T0235069

Roberto L. Benech-Arnold
Rodolfo A. Sánchez
Editors

Handbook
of Seed Physiology
Applications to Agriculture

Pre-publication
REVIEWS,
COMMENTARIES,
EVALUATIONS . . .

"This is a very interesting and timely book on seed physiology as it applies to agriculture. The range of topics is broad, but they fit neatly under several general headings: (1) the relationship between seeds and the soil in which they are planted, and strategies to improve seed performance in the field; (2) behavior of seeds in the field, emphasizing problems associated with dormancy, and lack of dormancy; (3) problems associated with seeds that can and cannot be stored in the dry state; and (4) the uses of commercially important seeds in an industrial context and the factors that influence their quality.

Several of these topics have not been comprehensively reviewed in recent times, making this an important and valuable addition to the seed literature. There is much to be learned from the chapters, as might be anticipated given the quality and expertise of the authors. The reviews related to seed behavior are particularly interesting since this area has rarely been covered in other books on seeds. Each chapter contains a very complete set of references, which is a useful guide to further reading. Many of the chapters will be extensively quoted for years to come."

J. Derek Bewley, PhD
Professor of Botany,
University of Guelph, Canada

More pre-publication
REVIEWS, COMMENTARIES, EVALUATIONS . . .

"Seeds are the beginning and the end of most agricultural practices. The ways in which seeds function—their physiology, biochemistry, molecular biology, and genetics—are critically important for agricultural success. But it is not only their use to humankind that makes seeds important objects for study; their biological properties, as agents for transmitting the legacy of one generation to the next, have long stimulated the intellect and investigative zeal of scientists.

The editors have judiciously chosen areas that reflect all of seed biology and have compiled an expert, authoritative team of seed scientists to write about them. This text brings together an exciting collection of articles covering virtually all of seed physiology important to agriculture, from seed germination, seed performance, and seedling establishment to dormancy, weed seeds, storage and longevity, and quality of cereals and oilseeds. The information is up to date, complete, and comprehensive. The book should attract and satisfy agriculturalists, seed scientists, and workers and students in related areas of biology. I warmly recommend this absorbing compendium for your study."

Michael Black, PhD
Emeritus Professor,
King's College, University of London, UK

"As the title indicates, *Handbook of Seed Physiology: Applications to Agriculture* updates several areas of seed biology and physiology related to the agricultural and industrial use of seeds. The book is divided into four sections that cover germination and crop establishment, the effects of seed dormancy in crop production and quality, seed longevity and conservation, and factors associated with seed quality and industrial uses of seeds.

This book covers a significant portion of current research related to the quality of seeds for both propagation and utilization. There is a good mix of physiological, genetic, biochemical, and modeling approaches that are applied to seed development, dormancy, germination, and composition. The integration of various levels of organization to understand how seeds behave in agricultural situations is an overall theme of the book. The coverage in these chapters offers enough detail for the book to be used in graduate courses in these topics, and also allows experts to update their knowledge of the current status of related fields."

Kent J. Bradford, PhD
Professor, Department of Vegetable Crops,
Director, Seed Biotechnology Center,
University of California, Davis

Food Products Press®
The Haworth Reference Press
Imprints of The Haworth Press, Inc.
New York • London • Oxford

Handbook
of Seed Physiology
Applications to Agriculture

FOOD PRODUCTS PRESS

Seed Biology, Production, and Technology
Amarjit S. Basra, PhD
Senior Editor

Heterosis and Hybrid Seed Production in Agronomic Crops
edited by Amarjit S. Basra

Seed Storage of Horticultural Crops by S. D. Doijode

Handbook of Seed Physiology: Applications to Agriculture
edited by Roberto L. Benech-Arnold and Rodolfo A. Sánchez

New, Recent, and Forthcoming Titles of Related Interest:

Wheat: Ecology and Physiology of Yield Determination
edited by Emilio H. Satorre and Gustavo A. Slafer

*Hybrid Seed Production in Vegetables: Rationale and Methods
in Selected Crops* edited by Amarjit S. Basra

Encyclopedic Dictionary of Plant Breeding and Related Subjects
by Rolf H. J. Schlegel

Handbook of Processes and Modeling in the Soil-Plant System
edited by D. K. Benbi and R. Nieder

*Biodiversity and Pest Management in Agroecosystems, Second
Edition* by Miguel A. Altieri and Clara I. Nichols

Molecular Genetics and Breeding of Forest Trees
edited by Sandeep Kumar and Matthias Fladung

Concise Encyclopedia of Plant Pathology by P. Vidhyasekaran

*Agrometeorology: Principles and Applications of Climate Studies
in Agriculture* by Harpal S. Mavi and Graeme J. Tupper

*Abiotic Stresses: Plant Resistance Through Breeding
and Molecular Approaches* edited by Muhammad Ashraf
and Philip John Charles Harris

Handbook
of Seed Physiology
Applications to Agriculture

Roberto L. Benech-Arnold
Rodolfo A. Sánchez
Editors

CRC Press
Taylor & Francis Group
Boca Raton London New York

CRC Press is an imprint of the
Taylor & Francis Group, an **informa** business

First published 2004 by The Haworth Press, Inc.

Published 2019 by CRC Press
Taylor & Francis Group
6000 Broken Sound Parkway NW, Suite 300
Boca Raton, FL 33487-2742

ISBN-13: 978-1-56022-929-2 (pbk)

Cover design by Marylouise E. Doyle.

Library of Congress Cataloging-in-Publication Data

Handbook of seed physiology : applications to agriculture / Roberto L. Benech-Arnold, Rodolfo A. Sánchez, editors.
    p. cm.
Includes bibliographical references and index.
ISBN 1-56022-928-4 (Case : alk. paper)—ISBN 1-56022-929-2 (Soft : alk. paper)
    1. Seeds—Physiology. 2. Seed technology. I. Benech-Arnold, Roberto L. II. Sánchez, Rodolfo A.
SB117.H27 2004
631.5'21—dc22
                                                                                    2003021276

**Visit the Taylor & Francis Web site at
http://www.taylorandfrancis.com**

**and the Psychology Press Web site at
http://www.psypress.com**

CONTENTS

SECTION III: SEED LONGEVITY AND STORAGE

SECTION IV: INDUSTRIAL QUALITY OF SEEDS

ABOUT THE EDITORS

Roberto L. Benech-Arnold, PhD, is Associate Professor of Grain Crops Production at the Department of Plant Production and Chairperson of the Plant Production Program of the School for Graduate Studies, both at the University of Buenos Aires, Argentina. He is the author or co-author of more than 100 professional papers, abstracts, and proceedings on various aspects of seed science. Dr. Benech-Arnold is Regional Representative for South America for the International Seed Science Society (ISSS) and speaks internationally on seed physiology, particularly in relation to the physiology and molecular biology of dormancy in grain crops. In addition, he serves as Independent Research Scientist for the National Council for Scientific and Technical Research in Argentina at IFEVA—the Agricultural Plant Physiology and Ecology Research Institute, and is also a member of many scientific and professional organizations.

Rodolfo A. Sánchez, PhD, is Professor of Plant Physiology at the Faculty of Agronomy and Chairperson of the Doctorate Program of the School for Graduate Studies, both at the University of Buenos Aires, Argentina. He is the author or co-author of more than 150 professional papers, abstracts, and proceedings. Professor Sánchez is a Guggenheim Fellow and an internationally recognized scientist in seed physiology and photobiology. He serves as Superior Research Scientist for the National Council for Scientific and Technical Research in Argentina. At present Professor Sánchez is the Director of IFEVA—the Agricultural Plant Physiology and Ecology Research Institute.

Miller B. McDonald, PhD, is Professor, Seed Biology Program, Department of Horticulture and Crop Science, The Ohio State University, Columbus, Ohio; e-mail: <mcdonald.2@osu.edu>.

R. Alejandra Mella, DrSci, is Research and Teaching Agronomist, Cátedra de Fisiología Vegetal, Facultad de Agronomía, Universidad de Buenos Aires, Argentina; e-mail: <mella@agro.uba.ar>.

José Luis Molina-Cano, PhD, is Scientist and Head of the Cereal Division, Centro Universitat de Lleida-Institut de Recerca i Tecnologia Agroalimentaries (UdL-IRTA), Lleida, Spain; e-mail: <Jose Luis.Molina@irta.es>.

Norman W. Pammenter, PhD, School of Life and Environmental Sciences, University of Natal, Durban, South Africa; e-mail <pammente @biology.und.ac.za>.

Valeria S. Passarella, is Research and Teaching Agronomist, Departamento de Producción Vegetal, Facultad de Agronomía, Universidad de Buenos Aires, Argentina; e-mail <vspasar@agro.uba. ar>.

Gary M. Paulsen, PhD, is Professor, Kansas State University; e-mail: <gmpaul@ksu.edu>.

Begoña Pérez-Vich, PhD, is Associate Scientist, Instituto de Agricultura Sostenible (CSIC), Alameda del Obispo s/n, Córdoba, Spain; e-mail: <bperez@cica.es>.

Roxana Savin, PhD, is Adjunct Professor, Departamento de Producción Vegetal, Facultad de Agronomía, Universidad de Buenos Aires, Argentina; e-mail: <savinrox@agro.uba.ar>.

Leonardo Velasco, PhD, is Associate Scientist, Instituto de Agricultura Sostenible (CSIC), Alameda del Obispo s/n, Córdoba, Spain; e-mail: <ia2veval@uco.es>.

Colin W. Wrigley, PhD, Food Science Australia and Value-Added Wheat CRC, North Ryde (Sydney), New South Wales, Australia; e-mail: <Colin.Wrigley@foodscience.afisc.csiro.au>.

Preface

Seeds have always caught the attention of both plant physiologists and agriculturists. Plant physiologists have been attracted by the multiplicity of processes that take place in such a small organ (i.e., desiccation tolerance, reserve deposition and utilization, dormancy, and germination); agriculturists, in turn, have been well aware from the beginning that the establishment of the "next crop" and the quality of their end product depend largely on "seed performance." Considerable progress has been made in recent decades in the field of seed physiology. The advancement made in some topics of this discipline is now sufficient to suggest approaches toward solving practical problems. On the other hand, attempts to solve these problems often raise issues or suggest approaches to more fundamental problems.

This book is a collection of chapters dealing with different aspects of seed physiology, each one having strong implications in crop management and utilization. The book has been divided in four major sections: (1) germination in the soil and stand establishment; (2) dormancy and the behavior of crops and weeds; (3) seed longevity and storage; and (4) industrial quality of seeds. Each section is composed of chapters dealing with specific aspects of an agricultural problem. Each chapter covers the most recent findings in the area, treated at a basic level (physiological, biochemical, and molecular level), but depicting the way in which that basic knowledge can be used for the development of tools leading to increase crop yield and/or improved industrial uses of the grain.

Section I addresses different aspects of crop germination and establishment. The physics of the seed environment, together with seed behavior in the soil in relation to seedbed preparation, are described in an introductory chapter of this section. The rest of the section is devoted to discussing seed responses to temperature and water availability, modeling crop emergence, breeding for germination at low temperatures and water availability, and suggesting techniques for improving crop germination performance in the field.

Section II covers dormancy problems in crop production. The first two chapters consider problems derived from the lack of control we have on the timing of exit from dormancy in grain crops: preharvest sprouting and the persistence of dormancy until the next sowing or seed industrial utilization. In Chapter 7 the termination of dormancy and the induction of germination is analyzed at a physiological and molecular level, mainly on the basis of the knowledge accumulated for two model species: tomato and *Datura ferox*. The section is completed with a chapter on dormancy in weedy species and the possibility of considering it in the generation of predictive models of weed emergence.

Section III presents an update in the field of seed longevity and conservation. The section is divided in two chapters dealing with orthodox and recalcitrant seeds, respectively.

Section IV considers aspects related to the industrial uses of seeds. The section has been divided in three chapters: one considering cereal grain quality for flour production, another dealing with industrial quality of oil crops, and a third devoted to discussing the development of good malting quality.

We attempted to give this book a different scope than other valuable works published recently in the area of seed biology. For example, the book *Seed Biology and the Yield of Grain Crops,* written by Dennis Egli (CAB International, 1998), covers only limited aspects of seed biology related to crop production (namely, those related to the determination of grain weight). On the other hand, the book *Seeds: Physiology of Development and Germination,* written by J. Derek Bewley and Michael Black (Plenum Press, 1994), is an excellent textbook on seed biology but is not focused on crop production. Similarly, the comprehensive *Seed Development and Germination,* edited by Jaime Kigel and Gad Galili (Marcel Dekker, 1995), sets the state of the art in seed science, without paying particular attention to the application of basic knowledge for the resolution of agricultural problems. The books *Seeds: The Ecology of Regeneration in Plant Communities,* edited by Michael Fenner (CAB International, 1992), and *Seeds: Ecology, Biogeography, and Evolution of Dormancy and Germination,* written by Carol and Jerry Baskin (Academic Press, 1998), discuss aspects of seed biology with the aim of understanding ecological processes. *Seed Quality: Basic Mechanisms and Agricultural Implications,* edited by Amarjit S. Basra (The Haworth Press, 1995) and *Seed Technology and Its Biological Basis,* edited by Michael Black and J. Derek Bewley (Sheffield Academic Press, 2000) are most closely related to this work; however, our book addresses aspects that are not covered in either *Seed Quality* or *Seed Technology* (i.e., dormancy of crops and weeds, models for predicting crop germination in the field, etc.).

We would like to thank all the authors who have contributed to this project. We are also indebted to our editorial assistant Juan Loreti who carried out very fine work. Our colleagues Antonio J. Hall and María E. Otegui acted as reviewers for some of the chapters and made comments and suggestions that greatly improved them.

SECTION I:
GERMINATION IN THE SOIL
AND STAND ESTABLISHMENT

Chapter 1

Seedbed Preparation—The Soil Physical Environment of Germinating Seeds

Amos Hadas

INTRODUCTION

Germination Processes in Seeds

Among the stages of the plant life cycle, seed germination and seedling establishment are the most vulnerable. The term *germination* includes sequences of complex processes that lead to the initiation of growth in the quiescent embryo in the seeds, seedling development, and emergence from the soil. During seed germination, various stored substrates are reactivated, repaired if damaged, and transformed into new building materials necessary for the initial growth of the embryo, its subsequent growth, and seedling establishment in its natural habitat (Koller and Hadas, 1982). To initiate the array of processes, the condensed, insoluble stored substrates must first be hydrated and then hydrolyzed to their basic forms before they can be reprocessed. The processes necessary to hydrate and reactivate enzymes, cell membranes, and cell organelles require much more respiratory energy than is required to maintain the dry seed (Bewley and Black, 1982).

The necessary sequential order of this complex array of processes, some of which may occur simultaneously and others in a serial, interdependent order, must be maintained to ensure its culmination in measurable and irreversible growth. To achieve this, the processes must be properly controlled, probably by endogenous growth regulators (Khan, 1975; Taylorson and Hendricks, 1977). Many of the metabolic events that are known to occur during germination may differ in their timing, both among the various organs of a particular seed and among seeds of different species (Mayer and Poljakoff-Mayber, 1989; Bewley and Black, 1982; Hegarty, 1978). Moreover, the transitions from one activity to another must be triggered by events that occur only when the appropriate thresholds, dictated and timed by endogenous regulators and/or varying environmental conditions, are reached.

3

The latter include environmental factors such as water availability, aeration, temperature, nutrients, and allelopathy caused by external toxins, e.g., allelochemicals (Currie, 1973; Come and Tissaoui, 1973; Koller and Hadas, 1982; Bewley and Black, 1982; Martin, McCoy, and Dick, 1990; Corbineau and Come, 1995; Bradford, 1995; Kigel, 1995).

Environmental Conditions

Proper germination of seeds and seedling emergence and establishment are critical processes in the survival and growth cycle of plant species in general. This is especially true in agriculture, since these processes determine uniformity, crop stand density, degree of weed infestation, and the efficient use of the nutrients and water resources available to the crop and ultimately affect the yield and quality of the crop (Hadas, 1997; Hadas, Wolf, and Rawitz, 1985; Hadas et al.,1990). Seed germination and stand establishment are especially critical under marginal environmental conditions. Under arid conditions (i.e., infrequent wetting, wide temperature fluctuations, and high evaporation rates), germinating seeds have to obtain their water from the rapidly diminishing soil water reserves and must overcome hardening soil seals formed at the soil surface. Many arid zone soils tend to slake upon wetting and then during the subsequent drying form hard crusts that impose mechanical obstacles to seed emergence and stand establishment, cause improper aeration, or lead to high-temperature injuries. Especially susceptible to these crusting conditions are minute seeds or seeds that are close to the soil surface, where the decrease of soil water content and the increase of soil seal resistance are fastest.

Where favorable ecological conditions prevail, other factors may decide the success or failure of an agricultural crop. Among these are seed development processes on the parent plants (Fenner, 1991; Gutterman, 1992), soil temperature (Probert, 1992), sensitivity to light (Scopel, Ballare, and Sanchez, 1991), seed burial, and depth regulation during dispersion and wetting (Koller and Hadas, 1982). Overgrazing, compaction caused by vehicular and animal traffic, irregular spatial dispersion and placement depth of seeds, and inadequate seedbed preparation are among adverse environmental factors. Obviously, knowledge of the specific physiological requirements of the various species of seeds and their physical interrelations with their environment, including climatic conditions, are of the utmost importance in ensuring successful seed germination and stand establishment.

This chapter is devoted to analyzing the soil physical environment of germinating seeds with the final aim of establishing the basis for optimization of seedbed preparation. To achieve this aim, the chapter has been structured in the following way: the first and second sections briefly analyze the environmental requirements for seed germination (i.e., water, temperature, aereation, and soil mechanical aspects) and the physics of the soil environment, respectively; the third section gives a characterization of seedbed attributes; the fourth section briefly discusses the biophysics of water uptake by seeds and seedlings; and the fifth section describes the physics of water movement from the soil matrix toward the germinating seed. Finally, and on the basis of all elements described in previous sections, the possibility of modeling seedbed attributes for optimization of stand establishment is discussed in the sixth section.

ENVIRONMENTAL REQUIREMENTS OF GERMINATING SEED

Seeds are self-contained units, in contrast to the plants that develop after germination, due to the materials stored in the seeds. Environmental requirements for germination are fewer and simpler than those for whole-plant development, so germination is relatively independent of the environment for a considerable period of seedling development. This assumption is based on the observation that a seedling does not photosynthesize; therefore, it requires neither light (except for regulatory or triggering functions) nor CO_2 for its proper development until the seedling breaks through the soil surface. Nevertheless, other environmental factors are needed, such as water, temperature, and oxygen.

Water Requirements

The effects of soil water on germinating seeds are difficult to define in biological terms, since soil water content and soil water potential are interdependent with soil constituents, their concentrations, and the scale and direction (draining or wetting) of the processes (Collis-George and Lloyd, 1979; Marshal et al., 1996). Water uptake by seeds is a prerequisite for proper germination, and under normal conditions, water uptake from the moist soil depends on the properties to water of the seed and the soil (Hegarty, 1978; Koller and Hadas, 1982).

The amount of water required by a seed for germination itself is very small. Water flow from the soil into the seed is driven by the water potential differences between the seed and the soil and is controlled by the soil conductivity to water. The total water potential of a dry seed, Ψ_{seed}, is very low compared to that of the soil, and the seed can draw water rapidly from the soil it comes in contact with. The driving force, water potential gradient, defined as ($\Delta\Psi = \Psi_{seed} - \Psi_{soil}$)/distance, is very large at the onset of imbibition and decreases as the imbibing seed reaches the required hydration level. Whether the amount of water taken up will suffice for germination depends on the water energy status of the seed and the adjacent soil total water potential, Ψ_{soil} (Koller and Hadas, 1982; Hadas, 1982; Bradford, 1995). Greater amounts of water are required for seedling development in the later part of the seedling growth than during the hydration stages because of the requirements of the radicles and root hairs (Hadas and Stibbe, 1973).

Species and cultivars may differ markedly in their water requirements for germination, and these differences have been attributed to the various endemic soil water regimes to which they were adapted (Bewley and Black, 1982; Koller and Hadas, 1982) and the differing soil physical conditions encountered during germination.

Temperature Requirements

Temperature affects both the soil properties with respect to water and the biological activity of seeds. Soil temperature varies greatly, both diurnally and seasonally, and is dependent on soil moisture, structure, layering, and soil color, as well as the site aspect and latitude (Marshal et al., 1996; van Wijk, 1963). The various effects of temperature on the rate of germination and the total germination have been discussed extensively (Mayer and Poljakoff-Mayber, 1989, Chapter 2). For germination to occur, the temperature of the seed environment should fall within a favorable, species-specific range.

Cardinal temperatures for germination are the base, maximum, and optimal temperatures, which are, respectively, the temperature below or above which no germination will occur and at which the faster germination rate is observed (see Chapter 2 in this book). Favorable temperature ranges, specific germination-enhancing conditions of diurnal or seasonal thermal periodicity, induction of secondary dormancy, and the combined effects of water stress and temperature vary among species (Kigel, 1995; Hegley, 1995; Benech-Arnold and Sanchez, 1995). Germination is greatly affected by the interactions between temperature, water potential, and water flow in the soil and by variations in the Q_{10} factors of the effective seed biological activity rates (Allrup, 1958; Bewley and Black, 1982; Meyer and Poljakoff-Mayber, 1989). The adverse effects of moisture stress on germinating seeds intensify as temperatures rise (McGinnies, 1960; Evans and Strickler, 1961) and may persist beyond the germination stages and extend into emergence and seedling growth stages through their strong effects on radicles and rootlet growth.

Aeration (Oxygen and CO_2) Requirements

The aeration regime (i.e., rates of gaseous exchange) greatly affects soil biological activity and the competition for oxygen with germinating seeds. However, their effects are complex and are difficult to define in biological terms (Currie, 1973; Collis-George and Lloyd, 1979). Such a definition requires knowledge of the interrelationships between complex diffusion processes (in air-filled pores in water films) that control oxygen supply and dissipation of respiratory and decomposition by-products (CO_2, N_2, NO_2, H_2S, ethylene, methane). Oxygen is required in germination as a terminal electron receptor in respiration and other oxidative processes of a regulatory nature (Roberts and Smith, 1977). Low oxygen availability reduces or even prevents germination in most species (Morinaga, 1926; Bewley and Black, 1982; Corbineau and Come, 1995). Oxygen supply to support the metabolic activity becomes decisive at a very early stage in germination, and oxygen-requiring metabolic activity is detected at an early stage of germination, indicated by a sharp rise in the respiration rate of seeds (Meyer and Poljakoff-Mayber, 1989). Another rise in respiration marks the beginning of the growth stage and radicle emergence. In between is a short period of constant respiration rate and oxygen consumption. Very often a conflict develops between oxygen supply and water supply to germinating seeds, which arises from the very low solubility and diffusivity of oxygen in water. Oxygen supply is greatly affected by the thickness of the water film covering the germinating seed and the hydrated seed coat (Come and Tissaoui, 1973), especially in seeds that have a swollen mucilaginous cover with very low diffusivity to oxygen (Heydecker and Orphanos, 1968; Witztum, Gutterman, and Evenari, 1969). Nevertheless, a few species, such as aquatic plants, are able to germinate under reduced oxygen or even anoxic conditions (Rumplo et al., 1984; Taylor, 1942). Seeds rich in fatty or starchy storage substances stop germinating when the oxygen level falls below 2 percent and lower, respectively (Al-Ani et al., 1982, 1985). Oxygen requirements increase with soil temperature and under light and/or water stress (Smoke et al., 1993; Gutterman, 1992).

Low CO_2 concentrations have been found to stimulate germination but may at times affect it in combination with ethylene (Corbineau and Come, 1995). Oxygen requirement and effects of oxygen and CO_2 concentrations on germination are rather complex and may be not fully understood (Bewley and Black, 1982; Meyer and Poljakoff-Mayber, 1989; Come and Corbineau, 1992).

Good aeration and gaseous exchange attained in well-structured, aggregated soil beds greatly assist germinating seeds, since the CO_2 and ethylene produced can easily diffuse out of the soil so that seed dormancy and germination retardation in CO_2-sensitive species are relieved (Bewley and Black, 1982; Corbineau and Come, 1995). The depth distributions of oxygen, CO_2, and ethylene concentration depend on soil temperature, soil air-filled porosity, and the exchange, consumption, and production of these gases (Smith and Dowdell, 1974; Marshall, Holmes, and Rose, 1996). Soil crusting and compaction may have deleterious effects on gas exchange and, in turn, on seed germination (Richard and Guerif, 1988a,b).

Soil Mechanical Impedance

Soil is a porous material made of particles of varied sizes and origins that form a matrix which exhibits a degree of resilience under mechanical stress, described as mechanical strength. Soil strength is a compound manifestation of soil mechanical properties (cohesion, angle of internal shear, compressibility) and depends on soil density, constituents, water content, and soil structure (Gill and van den Berg, 1967; Marshall, Holmes, and Rose, 1996). It increases with increasing bulk density (soil slaking, shrinking, and compaction) and decreases with increasing water content. Soils low in organic material or high in silt fractions tend to deform plastically and to compress easily, and to form seals under the impact and slaking action of raindrops or under instant flooding by water or irrigation (Marshall, Holmes, and Rose, 1996). Soil seals—thin, dense soil crusts—impede the germination and emergence of seedlings by restricting gaseous exchange and infiltration of water and by imposing a mechanical obstruction to emerging seedlings, or by any combination of these effects. Germination and final emergence are reduced as the seal strength increases and/or as the moisture content decreases (Arndt, 1965a,b; Richards, 1965; Hanks and Thorp, 1957; Hadas and Stibbe, 1977).

Adverse effects on seed germination and seedling development similar to those of seals are caused by soil compaction (Hadas, 1997; Hadas, Larson, and Allmaras, 1988; Hadas, Wolf, and Rawitz, 1983, 1985; Hadas et al., 1990). Increased soil strength is reflected not only in soil resistance to root proliferation, seed swelling, or tuber expansion, but also in restricted seedling emergence due to soil seals (Marshall, Holmes, and Rose, 1996). Nevertheless, under arid or semiarid situations, where soil moisture conditions are marginal, some soil compaction over the sown seeds has been found to improve germination and emergence (Hudspeth and Taylor, 1961; Dasberg, Hillel, and Arnon, 1966) and has been adopted as a common agronomic practice.

SOIL ENVIRONMENT—PHYSICAL ASPECTS

Soil is seldom an ideal environment, and it can be quite hostile to germinating seeds and emerging seedlings. Yet soils form the natural habitat in which most seeds germinate and the environment with which they interact and with which they establish themselves successfully, provided that the soil system and its constituents meet their requirements.

The Soil—A Three-Phased System

Soil is a three-phased system comprising solids (predominantly minerals, e.g., weathered primary parent materials, secondary particles—mainly clays—and organic matter), liquid (water and dissolved salts), and gases (a mixture in varying proportions).

The Solid Phase

Soil Constituents and Texture

The solid phase consists of (1) primary particles derived from the nonweathered rocks and deposits from which the soil is developed; (2) secondary minerals (clays) that are electrically charged, derived from weathered primary particles (Marshall, Holmes, and Rose, 1996); and (3) organic materials that consist of fully and partly decomposed organic residues and plant parts such as roots, fungal mycelia, and decomposed fauna.

Soil Structure

Solid soil particles of various origins, properties, and sizes, mixed in various proportions, define the textural soil types (e.g., sandy soils, clay soils). The solid soil particles are spatially arranged in various skeletal matrices that exhibit certain structural hierarchies (Tisdall, 1996; Tisdall and Oades, 1982; Hadas, 1987b; Dexter, 1988). These hierarchies are made of structural subunits of various sizes, in a variety of spatial arrangements, and have complex pore networks within and between the particles. In these pores, air and soil solution are found in varying proportions. The structural hierarchy follows a general pattern in which the smallest basic units, clay domains or tactoids (1 to ~20 μm in size) made of clay particles, are joined together by cation bonds, electrical attraction, and/or organic cements. These domains combine with larger particles and organic cementing substances to form microaggregates (50 to 200 μm in size) which, in turn, form larger units and so on (semi- and macroaggregates, clods, and blocks). The larger the soil unit is the coarser are its pores and the greater the number of interunit fissures. These structures recall the internal arrangement of smaller, denser units encapsulated in larger, more open ones (Tisdall and Oades, 1982; Oades and Watts, 1991). The pores are smallest in diameter within the domains and largest among the largest structural units. Soil structures, formed under natural conditions by wetting and drying, freezing and thawing, and swelling and shrinking cycles, or formed artificially by tillage operations, show great variability (Dexter, 1988, 1991; Marshall, Holmes, and Rose, 1996; Hadas, 1997).

Soil structure determines both total soil porosity and pore size and connectivity distributions. Intraparticle cohesion and interparticle adhesion are largest within and between the clay domains, and both decrease as the size and complexity of the structural units increase, because of the diminishing number of interparticle contact points and the increasing number of fissures and cracks. The structural stability of such a complicated matrix depends greatly on moisture content (which weakens cementing bonds and electrical attraction), internal stresses (caused by swelling, water surface tension, entrapped air pressure, and overburden), and external loads (vehicular traffic and animal tracking). Under these stresses soil structures will deform, fail, or collapse (slaking, compaction, seal forming) if the bonding forces (cohesion and adhesion) are weaker than the loads imposed on the structure (Hadas, 1987b; Dexter, 1988; Marshall, Holmes, and Rose, 1996). Both pore volume fraction and pore size distributions are closely related to water, thermal, and aeration regimes, as well as to the soil mechanical properties of natural soil environments or artificially produced seedbeds. Soil structural changes in response to climatic conditions (rain, freezing, and thawing events) or human activities (irrigation, tillage operations, compaction) cause great variations in soil density, total soil porosity, and pore size distribution and thus affect the water, thermal, and aeration regimes and soil strength (Marshall, Holmes, and Rose, 1996; Hadas, 1997).

Obviously, soil structure and its stability are of great importance to seed germination. Seeds may fall into natural fissures and cracks or be sown in between crumbs formed by tillage. Some may germinate; others may be entrapped by unstable, slaking structures or within fissures closed by swelling soil and their germination may be delayed or inhibited, or they may reenter secondary dormancy (Egley, 1995).

Soil Mechanical Behavior, Soil Crusts, and Soil Compaction

Soil mechanical behavior is determined by the intra- and interparticle bonds (cohesion and adhesion, respectively) which become stronger as the spacings between soil particle and unit diminish (i.e., as soil density increases). These forces are manifested in soil resistance to shear by tillage implements, compressibility under vehicular loads, impedance to penetration by fine needles, and tensile resilience. Soil impedance affects seed water uptake, thus, in turn impairing seed germination and stand establishment. Seed mechanical resistance can also diminish stand establishment by affecting elongation of radicles and roots and the emergence of coleoptiles and hypocotules through soil crusts (Bowen, 1981; Hadas and Stibbe, 1977; de Willingen and van Noordwijk, 1987; van Noordwijk and de Willingen, 1991; Unger and Kaspar, 1994; Marshal et al., 1996). Soil structure disintegration and slaking caused by fast soil surface wetting (because of low soil structural stability, raindrop impact, fast wetting, and implosion by entrapped air) and the subsequent formation and densification of soil seals reduce water infiltration and aeration. These crusts greatly impede seedling emergence, and this impedance increases as they become denser and drier (Bolt and Koening, 1972; Hadas and Stibbe, 1977; Dexter 1988; Bradford and Huang, 1992; Morin and Winkler, 1996).

Soil compaction results in soil densification caused either by shrinkage or external loads. Compaction, therefore, reduces the total soil porosity, pore size, gaseous exchange, and water infiltration, and increases soil impedance to penetration, impairing water spatial distribution and restricting seed germination and seedling establishment (Bowen, 1981; Hadas, Larson, and Allmaras, 1988; Gupta, Sharma, and De Franchi, 1989; Unger and Kaspar, 1994; Horn et al., 1994). Complete alleviation of an impaired soil physical environment depends on the processes that led to that impairment. The deleterious effects of crusts are rather easily alleviated by delicately fragmenting the newly formed crust, but complete rehabilitation of properties of compacted soil is almost impossible; great energy inputs are required to break up the dense soil into a favorable seedbed. Such efforts usually result in coarser seedbeds, improper stands, and lower yields (Hadas, Wolf, and Rawitz, 1983, 1985; Hadas et al., 1990; Wolf and Hadas, 1984).

Water Regimes In Soils

Water Content

In an air-dry soil, a minute amount of water is adsorbed on soil particles (hygroscopic water content), whereas in a saturated soil the pore system is completely filled with water. The water-filled volume fraction of the soil, termed the *volumetric soil water content*, θ_w, varies widely, especially in the upper soil layer. These variations depend on climatic and environmental conditions (e.g., rain, evaporation, drainage, vegetation, and human activity (irrigation). Integration of θ_w with respect to depth gives the total water amount held in the soil to a given depth. Periodic integration of θ_w with respect to soil depth leads to estimates of the soil water balance, i.e., the amounts of water added to or withdrawn from a given soil volume. Quantitative predictions of water movement into, within, and out of the soil can be derived from knowledge of the soil water energy status, i.e., the soil water potential, Ψ_{soil}, the water transport properties of the soil, and the appropriate physical equations governing water movement in soil.

Soil Water Potential

Various forces act on water adsorbed or held in the soil pores (e.g., gravitational, hydrostatic, matric forces derived from soil surface-water-air interactions, osmotic forces, and swelling forces derived from soil clay-water interactions). The influence of each of these forces, or their combinations, on soil water is given by the amount of work that must be done when a minute amount of water is transferred from a reference pool of water to the soil. That amount of work is termed the *soil water potential* (Kutilek and Nielsen, 1994; Marshall, Holmes, and Rose, 1996).

The soil water potential, Ψ_{soil}, is the algebraic sum of the several specific soil water potentials derived from the various forces acting on soil water. It is given in Equation 1.1, where Ψ_g, Ψ_p, Ψ_e, Ψ_m, and Ψ_{os} are the gravitational, hydrostatic, envelope (overburden, mechanical constraint), matric (derived from the adsorbed, interfacial soil-air-water tension), and osmotic soil water potential components, respectively. The gravitational and hydrostatic components (Ψ_g and Ψ_p) can be ignored when one deals with germinating seeds affected by a small volume of nonsubmerged, moist soil surrounding them (Equation 1.1a).

$$\Psi_{soil} = \Psi_m + \Psi_e + \Psi_{os} + \Psi_g + \Psi_p \qquad (1.1)$$
$$\Psi_{soil} = \Psi_m + \Psi_e + \Psi_{os} \qquad (1.1a)$$

When a soil is either saturated or submerged in pure water, Ψ_{soil} has negative values relative to pure water under the same conditions. In practical terms it means this water uptake by seeds or roots from unsaturated soil is carried out at the expense of metabolic energy. The osmotic component, Ψ_{os}, varies with salt concentrations and compositions, clay content, and clay type and requires a semipermeable membrane separating the soil water from the seed cells. The matric component, Ψ_m, exists in unsaturated soils and depends on water content, soil pore size, distributions, and soil structural stability. The relationship between θ_w and Ψ_m is known as the *soil characteristic* or *retention curve*. Assuming the soil pores resemble bundles of capillary tubes (Marshall, 1958, 1959), the first pores that will be drained under minute matric forces will be the large ones (the interaggregate pores, or fissures and cracks). As Ψ_m decreases, the smaller pores drain, with the narrowest pores (in the clay domains in which water is held by very strong matric forces) draining last. Upon wetting, the filling order of pores with water is the reverse of the draining order. Soil water characteristics curves are not unique and depend on the way they were obtained, either by draining a saturated soil or by wetting an unsaturated or an air-dry soil (Kutilek and Nielsen, 1994; Marshall, Holmes, and Rose, 1996). This phenomenon is called *soil water characteristic hysteresis*. The measured Ψ_m for a given θ_w will be lower for the draining characteristic curve than for the wetting one. This discrepancy results from irregular pore cross sections, bottlenecks connecting pores of differing radii, and smaller wetting angles than at draining (Kutilek and Nielsen, 1994; Marshall, Holmes, and Rose, 1996). In practical terms, this means that a seed embedded in a moist soil will start imbibing at a given value of Ψ_m, which will decrease instantaneously because of the abrupt change from wetting to draining characteristics. These variations in Ψ_m will be further aggravated if the seeds are placed in aggregated beds in which wide hysteretic variations are to be expected and are partly explained by the *pore exclusion principle* (Amemiya, 1965; Marshall, Holmes, and Rose, 1996).

Swelling Soils and Collapsed Soil Structure Matric Potential

During fast wetting, the soil structure deteriorates and breaks down (slakes), the structural fragments are reorganized, and, as in the case of clay soils, the soil undergoes volume change upon wetting or drying. In these cases, the pore system and the water characteristics of the soil change (Marshall, Holmes, and Rose, 1996). Upon draining, swollen or slaked soils remain saturated, although an appreciable volume change may be observed and Ψ_m becomes negative. Thus, seeds embedded in swollen or slaked soil may be subjected to oxygen deficiency during germination. Seeds germinating under external load or caught in a drying clay or compacted soil may be adversely affected by the confining pressure the external load or shrinking material may exert on them, reducing their ability to take up water and germinate (Collis-George and Williams, 1968; Hadas, 1985).

Water Movement in Soils

Water is forced to move in soil when there is a driving force resulting from a water potential gradient between two points in the soil or between seed and soil water. The rate of movement depends on the prevailing water potential gradient and the water conductivity of the matrix in which water movement occurs (e.g., seed or soil). As the water moves, the water content of a given soil volume may be depleted, remain unchanged, or increase. A general quantitative description of water flow, which accounts for water content variations (law of conservation of matter) that account for water potential and water content variations, is given in Equation 1.2, where q_w is the instantaneous water flux and K is the soil water conductivity. In saturated soil it will be termed the soil hydraulic conductivity, K_s, and in an unsaturated soil it is termed the soil capillary water conductivity, $K(\theta_w)$. K_s depends on the soil pore-size distribution, pore connectivity, and the total water content, whereas $K(\theta_w)$ depends on soil pore-size distribution and pore connectivity within the water-filled soil volume fraction,

$$\partial/\partial x(q_w) = \partial\theta_w/\partial t = \partial/\partial x[K(\theta_w)\partial\Psi_{soil}/\partial x] = \partial/\partial x[D(\theta_w)\partial\theta_w/\partial x] \quad (1.2)$$

where $(\partial\theta_w/\partial t)$ is the time variation of the volumetric water content and $D(\theta_w) = K(\theta_w)[\partial\Psi_m(\theta_w)/\partial\theta_w]$ is the soil diffusivity to water, and $[\partial\Psi(\theta_w)/\partial\theta_w]$ is the specific water yield or specific water capacity (Kutilek and Nielsen, 1994; Marshall, Holmes, and Rose, 1996). In aggregated beds K_s increases with increasing aggregate size, but as the soil matric potential decreases $K(\theta_w)$ increases as the aggregate size decreases (Amemiya, 1965).

As water is removed from an unsaturated soil, θ_w may change, sometimes causing changes in Ψ_m and Ψ_{os} and in the soil capillary conductivity to water, $K(\theta_w)$. These changes depend on the water amounts taken up and in which mode (wetting or drying). Solutions for Equation 1.2, derived for particular cases, e.g., water flow to seeds or roots, will be given and discussed in the following.

Temperature Effects on Soil Water Characteristics
and Transport in Soils

Soil water properties, i.e., water characteristics, conductivity, and diffusivity, are temperature dependent. Water characteristics are affected mostly by temperature because of changes in water surface tension and volume changes of entrapped air bubbles. The soil water matric potential decreases as the temperature increases, but changes observed in dry or saturated soils were smaller than those in moist to wet soils (Taylor and Stewart, 1960; Chahal, 1965). Water conductivity and diffusivity are affected by temperature-related changes in water viscosity and vapor diffusion, condensation, and evaporation in pores (de Vries, 1958, 1963). Temperature gradients cause water, in both liquid and vapor phases, to move from high to lower temperature zones in the soil (Philip and de Vries, 1957).

In moist soils the diurnal temperature wave will tend to reduce soil water loss to the atmosphere during the day by forcing the water to follow the heat wave into the soil. During the night the direction of water movement is reversed and losses to the atmosphere increase for a few days (Hadas, 1975) while the water is in the soil surface layer. Diurnal variations in water contents due to deposition and evaporation of water vapor condensation have been observed (Rose, 1968; Hadas, 1968; Jackson, 1973) and affect germination, as suggested by Collis-George and Melville (1975) and Wuest, Albrecht, and Skirvin (1999).

Soil Thermal Regime

The radiant energy, intercepted at the soil surface, governs the thermal regime of the soil. Its measure depends on latitude, land slope and relief, soil color, and vegetative cover (van Wijk, 1963). A rather small amount of the intercepted radiative energy heats the soil; a fraction of it is reflected, another fraction is reradiated as infrared radiation, a fraction directly heats the air in contact with the soil surface, and the rest is dissipated as latent heat by evaporating soil water. The soil heat flux, G, depends strongly on the thermal properties of the soil, namely soil heat capacity and thermal conductivity, which vary with soil texture, structure, bulk density, and water content (Buckingham, 1907; van Duin, 1956; van Wijk, 1963; de Vries, 1963).

Diurnal and annual radiation patterns result in diurnal and annual heat waves. The amplitude and phase shift of the diurnal wave strongly influence germinating seeds through (1) variations in Q_{10},* the rate of biological processes; (2) changes in the level of the competition for available oxygen with the surrounding microbiota; (3) changes in soil water properties (osmotic, matric and potentials, water conductivity, and diffusivity); and (4) coupled thermal, liquid, and vapor transport processes, vapor condensation, and evaporation (Philip and de Vries, 1957; de Vries, 1963; Hadas, 1968; Kutilek and Nielsen, 1994).

---

*Q_{10}: a factor for the change in reaction rate for a 10°C temperature increase

Volumetric Heat Capacity

The volumetric heat capacity of a soil, C, depends on the volumetric contents of the soil constituents, and their specific heat capacities and can be calculated from the sum of the respective products (the volumetric content of each soil constituent by its specific heat capacity). The greater the volumetric water and/or the solid fractions, the greater the soil volumetric heat capacity (de Vries, 1963). A larger volumetric heat capacity means that a greater amount of heat will be required for a given temperature increase of a given soil volume. In practical terms this means that a dry seedbed with a low volumetric heat capacity will tend to reach high temperatures during the day and low temperatures in the later part of the night. Such fluctuations expose young seedlings to risks of sun scorching during the summer and of frost damage in the early spring. These variations can be greatly moderated by increasing the volumetric heat capacity by increasing θ_w through irrigation and/or by compacting the soil.

Thermal Conductivity

The soil thermal conductivity, λ, depends strongly on soil constituents, i.e., the solids, air, and water. Whereas air is a poor conductor, the solid particles and water are good conductors; the heat conductivity of soil solids is four to five times greater than that of water, which is in turn about three orders of magnitude greater than that of air. In saturated soils and air-dry soils only two constituents contribute to λ, i.e., solids and water and solids and air, respectively. The thermal conductivity of a moist, unsaturated soil depends on all three constituents and can be calculated from their volume fractions, particles shape, and their respective thermal conductivities, as suggested by de Vries (1963), or it can be measured (van Wijk, 1963). The soil thermal conductivity increases as the solids and water volume fractions increase, due to better interparticle contacts.

Heat Transfer

The generalized heat transport relationship in soils is derived from Fourier's law, $G = -\lambda\, dT/dx$, in which G is the soil heat flux, dT/dx is the temperature gradient, and λ is the effective thermal conductivity. Under natural conditions in which the temperature of the soil surface varies constantly, the general heat transfer that takes account of temperature changes with time (law of energy conservation) is given, for a certain depth z, in Equation 1.3, in which $(\partial T/\partial t)$ is the rate of temperature change with time at a given point in the soil,

$$(\partial T/\partial t) = \partial/\partial z[\lambda/\rho c(\partial T/\partial z)] = \partial/\partial z[\kappa \partial T/\partial z] \qquad (1.3)$$

where $(\partial^2 T/\partial z^2)$ is the rate of change of temperature gradient with respect to distance, λ is the effective thermal conductivity, ρc is the volumetric heat capacity, and $\kappa = \lambda/\rho c$ is the thermal diffusivity of the soil (Marshall, Holmes, and Rose, 1996).

Diurnal and Annual Temperature Cycles in Soils

Although the diurnal and annual heat waves appear as a single, compound wave, it is possible to distinguish between the cycles by assuming the soil surface temperature to follow two different sinusoidal waves. The soil temperature dependence on time, t, and depth, z, for a homogeneous soil is given in Equation 1.4, where $T(z, t)$ is the soil temperature, $\omega = 2\pi/\tau$, τ is the period (day, year), and A_0 and A_{av} are the amplitude and mean temperature at the surface, respectively (Carslaw and Jaeger, 1959; Kirkham and Powers, 1972).

$$T(z,t) = A_{av} + A_0 \exp - \{(\omega/2\kappa)^{1/2}z\}\sin\{\omega t - (\omega/2\kappa)^{1/2}z\} \qquad (1.4)$$

The exponential term signifies the decay with depth of the temperature amplitude, and the argument of the sine term yields the lag between the times at which maximal or minimal temperature is reached at the soil surface and at depth z. The diurnal wave penetrates to a depth of 15 to 35 cm, and the annual wave to as much as 6 m, depending on the soil thermal properties. Normally, sown seeds are confined to the shallow layer below the soil surface, where they will be subjected to temperature amplitudes almost equal to those at the soil surface with a minimal time lag. Seeds sown into an aggregated seedbed may be scorched by extreme midday temperature amplitudes during the summer or be exposed to freezing hazards during late autumn and early spring. Kebreab and Murdoch (1999a) reported that inhibitory effects of temperature on germination were more evident under fluctuating than under constant temperatures; they found that the effect of high temperatures on germination was greatly influenced by the amplitude and thermoperiod of fluctuating temperature (Kebreab and Murdoch, 1999b; Stout, Brooke, and Hall, 1999).

Temperature Cycles in Layered Soils and Seedbeds

The temperatures around seeds are greatly affected by the aggregate size distribution, water content, and the existence of soil seals. Under natural situations, the soil thermal properties vary with time and soil depth; therefore, the heat wave becomes more complex with depth, and this is especially true in seedbeds in which great variations in soil structure and layering exist within short horizontal and vertical distances. These variations result in large temporal and spatial variations of λ, ρc, and κ; therefore, the thermal regimes of these soils cannot be predicted by simplified analytical equations as given previously (Hadas and Fuchs, 1973). Peerlkamp (1944), van Duin (1956), and van Wijk and Dirksen (1963) developed analytical models for predicting the changes in soil temperature in layered soils. However, where the soil properties and structure vary continuously, the analytical solutions fail and predictions of changes in soil temperature in layered soils require the use of computers and complex computer programs. Nevertheless, some conclusions can be drawn from the previous analysis. When an aggregated bed or a layer of dry soil lies over dense, moist soil, the soil surface temperature fluctuations will show increased amplitudes and the heat wave penetration will be shallower. This theoretical finding supports the practice of using dry mulches on the soil, either to keep the soil cooler in the summer or insulate it from cold in the late autumn.

Temperature Dependence of Q_{10} Coefficient

The diversity in the effects of temperature on germinating seeds caused by variations in the Q_{10} coefficient of enzymatic reactions is rather complicated. Q_{10} varies between 1.5 and about 3 for productive and synthesis reactions (which means that for each 10°C rise in temperature, the reaction rate increases by a factor of 1.5 to 3) and may be as high as 6 for denaturization processes (Voorhees, Allmaras, and Johnson, 1981). The Q_{10} biological reaction rate coefficient, defined by $R_1/R_2-Q_{10}^{[(T_2-T_1/10)]}$, is derived from a nonlinear relationship and its accumulated effect under fluctuating temperatures may be greater by 15 to 40 percent than that calculated from the Q_{10} value for the constant, mean temperature. This explains why seeds under a fluctuating temperature regime germinate faster then those under the constant mean temperature.

Aeration Regime

Water-free voids in the soil matrix contain mixture of gases, the proportions of which change with depth, temperature, root and microbial activity, water content, and void connectivity. Gaseous exchange occurs through air-filled pores and across water films. When the soil is air-dry the pore volume is air filled and free exchange is possible, whereas in a saturated soil, although the pores are water filled, some air bubbles are entrapped. Near the soil surface the soil air composition is very similar to that of the outer atmosphere (~79 percent N_2, 21 percent O_2, 0.03 percent CO_2, and other gases) (Marshall, Holmes, and Rose, 1996). Normally, the oxygen concentration decreases with soil depth while the concentrations of CO_2 and ethylene increase. During gaseous exchange, oxygen moves into the soil while CO_2 and ethylene move out of the soil. This exchange combines mass exchange driven by soil temperature and/or barometric pressure-related soil-air expansion, wind gusts at the surface, air displacement by rain or irrigation water, air entry resulting from soil desiccation, and gaseous diffusion. Diffusion is the most important gaseous exchange process in soils (Buckingham, 1904; Rommel, 1922; Marshall, Holmes, and Rose, 1996). Wind-generated air turbulence, which enhances diffusive exchange, is most effective in aggregated soil, where the gas exchange flux can be as high as 100 times the molecular diffusion flux, provided there are no seals (Farrell, Greacen, and Gurr, 1966; Farrell and Larson, 1973). For germinating seeds found next to the surface, oxygen supply by molecular movement, i.e., diffusion through the soil surface and within the air-filled soil pores, is of great importance. Seeds take up oxygen only after it has crossed the water films surrounding them under optimal situations, or after it has passed through water films, water-filled pores, or even saturated seals when adverse conditions prevail.

Molecular diffusion is described by the first Fick equation, in which the instantaneous oxygen flux, q_{ox} is driven by the oxygen gradient, dC/dx, and D_{ef} is the effective soil-air diffusion coefficient (Equation 1.5).

$$q_{ox} = -D_{ef} dC/dx \qquad (1.5)$$

In a medium such as soil, in which diffusion may occur only through interconnected air-filled pores or water films, D_{ef} equals a complex mean weighted value combining its value in water D_w and the effective continuous air-filled pore volume fraction (Buckingham, 1904; Currie, 1961, 1983, 1984; Currie and Rose, 1965; Marshall, Holmes, and Rose, 1996). Various simple relationships have been derived for the D_{ef}/D_{air}, e.g., $D_{ef}/D_{air} = b\theta_{air}$ (Pennman, 1940), or as power functions of θ_{air} (Marshall, Holmes, and Rose, 1996). These relationships point out the wide variations in D_{ef} with θ_{air} (volumetric air content). Greenwood (1975) and Wesseling (1974) have shown that for a volumetric water content $\theta_w \leq 0.10$ or 0.12, respectively, pore connectivity disappears and diffusion will occur only through the water in the soil pores. The oxygen diffusion coefficient in water is smaller than that in air by a factor of 10^4; therefore, it may well be that the oxygen supply rate to germinating seeds imbedded between moist soil crumbs in a well-prepared seedbed is controlled only by the water film covering the seeds. However, the O_2 supply in compacted or sealed soil will be reduced, because of the low porosity and high water content, respectively. The degree of impairment will be greater if, in addition to the factors just mentioned, the temperature is high, so that high biological activity will be enhanced (Glinski and Stepniewski, 1985; Glinski and Lipiec, 1990).

SEEDBED PREPARATION, CHARACTERIZATION OF SEEDBED ATTRIBUTES, AND SEEDBED ENVIRONMENT CONDITIONS AND SEED GERMINATION

The seedbed is the finely tilled, loose topsoil layer especially prepared to ensure fast, uniform germination and emergence into which seeds are sown (Keen, 1931; Slipher, 1932). Seedbed preparation requires a sequence of tillage operations aimed at fragmenting the bulk soil, manipulating the disturbed soil structure and improving soil tilth; it provides favorable air, water, and heat regimes and reduces mechanical resistance to seed germination, emergence, and root development (Slipher, 1932). Great importance is given to specifying and then attempting to produce a desired seedbed (Braunack and Dexter, 1989a,b; Dexter, 1991; Hadas, 1997); nevertheless, all the great effort, labor, and equipment invested in producing a specified seedbed may be wasted. The seedbed may, under optimal conditions, complete its usefulness within few days after sowing, once seedlings are established, but under the impact of adverse weather the soil structure may fail or collapse because of fast wetting and drop impact. As a result, the failed structure may impose mechanical constraints on seed germination and stand establishment.

Seedbed Preparation and Seedbed-Characterizing Indices

A bulk soil structure that has settled during previous seasons has to be fragmented and modified by one or more tillage operations and rearranged into a seedbed made of layers of aggregates that vary in their size ranges (Ojeniyi and Dexter, 1979a,b; Hadas and Shmulewich, 1990). Purposely modifying a soil bulk structure to form a desired seedbed is always a matter for compromises, aiming to minimize the risks of failures of seed germination, emergence, and stand establishment under anticipated future weather conditions, soil structure variations, and resultant seedbed environmental conditions, while reducing energy and labor investments. For each combination of soil, soil structure, structural stability, local climate uncertainties, and crop, several possible compromise solutions exist, derived by trial-and-error procedures (Hadas, 1997; Hadas, Wolf, and Meirson, 1978; Hadas, Wolf, and Stibbe, 1981; Unger, 1982).

Under arid conditions, the establishment of a uniform stand requires good seed-soil-water contact to ensure rapid water uptake and seedling emergence and to avoid the effects of fast soil-surface drying and crust hardening (Hillel, 1960; Hadas, 1997; Hadas and Stibbe, 1977). These requirements dictate a seedbed made of small, fine aggregates with narrow size distribution, so as to ensure good seed-soil contact and low evaporative losses (Russel, 1973; Hadas, 1975; Hadas and Russo, 1974a,b). However, a finely aggregated seedbed presents increased crusting hazards upon wetting and drying (Hadas, 1997). Under wet conditions, improved drainage, aeration, and enhanced soil warming in cold regions are sought, which stimulate efforts to produce coarse tilth and to shape the soil surface as ridges or benches to improve drainability and increase air-filled porosity. However, these latter benefits will be balanced by reduced heat capacity and thermal conductivity, which increase the risks of enhanced temperature variations near and at the soil surface in hot regimes, or even those of freezing in cold regimes (van Duin, 1956; van Wijk and Dirksen, 1963).

Soil Tillage and Seedbed Formation

Tillage implements exert external stresses on the soil bulk causing it to fail in several different modes (brittle, shear, compressive, and plastic deformation), depending on initial soil conditions (bulk density, water content and existing fissures, cracks, root channels), tillage implements, type, and modes of operation. The extent, mode, and fineness of soil failure or fragmentation determine the need for further tillage work and ultimately the quality of the produced seedbed.

Field soils that are compacted and tilled periodically consist of neatly arranged soil clods, macroaggregates, and blocks, differing in their size density and crack networks (Hadas, 1997). When tilled, soil units are separated, torn, or fragmented and moved sideward and upward. In dry soils, several tillage implements are applied sequentially to fragment the soil and produced the desired seedbed tilth (Hadas, Wolf, and Meirson, 1978). The number of tillage passes required diminishes as soil water content nears that of the plastic limit water content, which is approximately that of wilting point in many soils (Dexter, 1988, 1991; Hadas and Wolf, 1983; Hadas, Wolf, and Meirson, 1978). Current knowledge of soil fragmentation processes is very limited; therefore, exhaustive tillage trials are required to obtain the resulting soil fragmentation data (Dexter, 1977; Koolen, 1977; Hadas, Wolf, and Meirson, 1978; Hadas and Wolf, 1983; Gupta and Larson, 1982; Perdok and Kouwehoven, 1994; Guerif et al., 2001; Young, Crawford, and Rappoldt, 2001). The total inputs of energy and labor in tilling soils depend on (1) soil conditions and constituents (density, water content, fissures, pores and failure plane nets, surface energy of soils), (2) types of implements used, and (3) the soil structure fineness required (Gupta and Larson, 1982; Hadas, 1987a, 1997; Hadas and Wolf, 1983; Hadas, Wolf, and Meirson, 1978; Wolf and Hadas, 1987; McPhee et al., 1995; Roytenberg and Cheplin, 1995; Perfect, Zhai, and Belvins, 1997). It becomes obvious that any proposed seedbed preparatory procedure must be based on a huge database and must be formulated along a delicately balanced, compound probabilistic approach. That approach has to account for (1) annual and seasonal weather variability; (2) known probabilities of attaining the right tilth by using the right implements in the proper order and the best possible soil conditions; (3) soil tilth stability and probability of seedbed failure caused by weather events, traffic, etc.; and (4) known characteristics and behavior patterns of seed lots. In the light of the complexity of the processes involved, our current knowledge gaps, and our inability to assess the probabilities of the various system components, such an approach eludes us and any simple seedbed modeling and forecasting of its properties is precluded (Hadas, Wolf, and Meirson, 1978; Hadas, Larson, and Allmaras, 1988; Kuipers, 1984; Lal, 1991; Hadas, 1997; Guerif et al., 2001).

In order to standardize seedbed preparation, some physical indices, characterizing the preferred soil seedbed structure to be obtained, have to be defined, tested, and accepted as recognized and official indices. Suggested physical determinations of seedbed characteristics have appeared and have been discussed in the literature. The characteristics addressed included total porosity, aggregate size distribution, shear strength, infiltration rate, sorptivity, aggregate stability to water and wind abrasion, and resistance to penetration (Russell, 1973; Hadas and Russo, 1974b; Tennent and Humblin, 1987; Braunack and Dexter, 1989a,b; Thurburn, Hansen, and Glenville, 1987; Christiansen, Foley, and Glanville, 1987; Collis-George and Lloyd, 1979). The procedures to determine these indices have also been described (Hadas, Wolf, and Meirson, 1978; Tennent and Humblin, 1987; Braunack and McPhee, 1991), but so far none of the indices have been recognized as official indices, probably because their determinations are cumbersome and demand much time and labor, and once determined they may change in an instant by rainfall, irrigation, or traffic.

Seedbed Aggregate Size Distribution and Seed Germination

The most commonly used seedbed-defining index is the size distribution of the aggregates found at the seed placement depth. Russell (1973) has suggested that a seedbed consisting of aggregates larger then 0.5 mm but smaller then 5.0 to 6.0 mm will provide the ideal conditions for seed germination and emergence. In general, smaller aggregates reduce soil water evaporation and soil drying (Holmes, Greacen, and Gurr, 1960; Farrell, Greacen, and Gurr, 1966; Kimball and Lemon, 1971; Hadas, 1975). Nasr and Selles (1995) used two logistical models to predict wheat seed germination and concluded that final emergence rates were negatively affected by seedbed density and aggregate size. Beds made of 0.5 to 3.0 mm aggregates, 3.0 to 10.0 cm deep, were reported to maintain minimal water losses (Hillel and Hadas, 1972; Hadas, 1975). These seedbeds have to be prepared prior to soil wetting (Hillel and Hadas, 1972; Allmaras et al., 1977). Hadas and Russo (1974b) stated that a seedbed should consist of aggregates smaller than one-fifth to one-tenth of the seed size if seed-soil contact is the governing factor in water uptake and is crucial to seed germination. These recommendations seem to be right for medium to large seeds but fail when small seeds placed next to the soil surface are considered (e.g., celery, carrots, sugar beet). Under these situations, soil water content, aggregate size distribution, and good seed-soil contact are the germination controlling factors (Hadas and Russo, 1974b); frequent irrigation keeps the soil moist, improves seed-soil contact, and reduces the mechanical impedance of seals that may be formed.

Soil Surface Relief—Ridged Seedbed

Shaping the seedbed into ridges is a common practice for overcoming poor stand establishment on poorly drained soils or under low spring temperatures. Sowing on ridges enables the seedbed temperature on ridges to rise by 2 to 3°C above that of a flat seedbed, which promotes emergence in wet, cool regions (Spoor and Giles, 1973; Gupta et al., 1990). Ridging improves the utilization of winter-stored soil water by summer field crops and allows sowing a few days earlier than on flat seedbeds and thus enables crops to avoid pests (Hadas and Stibbe, 1973; Tisdall and Hodgson, 1990). However, yields on a flat seedbed may be the same as or higher than yields on ridges.

WATER UPTAKE BY SEEDS AND SEEDLINGS

Water uptake by seeds is an essential step toward rehydration of seed tissues and initiation of the metabolic processes in seeds, and the minute amounts of water required for germination depend on the seed genome and its individual constituents. The various organs (e.g., embryo, cotyledons) and tissues differ in their internal physical structure, biochemical properties, and chemical composition; therefore, they may differ in their water retention, distribution, and swelling properties (Stiles, 1948; Bewley and Black, 1982; Koller and Hadas, 1982).

Water uptake by dry seeds is characterized by three phases, controlled by one of of the following factors: (1) the seed properties with respect to water (e.g., seed water potential, diffusivity to water), (2) the soil-water properties (e.g., soil-water potential, diffusivity, and conductivity to water of the soil around the seed), and (3) the hydraulic properties of the seed-soil interface.

The initial phase in water uptake, the imbibition phase, is characterized by a saturation kinetics pattern, depending on soil-seed contact, seed composition, and the seed coat geometry and properties (Hadas, 1982). The second phase, the transition phase, is characterized by a low to negligible water uptake rate. The third phase, the growth phase, is characterized by a rapid, exponential increase in the water uptake rate, accompanied by the emergence of the radicle. The first two phases are observed in dead, inert, and viable seeds alike, whereas the growth phase is unique to viable, germinating seeds.

The Imbibition Phase

The imbibition phase, usually considered to be a passive one, starts with entry into the seed of water, which is distributed in crevices, cracks, and flaws in the seed cover and tissues and is absorbed by the seed colloids. Water uptake rate measurements toward the end of this phase have shown these rates to be temperature dependent and accompanied by observed increases in respiration rate and light sensitivity in some seed species (Pollock and Toole, 1966; Taylorson and Hendricks, 1972; Tobin and Briggs, 1969; Karssen, 1970; Berrie, Paterson, and West, 1974). These observations suggest that water uptake during imbibition is not passive at all but instead becomes an active process at a rather early stage of this phase. The end of the imbibition phase is generally marked by an asymptotic approach to a final water gain. The rate of approach to the final value of water gain and its value depend on soil water potential, soil hydraulic properties, and seed composition (Hadas, 1982; Bradford, 1995).

The Transition Phase

During the transition phase, also known as the *pause phase* (Haber and Luippold, 1960), the seed moisture content, respiration rate, and apparent morphology remain unchanged. Nevertheless, a variety of metabolic processes are activated (Koller and Hadas, 1982), and differences in activity levels of processes and the order of their occurrence have been observed among seeds of various species and among seeds differing in their hydration levels (Hegarty, 1978). Therefore, any adverse environmental conditions may lead to redrying of the seeds, so water stressing them and affecting their hydration levels may impair, retard, or even inhibit germination. If no damage resulted, no dormancy was induced, and no inhibitory processes were triggered, germination of these seeds upon rewetting would be enhanced due to the high concentrations of unused metabolite accumulated prior to drying (Boorman, 1968; Koller, 1970). These are the basis for seed priming, a technique known also as "chitting" (Hegarty, 1978) (see Chapter 4).

According to Bradford (1995), the transition phase can be considered as germination, as its duration influences the initiation time and the extent of radicle growth. Dormant seeds have been observed to reach the transition phase and to remain in it for long durations that extend to weeks or more before germination (Powell, Dulson, and Bewley, 1984; Bradford, 1995).

The Growth Phase

The growth phase starts with an increased respiratory rate, the initiation of cell division, and extension of the embryonic radicle cells and ends with radicle protrusion. The renewed water uptake rate depends on the water potential of the soil, adaptation of the seed water potential to soil environmental conditions, and the seed-soil contact properties (Hadas and Stibbe, 1973; Hadas, 1982; de Miguel and Sanchez, 1992; Ni and Bradford, 1992; Bradford, 1995). As pointed out earlier, the distinction between the phases is an arbitrary partitioning of the continuous, sequential order of processes that leads to germination. Actually, all the processes are interdependent and the interrelationships between them suggest that each phase greatly depends on the preceding phases, water uptake rates, and total water uptake (Hadas, 1977a; Hegarty, 1978).

In order to generalize the observations and conclusions brought up previously and to model germination in various seed-substrate systems, it is necessary to quantitatively define the physical properties of the substrate (e.g., soil) and the seed and their interactions. Practically, fulfillment of such a requirement is almost impossible; instead, one reverts to simple indices or characteristics such as critical seed hydration level or critical water potential (Hunter and Erickson, 1952).

It is generally accepted that to germinate, a seed must reach a minimal water content known as the *critical hydration level,* defined as the minimal amount of water taken up by a seed that will induce germination (Hadas, 1970; Koller and Hadas, 1982). It does not reflect the water distribution among the seed components, nor does it have an absolute value since it depends on the water uptake rate, variations in external soil water content, temperature, and seed adaptation to variations in these factors (Koller and Hadas 1982; Hadas 1982). The amount of water gained by the seed is the weighed mean water gains by the various parts of that seed. Blacklow (1973) has reported that whole corn seeds gained 75 percent of their initial weight, whereas the embryos, which form only 11 percent of the seed's weight, gained 261 percent, and the endosperm, which forms most of the seed mass, gained only 50 percent. The critical hydration level concept, developed for completely immersed seeds, fails in cases of partial seed wetting which occur when the wetted seed volume includes only the embryo and the adjacent storage tissues (Hydecker, 1968, personal communication).

Critical water potential is defined as the external water potential value at or below which seeds cannot reach their critical hydration level. Fully imbibed seeds can germinate and start growing even when the substrate or soil water potential is still decreasing and is far below that critical value (McDonough, 1975; Bradford, 1995). Hunter and Erickson (1952) determined critical water potential values of -1.25, -0.79, -0.66, and -0.35 MPa for corn, rice, pea, and clover seeds, respectively. Values of -1.52, -0.7, -1.2, -0.6, and -0.35 MPa were reported for sorghum, cotton, chickpea, pea, and clover seeds, respectively, by Hadas (1970), Hadas and Stibbe (1973), and Hadas and Russo (1974a,b). These values were determined under static equilibrium water potential conditions. In practical situations, in which external water potential varies with water uptake, soil evaporation, or drainage, these values may change as well. Computed critical water potential values of -1.4, -2.0, -0.45, -1.1, and -1.5 MPa, for corn, sorghum, clover, cotton, and chickpeas were reported for dynamic situations and perfect seed-soil-water contact by Hadas (1970), Hadas and Stibbe (1973), and Hadas and Russo (1974b). The values obtained for corn indicate that cotton, chickpea, sorghum, and corn seeds can probably germinate at lower critical water potentials than those observed for equilibrium conditions by Hunter and Erickson (1952) and Hadas (1970).

SEED-SOIL WATER RELATIONSHIPS

Water transport of water into, within, and out of the soil domain and into the imbibing seed depends on the soil water potential gradients and water transport properties (conductivity and diffusivity to water) of the various seed-soil system components (Marshall, Holmes, and Rose, 1996; Koller and Hadas, 1982; Hadas, 1982).

The water potential of dry seeds is extremely low compared to that of moist soils (Hegarty, 1978; Hadas, 1982). Seeds brought into contact with a moist soil will start taking in water at once, at a rate that depends on the water potential gradient between the seed and the soil. The seed water potential will increase in accordance with seed water characteristics, external water potential, seed storage materials, and ambient temperature (Mayer and Poljakoff-Mayber, 1989). Water will move first from the soil and then to the seed, and as the water uptake proceeds, water will be depleted from the soil farther away from the seed. The rate and degree of depletion will depend on the water flux into the seed and hydraullic properties of the soil-seed interface (Collis-George and Hector, 1966; Phillips, 1968; Hadas, 1969, 1970, 1982; Hadas and Russo, 1974a,b; Hadas and Stibbe, 1973; Shaykewich and Williams, 1971; Williams and Shaykewich, 1971). Changes in soil water content will induce changes in water potential gradients and water conductivity, and the seed water potential and diffusivity to water will change as well (because of seed metabolism and reconditioning of the seed membranes and seed coat).

Seed Water Potential

Air-dry seeds have an extremely low water potential, Ψ_{seed}, ranging between ~-50 and -100 MPa (Hegarty, 1978), but as the seed imbibes water, the water content of the seed organs and its water potential increase. Since the seed organs differ in their constituents and structure, their specific water potential characteristics will differ; nevertheless, the measured seed water potential, Ψ_{seed} reflects equilibrium water potential of the whole seed. The total water potential of a cell, Ψ_{cell}, in each of the seed organs equals the algebraic sum of the various water potential components, as given in Equation 1.6, where Ψ_{cell}, $\Psi_{os, cell}$, $\Psi_{m, cell}$ and $\Psi_{T\ cell}$ are the total, osmotic, matric, and turgor water potentials of the cells.

$$\Psi_{cell} = \Psi_{os, cell} = \Psi_{m, cell} + \Psi_{T\ cell} \qquad (1.6)$$

The osmotic cell water potential, $\Psi_{os,\ cell}$, reflects the osmotic potential contributions of the various cell constituents. It changes as the germination processes progress and adapt to the changing Ψ_{soil} near the seed. The seed matric water potential, $\Psi_{m,\ cell}$, reflects the matrical forces imposed on the cell water content by the cell wall structure and neighboring cells. The turgor component, $\Psi_{T\ cell}$, represents the counterpressure exerted by the stressed elastic cell wall structure in response to the $\Psi_{os,\ cell}$ and the swelling pressures of hydrated proteins and cell organelles. In general, Ψ_{cell} has a negative value except in fully turgid cells. By changing the concentrations of its constituents and by modification of its membrane activity and selectivity, a cell can regulate its water potential and its water uptake or loss. Therefore, the Ψ_{seed} changes during the various germination phases and as an adaptive response to varying environmental conditions, e.g., soil salinity or soil drying (Hadas and Stibbe, 1973), $\Psi_{T\ cell}$ may well also change when the seed membranes leak to the environment (Simon, 1974; Hegarty, 1978).

Specific Effects of Ψ_m, Ψ_{os}, and Ψ_e on Seed Germination

The two important soil water potential components, Ψ_m and Ψ_{os}, are directly involved in water transport to germinating seeds. Seeds have been reported to respond equally to equal changes in these two components, provided their membranes were intact and fully active (Ayers, 1952; Richards and Wadleigh, 1952; Hadas and Russo, 1974a; Manohar and Heydecker, 1974; Collis-George and Sands, 1962). Biological systems differ in their tissue permeability to water and salts and in their susceptibility to salt toxicity (Uhvits, 1946; Collis-George and Sands, 1959; Wiggans and Gardner, 1962; Bewley and Black, 1982; Meyer and Poljakoff-Mayber, 1989). Small reductions in soil matric potential were observed to affect germination to a greater extent than equal or even greater reductions in the soil osmotic potential (Uhvits, 1946; Ayers and Hayward, 1948; Collis-George and Sands, 1959; Wiggans and Gardner, 1962; Collis-George and Hector, 1966; Williams and Shaykewich, 1971; Hadas and Stibbe, 1973; Hadas and Russo, 1974a,b). This difference in response is due to the fact that a slight change in Ψ_m involves a change in soil water content, with corresponding reductions in both soil conductivity to water and seed soil contact (Sedgley, 1963; Collis-George and Hector, 1966; Hadas, 1970; Hadas and Russo, 1974a,b). An obvious corollary to that is that the critical water potential cannot always be taken as the sum of these two components when germination in soil is considered. The reason for this is that the presence of selective membranes will exclude salts from the water taken in by the seed and leave them outside the seed, so that Ψ_{os} will increase and will not be directly measured.

Values cited for critical water potential were determined under constant potential laboratory conditions; therefore, they might differ from those prevailing in real situations.

In natural situations, wetting of the soil by precipitation or irrigation will increase θ_w and Ψ_m, decrease Ψ_{os}, and improve environmental conditions for germinating seeds as long as aeration is not impaired. However, reduction in θ_w because of evaporation will decrease both Ψ_m and Ψ_{os} and may enforce changes in Ψ_{seed} and reduce the rate and final extent of germination (Hadas, 1976, 1977a,b). These responses will be further aggravated if seed-soil contact is impaired as well.

The possibility that the matric soil water potential affects germination by its direct contribution to the soil effective mechanical stress was examined by Collis-George and Hector (1966) and by Collis-George and Williams (1968). Their data suggested that the mechanical effective soil stress restricts seed swelling or even inhibits embryo development. Others (Hadas, 1970, 1977b; Shaykewich, 1973) found that under natural situations, normal stresses induced in seedbeds are too small to confine seeds or to impair their germination. A dry seed initially develops swelling pressures of up to ~400 MPa, but upon completion of imbibition the pressure decreases to ~0.1 MPa (Shaykewich, 1973). These values far exceed the normal soil stresses found in the field to be around 0.12 to 0.34 MPa (Williams and Shaykewich, 1971; Hadas, 1985). Observed poor germination in compacted soils and next to traffic lanes, or of seeds entrapped in shrinking soil, can result from greater mechanical constraints than those described earlier and imposed on the seeds.

MODELING SEED GERMINATION AND SEEDBED PHYSICAL ATTRIBUTES

Any attempt to model seed germination should address the relationships between the time course of seed germination, germination rate, final germination percentage, time lag in germination initiation, and the external factors affecting germination singly or in combination. Moreover, the model parameters should be quantifiable, have relevant biological significance, and be based on measured seed germination time patterns. Models should provide some forecasting capabilities. Determination of the relevant model parameters requires proper experimental procedures carried out under conditions that closely resemble actual conditions and that are aimed at minimizing uncertain results. Chapter 3 is fully devoted to analyzing existing germination models that account for changes in both germination rate and germination percentage in relation to environmental variables, mainly temperature and water availability. However, for completeness, modeling seed germination under field conditions should include quantification of the changing seedbed properties, the complex spatial and temporal variations in the soil structure, and the seed reactions to these variations. The final part of this chapter discusses quantifying the variables involved in these processes and developing models on the basis of such quantifications to complement germination models such as those described in Chapter 3.

Modeling Water Flow in Seed-Soil System and Germination

The dynamics of water uptake by seeds can be quantitatively calculated by applying solutions of Equation 1.2, the pertinent water potential gradients and water conduction properties, and the appropriate boundary and initial conditions to the system under considerations (e.g., determination of D_{seed}). In most reported experimental determination of seed water uptake data, the experimental procedure used precluded dynamic forecasting based on water flow. Furthermore, they do not permit any distinction to be made between specific effects on seed germination and those of changes in water potential components, conductivity or diffusivity to water, and seed-soil contact area. Water transport within the seed and from the soil to the seed can be simplified by (1) assuming seeds to resemble spheres or cylinders and (2) using water contents and mean weighted diffusivities to water as the flow equation parameters for solving Equation 1.2, (Phillips, 1968; Hadas, 1970; Hadas and Stibbe, 1973; Hadas and Russo, 1974a,b). This choice between using variations in water content and using mean weighted diffusivities to water allows one variable (Ψ_{seed} or Ψ_{soil}) to be dropped and simplifies computational complexities (Crank, 1956). Water flow from the bulk soil toward a germinating seed involves (1) water flow in the soil toward the seed surface, (2) flow across the soil-seed interface, (3) flow across the seed coat, and (4) flow into the seed itself (Phillips, 1968; Hadas, 1970, 1982; Koller and Hadas, 1982; Hadas and Russo, 1974a,b). To estimate water flow in each subsystem of the seed-soil system, Equation 1.2 has to be solved for each of the system components by using the specific boundary and initial conditions and the particular properties with respect to water (e.g., those of the seed, seed coat-soil, seed-soil interface)

Seed and Soil Diffusivities to Water

Several procedures, based on solving Equation 1.2 for the proper boundary and initial conditions, were used to calculate the seed mean diffusivity to water (for details see Phillips, 1968; Hadas, 1970; 1982). Reported seed diffusivity data for various seeds range from 1.5×10^{-5} to 1.6×10^3 m²/day (Phillips, 1968; Hadas, 1970; Shaykewich and Williams, 1971; Ward and Shaykewich, 1972; Hadas and Russo, 1974b). The reported values of soil diffusivity to water range between 4×10^4 and 5×10^7 m²/day for air-dry to near-saturated soils (Bruce and Klute, 1956; Rijtema, 1959; Kunze and Kirkham, 1962; Doering, 1965; Amemiya, 1965). The values for soils are higher than those reported for seeds. Hadas (1970) has shown that the seed radius increases as the amount of imbibed water increases. If seed swelling during imbibition tests for determining seed diffusivity to water was neglected, D_{seed} was found to increase with increasing seed mean water content (Phillips, 1968; Hadas, 1970; 1976, 1977a; Hadas and Russo,1974b; Ward and Shaykewich, 1972; Shaykewich and Williams, 1971). Collis-George and Melville (1975) used a solution of Equation 1.2 that accounted for seed swelling and found that the mean diffusivity of wheat seeds to water was 74 m²/day, a value which was practically the same as that reported by Ward and Shaykewich (1972), who used the simplified solution to Equation 1.2. Using reported data of soil diffusivity to water, Hadas (1970) showed that for a seed which maintains a low water potential and an active metabolic system, the soil can provide water to the seed at a greater rate than that observed experimentally. These calculations strongly suggest that seed water uptake and germination are controlled by seed coat impermeability apart from the low seed diffusivity to water.

Seed Coat and Seed-Soil Interface Diffusivity to Water

Water flow from the soil into a seed crosses the soil-seed interfacial zone, which consists of the seed coat and the seed-soil contact zone. Seed-soil contact is seldom perfect; therefore, a restriction is imposed on water flow from the soil into the seed. Seed coats vary in their permeability to water and may be impermeable, partially permeable, fully permeable, or even conditionally permeable in cases in which coat permeability varies in spots around the seed.

Seed Coat Permeability to Water

In general, seed coats are nonuniform in shape and roughness and present especially differentiated zones such as micropyle, hilum, chalaza, and areas covered with either hydrophilic or hydrophobic materials (Werker, Marbach, and Mayer, 1979). These variations in seed coat features (e.g., structure, ports of water entry) and properties affect seed coat permeability to water, seed water uptake rate, and seed-soil contact impedance to water flow (Christiansen and Moore, 1959; Manohar and Hydecker, 1974; Stone and Juhren, 1951; Quinlivan 1971). Morris, Campbell, and Wiebe (1968) reported seed coat permeability values ranging from 1.8×10^{-5} to 4.4×10^{-5} m/day for detached snapbean seed coats. Hadas (1976) calculated seed coat diffusivity to water ranging from 3×10^{1} to 3×10^{2}, 2.5×10^{-1} to 6×10^{0}, and 9×10^{-2} to 1.5×10^{0} m²/day for chickpea, pea, and vetch seeds, respectively. The lower values were for low seed coat hydration and increased with increasing coat hydration. Seed coat diffusivities to water, being much lower than those of a whole seed, may be considered to restrict imbibition. However, a decrease in water content, soil water conductivity, or seed-soil contact area, combined with low seed coat diffusivity to water, may restrict seed imbibition to a great extent (Dasberg and Mendel, 1971; Hadas, 1970; Hadas and Russo, 1974a; Williams and Shaykewich, 1971; Ward and Shaykewich, 1972).

Seed-Soil Interface Geometrical Configuration

The geometrical configuration of the seed-soil interface zone depends on a combination of seed coat surface properties, seed dimensions, and soil structure around the seed (Koller and Hadas, 1982; Hadas, 1982). Seeds placed on the soil surface or buried in a moist soil have partial contact with the soil particles; these contact points are few and of small area, thus their number and total area become negligible for seeds lying on the soil surface. The smaller the soil units are relative to the seeds, the greater the number of contact points and the total contact area will be (Hadas and Russo, 1974b; Hadas, Wolf, and Meirson, 1978). When these contact points are wetted with water, the contact area increases because water films and water collars form around the contact points; their shape and dimensions depend on the relative sizes of the seeds and the soil particles and on the water content (Collis-George and Hector, 1966; Hadas and Russo, 1974a,b). If the seeds are coated with a hairy cover, contact will be minimal unless the hairy cover is removed or the soil is compacted around the seeds. The wetted contact areas may be contiguous with impermeable sections of the seed coat, rendering these areas ineffective in transporting water to the imbibing seed. Sometimes, a minute contact point that touches a permeable area, identified as a *port of entry* (e.g., chalza, micropyle), can adequately supply the water needed by the imbibing seed (Berggren, 1963; Hyde, 1954; Manohar and Hydecker, 1974; Spurny, 1973). Most seeds, other than those sown in agricultural areas, once dispersed come to rest on the soil surface or fall into cracks, and their germination depends on: (1) enhanced seed-soil contact area because of soil surface roughness (Winkle, Roundy, and Box, 1991);

(2) wet priming (Finch-Savage and Pill, 1990); (3) swollen mucilaginous seed coats (Koller and Hadas, 1982); or (4) specific built-in burial mechanism (Gutterman, Witztum, and Evenari, 1967; Koller and Hadas, 1982; Meyer and Poljakoff-Mayber, 1989; Young and Evans, 1975).

Impedance to Water Flow Across the Seed-Soil Contact Zone

Under natural conditions, the hydraulic properties of the seed-soil contact zone vary during imbibition because of changing soil water content and seed-soil contact area. Moreover, these variations cannot be directly measured. Trials aimed at determining the effects of Ψ_{soil} and soil water conductivity on seed water uptake and germination have been carried out in porous substrates, soil plugs on sintered glass, or other materials. Either water flow was found to be restricted to a segment of the seed surface or the seeds were mechanically confined so that their swelling and imbibition were inhibited (Collis-George and Sands, 1959, 1962; Collis-George and Hector, 1966; Dasberg, 1971). The data interpretation in these studies was criticized by various researchers, who pointed out that the experimental procedures led to the observation of the combined effects of soil water content, soil mechanical stresses, water conductance, and seed-soil interface on germination (Hadas, 1970; Hadas and Russo, 1974a; Sedgley, 1963). Hadas and Russo (1974a,b) developed and used a procedure to enable determination of the separate effects of each of the water potential components, capillary conductivity to water, and seed-soil contact on seed germination. Their experimental results led to the conclusion that seed-soil contact impedance to flow increases with the decreasing seed wetted area, soil conductivity to water, or both. Contact impedance to water flow for a given size of seed and for a given Ψ_m increases with increasing coarseness of the soil texture, structure, or both. The final germination percentage was not affected by either Ψ_{soil} or $K(\theta_w)$, as long as Ψ_{soil} was higher than Ψ_b (Hadas and Russo, 1974a,b).

The model derived by Hadas and Russo (1974b) furnished a correlation between seed-soil water contact impedance and either wetted percentage of seed surface area or $K(\theta_w)$. Solutions of Equation 1.2, with the appropriate boundary and initial conditions for each part of the soil-seed system, enabled prediction of water uptake time courses that agreed well with data obtained in the laboratory and in small field plots. Those preliminary prediction capabilities prompted Hadas (1977a) to attempt to extend the model by correlating the predicted or measured imbibition time with the transition phase duration and final germination percentage. Good agreement was attained by using the extended model to predict the final germination and compare the forecast values with observed ones. These models, which have a physical basis, were found to be too cumbersome and unsuitable for practical application and their further refinement was abandoned.

These models, although not practical themselves, suggest some practical applications, namely that proper seedbed preparation which decreases the proportion of large-sized soil crumbs in the seedbed can control potential decreases in water uptake, rate of germination, and final germination percentage (Currie, 1973; Hadas, Wolf, and Meirson, 1978). Many seeds swell during imbibition, and the swollen seed imposes compacting forces on the particles around it, thus improving its contact with the soil and reducing seed-soil contact impedance. Concurrently, water uptake reduces soil water content next to the seed surface, while mechanical constraints and impedance to flow may increase. The combined effect of the two contrary trends may cause: (1) no change in the seed-soil interface impedance and, therefore, no delay of germination (Hadas, 1970, 1977b); (2) the effects of reduction in water content and water conductance to be greater than the effects of reduction in seed-soil contact impedance, thus impairing germination; or (3) the effects of reduced impedance to flow to be greater than those of reduced water content and conductance, so that water flow to the seed will not be restricted. It is obvious that the swelling of seeds lying on the soil surface will not compact the soil underneath, but rather will reduce their contact area, so that increased impedance to flow, reduced water uptake, and delayed germination are to be expected.

Under saline conditions, observed contact impedance effects may be partially obscured by the accumulation of excluded salts at the seed surface, which will reduce water uptake, germination rate, and final germination percentage, even though no changes in impedance to flow occurred (Hadas, 1970, 1976; Williams and Shaykewich, 1971).

The current knowledge of seed behavior during germination and their responses to changes in the soil environment has improved our understanding of the required conditions and properties of a seedbed. Although the various approaches to forecasting seed water uptake are, at best, good approximations, they do provide methods to be followed when planning seedbed preparations (Hadas and Russo, 1974a,b). It is evident that the crucial factors in seedbed preparation are control of the seed-soil contact area and the impedance to flow across the interface. This statement is based on the observations cited in the previous paragraphs. However, to date, the results of direct measurements of soil environmental conditions and of germination behavior have been either inconsistent or incomplete; therefore, we are left with large knowledge gaps which can be attributed to many factors. Probably the most important factors are (1) the extreme complexity of the soil system, including soil structure, stability, and hydraulic properties, and (2) theoretical aspects and experimental difficulties in the microscale analysis of flow around and into a seed and across the seed-soil interface.

Modeling Seedbed Structure: Temporal and Spatial Evolution of Seedbed Physical Properties

Any modeling effort aimed at characterizing seedbed physical properties and configuration has to rely on several criteria of soil structure characteristics and their spatial and temporal variations from initial preparation until seedling establishment. Such an endeavor first requires analytical presentation or modeling of the soil structure architecture, the physical properties derived from that architecture, the temporal and spatial evolution of soil structure changes, and the concurrent changes in the physical properties. Moreover, those procedures must be applied at various scales: soil structure stability must be considered at aggregate or subaggregate size, seed size, and on larger scales for a field stand (Hadas, 1997; Guerif et al., 2001).

In a detailed review Letey (1991) stated that soil structure does not lend itself to quantification. His statement was based on his recognition of the complexity involved in quantifying the heterogeneous soil structure. Tremendous knowledge gaps exist between what can be technically or experimentally obtained and the information and kind of data required for theoretical analysis. Young, Crawford, Rappoldt (2001), following Dexter (1988), who stated that spatial heterogeneity = spatial variability = soil structure, came to the conclusion that an explicit account of the heterogeneity inherent in the soil physical architecture has until recently been beyond experimental and theoretical insight.

These observations on the current state of the art indicate that special efforts are required to extend, improve, and create experimentally obtained databases which will yield empirical relationships between aggregate size, water regime, and structural stability. From these relationships, estimates of physical properties could be derived by means of currently existing models, operated at minute time and space steps. Taking heat, water, and air to be the major environmental factors involved in seed germination forecasting, the models chosen must be based on physical laws and must combine mass and energy fluxes and conservation principles. Moreover, although the numerical procedures will be complex, hard to follow, and difficult to handle, these models need to be validated. Partial efforts have already been made. Gupta and colleagues (1991) examined models used for predicting soil bulk density, water retention and conductivity, thermal conductivity, heat capacity, and gaseous diffusion with respect to their adaptation to fractured and compacted soils, and their critical examination identified shortcomings of the models. They pointed out knowledge gaps that require further research and indicated the difficulties to be expected in closing these gaps. Models of simultaneous heat and water transport have been developed for homogeneous soils (Nasser and Horton, 1992; Mullins et al., 1996), mulched soils or two-layered soils (Bristoe and Campbell, 1986), heterogeneous soils (Chung and Horton, 1987; Hares and Novak, 1992), and for ridged seedbeds (Benjamin, Ghaffarzadech, and Cruse, 1990; Gupta et al., 1990) and gaseous exchange (Richard and Guerif, 1988a,b). However, these models will have to be modified to include temporal changes in soil structure and the resulting variations in the physical properties.

Modeling soil fragmentation on the basis of soil dynamics, classical soil mechanics, and critical state theory, in an effort to predict the final soil structure, is only partly possible (Hettariachi, 1988), yet none of these models nor their extensions enable prediction of soil structure and seedbed tilth (Hadas, Larson, and Allmaras, 1988; Hadas, 1997). This statement is still valid (Guerif et al., 2001). These observations show that modeling of soil fragmentation is still a rather remote goal whose attainment will require tremendous efforts and much time.

CONCLUDING REMARKS

Timely, fast, and uniform seed germination, emergence, and final stand attainment are crucial for a successful crop and high yields, yet many field studies lack crucial information. Seeds deposited or sown respond individually to the microenvironment surrounding them. To specify the favorable soil physical properties, chemical constituents, microbiological population activity, and their interactions with one another and the climate would be an insurmountable task, in light of our current knowledge. Moreover, the microenvironment to which a seed responds tends to vary greatly and to induce great spatial variability across the field. Agricultural experience suggests that the soil physical properties are the major determinants of a successful seedbed conducive to optimal seed germination and stand attainment.

Although each seed reacts individually to its microenvironment, a field consists of a wide range of microenvironments. Since stand establishment under field conditions is our task, our approach to achieving that involves understanding how seeds germinate under field conditions. When cereals or grasses are considered, tillering is expected to correct the adverse effects of a poor stand, but for other crops, reduced emergence and low stand uniformity are associated with poor seedbed preparation (Perry, 1973; Hadas, Wolf, and Rawitz, 1983; Hadas and Wolf, 1984; Hadas et al., 1990).

There are great difficulties in tailoring recommended seedbeds, since field conditions present a great variety of soil structure stability, climatic uncertainties, and traffic history, all of which affect the performance of the next crop. Field conditions are difficult to reproduce under laboratory conditions, and, until recently, attempts to correlate laboratory studies (complex as they may be) failed to create a reliable database for field performance predictions. The material presented and discussed in the previous chapters presents a small variety of studies carried out to resolve that complex system and to furnish a reliable methodology for specifying the desired seedbed and the means to produce it. It is obvious that greater effort should be directed toward both the basic understanding of seed germination and the search for proven methodologies for specifying the proper seedbeds and recommending the means to achieve them.

REFERENCES

Al-Ani, A., Bruzau, F., Raymond P., Saint-Ges, V., Leblanc, J.M, and Pradet, A. (1985). Germination, respiration, and adenylate energy charge of seeds at various oxygen partial pressures. *Plant Physiology* 79: 885-890.

Al-Ani, A., Leblanc, J.M., Raymond, P., and Pradet, A. (1982). Effect de la pression particlle d'oxygene sur la vitesse de germination des semences a reserves lipidoques et amylacees: Role du metabolisme fermentaire. *Comptes Rendus de L'Academie des Sciences* 293: 271-274.

Allmaras, R.R., Hallauer, E.A., Nelson, W.W., and Evans, S.D. (1977). Surface energy balance and soil thermal property modifications by tillage-induced soil structure. *Agricultural Experimental Station University of Minnesota Technical Bulletin #306*.

Allrup, S. (1958). Effect of temperature on uptake of water in seeds. *Physiologia Plantarum* 11: 99-105.

Amemiya, M. (1965). The influence of aggregate size on moisture content-capillary conductivity relations. *Proceedings of the Soil Science Society of America* 29: 741-748.

Arndt, W. (1965a). The impedance of soil seals and the forces of emerging seedlings. *Australian Journal of Soil Research* 3: 55-68.

Arndt, W. (1965b). The nature of the mechanical impedance to seedlings by soil surface seals. *Australian Journal of Soil Research* 3: 45-54.

Ayers, A.D. (1952). Seed germination as affected by soil moisture and salinity. *Agronomy Journal* 44: 82-84.

Ayers, A.D. and Hayward, H.E. (1948). A method for measuring the effects of soil salinity on seed germination with observations on several crop plants. *Proceedings of the Soil Science Society of America* 13: 224-226.

Benech-Arnold, R.L. and Sanchez, R.A. (1995). Modeling weed seed germination. In Kigel, J. and Galili, G. (Eds.), *Seed Germination and Development* (pp. 545-566). New York: Marcel Dekker.

Benjamin, J.G., Ghaffarzadeh, M.R., and Cruse, R.M. (1990). Coupled water and heat transport in ridged soils. *Soil Science Society of America Journal* 54: 963-969.

Berggren, G. (1963). Is the ovula type of importance for the water absorption of the ripe seed? *Svensk Botanisk Tidskrift* 57: 377-395.

Berrie, A.M.M., Paterson, J., and West, H.P. (1974). Water content and responsivity of lettuce seeds to light. *Physiologia Plantarum* 31: 90-96.

Bewley, J.D. and Black, M. (1982). *Physiology and Biochemistry of Seeds*. Berlin: Springer Verlag.

Blacklow, W.M. (1973). Simulation model to predict germination and emergence of corn *Zea mays* L. in an environment of changing temperature. *Crop Science* 13: 604-608.

Bolt, G.H. and Koening, F.F.R. (1972). Physical and chemical aspects of the stability of sol aggregates. *Meded Fac Landbau Rijkuni Gent* 37: 955-973.

Boorman, L.A. (1968). Some aspects of the reproductive biology of *Limonium vulgare* Mill., and *Limonium humile* Mill. *Annals of Botany* (London) 32: 803-824.

Bowen, H.D. (1981). Alleviating mechanical impedance. In Arkin, G.F. and Taylor, H.M. (Eds.), *Modifying the Root Environment to Reduce Crop Stresses* (pp. 21-57). ASAE Monograph No. 4. St. Joseph, MI: American Society of Agricultural Engineers (ASAE).

Bradford, J.M. and Huang, C. (1992). Mechanics of crust formation: Physical components. In Sumner, M.E. and Stewart, B.A. (Eds.), *Soil Crusting: Chemical and Physical Processes* (pp. 55-72). Boca Raton, FL: Lewis Publishers.

Bradford, K.L. (1995). Water relations in seed germination. In Kigel, J. and Galili, G. (Eds.), *Seed Development and Germination* (pp. 351-396). New York: Marcel Dekker.

Braunack, M.V. and Dexter, A.R. (1989a). The effect of aggregate size in the seedbed on plant growth: A review. *Soil Tillage Research* 14: 281-291.

Braunack, M.V. and Dexter, A.R. (1989b). Soil aggregation in the seedbed—A review: II. Effect of aggregate sizes on plant growth. *Soil Tillage Research* 14: 181-298.

Braunack, M.V. and McPhee, J.E. (1991). The effect of the initial soil water content and tillage implement on seedbed formation. *Soil Tillage Research* 20: 5-17.

Bristow, K.L. and Campbell, G.S. (1986). Simulation of heat and moisture transfer through a surface-residue soil system. *Agriculture and Forest Meteorology* 36: 193-214.

Bruce, R.R. and Klute, A. (1956). The measurement of soil moisture diffusivity. *Proceedings of the Soil Science Society of America* 20: 458-462.

Buckingham, E. (1904). Contributions to our knowledge of the aeration of soils. U.S. Department of Agriculture Bureau Soils Bulletin #25. Washington, DC: U.S. Government Printing Office.

Buckingham, E. (1907). Studies in the movement of soil moisture. U.S. Department of Agriculture Soils Bulletin #38. Washington, DC: U.S. Government Printing Office.

Carslow, H.S. and Jaeger, J.E. (1959). *Conduction of Heat in Solids.* London: Oxford University Press.

Chahal, R.S. (1965). Effect of temperature and trapped air on matric suction. *Soil Science* 100: 262-266.

Christiansen, M.J. and Moore, P.P. (1959). Seed coat structure differences that influence water uptake and seed quality in hard seed cotton. *Agronomy Journal* 51: 582-584.

Christiansen, R.H., Foley, J., and Glanville, S.F. (1987). Effects of soil parameters on rain infiltration under several management systems. In Coughlin, K.J. and Troug, N.P. (Eds.), *Effects of Management Practices on Soil Physical Properties* (pp. 14-18). Toowoomba, Queensland, Australia: Proceedings of the National Workshop.

Chung, S.O. and Horton, R. (1987). Soil heat and water flow with a partial surface mulch. *Water Resources Research* 23: 2175-2186.

Collis-George, N. and Hector, J.B. (1966). Germination of seeds as influenced by matric potential and by area of contact between seed and soil water. *Australian Journal of Soil Research* 4: 145-164.

Collis-George, N. and Lloyd, J.E. (1979). The basis of a procedure to specify soil physical properties of a seedbed fir wheat. *Australian Journal of Agricultural Research* 30: 831-846.

Collis-George, N. and Melville, M.D. (1975). Water absorption by swelling seeds: I. Constant surface boundary conditions. *Australian Journal of Soil Research* 13: 141-158.

Collis-George, N. and Sands, J.E. (1959). The control of seed germination by moisture as a physical property. *Australian Journal of Agricultural Research* 10: 628-637.

Collis-George, N. and Sands, J.E. (1962). Comparison of the effects of physical and chemical components of soil water energy on seed germination. *Australian Journal of Agricultural Research* 13: 575-585.

Collis-George, N. and Williams, J. (1968). Comparison of the effects of soil matric potential and isotropic effective stress on the germination of *Lactuca sativa. Australian Journal of Soil Research* 6: 179-192.

Come, D. and Corbineau, F. (1992). Environmental control of seed dormancy and germination. In Jiarui, F. and Khan, A.A. (Eds.), *Advances in the Science and Technology of Seeds* (pp. 288-298). Beijing, New York: Science Press.

Come, D. and Tissaoui, T. (1973). Interrelated effects of imbibition, temperature and oxygen on seed germination. In Heydecker, W. (Ed.), *Seed Ecology* (pp. 157-168). London: Butterworth.

Corbineau, F. and Come, D. (1995). Control of seed germination and dormancy by the gaseous environment. In Kigel, J. and Galili, G. (Eds.), *Seed Development and Germination* (pp. 397-424). New York: Marcel Dekker.

Crank, J. (1956). *The Mathematics of Diffusion.* London: Oxford University Press.

Currie, J.A. (1961). Gaseous diffusion in porous media. Part 3: Wet granular materials. *British Journal of Applied Physics* 12: 275-281.

Currie, J.A. (1973). The seed-soil system. In Heydecker, W. (Ed.), *Seed Ecology* (pp. 463-479). London: Butterworth.

Currie, J.A. (1983). Gas diffusion through soil crumbs: The effects of wetting and swelling. *Journal of Soil Science* 34: 217-232.

Currie, J.A. (1984). Gas diffusion through soil crumbs: The effects of compaction and wetting. *Journal of Soil Science* 35: 1-10.

Currie, J.A. and Rose, D.A. (1965). Gas diffusion in structured materials: The effect of trimodal pore size distribution. *Journal of Soil Science* 36: 487-493.

Dasberg, S. (1971). Soil water movement to germinating seeds. *Journal of Experimental Botany* 22: 999-1008.

Dasberg, S., Hillel, D., and Arnon, I. (1966). Response of grain sorghum to seedbed compaction. *Agronomy Journal* 58: 199-201.

Dasberg, S. and Mendel, K. (1971). The effect of soil water and aeration on seed germination. *Journal of Experimental Botany* 22: 992-998.

de Miguel, L. and Sanchez, R.A. (1992). Phytochrome induced germination, endosperm softening and embryo growth potential in *Datura ferox* seeds: Sensitivity to low water potential and time to escape to FR reversal. *Journal of Experimental Botany* 45: 969-974.

de Vries, D.A. (1958). Simultaneous transfer of heat and water in porous media. *Transactions of the American Geophysical Union* 39: 909-916.

de Vries, D.A. (1963). Thermal properties of soils. In van Wijk W.R. (Ed.), *Physics of Plant Environment* (pp. 210-236). Amsterdam, the Netherlands: North-Holland.

de Willingen, P. and van Noorwijk, M. (1987). Root, plant production and nutrient use efficiency. PhD dissertation. Agricultural University, Wageningen, the Netherlands.

Dexter, A.R. (1977). Effect of rainfall on the surface microrelief of tilled soil. *Journal of Terramechanics* 14: 11-22.

Dexter, A.R. (1988). Advances in characterizaton of soil structure. *Soil Tillage Research* 11: 199-239.

Dexter, A.R. (1991). Soil amelioration by natural processes. *Soil Tillage Research* 20: 87-100.

Doering, E.J. (1965). Soil-water diffusivity by one-step method. *Soil Science* 99: 322-326.

Egley, G.H. (1995). Seed germination in soil: Dormancy cycles. In Kigel, Y. and Galili, G. (Eds.), *Seed Development and Germination* (pp. 397-424). New York: Marcel Dekker.

Evans, W.F. and Strickler, F.C. (1961). Grain sorghum seed germination under moisture and temperature stresses. *Agronomy Journal* 53: 369-372.

Farrell, D.A., Greacen, E.L., and Gurr, C.G. (1966). Vapor transfer in soil due to air turbulence. *Soil Science* 102: 303-315.

Farrell, D.A. and Larson, W.E. (1973). Effect of intra-aggregate diffusion on oscillatory flow dispersion in aggregated beds. *Water Resources Research* 9: 185-193.

Fenner, M. (1991). *Seeds: The Ecology of Regeneration in Plant Communities.* Wallingford, Oxon, United Kingdom: CAB International.

Finch-Savage, W.E. and Pill, W.G. (1990). Improvement of carrot crop establishment by combining seed treatments with increased seed-bed moisture availability. *Journal of Agricultural Science* 115: 75-81.

Gill, W.R. and van den Berg, G.E. (1967). *Soil Dynamics in Tillage and Traction.* Agricultural Handbook #316. Washington, DC: U.S. Department of Agriculture.

Glinski, J. and Lipiec, J. (1990). *Soil Physical Conditions and Plant Roots.* Boca Raton, FL: CRC Press.

Glinski, J. and Stepniewski, W. (1985). *Soil Aeration and Its Role for Plants.* Boca Raton, FL: CRC Press.

Greenwood, D.J. (1975). Measurement of soil aeration. In *Soil Physical Conditions and Crop Production* (pp. 261-272). Ministery of Agriculture, Fish and Food Techology Bulletin #29. London: Her Majesty's Stationery Office.

Guerif, J., Richard, G., Durr, C., Machet, J.M., Recous, S., and Roger-Estrade, J. (2001). A review of tillage effects on crop residue management, seedbed conditions and seedling establishment. *Soil Tillage Research* 61: 13-32.

Gupta, S.C. and Larson, W.E. (1982). Modeling soil mechanical behavior during tillage. In Unger, P.M. (Ed.), *Predicting Tillage Effects on Soil Physical Properties and Processes* (pp. 15-178). ASAE Special Publication #44. Madison, WI: American Society of Agronomy.

Gupta, S.C., Lowery, B., Moncrief, J.F., and Larson, W.E. (1991). Modelling tillage effects on soil physical properties. *Soil Tillage Research* 20: 293-318.

Gupta, S.C., Radke, J.K., Swan, J.B., and Moncrief, J.F. (1990). Predicting soil temperature under a ridge-farrow system in the U.S. corn belt. *Soil Tillage Research* 18: 145-165.

Gupta, S.C., Sharma, P.P., and De Franchi, S. (1989). Compaction effects on soil structure. *Advances in Agronomy* 42: 311-338.

Gutterman, Y. (1992). Maternal effects on seeds during development. In Fenner, M. (Ed.), *Seeds: The Ecology of Regeneration in Plant Communities* (pp. 27-59). Wallingford, Oxon, United Kingdom: CAB International.

Gutterman, Y., Witztum, A., and Evenari, M. (1967). Seed dispersal and germination in *Blepharis persica* (Burn) Kunze. *Israel Journal of Botany* 16: 213-234.

Haber, A.H. and Luippold, H.L. (1960). Effects of gibberelin, kinetin, thiourea and photomorphogenic radiation on mitotic activity in dormant lettuce seed. *Plant Physiology* 35: 486-494.

Hadas, A. (1968). Simultaneous flow of water and heat under periodic heat fluctuations. *Proceedings of the Soil Science Society of America* 32: 297-301.

Hadas, A. (1969). Effects of soil moisture stress on seeds germination. *Agronomy Journal* 61: 325-327.

Hadas, A. (1970). Factors affecting seed germination under soil moisture stress. *Israel Journal of Agricultural Research* 20: 3-14.

Hadas, A. (1975). Effect of external evaporative conditions on drying of soil. In *Transactions of the 10th International Congress of Soil Science* (Moscow, USSR) (pp. 1:136-142). Moscow: Nauka Publishing House.

Hadas, A. (1976). Water uptake and germination of leguminous seeds under changing external water potential conditions in osmotic solutions. *Journal of Experimental Botany* 27: 480-489.

Hadas, A. (1977a). A simple laboratory approach to test and estimate seed germination performance under field conditions. *Agronomy Journal* 69: 582-585.

Hadas, A. (1977b). Water uptake and germination of leguminous seeds under changing matric and osmotic water potential. *Journal of Experimental Botany* 28: 977-985.

Hadas, A. (1982). Seed-soil contact and germination. In Khan, A.A. (Ed.), *The Physiology and Biochemistry of Seed Development, Dormancy and Germination* (pp. 507-527). Amsterdam, the Netherlands: Elsevier.

Hadas, A. (1985). Water absorption by swelling leguminous seeds as affected by water potential and external mechanical constraints. *Israel Journal of Botany* 34: 7-16.

Hadas, A. (1987a). Dependence of true surface energy of soils on air entry pore size and chemical constituents. *Soil Science Society of America Journal* 51: 186-191.

Hadas, A. (1987b). Long term tillage practice effects on soil aggregation modes and strength. *Soil Science Society of America Journal* 51: 186-191.

Hadas, A. (1997). Soil tilth—The desired soil structural state obtained through proper soil fragmentation and reorientation processes. *Soil Tillage Research* 43: 7-40.

Hadas, A. and Fuchs, M. (1973). Prediction of the thermal regime of bare soils. In Hadas, A., Swartzendruber, D., Rijtema, P.E., Fuchs, M., and Yaron, B. (Eds.), *Physical Aspects of Soil Water and Salts in Ecosystems* (pp. 293-300). Ecological Studies 4. Berlin: Springer Verlag.

Hadas, A., Larson, W.E., and Allmaras, R.R. (1988). Advances in modelling machine-soil-plant interactions. *Soil Tillage Research* 11: 349-372.

Hadas, A. and Russo, D. (1974a). Water uptake by seeds as affected by water stress, capillary conductivity and seed soil contact: I. Experimental study. *Agronomy Journal* 66: 643-647.

Hadas, A. and Russo, D. (1974b). Water uptake by seeds as affected by water stress, capillary conductivity and seed soil contact: II. Analysis of experimental data. *Agronomy Journal* 66: 647-652.

Hadas, A. and Shmulewich, I. (1990). Spectral analysis of cone penetrometer data for detecting spatial arrangement of soil clods. *Soil Tillage Research* 18: 47-62.

Hadas, A., Shmulewich, I., Hadas, O., and Wolf, D. (1990). Forage wheat yields as affected by compaction and conventional vs. wide frame tractor traffic patterns. *Transactions of the ASAE* 33: 79-85.

Hadas, A. and Stibbe, E. (1973). Analysis of water uptake and growth patterns of seedlings of four species prior to emergence. In Hadas, A., Swartzendruber, D., Rijtema, P.E., Fuchs, M., and Yaron, B. (Eds.), *Physical Aspects of Soil Water and Salts in Ecosystems* (97-106). Ecological Studies 4. Berlin: Springer Verlag.

Hadas, A. and Stibbe, E. (1977). Soil crusting and emergence of wheat seedlings. *Agronomy Journal* 69: 547-550.

Hadas, A. and Wolf, D. (1983). Energy efficiency in tilling dry clod forming soils. *Soil Tillage Research* 3: 47-59.

Hadas, A. and Wolf, D. (1984). Refinement and reevaluation of the drop shatter soil fragmentation method. *Soil Tillage Research* 4: 237-249.

Hadas, A., Wolf, D., and Meirson, I. (1978). Tillage implements-soil structural relationships and their effects on crop stands. *Soil Science Society of America Journal* 42: 632-637.

Hadas, A., Wolf, D., and Rawitz, E. (1983). Zoning soil compaction and cotton stand under controlled traffic conditions. Paper No. 83-1042. ASAE 1983 Meeting, Bozeman, Montana.

Hadas, A., Wolf, D., and Rawitz, E. (1985). Residual compaction effects on cotton stand and yields. *Transactions of the ASAE* 28: 691-696.

Hadas, A., Wolf, D., and Stibbe, E. (1981). Tillage practices and crop response analysis of agro-ecosystems. *Agro-Ecosystems* 6: 235-248.

Hanks, R.J. and Thorp, F.C. (1957). Seedling emergence of wheat, grain sorghum and soy-beans as influenced by soil crust strength and moisture content. *Soil Science Society of America Proceedings* 21: 357-360.

Hares, M.A. and Novak, M.D. (1992). Simulation of energy balance and soil temperature under strip tillage: I. Model description. *Soil Science Society of America Journal* 56: 22-29.

Hegarty, T.W. (1978). The physiology of seed hydration and dehydration and the relation between water stress and the control of germination: A review. *Plant Cell and Environment* 1: 101-119.

Hegley, G.H. (1995). Seed germination in the soil: Dormancy cycles. In Kigel, J. and Galili, G. (Eds.), *Seed Development and Germination* (pp. 529-541). New York: Marcel Dekker.

Hettariachi, D.R.P. (1988). Theoretical soil mechanics and plant implement design. *Soil Tillage Research* 11: 325-348.

Heydecker, W. and Orphanos, P.I. (1968). The effect of excess moisture on the germination of *Spinacia oleracea* L. *Planta* 83: 237-247.

Hillel, D. (1960). Crust formation in loessial soil. *Transactions of the Seventh International Congress of Soil Science* I: 330-340.

Hillel, D. and Hadas, A. (1972). Isothermal drying of structurally layered soil columns. *Soil Science* 113: 65-73.

Holmes, J.W., Greacen, E.L., and Gurr, C.G. (1960). The evaporation of water from bare soils with different tilths. *Transactions of the Seventh International Congress of Soil Science* 1: 188-194.

Horn, R., Taubner, H., Wuttke, M., and Baumgartl, T. (1994). Soil physical properties related to soil structure. *Soil Tillage Research* 30: 187-216.

Hudspeth, E.B. and Taylor, H.M. (1961). Factors affecting seedling emergence of Blackwell switchgrass. *Agronomy Journal* 53: 331-335.

Hunter, J.R. and Erickson, A.E. (1952). Relation of seed germination to soil moisture tension. *Agronomy Journal* 44: 107-110.

Hyde, E.O.C. (1954). The function of the hilum in some Papilonaceae in relation to the ripening of the seed and the permeability of the testa. *Annals of Botany* (London) 18: 241-256.

Jackson, R.D. (1973). Diurnal changes in the soil water content during drying. In Bruce, R.R, Flach, K., and Taylor, H.M. (Eds.), *Field Soil Water Regime* (pp. 37-55). Special Publication #5. Madison, WI: Soil Science Society of America.

Karssen, C.M. (1970).The light promoted germination of the seeds of *Chenopodium album* L.: V. Dark reactions regulating quantity and the rate of response to red light. *Acta Botanica Neerlandica* 19: 187-196.

Kebreab, E. and Murdoch, A.J. (1999a). A model of the effects of a wide range of constant and alternating temperatures on seed germination of four Orobanche species. *Annals of Botany* 84: 549-557.

Kebreab, E. and Murdoch, A.J. (1999b). Modelling the effect of water stress and temperature on germination rate of *Orobanche aegyptiaca* seeds. *Journal of Experimental Botany* 50: 655-669.

Keen, B.A. (1931). *The Physical Properties of Soils.* London: Longmans and Green Co.

Khan, A.A. (1975). Primary, preventive and permissive roles of hormones in plant systems. *Botanical Review* 4: 391-420.

Kigel, J. (1995). Seed germination in arid and semi arid regions. In Kigel, J. and Galili, G. (Eds.), *Seed Development and Germination* (pp. 645-699). New York: Marcel Dekker.

Kimball, B.A. and Lemon, E.R. (1971). Air turbulence effects upon soil gas exchange. *Soil Science Society of America Journal* 35: 16-21

Kirkham, D. and Powers, W.L. (1972). *Advanced Soil Physics.* New York: Wiley-Interscience.

Koller, D. (1970). Analysis of the dual action of white light on germination of *Atriplex dimorphostegia* (Chenopodiceae). *Israel Journal of Botany* 19: 499-516.

Koller, D. and Hadas, A. (1982). Water relations in the germination of seeds. In Lange, O.L., Nobel, P.S., Osmond, C.B., and Zigler, H. (Eds.), *Encyclopedia of Plant Physiology,* Volume 12B (pp. 401-431). Berlin: Springer Verlag.

Kuipers, H. (1984). The challenge of soil cultivation and soil water problems. *Journal of Agricultural Engineering Research* 29: 177-190.

Kunze, R. and Kirkham, D. (1962). Simplified accounting for membrane impedance in capillary conductivity measurement. *Soil Science Society of America Proceedings* 26: 421-426.

Kutilek, M. and Nielsen, D.R. (1994). *Soil Hydrology.* Cremlingen-Destedt, Germany: Catena Verlag.

Lal, R. (1991). Tillage and agriculture sustainability. *Soil Tillage Research* 20: 133-145.

Letey, J. (1991). The study of soil structure: Science or art? *Australian Journal of Soil Science* 29: 699-707.

Manohar, M.S. and Heydecker, W. (1974). Effects of water potential on germination of pea seeds. *Nature* 202: 22-24.

Marshall, T.J. (1958). A relation between permeability and size distribution of pores. *Journal of Soil Science* 9: 1-8.

Marshall, T.J. (1959). Diffusion of gases through porous media. *Journal of Soil Science* 10: 79-82

Marshall, T.J., Holmes, J.W., and Rose, C.W. (1996). *Soil Physics.* Cambridge, United Kingdom: Cambridge University Press.

Martin, V.L., McCoy, E.L., and Dick, W.A. (1990). Allelopathy of crop residues influences corn seed germination and early growth. *Agronomy Journal* 82: 555-560.

Mayer, A.M. and Poljakoff-Mayber, A. (1989). *The Germination of Seeds,* Fourth Edition. London: Pergamon.

McDonough, W.T. (1975). Water potential of germinating seeds. *Botanical Gazette* 136: 106-108.

McGinnies, W.J. (1960). Effects of moisture stress and temperature on germination of six range grasses. *Agronomy Journal* 52: 159-163.

McPhee, J.E., Braunack, M.V., Garside, A.L., Reid, D.J., and Hilton, D.J. (1995). Controlled traffic for irrigated double cropping operations and energy use. *Journal of Agricultural Engineering Research* 60: 183-189.

Morin, J. and Winkler, J. (1996). The effect of raindrop impact and sheet erosion on infiltration rate and crust formation. *Soil Science Society of America Journal* 60: 1223-1227.

Morinaga, T. (1926). Effect of alternating temperatures on the germination of seeds. *American Journal of Botany* 13: 141-158.

Morris, I.J., Campbell, W.F., and Wiebe, H.H. (1968). A refractometric method for testing seed coat permeability. *Agronomy Journal* 60: 79-80.

Mullins, C.E., Townend, J., Mtakwa, P.W., Payne, C.A., Cowen, G., Simmonds, L.P., Daamen, C.C., Dunbabin,T., and Naylor, R.E.L. (1996). *Emergence User Guide: A Model to Predict Crop Emergence in the Semi-Arid Tropics.* Aberdeen, United Kingdom: Department of Plant and Soil Science, Aberdeen University.

Nasr, H.M. and Selles, F. (1995). Seedling emergence as influenced by aggregate size, bulk density, and penetration resistance of the seedbed. *Soil Tillage Research* 34: 61-76.

Nasser, I.N. and Horton, R. (1992). Simulation transfer of heat, water and solute in porous media: I. Theoretical development. *Soil Science Society of America Journal* 56: 1350-1357.

Ni, B.R. and Bradford, K.J (1992). Quantitative models characterizing seed germination responses to abscisic acid and osmoticum. *Plant Physiology* 98: 1057-1068.

Oades, J.M. and Watts, A. (1991). Aggregate hierarchy in soils. *Australian Journal of Soil Science* 29: 815-828.

Ojeniyi, S.O. and Dexter, A.R. (1979a). Soil factors affecting the macro-structure produced by tillage. *Transactions of the ASAE* 22: 339-343.

Ojeniyi, S.O. and Dexter, A.R. (1979b). Soil structural changes during multiple pass tillage. *Transactions of the ASAE* 22: 1068-1072.

Peerlkamp, P.K. (1944). Bodenmeteorologische onderzockingen te Wageningen. *Meded. Landbouwhogeschul Wageningen* 47: 1-96.

Penhman, H.L. (1940). Gas and vapour movement in soil: II. The diffusion of carbon dioxide through porous solids. *Journal of Agricultural Science* 30: 570-581.

Perdok, U.D. and Kouwehoven, J.K. (1994). Soil-tool interaction and field performance of implements. *Soil Tillage Research* 30: 283-326.

Perfect, E., Zhai, Q., and Belvins, R.L. (1997). Soil and tillage effects on the characteristic size and shape of aggregates. *Soil Science Society of America Journal* 61: 1459-1465.

Perry, D.A. (1973). Interacting effects of seed vigour and environment on seedling establishment. In Heydecker, W. (Ed.), *Seed Ecology* (pp. 463-479). London: Butterworth.

Philip, J.R. and de Vries, D.A. (1957). Moisture movement in porous materials under temperature gradients. *Transactions of the American Geophysics Union* 38: 222-232.

Phillips, R.E. (1968). Water diffusivity of germinating soy-bean, corn, and cotton seed. *Agronomy Journal* 60: 568-571.

Pollock, B.M. and Toole, V.K. (1966). Imbibition period as a critical temperature sensitive stage in germination of lima bean seeds. *Plant Physiology* 41: 221-229.

Powell, A.D., Dulson, J., and Bewley, J.D. (1984). Changes in germination and respiratory potential of embryos of dormant Grand Rapids lettuce seeds during long-term imbibed storage, and related changes in the endosperm. *Planta* 162: 40-45.

Probert, R.J. (1992). The role of temperature in germination ecophysiology. In Fenner, M. (Ed), *Seeds: The Ecology of Regeneration in Plant Communities* (pp. 285-325), Wallingford, Oxon, United Kingdom: CAB International.

Quinlivan, B.J. (1971). Seed coat impermeability in legumes. *Journal of the Australian Institute of Agricultural Science* 37: 283-295.

Richard, G. and Guerif, J. (1988a). Modelisation des transferts gazeux dans le lit semance: Application au diagnostic des conditions d'hypoxie des semaneces de beterave sucriere (*Beta vulgaris* L.) pendent la germination. I: Presentation du modele. *Agronomie* 8: 539-547.

Richard, G. and Guerif, J. (1988b). Modelisation des transferts gazeux dans le lit semence: Application au diagnostic des conditions d'hypoxie des semeneces de beterave sucriere (*Beta vulgaris* L.) pendent la germination. II: Resultats des simulations. *Agronomie* 8: 639-646.

Richards, L.A. (1965). Physical condition of water in soil. In Klute, A. (Ed.), *Methods of Soil Analysis,* Volume 1 (pp. 128-137). Agronomy Monograph No. 9. New York: Academic Press.

Richards, L.A. and Wadleigh, C.H. (1952). Soil water and plant growth. In Shaw, B.T. (Ed.), *Soil Physical Conditions and Plant Growth* (pp. 73-251). Agronomy Monograph No. 2. New York: Academic Press.

Rijtema, P.E. (1959). Calculation of capillary conductivity from pressure plate outflow data with non-negligible membrane impedance. *Netherlands Journal of Agricultural Science* 7: 209-216.

Roberts, E.H. and Smith, R.D. (1977). Dormancy and the pentose phosphate pathway. In Khan, A.A. (Ed.), *The Physiology and Biochemistry of Seed Dormancy and Germination* (pp. 385-411). Amsterdam, the Netherlands: Elsevier.

Rommel, L.G. (1922). Luftvaxlingen I marken som ekologisk faktor. *Medded Statens Skogsfarsoksanstalt* 19: no.2.

Rose, C.W. (1968). Water transport in soil with a daily temperature wave: 1. Theory and experiment. *Australian Journal of Soil Research* 6: 31-44.

Roytenberg, E. and Cheplin, J.S. (1995). Stochastic modeling of soil conditions during tillage. In Robert, P.C., Rust, R.H., and Larson, W.E. (Eds.), *Site Specific Management for Agricultural Systems* (pp. 581-599). Madison, WI: Second International Conference of the American Society of Agronomy.

Rumplo, M.E., Pradet, A., Khalik, A., and Kennedy, R.A. (1984). Energy change and emergence of the coleoptile and radicle at varying oxygen levels in *Echinocloa cruss galli. Physiologia Plantarum* 62: 133-139.

Russell, E.Y. (1973). *Soil Conditions and Plant Growth*, Tenth Edition. London: Longmans.

Scopel, A.L., Ballare, C.L., and Sanchez, R.A. (1991). Induction of extreme light sensitivity in buried weed seeds and its role in the perception of soil cultivation. *Plant, Cell and Environment* 14: 501-508.

Sedgley, R.H. (1963). The importance of liquid seed contact during the germination of *Medicago tribuloides. Australian Journal of Agricultural Research* 14: 646-654.

Shaykewich, C.F. (1973). Proposed method for measuring swelling pressure of seeds prior to germination. *Journal of Experimental Botany* 24: 1056-1061.

Shaykewich, C.F. and Williams, J. (1971). Resistance to water absorption in germinating rape seed (*Brassica napus* L.). *Journal of Experimental Botany* 22: 19-24.

Simon, E.W. (1974). Phospholipids and plant membrane permeability. *New Phytologist* 73: 377-420.

Slipher, J.A. (1932). The mechanical manipulation of soil as it affects structure. *Agricultural Engineering* 13: 7-10.

Smith, K.A. and Dowdell, R.J. (1974). Field studies of the soil atmosphere: 1. Relationships between ethylene, oxygen, soil moisture content and temperature. *Journal of Soil Science* 25: 217-230.

Smoke, M.A., Chojnowski, M., Corbineau, F., and Come, D. (1993). Effects of osmotic treatment on sunflower seed germination in relation with temperature and oxygen. In Come, D. and Corbineau, F. (Eds.), *Fourth International Workshop on Seeds: Basic and Applied Aspects of Seed Biology,* Volume 3 (pp. 1033-1038). Paris, France: ASFIS.

Spoor, G. and Giles, D.F.H. (1973). Effects of cultivation on raising spring soil temperature for germination with particular reference to maize. *Journal of Soil Science* 24: 392-398.

Spurny, M. (1973). The imbibition process. In Heydecker, W. (Ed.), *Seed Ecology* (pp. 367-389). London: Butterworth.

Stiles, W. (1948). Respiration. In Nobel, P.S., Osmond, C.B., and Zigler, H. (Eds.), *Seed Germination and Seedling Development, Encyclopedia of Plant Physiology,* Volume 12 (pp. 465-492). Berlin, Heidelberg: Springer Verlag.

Stone, E.C. and Juhren, G. (1951). The effect of fire on the germination of the seed of *Rhus ovata* Watts. *American Journal of Botany* 38: 368-372.

Stout, D.J., Brooke, B., and Hall, J.W. (1999). Effects of large diurnal temperature variation on alfalfa seed germination and hard seed content. *Seed Technology* 21: 5-14.

Taylor, D.L. (1942). Influence of oxygen tension on respiration, fermentation and growth in wheat and rice. *American Journal of Botany* 29: 721-738.

Taylor, S.A. and Stewart, G.L. (1960). Some thermodynamic properties of soil water. *Proceedings of the Soil Science Society of America* 24: 243-249.

Taylorson, R.B. and Hendricks, S.B. (1972). Rehydration of phytochrome in imbibing seeds of *Amaranthus retroflexus. Plant Physiology* 49: 663-665.

Taylorson, R.B. and Hendricks, S.B. (1977). Dormancy in seeds. *Annual Review of Plant Physiology* 28:551-354.

Tennent, D. and Humblin, A.P. (1987). Measurement of soil physical conditions in the seedbed under planting systems. In Coughlan, K.J. and Truong, N.P. (Eds.), *Effects of Management Practices on Soil Physical Properties* (pp. 43-47). Toowoomba, Queensland, Australia: Proceedings of the National Workshop. Queensland Department of Primary Industries, Brisbane.

Thurburn, P.J., Hansen, P.B., and Glenville, S.F. (1987). The effect of tillage and stuble management on the physical properties of a gray cracking clay. In Coughlan, K.J. and Truong, N.P. (Eds.), *Effects of Management Practices on Soil Physical Properties* (pp. 19-23). Toowoomba, Queensland, Australia: Proceedings of the National Workshop. Queensland Department of Primary Industries, Brisbane.

Tisdall, J.M. (1996). Crop establishment—A serious limitation to high productivity. *Soil Tillage Research* 40: 1-2.

Tisdall, J.M. and Hodgson, A.S. (1990). Ridge tillage in Australia: A review. *Soil Tillage Research* 18: 127-144.

Tisdall, J.M. and Oades, J.M. (1982). Organic matter and water stable aggregates in soil. *Journal of Soil Science* 19: 70-95.

Tobin, E.M. and Briggs, W.R. (1969). Phytochrome in embryos of *Pinus palustris. Plant Physiology* 44: 148-150.

Uhvits, R. (1946). Effect of osmotic pressure on water absorption and germination of alfalfa seeds. *American Journal of Botany* 33: 278-285.

Unger, P.W. (1982). Residual effects of soil profile modification on water infiltration, bulk density and wheat yields. *Agronomy Journal* 85: 656-659.

Unger, P.W. and Kaspar, T.C. (1994). Soil compaction and root growth: A review. *Agronomy Journal* 86: 759-766.

van Duin, R.H.A. (1956). *On the Influence of Heat, Diffusion of Air and Infiltration of Water in the Soil.* Wageningen, the Netherlands: Verlag Landbouwh.

van Noorwijk, M. and de Willingen, P. (1991). Root function in agricultural systems. In McMichaal, M.L. and Pearson, H. (Eds.), *Plant Roots and Their Function* (pp. 381-395). Amsterdam, the Netherlands: Elsevier.

van Wijk, W.R. (1963). *Physics of Plant Environment.* Amsterdam, the Netherlands: North Holland.

van Wijk, W.R. and Dirksen, W.J. (1963). Sinusoidal temperature variations in a layered soil. In van Wijk, W.R. (Ed.), *Physics of Plant Environment* (pp.171-210). Amsterdam, the Netherlands: North Holland.

Voorhees, W.B., Allmaras, R.R., and Johnson, C.E. (1981). Alleviating temperature stress. In Arkin, G.F. and Taylor, H.M. (Eds.), *Modifying the Root Environment to Reduce Crop Stresses* (pp. 217-266). ASAE Monograph No. 4. St. Joseph, MI: ASAE.

Ward, J. and Shaykewich, C.F. (1972). Water absorption by wheat seeds as influenced by hydraulic properties of the soil. *Canadian Journal of Soil Science* 52: 99-105.

Werker, E., Marbach, I., and Mayer, A.M. (1979). Relation between the anatomy of the testa, water permeability and the presence of phenolics in the genus *Pisum*. *Annals of Botany* (London). 43: 765-771.

Wesseling, J. (1974). Crop growth and wet soils. In Schilfgaarde, J. (Ed.), *Drainage for Agriculture* (pp. 7-37), ASAE Monograph No. 17. Madison, WI: ASAE.

Wiggans, S.C. and Gardner, E.B. (1962). Effectiveness of various solutions for simulating drought conditions as measured by germination and seedling growth. *Agronomy Journal* 51: 315-318.

Williams, J. and Shaykewich, C.F. (1971). Influence of soil matric potential and hydraulic conductivity on germination of rape seed (*Brassica napus* L.). *Agronomy Journal* 51: 315-318.

Winkle, V.K., Roundy, B.A., and Box, J.R. (1991). Influence of seedbed microsite characteristics on grass seedling emergence. *Journal of Range Management* 44: 210-214.

Witztum, A., Gutterman, Y., and Evenari, M. (1969). Integumentary mucilage as an oxygen barrier during germination of *Blepharis persica* (Burn) Kunze. *Botanical Gazette* 130: 238-241.

Wolf, D. and Hadas, A. (1984). Soil compaction effects on cotton emergence. *Transactions of the ASAE* 27: 655-659.

Wolf, D. and Hadas, A. (1987). Determining efficiencies of various mouldboard ploughs in fragmenting and tilling air-dry soil. *Soil Tillage Research* 10: 181-186.

Wuest, S.B., Albrecht, S.L., and Skirvin, K.W. (1999). Vapor transport vs. seed-soil contact in wheat germination. *Agronomy Journal* 91: 783-787.

Young, I.M., Crawford, J.W., and Rappoldt, C. (2001). New methods and models for characterising structural heterogeneity of soils. *Soil Tillage Research* 61: 33-45.

Young, J.A. and Evans, R.E. (1975). Mucilagenous seed coats. *Weed Science* 21: 52-54.

Chapter 2

The Use of Population-Based Threshold Models to Describe and Predict the Effects of Seedbed Environment on Germination and Seedling Emergence of Crops

William E. Finch-Savage

INTRODUCTION

Importance of Seedling Emergence to Crop Production

Seed germination and subsequent seedling growth to emergence from the soil are crucial steps in crop production. Although some field crops such as cereals can compensate for low stands by tillering, in many crop species no amount of effort and cost during plant growth can compensate for poor seedling establishment. A wide range of biotic and environmental factors interact with the potential performance of the seed lot to determine the success of seedling establishment (Hegarty, 1984). This chapter will use population-based threshold models to summarize current understanding of the interaction between the seedbed environment and the seed population from sowing to seedling emergence. The potential for these threshold models to predict seedling emergence in the field will then be discussed while describing the construction of an example simulation. In order to see the relevance and importance of the studies reviewed, it is necessary to briefly outline the consequences of nonoptimal seedling emergence in crops.

The timing, pattern, and extent of seedling emergence have a profound impact on crop yield and market value (Finch-Savage, 1995). Only part of

I would like to thank my colleagues Hugh Rowse, Kath Phelps, and Richard Whalley with whom I have collaborated over a number of years on crop establishment projects, in particular Hugh Rowse for use of his unpublished work. I thank the U.K. Department for Environment, Food and Rural Affairs (DEFRA) and, more recently, the Department for International Development (DFID) who have funded these collaborations.

the total biomass produced is harvested, and this component is crop and market specific. This economic yield is often determined by the whole plant population as a bulk weight per unit area, as in grain or sugar beet crops; in many horticultural crops economic yield is determined by individual plants within the population, for example, the number of plants within closely defined size grades (e.g., carrots, onions) or the number of plants that "mature" at a single harvest (e.g., lettuce). The effects of seedling emergence on economic yield are generalized in Figure 2.1. As the number of seedlings

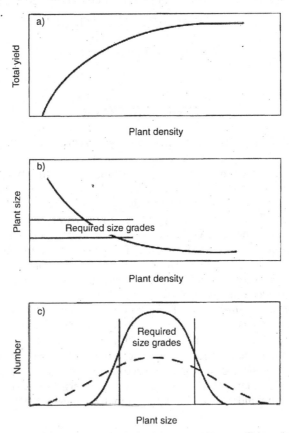

FIGURE 2.1. Schematic illustration of the effects of seedling emergence on marketable yield. Total yield increases asymptotically with increasing plant density (a) while there is a concurrent decrease in the size of individual plants (b). The uniformity of plant size at harvest determines the proportion of the crop in the required size grades (c, - - - - nonuniform crop, ——— uniform crop).

emerging per unit area (crop density) increases, yield increases asymptoti-cally (Figure 2.1a), but the size of individual plants decreases (Figure 2.1b). Thus target populations need to be achieved to grow bulk crops cost-effec-tively and to produce plants of the appropriate sizes for graded yields in hor-ticultural markets. Crop density can also influence time taken to reach ma-turity (e.g., onions, Mondal et al., 1986) and the uniformity of plants at maturity (e.g., cauliflower, Salter and James, 1975). It is also commonplace to oversow crops to avoid a limiting density, and this, under favorable con-ditions, results in densities that are too high. This situation can produce poor canopy structure which delays and reduces the uniformity of maturity.

Variation in the time to seedling emergence within the population can ac-count for much of the subsequent variation in plant size during crop growth to harvest (Benjamin and Hardwick, 1986; Benjamin 1990). The ranking of seedling size at the end of emergence changes little with time, and in many cases the difference between plants increases during growth (Benjamin and Hardwick, 1986). Thus more uniform seedling emergence can result in a greater proportion of the population falling within the required high-value size grade or maturation period to increase crop value (Figure 2.1c). Rapid and predictable emergence following sowing is particularly important in re-gions where season length limits yield (e.g., grain maize, Breeze and Milbourne, 1981; sweet corn, Cal and Obendorf, 1972), where water re-sources for irrigation are limited (Jordan, 1983), or when crops are grown in a programmed sequence of sowings (e.g., lettuce, Gray, 1976). Rapid emer-gence can also increase crop competitiveness with weeds emerging from the soil seedbank and facilitate earlier application of herbicides when weeds are more susceptible. Seedling emergence has an impact on many other aspects of crop management and its cost-effectiveness, not least be-cause many of the costs of production (e.g., fertilizers and disease and pest control) are likely to be similarly independent of the success of seedling emergence.

The time between sowing and seedling emergence can be conveniently divided into the phase before and after germination. Although these effects are confounded in most seedling-emergence studies, the two phases are each uniquely affected by adverse seedbed conditions (Finch-Savage, 1995). It is thought that the timing of germination can account for much of the vari-ation in seedling emergence time (Finch-Savage and Phelps, 1993; Finch-Savage, Steckel, and Phelps, 1998), whereas seedling losses and variation in the spread of seedling emergence times within the population occur largely in the postgermination growth phase (Hegarty and Royle, 1978; Durrant, 1981; Finch-Savage, Steckel, and Phelps, 1998). Therefore, to pre-dict the impact of seedbed environment on seedling emergence both phases must be considered.

Seedbed Environment

The seedbed environment provides a highly variable and often hostile environment for seedling emergence from crop seeds and those in the weed seed bank. For germination, most crop seeds require water, adequate temperature, and a favorable gaseous environment. Dormancy has little impact on seedling emergence of most commercial crops (Villiers, 1972; Maguire, 1984) but is a major factor in the emergence of weeds (Baskin and Baskin, 1998). For the weed seeds there are additional germination-promoting factors such as light and nitrate to consider (Hilhorst and Karssen, 2000). Modeling nondormant crop seed germination is therefore less complex and further aided because the crop seed is generally the same age and is sown at a narrow range of depths into the soil, so the environment for germination is more uniform within the population.

Crop seeds are sown close to the surface, and therefore soil water content and temperature can vary widely. Reduced oxygen availability can also have a major impact on germination and seedling emergence (Corbineau and Côme, 1995). This occurs when there is excessive water in the seedbed (e.g., Dasberg and Mendel, 1971; Hegarty and Perry, 1974; Perry, 1984), or when a soil crust forms to seal the seedbed surface or engulf the seed (Richard and Guérif, 1988a,b). Sensitivity to oxygen partial pressure (pO_2) differs among species, and linear relationships have been shown between germination rate and the logarithm of pO_2 (Al-Ani et al., 1985). This suggests that a threshold model, as discussed in the following for temperature and water potential, could be applied to this relationship. However, following good seedbed preparation, the oxygen concentration in the soil atmosphere in most cases does not fall below 19 percent (Richard and Boiffin, 1990). Crust formation and limiting oxygen environments occur only intermittently, and their effect can be minimized by good seedbed preparation (Chapters 1 and 3), whereas the variable strength of soil through which seedlings grow after germination is always a factor. The remainder of this chapter will be concerned with the effects of the three ubiquitous seedbed factors, water availability, temperature, and soil strength, that largely determine the patterns of germination and seedling emergence of crops observed in the field.

IMBIBITION

Water uptake by the seed generally occurs in three phases: rapid initial uptake, a lag phase with limited further uptake, and then a second phase of rapid water uptake associated with radicle emergence (Bewley and Black,

1994). Imbibition is identified with the first phase of water uptake and is regarded as a physical process, although metabolism is initiated before seeds reach full moisture content. Initial water uptake is driven by matric forces resulting from the hydration of cell walls, starch and protein bodies, etc. As the physiological range of water contents is approached there is a greater dependence on osmotic potential determined by the concentration of dissolved solutes. The rate of early water uptake can have a large negative impact on seed viability and the success of seedling emergence. If imbibition is too rapid, damage may be caused both directly and through a positive relationship with chilling injury. The extent of this damage is directly related to the integrity of the seed coat and other aspects of seed vigor (reviewed by Woodstock, 1988; Vertucci, 1989; Finch-Savage, 1995).

Imbibition can have an important influence on the prediction of germination and emergence times when seeds are sown into dry soils or when the contact between seed and soil is poor and therefore also likely to be variable in the seed population. The seed coat and other tissues can also have an important regulating affect on water uptake (e.g., soybean, McDonald, Vertucci, and Roos, 1988a) by controlling permeability. The movement of water into the seed is driven by gradients of water potential between the seed and the surrounding soil. Mechanistic models of imbibition have also been developed based on water concentration (diffusivity theory) rather than water potential gradients (hydraulic conductivity theory). This is a convenient simplification that can be used in homogeneous environments. When considered as a whole, the flow of water through the soil and into the seed is not a homogeneous system and is therefore considered here in terms of hydraulic flow. In this case, the rate of water uptake, in simple terms, is governed by the hydraulic conductivity of the seed and the soil and driven by the water potential gradient between them.

A reduction in water potential of the surrounding soil will therefore reduce the rate of water uptake by the seed because the gradient between them is less. However, the effect on rate is not directly proportional to changes in the gradient as hydraulic conductivity is also altered. Hydraulic conductivity is a function of the permeability of the seed and surrounding soil, the extent of contact between them, and temperature (reviewed by Bewley and Black, 1978; Vertucci, 1989). For example, rate of water uptake increases with temperature (Vertucci and Leopold, 1983). The situation is further complicated because there appears to be a wetting phase before hydraulic flow is initiated and hydraulic conductivity changes as the seed swells during imbibition (Vertucci, 1989). Soil water potential gradients may also form at the interface with the seed, and the relative importance of vapor transport of water to seed may be underestimated in many studies (Wuest, Albrecht, and Skirvin, 1999). In addition, seed coatings that are now commonly used in agriculture also influence imbibition (Schneider and Renault, 1997). Therefore, perhaps inevitably, seed imbibition under variable seedbed conditions is complex. Nevertheless, a number of models have been developed that make a range of different assumptions, and these have been reviewed in detail elsewhere (Hadas, 1970, 1982; Dasberg, 1971; Bruckler, 1983a,b; Bouaziz and Bruckler, 1989a,b; Vertucci, 1989; Schneider and Renault, 1997; Chapter 1).

Additional points to consider are that the different chemical compositions of seeds will affect the amount of water they take up; for example, equilibrium moisture content at any given water potential will always be greater in pea than soybean (Vertucci and Leopold, 1987). Equilibrium moisture contents also differ among seed tissues, often with the embryonic axis having a higher water content than the storage tissues (e.g., soybean, McDonald, Vertucci, and Roos, 1988b; maize, McDonald, Sulivan, and Lauer, 1994).

GERMINATION

The initiation of radicle growth at the end of the lag phase of imbibition terminates germination sensu stricto, and therefore germination is generally recorded when radicle growth is first observed. Following germination, desiccation tolerance is lost progressively during growth of the radicle, in most species, and so the initiation of growth is a critical step in the progression from sowing to seedling emergence. This critical step will occur at different times in each seed within the population, leading to a distribution of germination times and the characteristic sigmoidal cumulative germination curve. In agriculture, this spread of germination times can be very undesirable for the reasons described, but under natural conditions it presents a good strategy to cope with the highly variable conditions of temperature and water potential in the surface of the soil where seeds germinate. In the absence of significant disease, the interaction of this characteristic seedlot distribution of germination times with soil temperature and water potential largely determines the timing of seedling emergence in crops. Understanding this interaction and developing ways to model the outcome is essential to developing effective crop establishment practices. Population-based threshold models provide a useful framework for this purpose. Within these models the rate of development, such as progress toward germination or seedling growth, increases above a base (threshold) value for a given factor (temperature, water potential, hormone concentration, etc.). Below the base value, development ceases. The effect of the factor on rate of development above the base is described by an appropriate mathematical function. In many cases this function is linear. The base values may differ among individuals in the population and are therefore important in describing differences in their response to the factor concerned. The bases that are likely to have physiological importance (Welbaum et al., 1998; Meyer, Debaene-Gill, and Allen, 2000; Bradford, 2002) can be determined either explicitly by measuring the value at which development ceases or estimated implicitly in the case of linear relationships by extrapolation of the fitted line to the intercept.

Threshold Models: Effects of Temperature and Water Potential

For the purpose of modeling germination of nondormant seeds it is generally assumed that seeds germinate in a set order and that this order is not affected by germination conditions. Each seed can therefore be assigned a value of G, which is the fraction of the population at which it germinates (e.g., G_{10}, G_{50}, and G_{90} in Figure 2.2). The percentage of seeds that will germinate as well as germination time and spread of times within the seed population are all greatly influenced by temperature (reviewed by Roberts, 1988; Probert, 2000) and water potential (reviewed by Bradford, 1990, 1995, 2002).

Temperature

Seeds can germinate over a wide range of temperatures, but maximum percentage germination is typically reduced at the extremes of the range (Labouriau and Osborn, 1984; Roberts, 1988; Probert, 2000). Individual seeds within the population can therefore have different levels of tolerance (thresholds) at both high and low temperatures. For any individual seed in the population, germination rate, which is the reciprocal of germination time, increases from a base to an optimum temperature above which it decreases to a ceiling temperature that indicates the limit of its tolerance (Labouriau, 1970). In many cases this response to temperature can be described by linear relationships where the base and ceiling temperatures are defined by the intercepts on the temperature axis where rate tends to zero (Figure 2.2a, Labouriau, 1970; Bierhuizen and Feddes, 1973; Garcia-Huidobro, Monteith, and Squire, 1982a). A linear relationship at suboptimal temperatures has been shown for a wide range of species, for example, many temperate vegetables (Wagenvoort and Bierhuizen, 1977), other herbaceous species (Steinmaus, Prather, and Holt, 2000; Trudgill, Squire, and Thompson, 2000), subtropical crops (Covell et al., 1986), range grasses, and shrubs (Jordan and Haferkamp, 1989). At suboptimal temperatures a

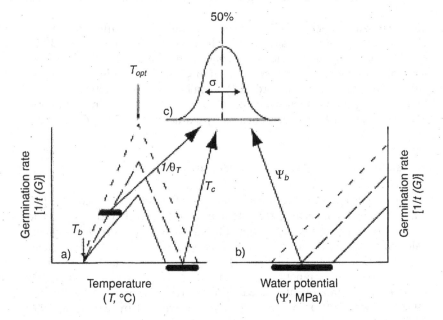

FIGURE 2.2. Schematic illustration of the effects of temperature (a) and water potential at suboptimal temperature (b) on the rate of germination. G_{10} (dotted lines), G_{50} (dashed lines), and G_{90} (solid lines) represent individual seeds in the population at percentiles 10, 50, and 90, respectively. Germination rate increases linearly with temperature above a base (T_b). The slopes of these lines are the reciprocal of the thermal times to germination ($1/\theta_T$). As temperature increases above an optimum (T_{opt}), rate of germination decreases to a ceiling temperature (T_c). Rate of germination also decreases linearly with water potential (Ψ) to a base (Ψ_b). T_b is common to all seeds in the population, but $1/\theta_T$, T_c, and Ψ_b vary among seeds in a normal distribution (c). For further explanation of these parameters see the text.

heat sum (Feddes, 1972; Bierhuizen and Feddes, 1973; Bierhuizen and Wagenvoort, 1974), more recently called thermal time (Garcia-Huidobro, Monteith, and Squire, 1982a), approach can therefore be used to predict germination time:

$$\theta_{Tl}(G) = [T\text{-}T_b(G)]\, t(G) \tag{2.1}$$

where, $\theta_{Tl}(G)$ is the thermal time to germination of percentile G, T is the temperature, $T_b(G)$ is the base temperature, and $t(G)$ is the time taken for germination of that percentile. In this context thermal times have been used most often to compare and predict time to 50 percent germination (G_{50}). In many cases little variation in $T_b(G)$ has been shown among individual seeds within the population, and T_b is therefore considered to be a constant although T_b differs greatly between species (Covell et al., 1986). Therefore, thermal time to germination of a given percentile is constant, but each percentile requires different thermal times to complete germination. Germination rate $1/t(G)$ is linearly related to temperature, but with a different slope for each percentile (e.g., G_{10}, G_{50}, and G_{90} in Figure 2.2a):

$$1/t(G) = (T-T_b)/\theta_{T1} (G) \qquad (2.2)$$

At supraoptimal temperatures the rate of germination declines in a series of parallel lines (Figure 2.2a). The intercepts therefore differ in the population and thus:

$$1/t(G) = [T_c (G) - T]/ \theta_{T2} \qquad (2.3)$$

where $T_c (G)$ is the ceiling temperature and θ_{T2} is thermal time. A natural extension of the thermal time approach allows the description of germination times of the whole population at a range of suboptimal temperatures by assuming T_b is constant and using a distribution to describe variation in θ_{T1} (G) (Covell et al., 1986; Ellis et al., 1986; Ellis, Simon, and Covell, 1987). If variation in $\theta_{T1} (G)$ is normally distributed (Figure 2.2c) then probits can be used:

$$1/t(G) = (T-T_b)/ \{[\text{probit } (G) - K]\sigma\} \qquad (2.4)$$

where σ is the standard deviation of $\theta_{T1} (G)$ and K is a constant. At supraoptimal temperatures θ_{T2} remains constant and variation in germination rate is accounted for by a normal distribution of $T_c (G)$ (Ellis et al., 1986; Ellis, Simon, and Covell, 1987) so that

$$1/t(G) = (\{[K_s - \text{probit } (G)]\sigma\} - T)/\theta_{T2} \qquad (2.5)$$

where σ is the standard deviation of $T_c (G)$ and K_s is a constant. Other distribution functions may be more appropriate to describe variation in thermal time for other species (e.g., Washitani, 1985; Covell et al., 1986; Ellis and Butcher, 1988). Equations 2.4 and 2.5 can be used to predict germination rate at any constant temperature for all seeds in the population.

In this work Ellis and colleagues used repeated probit regression analyses (Finney, 1971) of germination data from all the temperatures recorded to determine the best fit (least residual variance) to the data. Bradford

(1995) points out that a requirement of probit analysis is that samples at each time point should be independent (Finney, 1971). However, in usual practice, repeated measurements are made from a single sample to determine cumulative germination curves, rather than single measurements from a number of samples. Thus data do not conform to the criterion of independence. Bradford (1995) argues that for practical purposes, the results of the two methods are identical (Campbell and Sorensen, 1979). He continues, that although statistical comparisons based on probit analysis from cumulative scored data are invalid, other procedures developed specifically for the cumulative curve are available (Bliss, 1967). This same caveat applies where the use of probit analysis is mentioned in the following.

Water Potential

As with temperature, the rate of progress toward 50 percent germination has been shown to be linearly related to water potential (Hegarty, 1976). Gummerson (1986) was the first to consider germination in hydrotime, a scale analogous to thermal time that can be used to describe the response of seeds to different water potentials. The wider relevance of this concept was realized by Bradford who then developed and extended the use of hydrotime to provide insight into a wide range of seed behavior. A full review of the use of hydrotime is beyond the scope of this work but has been eloquently covered in detail elsewhere by Bradford and colleagues (Bradford, Dahal, and Ni, 1993; Bradford, 1995, 2002). Here, hydrotime will be considered only in relation to its contribution to the prediction of germination times for agricultural purposes.

The hydrotime (θ_H) approach considers germination rate as a function of the extent to which seed water potential exceeds a base water potential below which germination will not occur. It is analogous to thermal time and thus:

$$\theta_H = [\Psi - \Psi_b(G)] \, t(G) \qquad (2.6)$$

where $\Psi_b(G)$ is the base Ψ below which germination of percentile G will not occur. If θ_H is a constant, then the time required for germination [$t(G)$] of percentile G is inversely proportional to the amount by which seed Ψ exceeds its base water potential [$\Psi_b(G)$]. By analogy to Equation 2.2:

$$1/t(G) = [\Psi - \Psi_b(G)] / \theta_H \qquad (2.7)$$

Figure 2.2b shows that unlike T_b, but like T_c, Ψ_b is thought to differ among individual seeds in the population, and these differences result in the difference in the time seeds take to germinate (Gummerson, 1986; Bradford, 1990; Dahal and Bradford, 1990). Ψ_b is negatively related to germination rate, so slow-germinating seeds have the highest Ψ_b. In this case the difference between seed Ψ and $\Psi_b(G)$ is least, so hydrotime accumulates more slowly (Figure 2.3). With θ_H constant, differences in germination rate are therefore determined solely by the variation in Ψ_b that approximates to a normal distribution (Gummerson, 1986; Bradford, 1990; Dahal and Bradford, 1990). As seed Ψ decreases $\Psi - \Psi_b$ decreases and therefore hydrotime accumulates more slowly and the whole population of seeds take longer to germinate in clock time. When seed Ψ is reduced to less than its Ψ_b then it will not germinate. Therefore in experiments with fixed water potentials within the range of Ψ_b not all seeds will germinate. However, although the

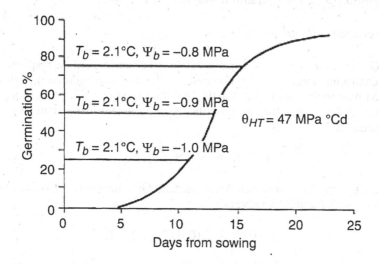

FIGURE 2.3. The influence of $\Psi_b(G)$ on the shape of the cumulative germination curve at suboptimal temperatures in carrot. Values of T_b and Ψ_b are shown for percentiles G_{25}, G_{50}, and G_{75}. According to the basic hydrothermal time concept, all seeds have the same T_b, but more rapidly germinating seeds have a lower $\Psi_b(G)$ and so $\Psi - \Psi_b(G)$ is greater and therefore more hydrothermal time (θ_{HT}) is accumulated per unit of clock time. As θ_{HT} to germination is the same for all seeds in the population (47 MPa °Cd), seeds with a lower Ψ_b will germinate first when their accumulated $\theta_{HT} = 47$ MPa °Cd. The shape of the cumulative germination curve is therefore determined by the distribution of $\Psi_b(G)$. (*Source:* Data from Finch-Savage, Steckel, and Phelps, 1998.)

seed will not complete germination and initiate radicle growth below Ψ_b, metabolism continues, and this has consequences for the timing of germination and therefore its prediction as discussed later in this chapter (section Seed Advancement Below Base Water Potential).

The repeated probit analysis technique used by Ellis and colleagues (1986) for thermal time was adapted by Bradford (1990) to describe the affect of water potential on germination for the whole population of seeds:

$$\text{Probit } (G) = \{\Psi - [\theta_H / t(G)] - \Psi_b(50)\} / \sigma_{\Psi_b} \qquad (2.8)$$

where $\Psi_b(50)$ is the median Ψ_b and θ_{Ψ_b} is the standard deviation of Ψ_b among seeds in the population. Following this analysis time courses of germination at a range of water potentials can be mapped onto a common hydrotime scale (Bradford, 1990, 1995; Dahal and Bradford, 1990; Ni and Bradford, 1992, 1993; Bradford and Somasco, 1994).

Water Potential and Temperature

Gummerson (1986) developed a combined description of the response of seeds to temperature and water potential in the theory of hydrothermal time. According to this theory, rates in thermal time are proportional to water potential and can therefore be described by an equation similar in form to Equation 2.2:

$$1/ \theta_T(G) = [\Psi - \Psi_b(G)] / \theta_{HT} \qquad (2.9)$$

where θ_{HT} (hydrothermal time) is a constant. Gummerson (1986) combined Equations 2.2 and 2.9 to give:

$$\theta_{HT} = [\Psi - \Psi_b(G)] (T - T_b) t(G) \qquad (2.10)$$

Consistent with the development of this theory, θ_{HT} and T_b are assumed constant and Ψ_b varies with (G). As pointed out by Gummerson (1986), it is possible that these assumptions are not entirely correct and this will be discussed further in the section Further Development of Threshold Models. Nevertheless, this approach has been shown to adequately describe germination curves produced in a wide range of combinations of constant temperature and water potential (Gummerson, 1986; Dahal and Bradford, 1994; Finch-Savage, Steckel, and Phelps, 1998; Roman et al., 1999; Shrestha et al., 1999; Allen, Meyer, and Khan, 2000). Accepting these assumptions, it is possible to describe the effect of suboptimal temperature and water potential of the whole population in a single equation by incorporating a suitable distribution (usually a normal distribution) of base water potentials within the population (Gummerson, 1986; Dahal and Bradford, 1994; Dahal, Bradford, and Haigh, 1993; Bradford, 1995). A form analogous to Equation 2.8 gives:

$$\text{Probit } (G) = \{[\Psi - \theta_{HT} / (T - T_b) \, t(G)] - \Psi_b(50)\} / \sigma_{\Psi_b} \qquad (2.11)$$

The best fit to the model can be obtained by repeated probit regressions varying the values of θ_{HT}. Following this approach used by Bradford (e.g., Bradford, 1995), the time courses of germination at suboptimal temperatures and water potentials can be mapped on to a common scale by multiplying time to germination [$t(G)$] by the factor $\{1 - [\Psi/\Psi_b(G)]\} (T - T_b)$. In a range of tomato seed lots the hydrothermal time model accounted for 73 to 93 percent of the variation in radicle emergence timing across a range of temperatures and water potentials (Cheng and Bradford, 1999).

To be useful for field predictions the hydrothermal time model must also be able to describe the reduction in germination rate and nongermination that occurs at supraoptimal temperatures. So far it has been assumed that the five parameters [θ_{HT}, T_b, $\Psi_b(50)$, σ_{Ψ_b}, and the fraction of viable seeds (G_m)], which can describe behavior of the whole seed population, are constant. However, Bradford (1995) suggested that progressive loss of dormancy, associated with increased percentage germination and germination rate, in a seed population may be related to a progressive decrease in $\Psi_b(50)$. Christensen, Meyer, and Allen (1996) demonstated that changes in the germination time courses of *Bromus tectorum* seeds during after-ripening could be fully accounted for by changes in $\Psi_b(50)$. Therefore as the time in storage (after-ripening) increased, the distribution of Ψ_b remained the same in the seed population, but their water potential thresholds were reduced below that of the ambient water potential to allow germination. As after-ripening continued, their Ψ_b was further reduced below that of ambient levels and germination rate increased. During thermoinhibition of lettuce *(Lactuca sativa)* the reverse occurred and the water potential thresholds increased as temperature approached the upper limit for germination (Bradford and Somasco, 1994). In this case, thresholds shifted above ambient water potential preventing germination. Subsequently, other studies have shown that as temperatures become supraoptimal and approach T_c, the Ψ_b distribution shifts progressively toward and above 0 MPa to reduce germination rate and eventually prevent germination (Figure 2.4a; Kebreab and Murdoch, 1999; Meyer, Debaene-Gill, and Allen, 2000; Bradford, 2002). In this way, germination time courses can be accounted for over the whole temperature range.

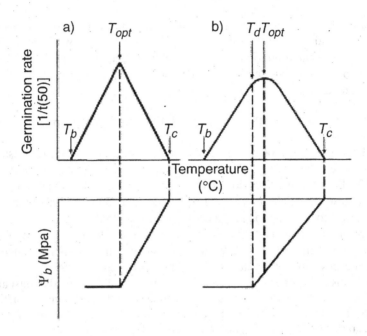

FIGURE 2.4. Schematic representation of the relationship between $\Psi_b(50)$ and temperature and their effect on the germination rate of percentile G_{50}. (a) $\Psi_b(G)$ initially remains constant as temperature increases above T_b, however $\Psi_b(G)$ increases at temperatures above the optimum (T_{opt}), as in Equation 2.12, to explain the reduction in germination rate. At T_c germination is prevented as $\Psi_b(G)$ increases above 0 MPa. (b) $\Psi_b(G)$ increases from the deviation temperature (T_d) in advance of (T_{opt}) as in the temperature modification described for Equations 2.15 and 2.16. In both situations, a and b, the distribution of $\Psi_b(G)$ around $\Psi_b(50)$ remains the same and this results in the same distribution for T_c (Equation 2.6).

The distribution of Ψ_b accounting for variation in germination times within the population remains the same, but above the optimum temperature (T_{opt}), $\Psi_b(G)$ increases linearly with T (Figure 2.4a). This in turn accounts for the parallel decreases in germination rate above the optimum and the distribution of ceiling temperatures discussed in relation to Equation 2.3 (Figure 2.2a). Therefore, Equation 2.10 was modified by Bradford (2002) to account for the germination response to supraoptimal temperatures:

$$\theta_{HT} = \left[\psi - \psi_b(G)_{opt} + k_T(T - T_{opt}) \right](T_{opt} - T_b)t(G) \quad (2.12)$$

where k_T is a constant (the slope of the $\Psi_b(G)$ versus T line when $T > T_{opt}$) and $\Psi_b(G)_{opt}$ is the threshold distribution at T_{opt}. This equation adjusts $\Psi_b(G)_{opt}$ to higher values as T increases above T_{opt}. Since the standard deviation of the $\Psi_b(G)$ distribution is not affected, the Ψ_b values of all of the seeds are adjusted upward by the same amount for each increment in T above T_{opt}. Equation 2.12 also stops the accumulation of thermal time at the value equivalent to that accumulated at T_{opt}. Thus, temperatures above T_{opt} do not contribute additional thermal time in the supraoptimal range. Instead, effects on germination are accounted for by the change in $\Psi_b(G)$.

By combining Equation 2.10 for suboptimal temperatures and Equation 2.12 for the supraoptimal temperatures, seed germination time courses can be described in hydrothermal time for all temperatures. Battaglia (1997) uses an alternative, and more flexible, linear predictor technique based on a similar conceptual framework that could incorporate responses to sub- and supraoptimal temperatures as well as other factors. However, the linear predictor assumes that variation in germination rate results from the distribution of the factor threshold (base or ceiling). It therefore accommodates Equations 2.3 (supraoptimal temperature) and 2.7 (water potential), but not Equation 2.3 (suboptimal temperature) where base temperature is assumed to be a constant.

Seed Advancement Below Base Water Potential

The models described previously have been used widely and successfully to describe data collected under laboratory conditions of constant temperature and water potential. Within the hydrotime and hydrothermal time models germination is arrested by water potentials that fall below Ψ_b. However, it is known from priming studies that metabolism continues below water potentials which prevent the completion of germination and radicle growth. After such a period of priming, germination is more rapid. The models described earlier do not take account of this advancement. Therefore under seedbed conditions where water potential varies above and below Ψ_b, germination time will be overestimated because no hydrotime or hydrothermal time is accumulated below Ψ_b. In addition to priming, successive wetting and drying of the seeds under field conditions can also significantly advance germination (e.g., Wilson, 1973; Koller and Hadas, 1982; Hegarty, 1978; Allen, White, and Markhart, 1993; Adams, 1999; González-Zertuche et al., 2001). Two approaches are currently being developed to account for seed advancement below Ψ_b.

In the first approach, the concept of hydropriming time was developed based upon the principles developed previously in this chapter (Tarquis and

Bradford, 1992). During priming, seeds accumulate hydropriming time in proportion to the difference between the water potential of the priming solution and the minimum required for metabolic advancement to occur during priming (Ψ_{min}). In this treatment the solution is always below Ψ_b. For lettuce, Tarquis and Bradford (1992) estimated that the median Ψ_b was -1.0 Mpa and Ψ_{min} was -2.4 MPa. A very similar Ψ_{min} has also been estimated for a range of tomato seed lots (Cheng and Bradford, 1999), but values between -5.0 and -8.0 Mpa were recorded for *Elymus elymoides* (Meyer, Debaene-Gill, and Allen, 2000). In the model, germination rates after priming are assumed to increase linearly with any increase in accumulated hydropriming time thus:

$$GR_{50} = GR_i + k \ (\Psi - \Psi_{min}) \ t_p \qquad (2.13)$$

where GR_{50} is the median germination rate ($1/t_{50}$) of primed seeds, GR_i is the initial median germination rate before priming, and t_p is the duration of the priming treatment at water potential Ψ, and k is a linear proportionality constant. (Tarquis and Bradford, 1992; Bradford and Haigh, 1994; Bradford, 1995). However, in the seedbed, such advancement at water potentials below Ψ_b will not occur at constant temperatures as they do in priming and so this concept was extended to include accumulated thermal time as hydrothermal priming time (Bradford and Haigh, 1994; Bradford, 1995). In this case germination rate after priming can be expressed as

$$GR_{50} = GR_i + k^1 \ (T - T_{min}) \ (\Psi - \Psi_{min}) \ t_p \qquad (2.14)$$

where T_{min} is the minimum temperature at which a priming effect will occur and k^1 is a proportionality constant. Hydrothermal priming time $[(T - T_{min}) \ (\Psi - \Psi_{min}) \ t_p]$ will accumulate in proportion to the extent by which Ψ exceeds Ψ_{min} and T exceeds T_{min}. Under laboratory conditions this approach was able to describe the effect of priming on germination rates (Cheng and Bradford, 1999; Meyer, Debaene-Gill, and Allen, 2000). In these priming models, a single value for Ψ_{min} has been used, but it may vary within the seed population with the effect of increasing variation in germination times. Bradford (1995) speculates that in a situation where water potential varies across the range of base and minimum water potentials hydrothermal and hydrothermal priming time could be used additively in some form to predict germination times.

A potential limitation to this approach was pointed out by Rowse, McKee, and Higgs (1999) in that it necessarily predicts that the increase in germination rate is proportional to t_p, whereas, in practice it tends to reach a maximum and then does not increase further with increased priming time. Rowse, McKee, and Higgs (1999) have developed an alternative model that can describe the effects of fixed and variable water potentials, both above and below Ψ_b. The model is loosely based on the idea that for a seed to initiate radicle growth its cells have to generate sufficient turgor pressure to exceed the yield threshold (Y) for growth. The model arbitrarily assumes that Y remains constant and turgor is determined in the cells by changes in osmotic potential and by changes in external water potential. Within the model, values of osmotic potential are empirical and they are therefore termed virtual (VOP) and assigned the symbol $\Psi_{\pi v}$. The model determines seed advancement to germination by integrating changes in $\Psi_{\pi v}$ that are proportional to the history of water potential experienced by the seed relative to minimum and base water potentials. The minimum (Ψ_{min}) defines the water potential below which there is no metabolic advancement (priming) and the base (Ψ_b) defines the water potential above which radicle growth can occur. According to the model, germination time for a given constant suboptimal temperature [$t(G, T)$] and water potential can be determined by

$$t(G,T) = \frac{1}{k_0(T)(1 - \psi / \psi_{min})} \ln\left(\frac{\psi_b(G) - Y}{\psi(G) - \psi}\right) \qquad (2.15)$$

where $k_0(T)$ is the rate constant when $\Psi = 0$. To fit the model, Ψ_b, as in the hydrotime model, is assumed to have a normal distribution. The VOP model can be used in finite difference simulation to calculate changes in $\Psi_{\pi v}$ and predict germination when Ψ varies thus:

$$d\psi_{\pi v}(G) / dt = k_0(T)(1 - \psi / \psi_{min})[\psi_b(G) - Y - \psi_{\pi v}(G)] \qquad (2.16)$$

The VOP model has now been extended to include temperature (Rowse, personal communication). At temperatures where the germination rate is proportional to ($T - T_b$), this is done by assuming that the rate constant is proportional to hydrothermal time [e.e., the terms $k_0(T)(1 - \Psi / \Psi_{min})$ in Equations 2.15 and 2.16 are replaced by $k(T/T_b - 1)(1 - \Psi / \Psi_{min})$]. Experiments on carrot and onion seed (Rowse, unpublished) have shown that above a critical temperature (T_d) the germination rate ceases to be linearly related to temperature (Figure 2.4b). This situation can be well accommodated by assuming that Ψ_b changes so that for any seed fraction the effective base water potential is given by $\Psi_b + m(T - T_d)$, where Ψ_b is the uncorrected base water potential and m is a coefficient. Thus Ψ_b increases at higher temperatures as described for hydrothermal time (Equation 2.12);

however, note that T_d is well below the temperature when the germination rate is a maximum (Figure 2.4b). For carrot and onion the fitted values are approximately 18 and 16°C, respectively, whereas maximum germination rate occurred close to 25°C. Thus increase in rate due to $T - T_b$ between T_d and the optimum is offset by an increase in Ψ_b reducing $\Psi - \Psi_b$. In this way a curved response results in contrast to that of hydrothermal time. Using this approach the model can be used to predict germination in conditions that vary above and below Ψ_b at the full range of temperatures. This same approach can also be used effectively to take account of supraoptimal temperatures in the hydrothermal time model (Rowse, pers comm).

The VOP model utilizes the concepts of base and minimum water potentials developed in hydrothermal time models; however, it has a differential formulation and does not assume that a seed must be in either a germinating or a priming state (water potential is treated as a continuous variable above the minimum water potential). Such a model is potentially very useful for predictions under variable seedbed conditions but has yet to be tested on a range of seed lots and conditions. In contrast, hydrothermal and hydrothermal priming time models have been tested and found to be descriptive on a wider range of seed lots but do not lend themselves so readily to prediction under variable conditions in their present form. The application of these models for field prediction is considered in the next section.

Further Development of Threshold Models

The threshold models described previously provide a robust framework in which to describe seed responses to the environment. Even though these models are fitted empirically, the thresholds determined appear to have a physiological basis. However, it is important to appreciate that at present these models do not account for all seed behavior and further development is necessary to incorporate sufficient flexibility to cover the extent of biological variability scientists have come to expect. Much of this variability may result from interactions between Ψ and T resulting in concurrent changes in thermal and hydrotime parameters. In addition, physiological adaptation occurring near both Ψ_b and T_b resulting in greater than expected germination rates and percentage germination may result from overlap of what we now consider to be separate priming and germination processes. As a greater range of species are investigated the discrete packaging of these model components, although convenient, may cease to be appropriate.

For example, the comprehensive data set and analysis conducted by Labouriau and Osborn (1984) on tomato seeds shows linear relationships between germination rate and temperature in both sub- and supraoptimal ranges. However, the optimum occurred over a range of temperatures between 25.9 and 29.5 rather than a sharply defined optimum at the convergence of the two linear relationships as used by Garcia-Huidobro, Monteith, and Squire (1982a) and Covell and colleagues (1986) for other species. This plateau can be accommodated in a further development of the thermal time model based on Gaussian curves which describes the germination response across both sub- and supraoptimal temperature ranges (Orozco-Segovia et al., 1996). However, the optimum temperature can also differ with water potential (Kebreab and Murdoch, 2000). Responses to both temperature and water potential can be accommodated within the hydrothermal time model by an increase in Ψ_b with temperature in advance of T_{opt}. In this case, the increased rate of hydrothermal time accumulation resulting from higher temperature (i.e., increase in $T - T_b$) as the optimum is approached would be offset by a concurrent increase in Ψ_b (reducing $\Psi - \Psi_b$). Data reported for fully after-ripened seeds of *Elymus elymoides* (Meyer, Debaene-Gill, and Allen, 2000) and observations in onion and carrot (Rowse, personal communication) can be explained in this way. For, example, in fully after-ripened *Elymus elymoides* seeds, Ψ_b increased linearly with temperature (10 to 30°C), resulting in little difference in germination rates over this range of temperature (Meyer, Debaene-Gill, and Allen, 2000). Kebreab and Murdoch (1999) have also shown that in *Orobanche aegyptiaca* seeds the underlying assumption of independance of Ψ and T effects within the current hydrothermal time model is not valid, and they give examples of work with other species where this is also the case. They found that Ψ_b varied with T, both above and below the optimum, and T_b varied with Ψ and developed a new and more general thermal time model that allows for the interaction of temperature and base water potential (Kebreab and Murdoch, 1999, 2000). Alternatively, the approach described by Battaglia (1997) can incorporate complex factor interactions and test them for significance in affecting the germination response. However, it has yet to be seen how well the hydrothermal time model, freed from the constraints of fixed thresholds, can account for the full range of seed responses to environment that are reported in the literature. Other current concerns, such as nonlinear relationships close to T_b outlined as follows, may also be reconciled in this way.

It is generally accepted that, when calculated by linear rate temperature relationships, there is a single base temperature below which germination of the whole population will not occur. Extrapolation of a linear relationship covering suboptimal temperatures of 10°C and above indicates a single base temperature in tomato (e.g., Dahal, Bradford, and Jones, 1990, and references within). Yet Labouriau and Osborn (1984), for example, show that percentage germination declines progressively over the range 10 to 6°C, indicating a range of thresholds within the seed population as seen in many studies. If this apparent dilemma is considered in terms of residual dormancy, expressed close to the base temperature, it could be accommodated in a modeling approach developed for species with seasonal changes in the range of temperatures which permit germination (Washitani and Takenaka, 1984; Washitani, 1987; Kruk and Benech-Arnold, 1998, 2000; Chapter 8). In this approach, T_b is used to calculate rates in thermal time as described previously, but seeds also have a lower temperature (T_l) below which germination is prevented by dormancy. T_l is assumed to have a normal distribution within the population. Thus percentage germination declines over a range of temperatures as T_b is approached. In weed species, T_l and an equivalent higher limit temperature (T_h) can change during the season as temperature changes to account for seasonal dormancy patterns. In genetically uniform crop seeds, produced without residual dormancy (i.e., $T_l = T_b$), a single base temperature may well be sufficiently accurate for predictions of germination, whereas in a seed lot from mixed populations or uncultivated species this is less likely to be the case. In fact, a normal distribution of minimum as well as maximum temperature thresholds is seen widely in the literature (e.g., Grundy et al., 2000). In addition, consistent deviations from the linear relationship between rate and temperature at suboptimal temperatures can occur in some crop species close to T_b. This behavior can severely affect the prediction of germination time at constant temperature close to the base (Marshall and Squire, 1996; Phelps and Finch-Savage, 1997).

Kebreab and Murdoch (2000) and Grundy and colleagues (2000) have developed separate modeling approaches to incorporate independent seed to seed variation in both minimum and maximum temperature thresholds within the general conceptual framework discussed here. These approaches suggest that thresholds and rates can behave independently, so they involve the separate determination of germination rates and final percentage germination. One advantage is that rate relationships within the threshold modeling approach are not constrained to be linear if this limits the precision required for field prediction (Grundy et al., 2000). Indeed, linear relationship between temperature and development have often been shown to occur within a limited temperature range only, and in many other biological systems rates are more often described by nonlinear relationships (Sharpe and DeMichelle, 1977; Schoolfield, Sharpe, and Magnuson, 1981). The mathematical approach adopted by Sharpe and DeMichelle (1977) closely fits observed data and accommodates the linearity in response over a limited temperature range that has been adopted in thermal time models. There is, however, a practical disadvantage. The determination of bases explicitly (final number that germinate) is time consuming, as germination inevitably takes a considerable time under conditions close to temperature or water potential thresholds and there is considerable risk of achieving a poor esti-mate by early termination of the experiment or the intervention of contami-nation. If curved responses between germination rate and temperature and variation in T_b can be accommodated within the hydrothermal time model by changes in Ψ_b these concerns could be eliminated. Further work is re-quired to determine whether the hydrothermal time model can be suffi-ciently flexible to accommodate the full range of seed behavior.

It is common knowledge that time spent below Ψ_b (e.g., Kahn, 1992) and T_b (Coolbear, Francis, and Grierson, 1984) can increase subsequent germi-nation rates when seeds are placed above these thresholds. Progress above and below these thresholds are not directly additive and there is not a clear predictive relationship between hydrothermal and hydrothermal priming time models (Cheng and Bradford, 1999), suggesting that they are separate processes. There is no obvious reason to consider these processes as mutu-ally exclusive. Indeed, seed characteristics can change in constant condi-tions above Ψ_b; for example, there is evidence that extended incubation of seeds between –0.5 MPa and Ψ_b results in a shift to lower values of Ψ_b (Ni and Bradford, 1992; Dahal and Bradford, 1994). This adaptation at constant low temperature may account for some observations of nonlinear behavior close to the threshold in laboratory experiments. In variable field environ-ments such prolonged exposure to a particular set of conditions is unlikely, so field prediction may not be affected by this behavior.

Garcia-Huidobro, Monteith, and Squire (1982b) point out that for thresh-old models developed from constant environments, several conditions need to be satisfied before they can be used in variable conditions. These include (1) the instantaneous rate of development should depend only on the current conditions and not their history of exposure and (2) values such as thermal time and base temperature should remain unaffected. In nondormant lentil seeds, at least at suboptimal temperatures, there was no effect of thermal history on germination rate and thermal time could be used to predict time required for germination at alternating temperatures (Ellis and Barrett, 1994). However, when seeds have residual dormancy then there can be sys-tematic deviation from predictions resulting from their history of exposure. For example, exposure to alternating temperatures can have a positive effect on rate of germination (Garcia-Huidobro, Monteith, and Squire, 1982b) and percentage germination (Murdoch, Roberts, and Goedert, 1989), whereas exposure to high temperatures can have a negative impact on these same measures (Garcia-Huidobro, Monteith, and Squire, 1982b). Although it is convenient to consider crop species as nondormant, deviation of their be-havior in some cases from that readily described by the commonly used hy-drothermal time model may result from limited residual dormancy. This is in keeping with the view that differences in crop seed performance (vigor) may be an extension of dormant behavior toward the end of a continuous scale (Hillhorst and Toorop, 1997).

The question to ask is, Do these current limitations of the hydrothermal time model have practical significance? For many purposes, such as comparison of genotypes or treatments (e.g., Covell et al., 1986; Dahal, Bradford, and Jones, 1990) this approach is very effective. For field prediction purposes, errors at low temperature are likely to have an impact in early season crops grown at suboptimal temperatures, but inaccuracy is likely to be limited, especially as ambient temperatures rise in the spring following sowing. However, errors in prediction close to the optimum temperature, when progress is rapid, can have a major impact on the prediction of germination and emergence in clock time.

OTHER GERMINATION MODELS

In the present work, population-based threshold models have been used to describe responses to the environmental because it is likely that they have physiological significance and provide a framework for developing a generic understanding of seed and seedling responses. However, for seedling emergence prediction in the field other modeling approaches can have significant merit, but here there is not space to do these models justice. A large number of models have been developed to describe germination and emergence responses (e.g., Wanjura, Buxton, and Stapleton, 1970; Blacklow, 1972; Scott, Jones, and Williams, 1984; Thornley, 1986; Forcella, 1993; King and Oliver, 1994; Hageseth and Young, 1994; Pemberton and Clifford, 1994; Gan, Stobbe, and Njue,1996). There are also more mechanistic approaches, for example, that of Bruckler (Bruckler, 1983a,b; Bouaziz and Bruckler, 1989a,b) and Dürr and colleagues (2001). A number of these modeling approaches have been reviewed by Forcella and colleagues (2000).

POSTGERMINATION SEEDLING GROWTH

Some monocot preemergent seedlings are resistant to desiccation due to their seminal root system that can readily replace damaged roots. However, in the majority of crop species, once the seed has initiated growth the growing seedlings progressively become desiccation sensitive and therefore are committed to continued growth (Bewley and Black, 1978). It is an obvious comment, but postgermination growth occurs in two directions; the pattern in which it does this is essential for survival and also for prediction of emergence time. Rapid downward growth is necessary to maintain contact with moisture in the seedbed as it dries from the surface. Growth upward, to reach light and establish an autotrophic seedling, usually occurs in a deteriorating seedbed (increasing impedance to growth) and must be completed before seed reserves are exhausted.

Close to the soil surface, germination tends to occur most often after rainfall. For example, not only is germination metabolism (as shown previously for priming) less sensitive to water potential than the initiation of growth, so is postgermination extension growth (Ross and Hegarty, 1979). Initiation of growth is therefore a moisture-sensitive, rate-limiting step that determines Ψ_b and ensures that in many species germination under variable soil conditions occurs only when sufficient moisture is likely to be available for subsequent seedling growth (Hegarty, 1977; Ross and Hegarty, 1979; Finch-Savage and Phelps, 1993, Finch-Savage, Steckel, and Phelps, 1998). Below Ψ_b, seed priming or advancement in the soil (Wilson, 1973; Allen, White, and Markhart, 1993; Rowse, McKee, and Higgs, 1999) means that germination can be rapid when water becomes available. In the absence of additional water, there is only a brief opportunity for the completion of germination and seedling growth before the surface soil layers dry again. Following germination, initial growth is downward in both epigeal and hypogeal seedlings, to maintain contact with soil moisture as it dries from the surface layers. As this drying occurs the hydraulic conductivity of soil in the surface layer quickly falls to a very low value, and this will tend to reduce the rate of water loss from deeper layers (e.g., Lascano and van Babel, 1986). The seedling root will therefore grow into increasingly wet soil and the seedling may become less dependent on moisture content of the surface layers (Bierhuizen and Feddes, 1973). This pattern may occur because hypocotyl extension is more sensitive than radicle extension to low matric potential which initially favors the growth of roots (Dracup, Davies, and Tapscott, 1993). The initial period of downward seedling growth following germination is therefore critical to successful seedling establishment. Upward growth often occurs in a deteriorating seedbed that has increasing soil strength. Even if water potential is not directly limiting, because the growing root maintains contact with adequate moisture, there can be a large indirect effect because soil strength above the seed will increase as water content decreases. Thus in practice, during the postgermination phase of crop emergence, mechanical impedance may have greater importance than water stress in delaying and reducing the number of seedlings emerging (Whalley et al., 1999). In addition, soil can become much stronger following rainfall even without subsequent drying (Hegarty and Royle, 1976, 1978).

A large number of studies have been made of the response of preemergence seedling growth to temperature (e.g., Wanjura, Buxton, and Stapleton, 1970; Blacklow, 1972; Hsu, Nelson, and Chow, 1984; Wheeler and

Ellis, 1991; Weaich, Bristow, and Cass, 1996; Vleeshouwers, 1997; Roman et al., 1999; Shrestha et al., 1999). Different methods have been used to describe growth data from different species, but in many cases a thermal time approach similar to that described for germination has been adopted which assumes growth rate is linearly related to temperature. However, the utility of thermal time and other techniques that account only for temperature have limited potential in practice for accurate crop emergence predictions because soil moisture and strength vary greatly in the surface layers of the soil. Fewer studies have been made on the interaction between temperature and water potential (e.g., Fyfield and Gregory, 1989; Choinski and Tuohy, 1991; Dracup, Davies, and Tapscott, 1993) or soil mechanical resistance to preemergence seedling growth (e.g., Mullins et al., 1996; Vleeshouwers, 1997; Vleeshouwers and Kropff, 2000). A review of this literature is not justified in the present work which has its emphasis on germination.

Recently a model was developed that incorporates the effects of the three ubiquitous seedbed factors, temperature, water potential, and soil impedance, on preemergence shoot growth (Whalley et al., 1999). The model (described in the Appendix) assumes a linear dependence on temperature and water potential and scales the basic thermal time model so that shoot elongation rate decreases proportionally as impedance increases and water potential decreases toward threshold values that will just stop elongation. The model has a single threshold for each factor and so describes mean shoot growth only (Whalley et al., 1999). Variation in elongation rates has been introduced into predictive models by assuming a normal distribution of rates within the population (Finch-Savage and Phelps, 1993; Vleeshouwers and Kropff, 2000). Finch-Savage and colleagues (2001) have developed a model that describes a distribution of temperature and water potential thresholds for preemergence growth to take account of variation within the population. For prediction, nonemergence can be modeled by incorporating seed weight and the exhaustion of reserves with time, in particular as emergence time is extended by soil resistance to growth (Whalley et al., 1999; Vleeshouwers and Kropff, 2000). Another factor to be considered is the effect of seedbed structure (Bouaziz and Bruckler, 1989b; Mullins et al., 1996; Dürr et al., 2001).

THRESHOLD MODELS: PREDICTION OF GERMINATION AND EMERGENCE PATTERNS IN THE FIELD

Forcella and colleagues (2000) have reviewed the use of thermal time in seedling emergence prediction and shown that in many circumstances it can be very effective. Bierhuizen and Feddes (1973) show that the heat sum approach can accurately predict time to 50 percent emergence of vegetable crops provided soil moisture content is taken into account. They suggest that in dry periods a quantity of 5 to 10 mm of irrigation should be given regularly, which will shorten the period to emergence and avoid the impact of crust formation. Indeed, heat sums can be used to time irrigation to better effect to achieve the same purpose (Finch-Savage, 1990a,b). However, in the absence of irrigation, soil moisture varies greatly in the surface layers of the soil and so thermal time has limited ability to predict emergence. In both untreated seed and seed after advancing treatments (e.g., priming) the timing of water availability in the surface layers of the seedbed and its effect on germination can be the main factor determining time to seedling emergence in crops with nondormant seeds (HåKansson and von Polgár, 1984; Finch-Savage, 1984a,b, 1987, 1990a). In the case of onions, seedling emergence was reduced or delayed by inadequate soil moisture on more than half of 45 sowings made over three years (Roberts, 1984). Following germination, particularly during upward growth of the shoot, it has been argued that soil impedance is likely to become the principal factor. Seedbed conditions are described in detail in Chapter 1 and will only be discussed here as they immediately relate to germination and preemergence seedling growth.

The relatively recent and continuing development of threshold models, describing both temperature and water potential effects on germination and preemergence growth, has so far resulted in few attempts to apply these techniques under variable field conditions. Hydrothermal time approaches have been used with some success to describe the timing of major flushes in crop seedling emergence (Finch-Savage and Phelps, 1993; Finch-Savage, Steckel, and Phelps, 1998; Finch-Savage et al., 2000) and that of weeds (Battaglia, 1997; Bauer, Meyer, and Allen, 1998; Roman, Murphy, and Swanton, 2000). However, the first section of this review illustrates the importance of being able to predict the uniformity and numbers of seedlings emerging as well. The remainder of this section will be used to illustrate seed responses in a field context to show how variation in seedling emergence is generated and how the models described can be applied to predict crop seedling emergence.

A Modeling Framework

Threshold models for germination and preemergence seedling growth can be used to simulate seed germination and seedling emergence by dividing time into manageable steps and applying the models to each step. The model time (e.g., thermal, hydrothermal time, VOP, etc.) is accumulated from each step to indicate progress toward seedling emergence in clock time within the seed population. A schematic illustration of the process of simulation is provided in Figure 2.5, showing separate submodels for seedbed conditions, germination, and preemergence seedling growth. The germination and preemergence growth submodels are driven by appropriate outputs from the seedbed model calculated from details of the soil and me-

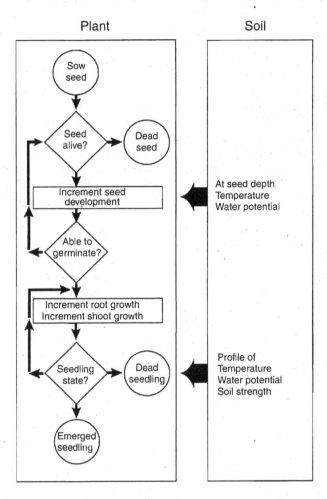

FIGURE 2.5. Flow diagram to illustrate the components of the simulation model used in Figure 2.6 (*Source:* Reproduced from W. E. Finch-Savage, 2003, Vegetable seed vigour: Looking to the future, *The Vegetable Farmer,* January, pp. 28-29, with permission from ACT Publishing.)

teorological data. The major drivers for germination are temperature and water potential, and as each seed in the population germinates, soil strength and its resistance to growth also become important. However, in the surface layers where the seeds are sown there can be steep profiles of all three of these variables, with temperature declining and moisture content increasing below the surface. In addition, fluctuations are dampened at increasing seedbed depths. To further complicate the situation, seeds are rarely sown at uniform depth, and irregularity in soil aggregate size will vary seed contact with the soil, so seeds within the population will experience different conditions. For simulation purposes this situation is accounted for by assigning depth and germination characteristics (e.g., base temperature and water potentials), drawn at random from the appropriate frequency distribution, to each seed. The models are then run for each seed using inputs appropriate for that depth. Increasing the number of individual seeds will generate a more reproducible prediction of seedling emergence times in the population. Following germination, as the seedlings grow, it is necessary to decide which temperature and water potential down the profile is most appropriate to use as inputs into the models. There is little guidance in the literature, so these decisions are largely pragmatic.

Although the measurement and simulation of seedbed conditions are described elsewhere (Chapter 1 and reviewed by Guérif et al., 2001), it is necessary to mention them here, in general terms, as the accuracy of these measurements determines the success of the simulation. These measurements arguably present the major limitation to predicting seedling emergence. For example, direct measurement of water potential at seed-scale resolution is not possible and so must itself be modeled from soil water content using a water release curve. When soils dry, water potential varies with water content according to a power law so that a small error in the measured water content can lead to a large error in the predicted water potential. Relevant soil water content measurement at seed-scale resolution is also difficult, even with more sophisticated techniques such as time-domain reflectometry (TDR). This equipment is sensitive to soil bulk density, so even in careful field experiments it is difficult to know which calibration to use (Whalley, 1993). Therefore, reliable water potential estimates are difficult to achieve and small differences can have a large impact upon the prediction of emergence time in a drying seedbed. Even for temperature, for which measurements at seed-scale resolution are possible, usually only a single soil or air temperature is available, so the temperature profile in the seedbed must be modeled. Prediction of seedling emergence will be only as good as these models of seedbed conditions.

Modeling the seedbed environment in any detail is very difficult, nevertheless models of differing complexity have been developed to provide relevant variables for seedling emergence models (e.g., Walker and Barnes, 1981; Forcella, 1993, 1998; Mullins et al., 1996; Brisson et al., 1998; Roman, Murphy, and Swanton, 2000; Dürr et al., 2001). These seedbed models can be coupled with threshold models that indicate how seeds and seedlings can respond to environment, to predict seedling emergence (Figure 2.5).

Germination

The following section is concerned with the prediction of crop seed germination and therefore the impact of environmental conditions on dormancy status will not be considered here (see Chapter 8). However, the application of hydrothermal time to the description and prediction of dormancy status in weeds has recently been reviewed elsewhere (Bradford, 2002). Under variable seedbed conditions, thresholds can be interpreted as switches for nondormant seeds: germination either proceeds or stops according to the value of ambient conditions relative to the threshold (Figure 2.6). Thus, rate of progress toward germination decreases and less physiological time (hydrothermal or hydrothermal priming time or VOP) is accumulated for a given clock time as Ψ_{min} and/or T_{min} and T_c are approached. Below Ψ_b or T_b, and above T_c radicle emergence will not occur.

It is likely that these thresholds have physiological significance relating to the initiation of radicle extension (Welbaum et al., 1998; Meyer, Debaene-Gill, and Allen, 2000; Bradford, 2002) and operate as rate-limiting steps in the progress of seedling emergence from the soil. For example, the interaction of Ψ_b and the changing amount of soil moisture can largely determine the timing of onion seedling emergence in the field (Finch-Savage and Phelps, 1993). Such a mechanism should avoid germination into seedbed conditions likely to be hostile to subsequent seedling growth.

The threshold models described can be applied in dynamic forms in finite difference simulation to predict progress toward germination above Ψ_{min}. The VOP model (Equation 2.16), with the modifications discussed for the effects of temperature, can now be applied directly to describe field germination (Rowse, personal communication; Figure 2.6), whereas a differential form of the hydrothermal time model covering both sub- and supraoptimal temperatures (Equations 2.10 and 2.12, respectively) must also somehow be coupled to hydrothermal priming time (Equation 2.14) to take account of progress that occurs below Ψ_b. One approach to the latter was attempted by Finch-Savage and colleagues (2000) who interpreted the equation suggested by Bradford (1995), and later modified (Bradford, 2002), to sum progress in both hydrothermal and hydrothermal priming time. However, the more flexible modeling approach used by Battaglia (1997) capable

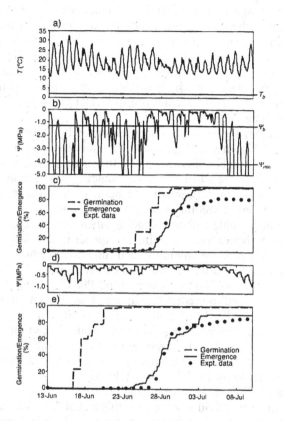

FIGURE 2.6. Simulation of the effects of soil temperature (a) and soil water potential (b, d) on cumulative germination and seedling emergence of onion (c, e) at two sowing depths (Rowse, Whalley, and Finch-Savage, unpublished). Seeds were sown at 9 ± 1 mm (b, c) or 19 ± 1mm (d, e). The sowing was made on June 13, 1996, at Wellesbourne, United Kingdom, in a sandy loam soil. Temperatures were measured at sowing depth and the water potential profiles were simulated from standard meteorological data and soil characteristics. There was some difference in temperature at the two depths, but for brevity only 9 mm is shown. Germination (dashed line) and emergence (solid line) times were predicted using models published by Rowse, McKee, and Higgs (1999) (Equation 2.16 with temperature modifications described previously) and Whalley and colleagues (1999, Appendix) respectively. Closed circles are observed seedling emergence. Temperature and water potential thresholds used in the simulation are shown. The predictions were made on 100 individual seeds each assigned a sowing depth, Ψ_b and T_b at random from distributions determined for the seed population. This simulation is being constructed with the future aim of demonstrating the consequences of environmental conditions and grower interventions on germination and seedling emergence.

of including a range of factors and their interactions may prove to be an effective way of applying threshold models to variable field conditions. Using this approach, reasonable prediction of *Eucalyptus delegatensis* seed germination in the field was possible (Battaglia, 1997). It is relevant to point out that in contrast to crop species, germination of the *Eucalyptus delegatensis* seed population was spread over months in these experiments. Under these circumstances the accuracy of soil water prediction discussed earlier has less importance.

The median Ψ_b for lettuce has been estimated by Tarquis and Bradford (1992) to be -1.0 MPa and Ψ_{min} as -2.4 MPa. Similar values have now been determined for a number of crop species. In practice, water potential in the surface layers of the seedbed can change quickly (e.g., Figure 2.6) so that little clock time is spent between these water potential thresholds. Thus reasonable predictions of seedling emergence time, under many seedbed conditions, may be possible without including advancement below Ψ_b. Indeed, the timing of flushes of onion seedling emergence can be described reasonably well using a modified hydrothermal time model which assumed that germination progressed in thermal time (unaffected by Ψ) above Ψ_b and progress ceases below it (Finch-Savage and Phelps, 1993). A similar approach has been used to predict the timing of weed seedling flushes with some success (Forcella, 1998; Forcella et al., 2000). This approach works because soil water potential rises instantly upon rain and then, at shallow sowing depth, falls rapidly to levels below which metabolic advancement occurs in the seed. This clear pattern of soil water potential can mask the inaccuracies of the simple model used, but the success of the model illustrates the importance of the rate-limiting moisture-sensitive step (Ψ_b). Nevertheless, in a further set of field experiments, this model described carrot germination more accurately than the hydrothermal model which overestimated time to germination (Finch-Savage, Steckel, and Phelps, 1998) when soil moisture was limiting. Similarly, Roman, Murphy, and Swanton (2000) accurately predicted seedling emergence of *Chenopodium album* in spring under no-till conditions using hydrothermal time to account for germination in the model. However, in seedbeds that were cultivated and therefore more subject to drying, the model overestimated time to emergence. One reason for overestimation could have been that progress below Ψ_b was not considered. An attempt to account for hydrothermal priming time in the carrot experiments improved prediction of germination times in dry conditions, but time to germination was still overestimated (Finch-Savage et al., 2000).

It has been argued previously that the most likely cause of overestimation of the time to germination results from inaccurate estimates of soil water potential at sowing depth. However, there are a number of other reasons why this may have occurred. There may have been a catastrophic effect; for example, rapid drying imposed shortly before radicle emergence can alter germination rates upon return to water (Debaene-Gill, Allen, and White, 1994). Alternatively, overestimation of germination time suggests that seeds progressed toward germination faster in variable field conditions than would be expected from laboratory experiments under constant conditions. As discussed earlier, seeds have been shown to adapt physiologically to prolonged exposure to low water potentials by lowering Ψ_b (Ni and Bradford, 1992). There is also an implicit assumption in the models that seeds wet up and dry as rapidly and to the same extent as the surrounding soil and that the impact of this on the rate of progress toward germination, including radicle emergence, is equal throughout the germination process. However, seeds may resist water loss so that they do not dry as quickly as the surrounding soil or they may acquire water that condenses on them in a cycling temperature environment. Moisture in the vapor phase can also be an important component of seed germination in the field (Bruckler, 1983a,b; Wuest, Albrecht, and Skirvin, 1999). The seed response may not remain the same throughout all stages of germination. It is also possible that once the seeds have become sufficiently hydrated they subsequently germinate faster at suboptimal water potentials than if they were imbibed and remained at that water potential, as in constant laboratory conditions. These possibilities could be resolved by experimentation using controlled changes in external water potential and temperature; however, such experimental data is lacking in the literature.

In horticulture, it is reasonable to assume that seeds are sown into adequate moisture (Finch-Savage and Phelps, 1993) for initial imbibition. However, this will not be the case in all situations or with other crops. A further potential error is that the models assume instantaneous reaction to changes in seedbed conditions; as discussed previously, this may not be true in the case of water potential changes. For prediction, imbibition is important when seeds are sown into dry soils or when the contact between seed and soil is poor (Bruckler, 1983a,b; Bouaziz and Bruckler, 1989a,b) and it is therefore also likely to be variable in the seed population. Therefore, a future improvement to simulations using threshold models for prediction, in variable conditions of moisture, may be the incorporation of a suitable imbibition model. One approach would be to apply an imbibition submodel to cover water uptake to Ψ_{min} and water loss below Ψ_{min}. As discussed previously, imbibition of water below Ψ_{min} can be considered a physical process. Above Ψ_{min} the seeds become metabolically active and progress toward germination should be predicted using a germination model. Consideration should also be given to the incorporation of seedbed structure effects on seed-soil contact area and water uptake in the vapor phase (Hadas, 1982;

Bruckler, 1983a,b; Bouaziz and Bruckler, 1989a,b; Wuest, Albrecht, and Skirvin, 1999).

Postgermination Seedling Growth and Emergence

There are comparatively few field studies in which the fate of nonemerging seedlings has been studied. In the absence of disease, most viable seeds are thought to germinate and seedling losses occur during postgermination seedling growth (e.g., Hegarty and Royle, 1978; Durrant, 1981; Finch-Savage, Steckel, and Phelps, 1998). It is therefore necessary to consider and model the exhaustion of seed reserves, especially in small-seeded crops. Limiting conditions of temperature, water, and soil strength all increase the time to seedling emergence when reserves can be used. However, under limiting water and temperature respiration rate also decreases. Recent experiments have found that respiration rate in onion seedlings also decreases as resistance to postgermination growth increases (Finch-Savage and Peach, unpublished data). In this case the total CO_2 evolved for a given increment of seedling growth was very similar over a range of resistances under nonlimiting temperature and water potential. Nevertheless, seedling growth, and therefore seedling emergence, can be reduced in this crop by increased soil resistance, probably from allocation of resources to additional structural components (e.g., Whalley et al., 1999).

The relative success of the model described by Finch-Savage and Phelps (1993) to describe emergence patterns implies that, once germination has occurred, seedling growth under a wide range of conditions does not experience significant water stress even though the soil surface becomes very dry. This view is supported by the work of Vleeshouwers and Kropff (2000) who show that if the germination percentage in the soil is known, accurate prediction of numbers of seedlings emerging is possible using their model which considers only temperature, soil penetration resistance, and seed weight. Few attempts have been made to include the effects of seedbed structure (Bouaziz and Bruckler, 1989b; Mullins et al., 1996; Dürr et al., 2001), which is likely to further improve seedling growth predictions. A model has now been developed that includes the effect of aggregate size and organization in the seedbed and crust development on hypocotyl growth but does not yet include the effects of moisture content (Dürr et al., 2001). However, reasonable predictions are possible from a simulation that accounts for soil moisture, temperature, soil resistance to growth, and time (Figure 2.6).

Simulations can be used to understand more about how seedling emergence patterns are developed. In the simple example shown (Figure 2.6), onion seeds were sown at the same time but at two depths in a randomized plot experiment. Seeds that were sown more deeply germinated faster and more uniformly (Figure 2.6e) as they were exposed to greater water potentials (Figure 2.6d) than those sown shallow (Figure 2.6 b and c). Soil water potential at the shallow sowing depth was much more variable and spent time below Ψ_b and Ψ_{min} (Figure 2.6b), and therefore seeds germinated later (Figure 2.6c). However, the period of seedling growth was greater from deeper-sown seeds with a greater influence of soil impedance. The recorded emergence shows that under the conditions following this sowing, despite the different germination times, emergence times were very similar from shallow and more deeply sown seeds (Figure 2.6c and e). Under the drier and more variable conditions experienced at shallow sowings the prediction of seedling numbers was less accurate, underlining the difficulties described in the previous section. An additional interesting point is that sowing depth varies in the seed population following sowing and therefore, in simulation, seeds are assigned to different depths and characteristic base temperatures and water potentials at random. As conditions differ in the seedbed profile, seeds may not germinate in the same set order that they are assumed to under constant laboratory conditions. For example, a faster germinating seed (i.e., low Ψ_b) may be exposed to a lower water potential than a slow germinating seed (i.e., high Ψ_b) sown deeper in the seedbed profile. Thus $\Psi - \Psi_b$ may be greater in the deeper sown, slower germinating seed, causing it to germinate faster in practice. This is accounted for in the simulation because the models are effectively run for each seed separately using Monte Carlo simulation principles.

SUMMARY AND CONCLUSIONS

The complex interactions between germination and preemergence growth characteristics in the seed population and seedbed conditions that determine seedling emergence present a challenging subject. Threshold models can accurately describe the range of responses from individuals within the population to constant environmental conditions in the laboratory and protocols are being developed to extend this to field conditions that vary. To date there have been few attempts to use these population-based models to simulate and predict germination and emergence in the field. The prospects look good, but further development of the models to include greater flexibility to account for interactions between temperature and water potential effects will be required along with testing in laboratory and field conditions that vary. However, the difficulty in obtaining accurate seed-scale environmental measurements may be the factor that limits accurate predictions of the numbers and spread of germination and emergence time in the population. Nevertheless, accurate prediction of mean germination times and the timing of seedling flushes seems possible.

An important application of simulation models is to understand the apparent contradictions that can result from field experimentation. They also form a powerful vehicle for the extension of scientific research to the farmer. An immediate use for these models may be for the selection of suitable seed sources for the site to be sown, sowing times, and suitable sites in natural populations (Battaglia, 1997). In agriculture, they could have practical application, such as determining the relative timing of crop and weed populations to develop strategies for reducing competition. They can also have an educational role in developing an improved appreciation of how variations in germination and emergence times are generated. A further use of field simulation models is to determine potential improvements in grower practice through exhaustive scenario testing, under a wide range of weather conditions, which is not practical by experimentation. For example, delayed emergence allows more time for seedbed deterioration and consequential effects on the uniformity and numbers emerging from the population. Delayed emergence can also result in reduced vigor and reduced photosynthetic efficiency of individual seedlings (Tamet et al., 1996). Scenario testing can be used to develop protocols that keep seedling emergence time to a minimum, such as timing irrigation in relation to seed development (physiological time) rather than clock time (Finch-Savage, 1990a,b). In this way, even if accurate prediction of seedling emergence under more extreme conditions is not possible, the models may be used to avoid these extremes by determining the timing of farmer intervention at crucial stages to provide predictable seedling emergence.

APPENDIX

A Threshold Model for Postgermination Seedling Growth

Whalley and colleagues (1999) developed a model to describe the elongation rate of a shoot that was based on the monomolecular function written as

$$\frac{dL}{dt} = b\left(A - L \right) \tag{2.17}$$

where t is the thermal time, L is the shoot length, and A and b are constants. To take into account the effect of water stress and mechanical impedance on elongation rate, both A and b were scaled by the following factor:

$$\left[1-\left(\frac{q}{q_L}\right)^n\right]\left[1-\left(\frac{\Psi}{\Psi_L}\right)\right] \tag{2.18}$$

where q is the penetrometer pressure (i.e., proportional to mechanical impedance) and Ψ is the water stress, and q_L, Ψ_L, and n are all constants. q_L is a conceptual value of penetrometer pressure that will just stop elongation and Ψ_L, a water potential that will just stop elongation. In this form, the model gave a reasonable description of the response of carrot and onion shoots to constant levels of mechanical impedance; however, it was poor at describing how shoots recovered in a stress-free environment, following prolonged exposure to combinations of water stress and mechanical impedance. A practical example of this situation would be irrigation of a dry and strong soil. To improve this aspect of the model, Whalley and colleagues (1999) allowed A to decline with thermal time according to a logistic function written as

$$A = \left[\frac{c}{1+\left(\frac{t}{d}\right)^m}\right] \tag{2.19}$$

where t is thermal time accumulated by a seedling and c, d, and m are constants. To make predictions with the model it needs to be solved numerically. Some analogy can be drawn between this model for shoot elongation rate and the threshold models used to describe germination. Both models have a value of water potential below which no elongation can occur. In addition, the shoot elongation model has an upper limit to mechanical impedance which stops elongation. However, the shoot elongation model is different because the rate of elongation depends on the length of the shoot (L) and the thermal time that the seedling has accumulated (t) in addition to the difference between the level of a physical stress (mechanical impedance, q or/and water stress, Ψ) and its threshold value (q_L and/or Ψ_L). In contrast, the rate of germination at a given temperature depends only on the difference between water potential and the appropriate value of base water potential.

REFERENCES

Adams, R. (1999). Germination of *Callitris* seeds in relation to temperature, water stress, priming, and hydration-dehydration cycles. *Journal of Arid Environments* 43: 437-448.

Al-Ani, A., Bruzau, F., Raymond, P., Saint-Ges, V., Leblac, J.M., and Pradet, A. (1985). Germination, respiration, and adenylate energy charge of seeds at various oxygen partial pressure. *Plant Physiology* 79: 885-890.

Allen, P.S., Meyer, S.E., and Khan, M.A. (2000). Hydrothermal time as a tool in comparative germination studies. In Black, M., Bradford, K.J., and Vázquez-Ramos, J. (Eds.), *Seed Biology: Advances and Applications* (pp. 401-410). Wallingford, UK: CAB International.

Allen, P.S., White, D.B., and Markhart, A.H. (1993). Germination of perennial ryegrass and annual blue grass seeds subjected to hydration-dehydration cycles. *Crop Science* 33: 1020-1025.

Baskin, C.C. and Baskin, J.M. (1998). *Seeds: Ecology, Biogeography, and Evolution of Dormancy and Germination.* New York: Academic Press.

Battaglia, M. (1997). Seed germination model for *Eucalyptus delegatensis* provenances germinating under conditions of variable temperature and water potential. *Australian Journal of Plant Physiology* 27: 69-79.

Bauer, M.C., Meyer, S.E., and Allen, P.S. (1998). A simulation model to predict seed dormancy loss in the field for *Bromus tectorum* L. *Journal of Experimental Botany* 49: 1235-1244.

Benjamin, L.R. (1990). Variation in time of seedling emergence within populations: A feature that determines individual growth and development. *Advances in Agronomy* 44: 1-25.

Benjamin, L.R. and Hardwick, R.C. (1986). Sources of variation and measures of variability in even-aged stands of plants. *Annals of Botany* 58: 527-537.

Bewley, J.D. and Black, M. (1978). *Physiology and Biochemistry of Seeds.* New York: Springer-Verlag.

Bewley, J.D. and Black, M. (1994). *Seeds: Physiology of Development and Germination,* Second Edition. New York: Plenum Press.

Bierhuizen, J.F. and Feddes, R.A. (1973). Use of temperature and short wave radiation to predict the rate of seedling emergence and harvest date. *Acta Horticulturae* 27: 269-274.

Bierhuizen, J.F. and Wagenvoort, W.A. (1974). Some aspects of seed germination in vegetables: I. The determination and application of heat sums and minimum temperature for germination. *Scientia Horticulturae* 2: 213-219.

Blacklow, W.M. (1972). Simulation model to predict germination and emergence of corn (*Zea mays* L.) in an environment of changing temperature. *Crop Science* 13: 604-608.

Bliss, C.I. (1967). *Statistics in Biology,* Volume 1. New York: McGraw-Hill.

Bouaziz, A. and Bruckler, I. (1989a). Modeling of wheat imbibition and germination as influenced by soil physical properties. *Soil Science Society of America* 53: 219-227.

Bouaziz, A and Bruckler, I. (1989b). Modeling of wheat seedling growth and emergence: II. Comparison with field experiments. *Soil Science Society of America* 53: 1838-1846.

Bradford K.J. (1990). A water relations analysis of seed germination rates. *Plant Physiology* 94: 840-849.

Bradford K.J. (1995). Water relations in seed germination. In Kigel, J. and Galili, G. (Eds.), *Seed Development and Germination* (pp. 351-396). New York: Marcel Dekker, Inc.

Bradford, K.J. (2002). Applications of hydrothermal time to quantifying and modelling seed germination and dormancy. *Weed Science* 50: 248-260.

Bradford K.J., Dahal, P., and Ni, B.R. (1993). Quantitative models describing germination responses to temperature, water potential, and growth regulators. In Côme, D. and Corbineau, F. (Eds.), *Fourth International Workshop on Seeds: Basic and Applied Aspects of Seed Biology*, Volume 1 (pp. 239-248). Paris: Association pour la Formation Professionnelle de l'Interprofession Semences.

Bradford, K.J. and Haigh, A.M. (1994). Relationship between accumulated hydrothermal time during seed priming and subsequent seed germination rates. *Seed Science Research* 4: 1-10.

Bradford, K.J. and Somasco, O.A. (1994). Water relations of lettuce seed thermoinhibition: I. Priming and endosperm effects on base water potential. *Seed Science Research* 4: 1-10.

Breeze V.G. and Milbourn, G.M. (1981). Inter-plant variation in temperate crops of maize. *Annals of Applied Biology* 99: 335-352.

Brisson, N., Mary, B., Ripoche, D., Jeuffroy, M.H., Ruget, F., Nicoullaud, B., Gate, P., Devienne-Barret, F., Antonioletti, R., Durr, C., et al. (1998). STICS: A generic model for the simulation of crops and their water and nitrogen balances: I. Theory and parameterization applied to wheat and corn. *Agromonie* 18: 311-346.

Bruckler, L. (1983a). Rôle des propriétés physiques du lit de semences sur l'imbibition et la germination: I. Elaboration d'un modèle du système "terre-graine." *Agronomie* 3: 213-222.

Bruckler, L. (1983b). Rôle des propriétés physiques du lit de semences sur l'imbibition et la germination: II. Contrôle expérimental d'un modéle d'imbibition et possibilitiés d'application. *Agronomie* 3: 223-232.

Cal, J.P. and Obendorf, R.L. (1972). Imbibitional chilling injury in *Zea mays* L. altered by initial kernel moisture and maternal parent. *Crop Science* 12: 369-373.

Campbell, R.K. and Sorensen, F.C. (1979). A new basis for characterizing germination. *Journal of Seed Technology* 4: 24-34.

Cheng, Z. and Bradford, K.J. (1999). Hydrothermal time analysis of tomato seed germination responses to priming treatments. *Journal of Experimental Botany* 50: 89-99.

Choinski, J.S., Jr. and Tuohy, J.M. (1991). Effect of water potential and temperature on the germination of four species of African savana trees. *Annals of Botany* 68: 227-233.

Christensen, M., Meyer, S.E., and Allen, P.S. (1996). A hydrothermal time model of seed after-ripening in *Bromus tectorum* L. *Seed Science Research* 6: 155-163.

Coolbear, P., Francis, A., and Grierson, D. (1984). The effect of low temperature pre-sowing treatment on the germination performance and membrane integrity of artificially aged tomato seeds. *Journal of Experimental Botany* 35: 1609-1617.

Corbineau, F. and Côme, D. (1995). Control of seed germination and dormancy by the gaseous environment. In Kigel, J. and Galili, G. (Eds.), *Seed Development and Germination* (pp. 397-424). New York: Marcel Dekker, Inc.

Covell, S., Ellis, R.H., Roberts, E.H., and Summerfield, R.J. (1986). The influence of temperature on seed germination rate in grain legumes: I. A comparison of chickpea, lentil, soybean, and cowpea at constant temperatures. *Journal of Experimental Botany* 37: 705-715.

Dahal, P. and Bradford, K.J. (1990). Effects of priming and endosperm integrity on seed germination of tomato genotypes: II. Germination at reduced water potential. *Journal of Experimental Botany* 41: 1441-1453.

Dahal, P. and Bradford, K.J. (1994). Hydrothermal time analysis of tomato seed germination at suboptimal temperature and reduced water potential. *Seed Science Research* 4: 71-80.

Dahal, P., Bradford, K.J., and Haigh, A.M. (1993). The concept of hydrothermal time in seed germination and priming. In Côme, D. and Corbineau, F. (Eds.), *Fourth International Workshop on Seeds: Basic and Applied Aspects of Seed Biology*, Volume 3 (pp. 1009-1004). Paris: Association pour la Formation Professionnelle de l'Interprofession Semences.

Dahal, P., Bradford, K.J., and Jones, R.A. (1990). Effects of priming and endosperm integrity on seed germination rates of tomato genotypes: I. Germination at suboptimal temperature. *Journal of Experimental Botany* 41: 1431-1439.

Dasberg, S. (1971). Soil water movement to germinating seeds. *Journal of Experimental Botany* 22: 999-1008.

Dasberg, S. and Mendel, K. (1971). The effect of soil water and aeration on seed germination. *Journal of Experimental Botany* 22: 992-998.

Debaene-Gill, S.B., Allen, P.S., and White, D.B. (1994). Dehydration of germinating perennial ryegrass seeds can alter the rate of subsequent radicle emergence. *Journal of Experimental Botany* 45: 1301-1307.

Dracup, M., Davies, C. and Tapscott, H. (1993). Temperature and water requirements for germination and emergence of lupin. *Australian Journal of Experimental Agriculture* 33: 759-766.

Dürr, C., Aubertot, J.N., Richard, G., Dubrulle, P., Duval, Y., and Boiffin, J. (2001). SIMPLE: A model for simulation of plant emergence predicting the effects of soil tillage and sowing operations. *Soil Science Society of America Journal* 65: 414-423.

Durrant, M.J. (1981). Some causes of the variation in plant establishment. In *Proceedings IIRB 44th Winter Congress 1981* (pp. 7-20). Brussels: Institut International de Recherche Betteraviére.

Ellis, R.H. and Barrett, S. (1994). Alternating temperatures and rate of seed germination in lentil. *Annals of Botany* 74: 519-524.

Ellis, R.H. and Butcher, P.S. (1988). The effects of priming and "natural" differences in quality amongst onion seed lots on the response of the rate of germination to temperature and the identification of the characteristics under genotypic control. *Journal of Experimental Botany* 39: 935-950.

Ellis, R.H., Covell, S., Roberts, E.H., and Summerfield, R.J. (1986). The influence of temperature on seed germination rate in grain legumes: II. Intraspecific variation in chickpea at constant temperatures. *Journal of Experimental Botany* 37: 1503-1515.

Ellis, R.H., Simon, G., and Covell, S. (1987). The influence of temperature on seed germination rate in grain legumes: III. A comparison of five faba bean genotypes at constant temperatures using a new screening method. *Journal of Experimental Botany* 38: 1033-1043.

Feddes, R.A. (1972). Effects of water and heat on seedling emergence. *Journal of Hydrology* 16: 341-359.

Finch-Savage, W.E. (1984a). Effects of fluid drilling germinating onion seeds on seedling emergence and subsequent plant growth. *Journal of Agricultural Science* 102: 461-468.

Finch-Savage, W.E. (1984b). The effects of fluid drilling germinating seeds on the emergence and subsequent growth of carrots in the field. *Journal of Horticultural Science* 59: 411-417.

Finch-Savage, W.E. (1987). A comparison of seedling emergence and early seedling growth from dry-sown natural and fluid-drilled pregerminated onion (*Allium cepa* L.) seeds in the field. *Journal of Horticultural Science* 62: 39-47.

Finch-Savage, W.E. (1990a). Effects of osmotic seed priming and the timing of water availability in the seedbed on the predictability of carrot seedling establishment in the field. *Acta Horticulturae* 267: 209-216.

Finch-Savage, W.E. (1990b). Estimating the optimum time of irrigation to improve vegetable crop establishment. *Acta Horticulturae* 278: 807-814.

Finch-Savage, W.E. (1995). Influence of seed quality on crop establishment, growth and yield. In Basra, A.S. (Ed.), *Seed Quality: Basic Mechanisms and Agricultural Implications* (pp. 361-384). Binghamton, NY: Food Products Press.

Finch-Savage, W.E. and Phelps, K. (1993). Onion (*Allium cepa* L.) seedling emergence patterns can be explained by the influence of soil temperature and water potential on seed germination. *Journal of Experimental Botany* 44: 407-414.

Finch-Savage, W.E., Phelps, K., Peach, L., and Steckel, J.R.A. (2000) Use of threshold germination models under variable field conditions. In Black, M., Bradford, K.J., and Vázquez-Ramos, J. (Eds.), *Seed Biology: Advances and Applications* (pp. 489-497). Wallingford, UK: CAB International.

Finch-Savage, W.E., Phelps, K., Steckel, J.R.A., Whalley, W.R., and Rowse, H.R. (2001). Seed reserve-dependant growth responses to temperature and water potential in carrot (*Daucus carota* L.). *Journal of Experimental Botany* 52: 2187-2197.

Finch-Savage, W.E., Steckel, J.R.A., and Phelps, K. (1998). Germination and postgermination growth to carrot seedling emergence: Predictive threshold models and sources of variation between sowing occasions. *New Phytologist* 139: 505-516.

Finney, D.J. (1971). *Probit Analysis,* Third Edition. Cambridge, UK: Cambridge University Press.

Forcella, F. (1993). Seedling emergence model for velvetleaf. *Agronomy Journal* 85: 929-933.

Forcella, F. (1998). Real-time assessment of seed dormancy and seedling growth for weed management. *Seed Science Research* 8: 201-209.

Forcella, F., Benech-Arnold, R.L., Sanchez, R., and Ghersa, C.M. (2000). Modeling seedling emergence. *Field Crops Research* 67: 123-139.

Fyfield, T.P. and Gregory, P.J. (1989). Effects of temperature and water potential on germination, radicle elongation and emergence of mungbean. *Journal of Experimental Botany* 40: 667-674.

Gan, Y., Stobbe, E.H., and Njue, C. (1996). Evaluation of selected nonlinear regression models in quantifying seedling emergence rate of spring wheat. *Crop Science* 36: 165-168.

Garcia-Huidobro, J., Monteith, J.L., and Squire, G.R. (1982a). Time, temperature and germination of pearl millet (*Pennisetum typhoides* S. & H.): I. Constant temperature. *Journal of Experimental Botany* 33: 288-296.

Garcia-Huidobro, J., Monteith, J.L., and Squire, G.R. (1982b). Time, temperature and germination of pearl millet (*Pennisetum typhoides* S. & H.): II. Alternating temperature. *Journal of Experimental Botany* 33: 297-302.

González-Zertuche, L., Vázquez-Yanes, C., Gamboa, A., Sánchez-Coronado, M.E., Aguilera, P., and Orozco-Segovia, A. (2001). Natural priming of *Wigandia urens* seeds during burial: Effects on germination, growth and protein expression. *Seed Science Research* 11: 27-34.

Gray, D. (1976). The effect of time to emergence on head weight and variation in head weight at maturity in lettuce (*Lactuca sativa*). *Annals of Applied Biology* 82: 569-575.

Grundy, A.C., Phelps, K., Reader, R.J., and Burston, S. (2000). Modeling the germination of *Stellaria media* using the concept of hydrothermal time. *New Phytologist* 148: 433-444.

Guérif, J., Richard, G., Durr, C., Machet, J.M., Recous, S., and Roger-Estrade, J. (2001). A review of tillage effects on crop residue management, seedbed conditions and seedling establishment. *Soil and Tillage Research* 61: 13-32.

Gummerson, R.J. (1986). The effect of constant temperatures and osmotic potential on the germination of sugar beet. *Journal of Experimental Botany* 37: 729-958.

Hadas, A. (1970). Factors affecting seed germination under soil moisture stress. *Israel Journal of Agricultural Research* 20: 3-14.

Hadas, A. (1982). Seed-soil contact and germination. In Khan, A.A. (Ed.), *The Physiology and Biochemistry of Seed Development, Dormancy, and Germination* (pp. 507-527). Amsterdam: Elsevier.

Hageseth, G.T. and Young, C.W. (1994). The four-compartment thermodynamic energy-level diagram for isothermal seed germination. *Journal of Thermal Biology* 19: 1-11.

HåKansson, I. and von Polgár, J. (1984). Experiments on the effects of seedbed characteristics on seedling emergence in a dry weather situation. *Soil and Tillage Research* 4: 115-135.

Hegarty, T.W. (1976). Effect of fertilizer on the seedling emergence of vegetable crops. *Journal of the Science of Food and Agriculture* 27: 962-968.

Hegarty, T.W. (1977). Seed activation and seed germination under moisture stress. *New Phytologist* 78: 349-359.

Hegarty, T.W. (1978). The physiology of seed hydration and dehydration, and the relation between water stress and the control of germination: A review. *Plant, Cell and Environment* 1: 101-119.

Hegarty, T.W. (1984). The influence of environment on seed germination. *Aspects of Applied Biology* 7: 13-31.

Hegarty, T.W. and Perry, D.A. (1974). Predictable stands: Problems and prospects. *Arable Farming* 14-17.

Hegarty T.W. and Royle, S.M. (1976). Impedance of calabrese seedlings from light soils after rainfall. *Horticultural Research* 16: 107-114.

Hegarty T.W. and Royle, S.M. (1978). Soil impedance as a factor reducing crop seedling emergence, and its relation to soil conditions at sowing, and to applied water. *Journal of Applied Ecology* 15: 897-904.

Hilhorst, H.W.M. and Karssen, C.M. (2000). Effect of chemical environment on seed germination. In Fenner, M. (Ed.), *Seeds: The Ecology of Regeneration in Plant Communities,* Second Edition (pp. 293-309). Wallingford, UK: CAB International.

Hilhorst, H.W.M. and Toorop, P.E. (1997). A review on dormancy, germinability and germination in crop and weed seeds. *Advances in Agronomy* 61: 111-166.

Hsu, F.H., Nelson, C.J., and Chow, W.S. (1984). A mathematical model to utilize the logistic function in germination and seedling growth. *Journal of Experimental Botany* 35: 1629-1640.

Jordan, G.L. (1983). Planting limitations for arid, semi-arid and salt-desert shrublands. In Monsen, S. and Shaw, N. (Eds.), *Managing Intermountain Rangelands,* USDA Forest Service, Intermountain Research Station General Technical Report INT 157 (pp. 11-16). September 15-17, 1981, Twin Falls, Idaho, and June 22-24, 1982, Elko, Nevada: USDA Forest Service.

Jordan, G.L. and Haferkamp, M.R. (1989). Temperature responses and calculated heat units for germination of several range grases and shrubs. *Journal of Range Management* 42: 41-45.

Kebreab, E. and Murdoch, A.J. (1999). Modeling the effects of water stress and temperature on germination rate of *Orobanche aegyptiaca* seeds. *Journal of Experimental Botany* 50: 655-644.

Kebreab, E. and Murdoch, A.J. (2000). The effect of water stress on the temperature range for germination of *Orobanche aegyptiaca* seeds. *Seed Science Research* 10: 127-133.

Khan, A.A. (1992). Preplant physiological seed conditioning. *Horticultural Reviews* 14: 131-181.

King, C.A. and Oliver, L.R. (1994). A model for predicting large crabgrass *(Digitaria sanguinalis)* emergence as influenced by temperature and water potential. *Weed Science* 42: 561-567.

Koller, D. and Hadas, A. (1982). Water relations in the germination of seeds. In Lange. O.L., Nobel, P.S., Osmond, C.B., and Ziegler, H. (Eds.), *Encyclopedia of Plant Physiology,* New Series, Volume 12B (pp. 401-431). Berlin: Springer-Verlag.

Kruk, B.C. and Benech-Arnold, R.L. (1998). Functional and quantitative analysis of seed thermal responses in prostrate knotweed *(Polygonum aviculare)* and common purslane *(Portulaca oleracea). Weed Science* 46: 83-90.

Kruk, B.C. and Benech-Arnold, R.L. (2000). Evaluation of dormancy and germination responses to temperature in *Cardus acanthoides* and *Anagallis arvensis* using a screening system, and relationship with field observed emergence patterns. *Seed Science Research* 10: 77-88.

Labouriau, L.G. (1970). On the physiology of seed germination in *Vicia graminea* Sm. I. *Annales Academia Brasilia Ciencia* 42: 235-262.

Labouriau, L.G. and Osborn, J.H. (1984). Temperature dependence of the germination of tomato seeds. *Journal of Thermal Biology* 9: 285-294.

Lascano, R.J. and van Babel, C.H.M. (1986). Simulation and measurement of evaporation from bare soil. *Soil Science Society of America Journal* 50: 1127-1132.

Maguire, J.D. (1984). Dormancy in seeds. *Advances in Research and Technology of Seeds* 9: 25-60.

Marshall, B. and Squire, G.R. (1996). Non-linearity in rate-temperature relationships of germination in oilseed rape. *Journal of Experimental Botany* 47: 1369-1375.

McDonald, M.B., Jr., Sulivan, J., and Lauer, M.J. (1994). The pathway of water uptake in maize seeds. *Seed Science and Technology* 22: 79-90.

McDonald, M.B., Jr., Vertucci, C.W., and Roos, E.E. (1988a). Seed coat regulation of soybean seed imbibition. *Crop Science* 28: 987-992.

McDonald, M.B., Jr., Vertucci, C.W., and Roos, E.E. (1988b). Soybean seed imbibition: Water absorption by seed parts. *Crop Science* 28: 993-997.

Meyer, S.E., Debaene-Gill, S.B., and Allen, P.S. (2000). Using hydrothermal time concepts to model seed germination response to temperature, dormancy loss, and priming effects in *Elymus elymoides. Seed Science Research* 10: 213-223.

Mondal, M.F., Brewster, J.L., Morris, G.E.L., and Butler, H.A. (1986). Bulb development in onion *(Allium cepa* L.): I. Effects of plant density and sowing date in field conditions. *Annals of Botany* 58: 187-195.

Mullins, C.E., Townend, J., Mtakwa, P.W., Payne, C.A., Cowan, G., Simmonds, L.P., Daamen, C.C., Dunbabin, T., and Naylor, R.E.L. (1996). *EMERGE User Guide: A Model to Predict Crop Emergence in the Semi-Arid Tropics.* Aberdeen: Department of Plant and Soil Science, University of Aberdeen.

Murdoch, A.J., Roberts, E.H., and Goedert, C.O. (1989). A model for germination responses to alternating temperatures. *Annals of Botany* 63: 97-111.

Ni, B.R. and Bradford, K.J. (1992). Quantitative models characterizing seed germination responses to abscisic acid and osmoticum. *Plant Physiology* 98: 1057-1068.

Ni, B.R. and Bradford, K.J. (1993). Germination and dormancy of abscisic acid- and gibberellin-deficient mutant tomato seeds: Sensitivity of germination to abscisic acid, gibberellin, and water potential. *Plant Physiology* 101: 607-617.

Orozco-Segovia, A., González-Zertuche, L., Mendoza, A., and Orozco, S. (1996). A mathematical model that uses Gaussian distribution to analyze the germination of *Manfreda brachystachya* (Agavaceae) in a thermogradient. *Physiologia Plantarum* 98: 431-438.

Pemberton, M.R. and Clifford, H.T. (1994). Seed germination models. *Seed Science and Technology* 22: 209-221

Perry, D.A. (1984). Factors influencing the establishment of cereal crops. *Aspects of Applied Biology* 7: 65-83.

Phelps, K. and Finch-Savage, W.E. (1997). A statistical perspective on threshold type models. In Ellis, R.H., Black, M., Murdoch, A.J., and Hong, T.D. (Eds.), *Basic and Applied Aspects of Seed Biology* (pp. 361-368). Dordrecht, the Netherlands: Kluwer.

Probert, R.J. (2000). The role of temperature in the regulation of seed dormancy and germination. In Fenner, M. (Ed.), *Seeds: The Ecology of Regeneration in Plant Communities*, Second Edition (pp. 261-292). Wallingford, UK: CAB International.

Richard, G. and Boiffin, J. (1990). Effets de l'état structural du lit de semences sur la germination et la levée des cultures. In Boiffin, J. and Marin-Laflèche, A. (Eds.), *La structure du sol et son évolution* (pp. 112-136). Paris: Institut National de la Recherche Agronomique (INRA).

Richard, G. and Guérif, J. (1988a). Modélisation des transferts gazeux dans le lit de semence: Application au diagnostic des conditions d'hypoxie des semences de betterave sucrière (*Beta vulgaris* L.) pendent la germination: I. Présentation du modèle. *Agronomie* 8: 539-547.

Richard, G. and Guérif, J. (1988b). Modélisation des transferts gazeux dans le lit de semence: Application au diagnostic des conditions d'hypoxie des semences de betterave sucrière (*Beta vulgaris* L.) pendent la germination: II. Résultats des simulations. *Agronomie* 8: 639-646.

Roberts, E.H. (1988). Temperature and seed germination. In Long, S.P. and Woodward, F.I. (Eds.), *Plants and Temperature* (pp. 109-132). Cambridge, UK: Society for Experimental Biology.

Roberts, H.A. (1984). Crop and weed emergence patterns in relation to time of cultivation and rainfall. *Annals of Applied Biology* 105: 263-275.

Roman, E.S., Murphy, S.D., and Swanton, C.J. (2000). Simulation of *Chenopodium album* seedling emergence. *Weed Science* 48: 217-224.

Roman, E.S., Thomas, A.G., Murphy, S.D., and Swanton, C.J. (1999). Modeling germination and seedling elongation of common lambsquarters (*Chenopodium album*). *Weed Science* 47: 149-155.

Ross, H.A and Hegarty, T.W. (1979). Sensitivity of seed germination and seedling radicle growth to moisture stress in some vegetable crop species. *Annals of Botany* 43: 241-243.

Rowse, H.R., McKee, J.M.T., and Higgs, E.C. (1999). A model of the effects of water stress on seed advancement and germination. *New Phytologist* 143: 273-279.

Salter, P.J. and James, J.M. (1975). The effect of plant density on the initiation, growth and maturity of curds of two cauliflower varieties. *Journal of Horticultural Science* 50: 239-248.

Schneider, A. and Renault, P. (1997). Effects of coating on seed imbibition: I. Model estimates of water transport coefficient. *Crop Science* 37: 1841-1849.

Schoolfield, R.M., Sharpe, P.J.H., and Magnuson, C.E. (1981). Non-linear regression of biological temperature-dependant rate models based on absolute reaction-rate theory. *Journal of Theoretical Biology* 88: 719-731.

Scott, S.J., Jones, R.A., and Williams, W.A. (1984). Review of data analysis methods for seed germination. *Crop Science* 24: 1192-1199.

Sharpe, P.J.H. and DeMichele, D.W. (1977). Reaction kinetics of poiklotherm development. *Journal of Theoretical Biology* 64: 649-670.

Shrestha, A., Roman, E.S., Thomas, A.G., and Swanton, C.J. (1999). Modeling germination and shoot-radicle elongation of *Ambrosia artemisiifolia*. *Weed Science* 47: 557-562.

Steinmaus, S.J., Prather, T.S., and Holt, J.S. (2000). Estimation of base temperatures for nine weed species. *Journal of Experimental Botany* 51: 275-286.

Tamet, V., Boiffin, J., Durr, C., and Souty, N. (1996). Emergence and early growth of an epigeal seedling (*Daucus carota* L.): Influence of soil temperature, sowing depth, soil crusting and seed weight. *Soil and Tillage Research* 40: 25-38.

Tarquis, A. and Bradford, K.J. (1992). Prehydration and priming treatments that advance germination also increase the rate of deterioration of lettuce seed. *Journal of Experimental Botany* 43: 307-317.

Thornley, J.H.M. (1986). A germination model: Responses to time and temperature. *Journal of Theoretical Biology* 123: 481-492.

Trudgill, D.L., Squire, G.R., and Thompson, K. (2000). A thermal time basis for comparing the germination requirments of some British herbaceous plants. *New Phytologist* 145:107-114.

Vertucci, C.W. (1989). The kinetics of seed imbibition: Controlling factors and relevance to seedling vigor. In Stanwood, P.C. (Ed.), *Seed Moisture* (pp. 93-115). Special Publication No 14. Madison, WI: Crop Science Society of America

Vertucci, C.W. and Leopold, A.C. (1983). Dynamics of imbibition in soybean embryos. *Plant Physiology* 72: 190-193.

Vertucci, C.W. and Leopold, A.C. (1987). Water binding in legume seeds. *Plant Physiology* 85: 224-231.

Villiers, T.A. (1972). Seed dormancy. In Kozlowski, T.T. (Ed.), *Seed Biology*, Volume II (pp. 220-281). New York: Academic Press.

Vleeshouwers, L.M. (1997). Modeling the effect of temperature, soil penetration resistance, burial depth and seed weight on preemergence growth of weeds. *Annals of Botany* 79: 553-563.

Vleeshouwers, L.M. and Kropff, M.J. (2000). Modeling field emergence patterns in arable weeds. *New Phytologist* 148: 445-457.

Wagenvoort, W.A. and Bierhuizen, J.F. (1977). Some aspects of seed germination in vegetables: II. The effect of temperature fluctuation, depth of sowing, seed size and cultivar, on heat sum and minimum temperature for germination. *Scientia Horticulturae* 6: 259-270.

Walker, A. and Barnes, A. (1981). Simulation of herbicide persistence in soil, a revised computer model. *Pesticide Science* 12: 123-132.

Wanjura, D.F., Buxton, D.R., and Stapleton, H.N. (1970). A temperature model for predicting initial cotton emergence. *Agronomy Journal* 62: 741-743.

Washitani, I. (1985). Germination-rate dependency on temperature of *Geranium carolinianum* seeds. *Journal of Experimental Botany* 36: 330-337.

Washitani, I. (1987) A convenient screening test system and a model for thermal germination responses of wild plant seeds: Behavior of model and real seeds in the system. *Plant, Cell and Environment* 10: 587-598.

Washitani, I. and Takenaka, A. (1984). Mathematical description of the seed germination dependency on time and temperature. *Plant, Cell and Environment* 7: 359-362.

Weaich, K., Bristow, K.L., and Cass, A. (1996). Modeling preemergent maize shoot growth: I. Physiological temperature conditions. *Agronomy Journal* 88: 391-397.

Welbaum, G.E., Bradford, K.J., Yim, K.O., Booth, D.T., and Oluoch, M.O. (1998). Biophysical, physiological and biochemical processes regulating seed germination. *Seed Science Research* 8: 161-172.

Whalley, W.R. (1993). Considerations on the use of time-domain reflectometry (TDR) for measuring soil water content. *Journal of Soil Science* 44: 1-9.

Whalley, W.R., Finch-Savage, W.E., Cope, R.E., Rowse, H.R., and Bird, N.R.A. (1999). The response of carrot (*Daucus carota* L.) and onion (*Allium cepa* L.) seedlings to mechanical impedance and water stress at suboptimal temperatures. *Plant, Cell and Environment* 22: 229-242.

Wheeler, T.R. and Ellis, R.H. (1991). Seed quality, cotyledon elongation at suboptimal temperatures, and the yield of onion. *Seed Science Research* 1: 57-67.

Wilson, A.M. (1973). Responses of crested wheatgrass seeds to environment. *Journal of Range Management* 26: 43-46.

Woodstock, L.W. (1988). Seed imbibition: A critical period for successful germination. *Journal of Seed Technology* 12: 1-15.

Wuest, S.B., Albrecht, S.L., and Skirvin, K.W. (1999). Vapor transport vs. seed-soil contact in wheat germination. *Agronomy Journal* 91: 783-787.

Chapter 3

Seed and Agronomic Factors Associated with Germination Under Temperature and Water Stress

Mark A. Bennett

INTRODUCTION

Substantial progress has been made in our understanding of physiological mechanisms in seeds that confer the ability to germinate under stress conditions. Parallel to this progress is a series of agronomic changes, including (1) shifts to earlier planting dates and tillage practices; (2) greater expectations of precision and uniformity in seedling establishment; and (3) double-cropping systems that require continued seed research and new strategies for reliable crop production.

The objective of this chapter is to describe and review present knowledge on physiological, morphological, and cultural factors involved in germination under stress conditions. Although not intended to be a comprehensive literature review of this wide-ranging subject, references to related reviews, proceedings, and books are made in connection with several sections of this chapter.

The overarching goal of crop establishment is to achieve rapid and uniform germination, followed by rapid and uniform seedling emergence plus autotrophy (Covell et al., 1986). Seeds are particularly vulnerable to stress(es) encountered between sowing and seedling establishment (Carter and Chesson, 1996). Germination and seedling establishment in crop species are the end result of a complex and interactive process, involving a number of physiological, morphological, environmental, and cultural fac-

The assistance of Julie Hering in preparing the manuscript and Jimmie Jones in figure design is appreciated. Salaries and research support provided in part by state and federal funds appropriated to the Ohio Agricultural Research and Development Center, The Ohio State University.

tors (Figure 3.1). Insights into the physiological mechanisms and cultural practices that increase the ability of seeds to perform optimally under stressful conditions will be useful for (1) sowing on atypical dates and (2) when introducing crops into new production areas or systems (Covell et al., 1986; Thiessen Martens and Entz, 2001).

SEED COATS

The seed coat, or testa, has an important role in germination under stress conditions. An intact seed coat is essential for controlled water uptake and protection from injury to the embryo or other tissues (Chachalis and Smith, 2000; Baskin and Baskin, 1998). Seeds of various Fabaceae species have been studied to compare traits of permeable versus water-impermeable genotypes. Studies with 'Williams 82' soybean seedlots demonstrated the ability of seed coats to (1) direct water penetration to the embryo and (2) serve as a reservoir of water for the developing axis (McDonald, Vertucci, and Roos, 1988a). The testa can also decrease levels of solute leakage re-

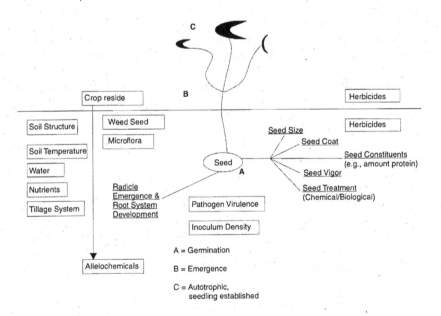

FIGURE 3.1. Schematic model of physiological, morphological, and cultural factors involved in germination and seedling establishment

sulting from seed water uptake and imbibitional damage. A comparative study of 19 soybean accessions with a wide range of seed size (60 to 257 mg/seed) and testa color showed a testa dry weight range (5.8 to 18.3 mg/seed) that was closely correlated to seed dry weight (Chachalis and Smith, 2000). Total dry weight per unit area ranged from 0.075 to 0.150 mg·mm^{-2} and was negatively correlated with total seed dry weight. Rates of water uptake and testa dry weight:dry weight ratios (6.5 percent to 13.8 percent) were not correlated (Chachalis and Smith, 2000). Lupin seeds (especially lines of *Lupinus pilosus*) had thinner coats when produced in a dry season with about 50 percent of average rainfall (Miao, Fortune, and Gallagher, 2001). Genetic characteristics and production environment effects on seed coat structure can interact significantly.

Seed coat surface deposits, phenolic materials, and pore development patterns have also been studied in relation to water uptake (Mayer and Poljakoff-Mayber, 1989). Phenolic materials in permeable and impermeable legume seeds have not been strongly linked to water uptake patterns (Slattery, Atwell, and Kuo, 1982; Chachalis and Smith, 2001). Developmental studies of four soybean genotypes from maturity groups III through V showed pores formed first around the hilum (approximately 36 days after flowering). Pore development next encircled the seed parallel with the axis and then formed on the abaxial surface, i.e., the area covering the round face of the cotyledon (Yaklich, Vigil, and Wergin, 1986). Soybean imbibition studies with four permeable and three impermeable seed lines detected a lack of pores in the abaxial region of the seed coat in VLS-1, a delayed-permeability genotype. In two lines possessing a rapid-permeability seed coat characteristic, pores were observed to be deep, wide open, and densely distributed (Chachalis and Smith, 2001). The VLS-1 (delayed-permeability) line is black seeded; if the associated pure characteristic is inheritable, breeders could transfer this trait to yellow-seeded genotypes. Alternatively, genotypes could be selected based on pore characteristics to provide more resistance to imbibition damage from waterlogged soil conditions, etc.

SEED SIZE

Large seed size is widely thought to improve the chances for crop emergence under a wide range of environments. It is also generally considered that, within a seedlot, seeds with a greater seed weight have greater storage reserves and thereby have increased seed vigor (Powell, 1988). Studies of seed size effects on stand establishment are conflicting, however, and several possible explanations exist for the mixed findings. Seed size classes should be kept distinct from seed quality (vigor) assessments. Among red clover seedlots, for example, the relationship between thousand seed weight (TSW) and seed vigor was weak or nonexistent (Wang and Hampton, 1989). Although many reports suggest that larger seeds produce seedlings with greater early growth and increased competitive ability against weeds and pests (Chastain, Ward, and Wysocki, 1995; Douglas, Wilkins, and Churchill, 1994; Mian and Nafziger, 1992), the sheer range of conditions examined in the literature is cause for careful interpretation of results.

Plants grown from smaller spring wheat (*Triticum aestivum* L.) seeds emerged faster but accumulated less shoot weight than plants grown from large seeds (Lafond and Baker, 1986a). Seed size accounted for approximately 50 percent of the variation in seedling shoot dry weight for the nine cultivars tested over two years in Saskatchewan, Canada. A large survey of winter wheat stand establishment in the southern Great Plains of the United States showed reduced percent emergence (45 percent) for the smallest seed class (<19.8 g, TSW). This compared to approximately 60 percent emergence for seed classes of 19.8 to 22.7, 25.5 to 28.0, and >28 g per TSW (Stockton et al., 1996). Seed size studies are interesting in that the cause(s) of smaller seed can be quite diverse. Mian and Nafziger (1992) examined the effect of three seed sizes of soft red winter wheat, producing seed size classes with sequential harvests of 21, 28, and 35 days after anthesis (DAA). All seed sizes emerged equally well in the two-year study in Illinois, even though the seed size range for 21 to 35 DAA lots varied considerably by year of harvest (Mian and Nafziger, 1992). Seed size differences in other studies are often established by screening or other sorting procedures which may confound effects from (1) drought, (2) disease or insect damage, (3) ear or head position, and (4) seed size and dormancy physiology interactions.

Corn seed size classes have also shown differences in imbibition, with small flat (SF) seeds having a faster rate of water uptake than large round (LR) kernels during initial stages of germination (Shieh and McDonald, 1982). First counts of the standard germination test showed smaller seeds germinated more rapidly than large seeds for the two inbreds tested. Similar findings were reported by Muchena and Grogan (1977), who noted that smaller seeds may require less water due to less seed volume. They also speculate that small-seeded corn lines could provide more rapid and improved germination under conditions of limited soil moisture. Seed size studies with a sweet corn inbred divided a composite lot into LF (large flat), LR, SF, and SR (small round) classes. Under greater crop establishment stress, the SF lot performed well in seedling dry weight accumulation (four-leaf stage) relative to the other classes (Bennett, Waters, and Curme, 1988). Seedlot nitrogen (N) concentration, TSW, and age all had significant ($P < 0.001$) effects on perenial ryegrass (*Lolium perenne* L.) seed vigor (Cookson, Rowarth, and Sedcole, 2001). N concentration accounted for more variability in laboratory emergence and perennial ryegrass seedling dry weight than TSW in these studies, but individual effects of N and TSW are often confounded (Bennett, Rowarth, and Jin, 1998; Lowe and Ries, 1972).

Seed size may also be linked to emergence through soil crusts. Research comparing crust-tolerant and crust-susceptible sorghum genotypes indicated differences in seed-seedling conversion efficiency. Susceptible genotypes used about 60 percent of their initial seed weight for forming seedling tissue, while the tolerant genotypes used only 40 percent (Soman, Jayachandran, and Peacock, 1992). Tolerant genotypes had longer mesocotyls with faster growth rates, allowing an avoidance mechanism for soil crusting situations.

Carrying seed size studies to yield data comparisons is generally suspect since many intervening factors may outweigh the original factor. The research by Mian and Nafziger (1992) noted a higher wheat yield in one year from small seed plots, but it was largely due to reduced lodging. Many other factors (planting depth, planting date, optimal versus suboptimal growing conditions or cultural practices) can intervene between seed size at sowing and eventual yield. In other species, such as kura clover (*Trifolium ambiguum* M. Bieb.), shoot weights may serve as a better indicator of seedling vigor than seed size since the plant allocates a majority of its reserves to root and rhizome development during seedling establishment (DeHaan, Ehlke, Sheaffer, 2001).

SEED WATER UPTAKE

Available soil water is an essential factor for seed germination. Uptake typically follows a triphasic curve of (1) rapid uptake, (2) lag phase, and (3) additional hydration from cell expansion and radicle growth (Obroucheva and Antipova, 1997). Seeds have the same general response to water supply as to temperature fluctuations. An optimal seed substrata water status exists for germination percentage or rates, with lower germination values on either side of the optimum (Gulliver and Heydecker, 1973; Chatterjee, Das, and Deb, 1981). Indirect effects of water status may be on (1) leaching of endogenous inhibitors, (2) soil crusting from flooded soils followed by rapid drydown, (3) decreased oxygen availability, or (4) increased competition from microbes favored by super- or suboptimal water supply.

In undamaged seed, phase I water uptake is closely linked to colloidal or physical properties. Nitrogen and protein content are major colloidal constituents of many important crop species (Cardwell, 1984; Vertucci and Leopold, 1987). Seeds rich in protein imbibe more water than fat-storing seeds. Water-insoluble carbohydrates from soybean seeds were found to hold tenfold their weight in water, while protein held only twice its weight in water (Smith and Circle, 1972). Soybean seed parts (axes versus cotyledons) differed substantially in moisture content when whole seeds were allowed to imbibe for 72 h. Axes contained 800 g water/kg fresh weight at 48 h (germination) versus 550 to 600 g water/kg fresh weight for cotyledon tissue with differences largely due to lipid versus carbohydrate contents (McDonald, Vertucci, and Roos, 1988b). Studies with seed size of soft white winter wheat indicated no effect on emergence at soil water contents of 0.12 to 0.16 g water/g. However, light (small) seeds, which also had the highest percent protein content, were observed to emerge more rapidly at the lowest soil water status (0.10 $g \cdot g^{-1}$) tested (Douglas, Wilkins, Churchill, 1994).

Seeds may become hydrated by water in the liquid or vapor phases. Under conditions of severe low soil water stress, seeds may be suspended in stages of incomplete hydration (Chatterjee, Das, and Deb, 1982; Hegarty, 1978). Metabolic changes and associated shifts in seed storage materials may allow water uptake to later resume or rapidly germinate upon subsequent rainfall or irrigation (Hadas, 1982). The concept of natural priming of seeds may also apply in these and semiarid or low water stress situations (González-Zertuche et al., 2001). Partial imbibition, inadequate for germination per se, may result in a type of priming with rapid and more uniform germination plus emergence occurring with subsequent rainfall (Kigel, 1995). Seed metabolism will also vary with respect to critical hydration levels (Vertucci and Farrant, 1995). Exposure of perennial ryegrass and annual bluegrass seeds to controlled hydration-dehydration cycles resulted in delayed but more uniform germination (Allen, White, and Markhart, 1993). Cycled seeds required fewer hours in contact with liquid water to germinate than continuously hydrated seeds. A more thorough understanding of seed germination patterns in crop species after wetting and drying cycles will benefit seedling establishment under stressful field conditions (Koller and Hadas, 1982).

Initial radicle protrusion is dependent upon cell expansion, not cell division (Haigh, 1988; Gornik et al., 1997). Minimum seed moisture contents required for turgor pressure and base water potential (Ψ_b) for germination may vary slightly for individual seeds within a seedlot, but estimated minimum values for seeds of several species are reported in the literature (Table 3.1). Moisture stress of -1.0 to -1.3 MPa is known to delay lettuce germination (Haber and Luippold, 1960). Cultivar differences in water requirements for tomato seed germination were noted by Liptay and Tan (1985). Using different available soil moisture (ASM) treatments of 5, 35, 60, 75, and 100 percent of a loamy sand, one cultivar germinated well at 60 percent ASM or greater while the other cultivar required 100 percent ASM for optimal germination. Differences in seed moisture contents required for germi-

TABLE 3.1. Estimated minimum water potential (Ψ_{min}) values and base water potential values (Ψ_b) of various seeds. Minimum water potential is considered the minimum seed hydration level at which metabolic advancement will occur, and Ψ_b is the threshold water potential below which germination will not proceed.

Species	Ψ_{min} (MPa)	Source
Tomato (*Lycopersicon esculentum* Mill.)	−2.45	Cheng and Bradford, 1999
Tomato	−2.50	Bradford and Haigh, 1994
Lettuce (*Lactuca sativa* L.)	−2.40	Tarquis and Bradford, 1992
Nemophila menziesii Aggr.	−2.0	Cruden, 1974
	Ψ_b (MPa)	
Lettuce	−1.3	Haber and Luippold, 1960
Wheat (*Triticum aestivum* L.)	< −1.5	Owen, 1952
Wheat	< −2.0	Lindstrom, Papendick, and Koehler, 1976

nation have also been reported among sorghum cultivars (Mali, Varade, and Musande, 1979). In studies comparing commercial tomato *(Lycopersicon esculentum)* with two wild tomato species, *L. chilense* and *Solanum pennellii,* germination of *L. esculentum* appeared less sensitive to water deficits (−0.2 to −0.8 MPa) than did the two wild species (Taylor, Motes, and Kirkham, 1982). Germination of all species was inhibited more by water stress, as compared to seedling growth responses. This conclusion was also reached by Bhatt and Srinivasa Rao (1987) studying four *L. esculentum* cultivars and the wild tomato species *L. pimpinellifolium.* Recent work on the genetic basis of tomato seed germination rates at reduced water potential will be useful in understanding the physiological determinates of Ψ_b (Foolad and Lin, 1997; Foolad et al., 1997). Base water potentials can also change with seed maturity. As broccoli seed matured, germination response to osmotic stress decreased (Still, 1999). Seeds harvested at 38 days after flowering (DAF) had a Ψ_b of −0.6MPa, but the Ψ_b decreased to −0.8 MPa for seeds harvested at 49 DAF, and reached an intermediate level (−0.7 MPa) for 56 DAF seeds (Still, 1999).

Mucilaginous seeds are able to establish better seed-soil contact, which assists in water uptake. Myxospermy (mucilaginous seeds) is more common for seeds of plant families native to arid/semiarid regions (e.g., Brassicaceae, Euphorbiaceae, Plantaginaceae, Labiateae) (van der Pijl, 1982). In an attempt to mimic nature, the application of hydrophilic polymers as seed coatings or seed furrow amendments to absorb water have given inconsistent results. Hydrolyzed starch-graft-polyacrylonitrile (H-SPAN) applied at 2 to 5 g·kg⁻¹ of sweet corn seed showed improved emergence, while similar tests reduced cowpea *(Vigna unguiculata)* emergence and seedling dry weight (Baxter and Waters, 1986a). These results were likely linked to differences in seed storage materials and rates of water uptake. Laboratory experiments with H-SPAN on sweet corn planted into a silt loam soil at matric potentials of −0.01, −0.40, −1.0, and −1.5 MPa were also conducted. Treated seeds had greater imbibition, respiration, and emergence at −0.01 and −0.40 MPa than control seeds, but the H-SPAN coating had a deleterious effect as soil water matric potential decreased to −1.0 and −1.5 MPa (Baxter and Waters, 1986b). This effect was also reported for Russian wildrye seeds (Berdahl and Barker, 1980). For hydrophilic coatings to be most effective, soil water potential should be near field capacity initially. This water can first hydrate the coating sufficiently to cause germination. Coatings may be useful in nonirrigated dryland conditions where seeds are planted just prior to or immediately after rainfall. Coatings could then trap water around the seed for improved germination.

RADICLE EMERGENCE
AND ROOT SYSTEM DEVELOPMENT

Seed reserves and environmental factors largely determine the initial patterns of germination and seminal root growth. After seed reserves are exhausted, however, the size and activity of the young root system plays a major role in determining the rate of early seedling shoot growth and dry matter accumulation (Hoad et al., 2001). In the small grains, primary (seminal) roots develop from the radicle and comprise approximately 5 to 10 percent of the total root volume at full growth. Secondary roots (also referred to as nodal, adventitious, or crown roots) arise from nodes at the stem or tiller base (Hoad et al., 2001). Soil compaction, greater bulk density values, high seeding rates, and moisture stress can reduce root development and seedling establishment. However, radicle lengths did not differ between soil crusting-tolerant and -susceptible lines of sorghum, and no effect was detected of crusting treatments on radicle length (Soman, Jayachandran, and Peacock, 1992; Dexter and Hewitt, 1978).

Selection for longer pearl millet seedling root length in greenhouse sand culture correlated well with field seedling emergence and shoot height (McGrath et al., 2000). Seedlings with longer roots tolerated or avoided moisture stress better than five other populations (with short roots, long coleoptile, or short coleoptiles) tested. Under flooding stress (hypoxia), corn genotypes were shown to differ in adventitious root production, with some hybrids producing more root weight per seedling at hypoxic levels (10.5 and 14.0 KPa O_2) than at ambient (20.9 KPa O_2) control levels (VanToai, Fausey, and McDonald, 1988).

Critical O_2 concentration (COC), or the O_2 concentration below which a process becomes dependent on $[O_2]$, was 4.8 KPa for most of the ten corn genotypes evaluated, but 10.5 KPa for flood-susceptible genotypes Mo17 and B37 (VanToai, Fausey, and McDonald, 1988). Correlations between germination and adventitious root formation across the ten genotypes tested was not very high (at 2.5 KPa O_2, correlations of 0.6 and 0.4 for inbreds and hybrids, respectively) (VanToai, Fausey, and McDonald, 1988). Reported COC values for tomato root growth (0.14 mol·m³) and barley (0.6 mol·m³ or 0.13 KPa O_2) indicate the wide range among species (Benjamin and Greenway, 1979).

GENETIC LINKS TO GERMINATION
TEMPERATURE LIMITS

In temperate regions, low temperatures at the time of planting are often the most limiting environmental component for germination and seedling growth under early spring field conditions. Optimum temperatures for seedling growth and for radicle emergence (germination) are likely to differ for most species (Gulliver and Heydecker, 1973; Marsh, 1992; Nomura et al., 2001). Germination base temperatures for chickpea (0°C) and cowpea (8°C) were quite consistent, while soybean varied widely for temperate-origin genotypes (4°C) versus those with tropical origins (10°C) (Covell et al., 1986). Base temperatures for germination have also been shown to be unaffected by seed age in barley and wheat (Ellis, Hong, and Roberts, 1987; Khah, Ellis, and Roberts, 1986).

Earlier planting to achieve (1) longer growing seasons, (2) better use of sunlight and rainfall, and (3) enhanced yield potential has placed a premium on selecting cold-tolerant (CT) populations for major crop species (Yu and Tuinstra, 2001). An example is maize, which is now grown at 55° latitude (North), despite its warm-season characteristics and subtropical origins (Shaw, 1988). As for most species examined to date, at least two mechanisms appear to be involved in CT of maize—one for germination and emergence and another for seedling growth (Revilla et al., 2000). For cold-tolerance breeding in tomato, Foolad and Lin (2000) suggest that each stage of plant development may require evaluation and selection. No cold-tolerant tomato cultivar has yet been developed or released for commercial use. A more thorough understanding of CT at different growth stages in tomato (and other species) will likely require genetic mapping, cloning, and characterization of the functional genes that confer tolerance at each stage (Foolad and Lin, 2001).

An oilseed rape cultivar (Martina) showed potential to give rise to genetically distinct populations that could exploit different environments (Squire et al., 1997). Early germinator seeds (5°C) and viable, but nongerminating seeds at 5°C were hand selected and selfed for testing of progeny seedlots. Germination differences were small at 19°C but large at <10°C (Squire et al., 1997). It has been noted that acceptable germplasm screening systems must distinguish between temperature responses due to genotypes and responses due to presowing environment (i.e., seed production effects on vigor) (Ellis et al., 1986). A streamlined screening protocol for faba bean (*Vicia faba* L.) genotype germination response to sub- and supraoptimal temperatures was proposed by Ellis, Simon, and Covell (1987). Four temperatures (10°C, 20°C, 27°C, 30°C) were used with repeated probit analyses to compare genotypes for base and optimum temperature values. Garcia-Huidobro, Monteith, and Squire (1982a) used a wide range of constant temperatures (12°C to 47°C) and showed pearl millet rate of germination increased linearly from a base temperature to a clearly defined optimum. Beyond the optimum, germination rate decreased linearly with increased temperature to a T_{max} and no germination.

Use of alternating or fluctuating temperatures may be more appropriate for predicting or interpreting field germination and emergence data. Beyond the seed dormancy breaks associated with alternating temperature regimes, fluctuating temperatures are thought to (1) increase the maximum fraction that will germinate in a seed population and (2) possibly increase the rate of germination (Garcia-Huidobro, Monteith, and Squire, 1982b). In studies with pearl millet, large diurnal temperature amplitudes (8°C) and temperatures from 15° to < 42°C accelerated the germination rate. Differences were small relative to comparable average constant temperatures but may be important for seeds germinating in field environments.

SEED PRODUCTION AND SEED VIGOR

Among the many cultural factors and management decisions that impact germination under environmental stress, an important early indicator of crop establishment is seedlot vigor. A high quality or vigorous seedlot possesses the ability to germinate and emerge uniformly and quickly under the wide range of conditions (temperature, moisture, biotic stresses, etc.) commonly encountered in field settings (Association of Official Seed Analysts [AOSA], 1983). Environmental conditions during seed development can have major effects on seed quality (Wulff, 1995; Syankwilimba, Cochrane, and Duffus, 1997). Seed production per se will not be covered extensively in this section, but implications for better germination under stressful conditions will be addressed.

Seed development and seed production of *Brassica* species is a useful starting point for examining the development of seed vigor. Red cabbage and rapeseed were selected from among the brassica crops for studies conducted by Still and Bradford (1998) and Still (1999). The indeterminate growth pattern and extended flowering period (typically 35 d) of brassicas force seed producers into a compromise over mature seed yield potential versus shattering losses.

Rapeseed maximum seed dry weight was achieved by 33 days after flowering, while for red cabbage this stage was reached by 54 DAF (Still and Bradford, 1998). Different sensitivities to water stress (reduced water potential) (Still and Bradford, 1998) and exogenous abscisic acid (ABA) treatments (Benech-Arnold, Fenner, and Edwards, 1991) were linked to the environment experienced by the mother plant during seed development. Temperature thresholds may also be linked to different seed maturity stages (Ellis et al., 1986).

For many species and seed crops, it is difficult to determine when physiological maturity (PM) or maximum seed quality has occurred and whether this developmental stage is synchronized with maximum seed dry weight. In rapeseed, PM was reached 4 to 9 d after maximum seed dry weight; in red cabbage, PM was observed at 6 or 7 d later than maximum dry weight (Still and Bradford, 1998). Best seed quality (PM) is reported to occur after maximum seed dry weight accumulation for many other crops, including sweet corn (Wilson and Trawatha, 1991), barley (Pieta-Filho and Ellis, 1991), and *Phaseolus vulgaris* (Sanhewe and Ellis, 1996). Seed moisture can also be a useful indicator of seed quality development in field bean, with seed quality assessment (standard germination, controlled deterioration, and conductivity) values leveling off at about 0.4 g water/g fresh weight. This seed moisture content did not differ across years, anthesis dates, or pod locations on the plant (Coste, Ney, and Crozat, 2001).

Shattering is not a critical factor in *sh2* sweet corn seed production, but slow drydown in the field has lead to studies on earlier harvests (0.45 to 0.65 g water/g fresh weight) for this unique endosperm type (Borowski, Fritz, and Waters, 1991). Sweet corn seed can be harvested at higher than normal (0.35 to 0.45 g water/g fresh weight) moisture levels with proper attention to harvesting, handling, and drying operations (Borowski, Fritz, and Waters, 1995). Continued research on membrane and pericarp integrity changes during seed production will help to provide flexibility in seed harvest windows and supply good yields of high-quality seed.

An early indicator of loss of seed vigor is a narrowing of the range of conditions (e.g., temperatures, water) in which seeds will germinate (Abdul-Baki and Anderson, 1972). New seedling imaging systems for quick assessment of seed vigor will be useful in seed production and seed inventory decisions (Sako et al., 2001).

SOWING DEPTHS AND PLANTER TECHNOLOGY

Seedling establishment is often compromised by wide-ranging soil moisture conditions (near field capacity to levels too dry for germination), planting depths (<1 cm to 15 cm or more), and seedbed temperatures (0.25°C to 0.5°C). Delays in germination and emergence subjects seedlings to greater risk from soil crusting impedance and greater competition or damage from various pathogens, insects, and weeds. Deep planting (8 to 10 cm or more) is often required to place seeds of barley, winter wheat, and other crops in moist soil (Lindstrom, Papendick, and Koehler, 1976; Radford, 1987).

It is useful, especially in deep planting situations, to separately consider the processes of (1) germination and radicle emergence and (2) subsurface seedling elongation. The emergence phase (subsurface coleoptile elongation, etc.) is generally more sensitive to marginal seedbed conditions (Lindstrom, Papendick, and Koehler, 1976). Long coleoptile length (usually highly correlated with seed weight) is clearly desirable when deep sowing is required in crop production. Within most barley, oat, and wheat cultivars, larger seed with good germinability produced longer coleoptiles (Kaufman, 1968). The extra seed reserves for emergence in larger seed plus longer coleoptiles were both linked to more successful seedling establishment. In studies with seven barley cultivars, constant temperatures of <10°C and >20°C reduced coleoptile lengths for all genotypes (Radford, 1987). At 10°C, coleoptile length ranged from 64 to 106 mm, while at 25°C the lengths dropped to 58 to 80 mm. Optimal barley seed zone temperatures varied by cultivar. One line showed optimal coleoptile growth at 10°C, 10 or 15°C, and 15 or 20°C, while four cultivars produced optimal coleoptiles anywhere across the 10 to 20°C range (Radford, 1987). It is recommended in deep-sowing situations that furrows be formed over the seed rows to minimize the actual depth of soil covering, e.g., deep sow at 110 mm and firm soil directly above the seed with a press wheel to leave 75 to 80 mm of soil actually over the seed. (Radford, 1987).

Precision agriculture techniques may also be useful for sowing depth and variable seed placement decisions. Within-field variability leads to substantial ranges of soil temperatures and moisture, and refinements in planter engineering show promise for dealing with these key variables (Carter and

Chesson, 1996; Price and Gaultney, 1993). The use of global positioning systems (GPS) and geographic information systems (GIS) allow the mapping of fields for many applications, including seed placement, for improved stand establishment. Field studies with several *shrunken-2* sweet corn cultivars showed that seedling emergence for an entire field was greater using variable planting depths (2 to 4 cm) based on mapped soil type differences versus a single planting depth of 2 cm (Barr, Bennett, and Cardina, 2000). Additional mapping data on soil compaction (Hakansson, Voorhees, and Riley, 1988; Wolfe et al., 1995) and a better understanding of cultivar interactions will improve the accuracy and utility of precision planting techniques for a wider range of crop species and field environments.

TILLAGE SYSTEMS AND SOIL STRUCTURE EFFECTS

Worldwide concerns about soil erosion and deteriorating soil structure have spurred research and use of various conservation tillage systems that preserve more crop residue at or near the soil surface. Germination and emergence can be impacted by increased residues in many ways, including (1) cooler, wetter microclimates, (2) decreased seed-soil contact for water uptake, (3) allelochemical interactions, and (4) modified levels of ethylene production and removal (Douglas, Wilkins, Churchill, 1994; Creamer, Bennett, and Stinner, 1996; Hadas and Russo, 1974a,b; Karssen and Hilhorst, 1992; Chase, Nair, and Putnam, 1991; Arshad and Frankenberger, 1990). Many crop producers also feel pressured to plant earlier in order to meet market windows or optimize light interception, and earlier plantings are often made into cold, wet soils regardless of the tillage system employed (Hakansson, Voorhees, and Riley, 1988). Soil compaction is commonly caused by vehicle traffic on wet soil, which puts additional stress on germination and seedling emergence. Systems or environments that slow stand establishment also prolong the period of seedling vulnerability to soil impedance, diseases, insects, and weed competition (Wolfe et al., 1995; Mohler and Galford, 1997).

Soil attributes and critical threshold values for a number of variables have recently been proposed by Pilatti and deOrellana (2000) for mollisols in Argentina. Among the many attributes considered, at least four are linked to germination and seedling establishment concerns. They are (1) root penetration resistance/impedance, (2) surface crusting potential, (3) water storage capacity, and (4) total biological activity. The effort to describe critical values of an "ideal soil" and establish threshold values for 25 or more attributes seems promising for more accurate assessment and crop decision-making processes. Combining this information with GPS/GIS precision farming systems (Barr, Bennett, and Cardina, 2000) should aid in overcoming many crop establishment obstacles in coming decades.

Genotypes of important crop species, including corn, are known to differ in their germination and seedling growth response to low oxygen concentrations (VanToai, Faussey, and McDonald, 1988). While O_2 content may be more closely linked to soil drainage systems, soil type, and topography than to tillage systems, higher crop residues can also slow the loss of water. In the study by VanToai, Faussey, and McDonald (1988), only high-vigor lots of inbred and hybrid corn lines were assessed to avoid any confounding of seed vigor with hypoxia or anoxia responses. Low O_2 levels are usually more limiting during germination than after radicle protrusion, which likely facilitates at least some increase in O_2 availability (Al-Ani et al., 1985; Wuebker, Mullen, and Koehler, 2001). Fluctuating from high (content of 20.0 KPa O_2) to low O_2 concentration was most damaging to corn germination and seedling growth, especially when a period of true anoxia was imposed (VanToai, Faussey, and McDonald, 1988). In species which are extremely tolerant to flooding, such as rice (*Oryza sativa* L.) and barnyardgrass (*Echinochloa crus-galli* L.), low O_2 actually stimulates coleoptile growth while inhibiting root development (Rumpho et al., 1984; Alpi and Beevers, 1983). For corn, moderate levels of hypoxia (10 to 14 KPa O_2) also stimulated shoot growth of the five hybrids tested, but not the inbreds (VanToai, Faussey, and McDonald, 1988).

The occurrence of ethylene (C_2H_4) in soils is also important due to its many effects on plant development, from seed germination to senescence (Raven, Evert, and Eichhorn, 1997). Changes in levels of organic matter and associated soil microorganisms with various tillage and soil management systems can be expected to affect ethylene production, removal, and stability (Arshad and Frankenberger, 1990). The biologically active rhizosphere and spermosphere are likely to be very active sites for C_2H_4 generation and consumption, with possible effects on crop and weed seed germination plus seedling establishment (Karssen and Hilhorst, 1992).

INTERACTIONS WITH SEED TREATMENTS
AND OTHER CROP PROTECTION CHEMICALS

Fungicide and insecticide seed treatments are often employed to protect crops from biotic stress. Emergence from cold (2 to 7°C), wet soils is often slow and incomplete, with stand establishment appearing to differ for various seed treatments (Smiley, Patterson, and Shelton, 1996). Changes in tillage operations (e.g., increased use of conservation tillage or stubble-mulch systems) have led to more research or optimal treatments for these modified microenvironments (Bradley et al., 2001). Planting depth can also affect the recommended treatments, with some products not suggested for seedings made deeper than 5 cm. Three greenhouse and seven field experiments were conducted with deeply planted winter wheat to compare the efficacy of five seed fungicide products (Smiley, Patterson, and Rhinhart, 1996). In the greenhouse studies, treated seed was planted 2.5 cm deep into moist (7, 10, or 15 percent water) and warm (24°C) silt loam soil, then topped off with 10 cm of dry soil to simulate planting 12.5 cm deep into a stubble-mulch fallow system. Field study plantings were at 2.5 to 12.7 cm deep into warm soils (21 to 27°C at seed zone) with seed zone water contents of 5 to 17 percent. Three of the seed fungicide treatments evaluated had variable effects on seedling emergence or established stand density values (Smiley, Patterson, and Rhinhart, 1996). Coleoptile lengths were not affected by the fungicide treatment. Seed fungicide treatment decisions can interact with (1) planting depth, (2) irrigation availability, (3) planting season and likelihood of soil crusting, (4) species or class (i.e., hard-red versus soft-white wheat), and (5) key pathogens associated with given fields or planting season. Natural resistance to various diseases and pests has been linked to colored (pigmented) seedcoats. Red pericarps have been linked to grain mold resistance in sorghum (Esele, Frederiksen, and Miller, 1993). It is also believed that general resistance to pathogens is associated with phytoalexin (pigment) accumulation in sorghum plant tissues in response to pathogen infection (Nicholson et al., 1987). Recent work by Pedersen and Toy (2001) tested the combined effects of plant and seed color on sorghum germination, emergence, and other agronomic factors. Using 20 near-isogenic lines, seedling emergence was higher for red-seed versus white-seed phenotypes (Pedersen and Toy, 2001). Grain sorghum markets, however, often prefer white grain, which is free of pigment stains. Purple plant phenotypes produced seed with (1) higher cold germination and accelerated aging values and (2) greater seedling elongation at 10 d versus results from tan phenotypes, although standard (warm) germination values were not different (Pedersen and Toy, 2001). Higher grain yields were associated with white seeded, purple plant types.

Unexpected losses in seedling establishment (and eventual yields) can also occur from crop responses to multiple pesticide applications. Interactions among fungicides, insecticides, and herbicides can be complicated further by soil characteristics (Morton et al., 1993). Soil moisture, pH levels, and organic matter content may all influence the actual amount of chemical taken up by a young plant. For example, if a systemic soil-applied insecticide such as terbufos is taken up in greater than normal amounts and distributed at high levels throughout the young plant, its presence can reduce the metabolism of later herbicide applications (Morton et al., 1993). Herbicide rates that are normally safe and free of phytotoxic effects can then cause foliar injury and stand losses. Cold stress, seedling size, and endosperm class were also shown to influence sweet corn response to four herbicide treatments in field and controlled environment studies (Bennett and Gorski, 1989). Introduction of new crop protection chemistry and new germplasm call for careful compatibility studies, especially for seedling establishment in stress environments.

SCREENING PROTOCOLS FOR GERMINATION TOLERANCE TO LOW TEMPERATURE AND WATER STRESS

The study of germination stress tolerance in field settings is difficult. Soil temperature and moisture ranges needed for careful cultivar or germplasm evaluations are often lacking or are unpredictable (Schell et al., 1991; Blacklow, 1972; Washitani, 1987). Many researchers and crop practitioners have noted that use of controlled environment settings would be a more efficient strategy for examining genotype differences in germination and seedling emergence (Heydecker and Coolbear, 1977; Khan, 1992; Tadmor, Cohen, and Harpaz, 1969; McGrath et al., 2000).

Changes in alfalfa (*Medicago sativa* L.) emergence and seedling height after laboratory selection at suboptimal temperatures (<10°C) successfully improved seedling heights in the field for some populations without changing other agronomic and forage quality traits.

Seedling height appeared to be a better trait than germination time on which to base predicted field performance if traits are measured in lab or greenhouse studies (Klos and Brummer, 2000). The consistence of laboratory and field responses to recurrent selection varied considerably within the six alfalfa cultivars assessed. A field location and population interaction was also observed for seedling height, due to both rank and magnitude differences (Klos and Brummer, 2000). Future evaluations of response to selection for such traits should therefore be performed at multiple locations.

It has also been noted that at suboptimal temperature ranges for a given crop, thermal time and germination of different individuals and fractions of the seed lot (population) are normally distributed. Less variation is observed when supraoptimal temperatures are imposed (Covell et al., 1986; Ellis, Simon, and Covell, 1987). Screening procedures for selecting grain legume germplasm (chickpea, lentil, cowpea, soybean) tolerant to suboptimal temperatures, based on cumulative germination and thermal time patterns, are well described by Covell and colleagues (1986). It is also useful to distinguish between (1) genotype × temperature responses and (2) presowing or seed production environment effects linked to temperature responses if a truly acceptable germplasm screening protocol is desired (Ellis et al., 1986).

Higher catalase activity, lower lipoperoxidation, higher total oxygen consumption at 3°C, and a doubling of fructan content were all correlated with the improved cold tolerance of oat cultivar OT220 versus the cold-sensitive cultivar America (Massardo, Corcuera, and Alberdi, 2000). Oxygen-scavenging enzymes, such as catalase, provide one mechanism for reducing oxidative injury due to cold stress. Lipoperoxidation in 'America' oat embryos increased 25 percent when germinated at 30°C versus 17°C, while lipoperoxidation did not increase with cold treatment of the cold-tolerant cultivar OT220 (Massardo, Corcuera, and Alberdi, 2000). These and other physiological responses to cold described previously are correlative evidence that may be important links to genetic differences. These responses also have potential use in broader germplasm screening programs for germination tolerance to low temperatures. Embryo adenosine triphosphate (ATP) levels of two corn hybrids imbibed for 64 h were different at 10°C but not at 20°C (Schell et al., 1991). Cold test germination, emergence index, field emergence and dry weight (30 days after planting) values showed good agreement with embryo ATP levels for these hybrids. Schell and colleagues (1991) observed that imbibition times of 16 h may be used if ATP accumulation rates, rather than ATP content/embryo values, are analyzed.

Lafond and Baker (1986b) assessed the germination responses of nine spring wheat cultivars to varying levels of temperature and moisture stress. Temperature ranges (5 to 30°C) and moisture stress using polyethylene glycol (PEG8000) solutions with osmotic potentials of 0.0, –0.4, and –0.8 MPa (at 10°C and 20°C) gave final germination values of over 90 percent for all environments tested. Consistent cultivar rankings and differences (although magnitude decreased) were reported across the range of 5 to 30°C. Increasing the (osmotic) water stress from –0.0 to –0.8 MPa caused median germination time to increase from 90 h to 156 h at 10°C, and from 36 h to 64 h at 20°C. Relative ranking of germination times for the nine wheat cultivars was consistent over the levels of moisture stress. Seed and seedling tolerance to soil moisture stress is another important trait to test, but it generally receives less attention than low-temperature tolerance (Hegarty, 1977; Bradford, 1995). Various systems for controlled water stress have been used for seed germination and priming studies (Pavmar and Moore, 1968; Bennett and Waters, 1984; Bradford, 1997) and are again more reliable than using a range of field experiments. Water potential has also been shown to affect the temperature range over which optimal germination was observed (Sharma, 1976; Kebreab and Murdoch, 2000). Optimal germination of *Orobanche aegyptiaca* at 0.0 MPa occurred over 17 to 26°C (9°C range) compared with 17 to 20°C (3°C range) at −1.25 MPa. Optimum germination temperature for this parasitic weed also tended to decrease with decreasing water potential (Kebreab and Murdoch, 2000), and these points should be considered if combining temperature and water stress assessments (Gummerson, 1986). Drought-tolerance assessments used for whole plants may also hold promise for screening seeds and seedlings (Ali Dib et al., 1994; Bajji, Lutts, and Kinet, 2001).

Germination of sugarbeet (*Beta vulgaris* L.) seed submerged in hydrogen peroxide and water has recently been proposed for screening cultivar and seedlot vigor (McGrath et al., 2000). Thirty-nine commercial seedlots representing 24 cultivars were tested in a range of laboratory and field experiments. Total germination (96 h) in 0.3 percent H_2O_2 was identified as the best laboratory screen. McGrath and colleagues (2000) observe that although it is unlikely a water germination test can be developed to fully mimic field conditions, it should be useful in evaluating relative emergence potential for species that tolerate immersion for several days. Physiological and agronomic information from the germination tests will be used to identify target genes for use as markers in breeding for improved field emergence.

CONCLUDING REMARKS

Crop physiology and management studies often describe and quantify the changes plant breeders and geneticists have delivered in new germplasm but rarely address the specific changes needed to advance crop establishment, yield potential, or other agronomic goals (Snape, 2001). As discussed throughout this chapter, germination and seedling establishment in the field is a complex process influenced by many interacting factors. Advances in genetics and genomics will contribute much precision to the next wave of crop physiology and seedling establishment research. Extreme environmental stresses will always pose limitations for crop establishment, but continued progress in germplasm screening protocols and crop management research should also lead to new varieties with a tailored set of agronomic practices for given environments and cultural practices.

REFERENCES

Abdul-Baki, A.A. and Anderson, J.D. (1972). Physiological and biochemical deterioration of seeds. In Kozlowski, T.T. (Ed.), *Seed Biology,* Volume II (pp. 283-315). New York: Academic Press.

Al-Ani, A., Bruzan, F., Raymond, P., Saint-Ges, V., Leblanc, J.M., and Pradet, A. (1985). Germination, respiration and adenylate energy charge of seeds at various oxygen partial pressures. *Plant Physiology* 79: 885-890.

Ali Dib, T., Monneveux, P., Acevedo, E., and Naeliot, M.M. (1994). Evaluation of proline analysis and chlorophyll fluorescence quenching experiments as drought tolerance indicators in durum wheat (*Triticum turgidum* L. var. *durum*). *Euphytica* 79: 65-73.

Allen, P.S., White, D.B., and Markhart, A.H. (1993). Germination of perennial ryegrass *(Lolium perenne)* and annual bluegrass *(Poa annua)* seeds subjected to hydration-dehydration cycles. *Crop Science* 33: 1020-1025.

Alpi, A. and Beevers, H. (1983). Effects of oxygen concentration on rice seedlings. *Plant Physiology* 71: 30-34.

Arshad, M. and Frankenberger, W.T. Jr. (1990). Ethylene accumulation in soil in response to organic amendments. *Soil Science Society of America Journal* 54: 1026-1031.

Association of Official Seed Analysts (1983). *Seed Vigor Testing Handbook.* AOSA Handbook 32. Lincoln, NE: Association of Official Seed Analysts.

Bajji, M., Lutts, S., and Kinet, J.M. (2001). Water deficit effects on solute contribution to osmotic adjustment as a function of leaf ageing in three durum wheat (*Triticum durum* Desf.) cultivars performing differently in arid conditions. *Plant Science* 160: 669-681.

Barr, A., Bennett, M., and Cardina, J. (2000). Geographic information systems show impact of field placement of *sh2* sweet corn stand establishment. *HortTechnology* 10: 341-350.

Baskin, C.C. and Baskin, J.M. (1998). *Seeds: Ecology, Biogeography and Evolution of Dormancy and Germination.* San Diego, CA: Academic Press.

Baxter, L. and Waters, L., Jr. (1986a). Effect of a hydrophilic polymer seed coating on the field performance of sweet corn and cowpea. *Journal of the American Society of Horticultural Science* 111: 31-34.

Baxter, L. and Waters, L., Jr. (1986b). Effect of a hydrophilic polymer seed coating on the imbibition, respiration and germination of sweet corn at four matric potentials. *Journal of the American Society of Horticultural Science* 111: 517-520.

Benech-Arnold, R.L., Fenner, M., and Edwards, P.J. (1991). Changes in germinability, ABA content and ABA embryonic sensitivity in developing seeds of *Sorghum bicolor* (L.) Moench. induced by water stress during grain filling. *New Phytologist* 118: 339-347.

Benjamin, L.R. and Greenway, H. (1979). Effects of a range of oxygen concentration on porosity of barley roots and on their sugar and protein concentrations. *Annals of Botany* 43: 383-391.

Bennett, J.S., Rowarth, J.S., and Jin, Q.F. (1998). Seed nitrogen and potassium nitrate influence browntop (*Agrostis capillaris* L.) and perennial ryegass (*Lolium perenne* L.) vigour. *Journal of Applied Seed Production* 16: 77-81.

Bennett, M.A. and Gorski, S.F. (1989). Response of sweet corn *(Zea mays)* endosperm mutants to chloracetamide and thiocarbamate herbicides. *Weed Technology* 3: 475-478.

Bennett, M.A. and Waters, L., Jr. (1984). Influence of seed moisture on lima bean stand establishment and growth. *Journal of the American Society of Horticultural Science* 109: 623-626.

Bennett, M.A., Waters, L., Jr., and Curme, J.H. (1988). Kernel maturity, seed size, and seed hydration effects on the seed quality of a sweet corn inbred. *Journal of the American Society of Horticultural Science* 113: 348-353.

Berdahl, J.D. and Barker, R.E. (1980). Germination and emergence of Russian wild rye seeds coated with hydrophilic materials. *Agronomy Journal* 72: 1006-1008.

Bhatt, R.M. and Srinivasa Rao, N.K. (1987). Seed germination and seedling growth responsees of tomato cultivars to imposed water stress. *Journal of Horticultural Science* 62: 221-225.

Blacklow, W.M. (1972). Mathematical description of the influence of temperature and seed quality on imbibition of seeds of corn. *Crop Science* 12: 643-646.

Borowski, A.M., Fritz, V.A., and Waters, L., Jr. (1991). Seed maturity influences germination and vigor of two *shrunken-2* sweet corn hybrids. *Journal of the American Society of Horticultural Science* 116: 401-404.

Borowski, A.M., Fritz, V.A., and Waters, L., Jr. (1995). Seed maturity and desiccation affect carbohydrate composition and leachate conductivity in *sh2* sweet corn. *HortScience* 30: 1396-1399.

Bradford, K.J. (1995). Water relations in seed germination. In Kigel, J. and Galili, G. (Eds.), *Seed Development and Germination* (pp. 351-396). New York: Marcel Dekker, Inc.

Bradford, K.J. (1997). The hydrotime concept in seed germination and dormancy. In Ellis, R.H., Black, M., Murdoch, A. J., and Hong, T.D. (Eds.), *Basic and Applied Aspects of Seed Biology* (pp. 349-360). Dordrecht, the Netherlands: Kluwer Academic Publishers.

Bradford, K.J. and Haigh, A.M. (1994). Relationship between accumulated hydrothermal time during seed priming and subsequent seed germination rates. *Seed Science Research* 4: 63-69.

Bradley, C.A., Wax, L.M., Ebelhar, S.A., Bollero, G.A., and Pedersen, W.L. (2001). The effect of fungicide seed protectants, seeding rates, and reduced rates of herbicides on no-till soybean. *Crop Protection* 20: 615-622.

Cardwell, V.B. (1984). Seed germination and crop production. In Tesar, M.B. (Ed.), *Physiological Basis of Crop Growth and Development* (pp. 53-91). Madison, WI: American Society of Agronomy.

Carter, L.M. and Chesson, J.H. (1996). Two USDA researchers develop a moisture-seeking attachment for crop seeders that is designed to help growers plant seed in soil sufficiently moist for germination. *Seed World* 134 (March): 14-15.

Chachalis, D. and Smith, M.L. (2000). Imbibition behavior of soybean [*Glycine max* (L.) Merrill] accessions with different testa characteristics. *Seed Science and Technology* 28: 321-331.

Chachalis, D. and Smith, M.L. (2001). Seed coat regulation of water uptake during imbibition in soybeans [*Glycine max* (L.) Merr.]. *Seed Science and Technology* 29: 401-412.

Chase, W.P., Nair, M.G., and Putnam, A.R. (1991). 2,2'-1,1'-azobenzene: Selective toxicity of rye (*Secale cereale* L.) allelochemicals to weed and crop species. II. *Journal of Chemical Ecology* 17: 9-19.

Chastain, T.G., Ward, K.J., and Wysocki, D.J. (1995). Stand establishment responses of soft white winter wheat to seedbed residue and seed size. *Crop Science* 35: 213-218.

Chatterjee, D., Das, D.K., and Deb, A.R. (1981). Water uptake and diffusivities of germinating gram, cotton, soybean and cowpea seeds. *Seed Research* 9: 109-221.

Chatterjee, D., Das, D.K., and Deb, A.R. (1982). Water absorption, diffusivity and seed-surface soil water matric potentials of germinating maize seeds. *Seed Research* 10: 46-52.

Cheng, Z. and Bradford, K.J. (1999). Hydrothermal time analysis of tomato seed germination responses to priming treatments. *Journal of Experimental Botany* 50: 89-99.

Cookson, W.R., Rowarth, J.S., and Sedcole, J.R. (2001). Seed vigor in perennial ryegrass (*Lolium perenne* L.): Effect and cause. *Seed Science and Technology* 29: 255-270.

Coste, F., Ney, B., and Crozat, Y. (2001). Seed development and seed physiological quality of field grown beans (*Phaseolus vulgaris* L.). *Seed Science and Technology* 29: 121-136.

Covell, S., Ellis, R.H., Roberts, E.H., and Summerfield, R.J. (1986). The influence of temperature on seed germination rate in grain legumes: I. A comparison of chickpea, lentil, soybean and cowpea at constant temperatures. *Journal of Experimental Botany* 37: 705-715.

Creamer, N.G., Bennett, M.A., and Stinner, B.R. (1996). Mechanisms of weed suppression in cover crop-based production systems. *HortScience* 31: 410-413.

Cruden, R.W. (1974). The adaptive nature of seed germination in *Nemophila menziesii* Aggr. *Ecology* 55: 1295-1305.

DeHaan, L.R., Ehlke, N.J., and Sheaffer, C.C. (2001). Recurrent selection for seedling vigor in kura clover. *Crop Science* 41: 1034-1041.

Dexter, A.R. and Hewitt, J.S. (1978). The deflection of plant roots. *Journal of Agricultural Engeneering Research* 23: 17-22.

Douglas, C.C., Jr., Wilkins, D.E., and Churchill, D.B. (1994). Tillage, seed size, and seed density effects on performance of soft white winter wheat. *Agronomy Journal* 86: 707-711.

Ellis, R.H., Covell, S., Roberts, E.H., and Summerfield, R.J. (1986). The influence of temperature on seed germination rate in grain legumes: II. Intraspecific variation in chickpea (*Cicer arietinum* L.) at constant temperatures. *Journal of Experimental Botany* 37: 1503-1515.

Ellis, R.H., Hong, T.D., and Roberts, E.H. (1987). Comparison of cumulative germination and rate of germination of dormant and aged barley seed lots at different constant temperatures. *Seed Science and Technology* 15: 717-727.

Ellis, R.H., Simon, G., and Covell, S. (1987). The influence of temperature on seed germination rate in grain legumes: III. A comparison of five faba bean genotypes at constant temperatures using a new screening method. *Journal of Experimental Botany* 38: 1033-1043.

Esele, J.P., Frederiksen, R.A., and Miller, F.R. (1993). The association of genes controlling caryopsis traits with grain mold resistance in sorghum. *Phytopathology* 83: 490-495.

Foolad, M.R. and Lin, G.Y. (1997). Genetic potential for salt tolerance during germination in *Lycopersicon* species. *HortScience* 32: 296-300.

Foolad, M.R. and Lin, G.Y. (2000). Relationship between cold tolerance during seed gemination and vegetative growth in tomato: Germplasm evaluation. *Journal of the Amererican Society of Horticultural Science* 125: 679-683.

Foolad, M.R. and Lin, G.Y. (2001). Relationship between cold tolerance during seed gemination and vegetative growth in tomato: Analysis of response and correlated response to selection. *Journal of the American Society of Horticultural Science* 126: 216-220.

Foolad, M.R., Stoltz, T., Dervinis, C., Rodriquez, R.L., and Jones, R.A. (1997). Mapping QTLs conferring salt tolerance during germination in tomato by selective genotyping. *Molecular Breeding* 3: 269-277.

Garcia-Huidobro, J., Monteith, J.L., and Squire, G.R. (1982a). Time, temperature and germination of pearl millet (*Pennisetum typhoides* S. and H.): I. Constant temperatures. *Journal of Experimental Botany* 33: 288-296.

Garcia-Huidobro, J., Monteith, J.L., and Squire, G.R. (1982b). Time, temperature and germination of pearl millet (*Pennisetum typhoides* S. and H.): II. Alternating temperature. *Journal of Experimental Botany* 33: 297-302.

González-Zertuche, L., Vázquez-Yanes, C., Gamboa, A., Sánchez-Coronado, M.E., Aguilera, P., and Orozco-Segovia, A. (2001). Natural priming of *Wigandia urens* seeds during burial: Effects on germination, growth and protein expression. *Seed Science Research* 11: 27-34.

Gornik, K., Castro, R.D., Lin, Y., Bino, R.J., and Groot, P.C. (1997). Inhibition of cell division during cabbage (*Brassica oleraceae* L.) seed germination. *Seed Science Research* 7: 333-340.

Gulliver, R.L. and Heydecker, W. (1973). Establishment of seedlings in a changeable environment. In Heydecker, W. (Ed.), *Seed Ecology* (pp. 433-462). University Park: Penn State University Press.

Gummerson, R.J. (1986). The effect of constant temperatures and osmotic potential on the germination of sugar beet. *Journal of Experimental Botany* 37: 729-741.

Haber, A.H. and Luippold, H.J. (1960). Separation of mechanisms initiating cell division and cell expansion in lettuce seed germination. *Plant Physiology* 35: 168-173.

Hadas, A. (1982). Seed-soil contact and germination. In Khan, A.A. (Ed.), *The Physiology and Biochemistry of Seed Development, Dormancy and Germination* (pp. 507-527). Amsterdam: Elsevier Biomedical Press.

Hadas, A. and Russo, D. (1974a). Water uptake by seeds as affected by water stress, capillary conductivity, and seed-soil water contact: I. Experimental study. *Agronomy Journal* 66: 643-647.

Hadas, A. and Russo, D. (1974b). Water uptake by seeds as affected by water stress, capillary conductivity, and seed-soil water contact: II. Analysis of experimental data. *Agronomy Journal* 66: 647-652.

Haigh, A.M. (1988). Why do tomato seeds prime? Physiological investigations into the control of tomato seed germination and priming. PhD dissertation, Macquarie University, Sydney, Australia.

Hakansson, I., Voorhees, W.B., and Riley, H. (1988). Vehicle and wheel factors influencing soil compaction and crop response in different traffic regimes. *Soil Tillage Research* 11: 239-282.

Hegarty, T.W. (1977). Seed and seedling susceptibility to phased moisture stress in soil. *Journal of Experimental Botany* 28: 659-668.

Hegarty, T.W. (1978). The physiology of seed hydration and dehydration, and the relation between water stress and the control of germination: A review. *Plant, Cell and Environment* 1: 101-119.

Heydecker, W. and Coolbear, P. (1977). Seed treatments for improved performance—Survey and attempted prognosis. *Seed Science and Technology* 13: 299-355.

Hoad, S.P., Russell, G., Lucas, M.E., and Bingham, I.J. (2001). The management of wheat, barley and oat root systems. *Advances in Agronomy* 74: 193-246.

Karssen, C.M. and Hilhorst, H.W.M. (1992). Effect of chemical environment on seed germination. In Fenner, M. (Ed.), *Seeds: The Ecology of Regeneration in Plant Communities* (pp. 327-348). Wallingford, UK: CAB International.

Kaufmann, M.L. (1968). Coleoptile length and emergence in varieties of barley, oats and wheat. *Canadian Journal of Plant Science* 48: 357-361.

Kebreab, E. and Murdoch, A.J. (2000). The effect of water stress on the temperature range for germination of *Orabanche aegyptiaca* seeds. *Seed Science Research* 10: 127-233.

Khah, E.M., Ellis, R.H., and Roberts, E.H. (1986). Effects of laboratory germination, soil temperature and moisture content on the emergence of spring wheat. *Journal of Agricultural Science* 107: 431-438.

Khan, A.A. (1992). Preplant physiological seed conditioning. *Horticultural Review* 14: 131-181.

Kigel, J. (1995). Seed germination in arid and semiarid regions. In Kigel, J. and Galili, G. (Eds.), *Seed Development and Germination* (pp. 645-649). New York: Marcel Dekker, Inc.

Klos, K.L.E. and Brummer, E.C. (2000). Field response to selection in alfalfa for germination rate and seedling vigor at low temperatures. *Crop Science* 40: 1227-1232.

Koller, D. and Hadas, A. (1982). Water relations in the germination of seeds. In Lange, O.L., Nobel, P.S., Osmond, C.B., and Ziegler, H. (Eds.), *Encyclopedia of Plant Physiology,* New Series, Volume 12B (pp. 401-431). Berlin: Springer-Verlag.

Lafond, G.P. and Baker, R.J. (1986a). Effects of genotype and seed size on speed of emergence and seedling vigor in nine spring wheat cultivars. *Crop Science* 26: 341-346.

Lafond, G.P. and Baker, R.J. (1986b). Effects of temperature, moisture stress, and seed size on germination of nine spring wheat cultivars. *Crop Science* 26: 563-567.

Lindstrom, M.J., Papendick, R.I., and Koehler, F.E. (1976). A model to predict winter wheat emergence as affected by soil temperature, water potential, and depth of planting. *Agronomy Journal* 68: 137-141.

Liptay, A. and Tan, C.S. (1985). Effect of various levels of available water on germination of polyethylene glycol (PEG) pretreated or untreated tomato seeds. *Journal of the American Society of Horticultural Science* 110: 748-751.

Lowe, L.B. and Ries, S.K. (1972). Effects of environment on the relation between seed protein and seedling vigor in wheat. *Canadian Journal of Plant Science* 52: 157-164.

Mali, C.V., Varade, S.B., and Musande, V.G. (1979). Water absorption of germinating seeds of sorghum varieties at different moisture potentials. *Indian Journal of Agricultural Science* 49: 22-25.

Marsh, L. (1992). Emergence and seedling growth of okra genotypes at low temperatures. *HortScience* 27: 1310-1312.

Massardo, F., Corcuera, L., and Alberdi, M. (2000). Embryo physiological responses to cold by two cultivars of oat during germination. *Crop Science* 40: 1694-1701.

Mayer, A.M. and Poljakoff-Mayber, A. (1989). *The Germination of Seeds*. Oxford, UK: Pergamon Press.

McDonald, M.B., Jr., Vertucci, C.W., and Roos, E.E. (1988a). Seed coat regulation of soybean seed imbibition. *Crop Science* 28: 987-992.

McDonald, M.B., Jr., Vertucci, C.W., and Roos, E.E. (1988b). Soybean seed imbibition: Water absorption by seed parts. *Crop Science* 29: 993-997.

McGrath, J.M., Derrico, C.A., Morales, M., Copeland, L.O., and Christenson, D.R. (2000). Germination of sugar beet (*Beta vulgaris* L.) seed submerged in hydrogen peroxide and water as a means to discriminate cultivar and seedlot vigor. *Seed Science and Technology* 28: 607-620.

Mian, A.R. and Nafziger, E.D. (1992). Seed size effects on emergence, head number, and grain yield of winter wheat. *Joural of Production and Agriculture* 5: 265-268.

Miao, Z.H., Fortune, J.A., and Gallagher, J. (2001). Anatomical structure and nutritive value of lupin seed coats. *Australian Journal of Agricultural Research* 52: 985-993.

Mohler, C.C. and Galford, A.E. (1997). Weed seedling emergence and seed survival: Separating the effects of seed position and soil modification by tillage. *Weed Research* 37: 147-155.

Morton, C.A., Harvey, R.G., Kells, J.J., Landis, D.A., Lueschen, W.E., and Fritz, V.A. (1993). In-furrow terbufos reduces field and sweet corn *(Zea mays)* tolerance to nicosulfuron. *Weed Technology* 7: 934-939.

Muchena, S.C. and Grogan, C.O. (1977). Effect of seed size on germination of corn (*Zea mays* L.) under simulated water stress conditions. *Canadian Journal of Plant Science* 57: 921-923.

Nicholson, R.L., Kollipara, S.S., Vincent, J.R., Lyons, P.C., and Cadena-Gomez, G. (1987). Phytoalexin synthesis by the sorghum mesocotyl in response to infection by pathogenic and nonpathogenic fungi. *Proceedings of the National Academy of Sciences* 84: 5520-5524.

Nomura, K., Endo, I., Tateishi, A., Inoue, H., and Yoneda, K. (2001). A chilling-insensitive stage in germination of a low-temperature-adapted radish, rat's tail radish (*Raphanus sativus* L.) cv. "Pakki-hood." *Scientia Horticulturae* 90: 209-218.

Obroucheva, N.V. and Antipova, O.V. (1997). Physiology of the initiation of seed germination. *Russian Journal of Plant Physiology* 44: 250-264.

Owen, P.C. (1952). The relation of germination of wheat to water potential. *Journal of Experimental Botany* 3: 188-203.

Pavmar, M.T. and Moore, R.P. (1968). Carbowax 6000, mannitol, and sodium chloride for simulating drought conditions in germination studies of corn (*Zea mays* L.) of strong and weak vigor. *Agronomy Journal* 60: 192-195.

Pedersen, J.F. and Toy, J.J. (2001). Germination, emergence and yield of 20 plant-color, seed-color near-isogenic lines of grain sorghum. *Crop Science* 41: 107-110.

Pieta-Filho, C.P. and Ellis, R.H. (1991). The development of seed quality in spring barley in four environments: II. Field emergence and seedling size. *Seed Science Research* 1: 179-185.

Pilatti, M.A. and deOrellana, J.A. (2000). The ideal soil: II. Critical values of an "ideal soil," for mollisols in the north of the Pampean region in Argentina. *Journal of Sustainable Agriculture* 17: 89-111.

Powell, A.A. (1988). Seed vigour and field establishment. *Advances in Research and Technology of Seeds* 16: 419-426.

Price, R.R. and Gaultney, L.D. (1993). Soil moisture sensor for predicting seed planting depth. *Transactions of the American Society of Agricultural Engineers* 36: 1703-1711.

Radford, B.J. (1987). Effect of cultivar and temperature on the coleoptile length and establishment of barley. *Australian Journal of Experimental Agriculture* 27: 313-316.

Raven, P.H., Evert, R.F., and Eichhorn, S. (1997). *Biology of Plants*, Sixth Edition. New York: Worth Publishing.

Revilla, P., Malvar, R.A., Cartea, M.E., Butrón, A., and Ordás, A. (2000). Inheritance of cold tolerance at emergence and during early season growth in maize. *Crop Science* 40: 1579-1585.

Rumpho, M.E., Pradet, A., Khalik, A., and Kennedy, R.A. (1984). Energy charge and emergence of the coleoptile and radicle at varying oxygen levels in *Echinocloa crus-galli*. *Physiologia Plantarum* 62: 133-138.

Sako, Y., McDonald, M.B., Fujimura, K., Evans, A.F., and Bennett, M.A. (2001). A system for automated seed vigor assessment. *Seed Science and Technology* 29: 625-629.

Sanhewe, A.J. and Ellis, R.H. (1996). Seed development and maturation in *Phaseolus vulgaris:* II. Post-harvest longevity in air-dry storage. *Journal of Experimental Botany* 47: 959-965.

Schell, L.P., Danehower, D.A., Anderson, J.R., Jr., and Patterson, R.P. (1991). Rapid isolation and measurement of adenosine triphosphate levels in corn embryos germinated at suboptimal temperatures. *Crop Science* 31: 425-430.

Sharma, M.L. (1976). Interaction of water potential and temperature effects on germination of three semi-arid plant species. *Agronomy Journal* 68: 390-394.

Shaw, R.H. (1988). Climate requirement. In Sprague, G.F. and Dudley, J.W. (Eds.), *Corn and Corn Improvement,* Third Edition (pp. 609-638). Madison, WI: American Society of Agronomy.

Shieh, W.J. and McDonald, M.B., Jr. (1982). The influence of seed size, shape and treatment on inbred seed corn quality. *Seed Science and Technology* 10: 307-313.

Slattery, H.D., Atwell, B.J., and Kuo, J. (1982). Relationship between colour, phenolic content and impermeability in the seed coat of various *Trifolium subterraneum* L. genotypes. *Annals of Botany* 50: 373-378.

Smiley, R.W., Patterson, L.M., and Rhinhart, K.E.L. (1996a). Fungicide seed treatment effects on emergence of deeply planted winter wheat. *Journal of Production and Agriculture* 9: 564-570.

Smiley, R.W., Patterson, L.M., and Shelton, C.W. (1996b). Fungicide seed treatments influence emergence of winter wheat in cold soil. *Journal of Production and Agriculture* 9: 559-563.

Smith, A.K. and Circle, S.J. (1972). Chemical composition of the seed. In Smith, A.K. and Circle, J.J. (Eds.), *Soybeans: Chemistry and Technology* (pp. 61-93). Westport, CT: AVI Publ. Co.

Snape, J. (2001). The influence of genetics on future crop production strategies: From traits to genes, and genes to traits. *Annals of Applied Biology* 138: 203-206.

Soman, P., Jayachandran, R., and Peacock, J.M. (1992). Effect of soil crusting on seedling growth in contrasting sorghum lines. *Experimental Agriculture* 28: 49-55.

Squire, G.R., Marshall, B., Dunlop, G., and Wright, G. (1997). Genetic basis of rate-temperature characteristics for germination in oilseed rape. *Journal of Experimental Botany* 48: 869-875.

Still, D.W. (1999). The development of seed quality in brassicas. *HortTechnology* 9: 335-340.

Still, D.W. and Bradford, K. J. (1998). Using hydrotime and ABA-time models to quantify seed quality of Brassicas during development. *Journal of the American Society of Horticultural Science* 123: 692-699.

Stockton, R.D., Krenzer, E.G., Jr., Solie, J., and Payton, M.E. (1996). Stand establishment of winter wheat in Oklahoma: A survey. *Journal of Production and Agriculture* 9: 571-575.

Syankwilimba, I.S.K., Cochrane, M.P., and Duffus, C.M. (1997). Effect of elevated temperatures during grain development on seed quality of barley (*Hordeum vulgare* L.). In Ellis, R., Black, M., Murdoch, A., and Hong, T. (Eds.), *Basic and Applied Aspects of Seed Biology* (pp. 585-592). Dordrecht, the Netherlands: Kluwer Academic Publishers.

Tadmor, N.H., Cohen, Y., and Harpaz, Y. (1969). Interactive effects of temperature and osmotic potential on the germination of range plants. *Crop Science* 9: 771-773.

Tarquis, A. and Bradford, K.J. (1992). Prehydration and priming treatments that advance germination also increase the rate of deterioration in lettuce seed. *Journal of Experimental Botany* 43: 307-317.

Taylor, A.G., Motes, J.E., and Kirkham, M.B. (1982). Germination and seedling growth characteristics of three tomato species affected by water deficits. *Journal of the American Society of Horticultural Science* 107: 282-285.

Thiessen Martens, J.R. and Entz, M.H. (2001). Availability of late-season heat and water resources for relay and double cropping with winter wheat in prairie Canada. *Canadian Journal of Plant Science* 81: 273-276.

van der Pijl, L. (1982). *Principles of Dispersal in Higher Plants*, Third Edition. Berlin: Springer-Verlag.

VanToai, T., Fausey, N., and McDonald, M., Jr. (1988). Oxygen requirements for germination and growth of flood-susceptible and flood tolerant corn lines. *Crop Science* 28: 79-83.

Vertucci, C.W. and Farrant, J.M. (1995). Acquisition and loss of desiccation tolerance. In Kigel, J. and Galili, G. (Eds.), *Seed Development and Germination* (pp. 237-271). New York: Marcel Dekker, Inc.

Vertucci, C.W. and Leopold, A.C. (1987). Water binding in legume seeds. *Plant Physiology* 85: 224-231.

Wang, Y.R. and Hampton, J.G. (1989). Red clover (*Trifolium pratense* L.) seed quality. *Proceedings of the Agronomy Society of New Zealand* 19: 63-70.

Washitani, I. (1987). A convenient screening test system and a model for thermal germination responses of wild plant seeds: Behaviour of model and real seeds in the system. *Plant, Cell and Environment* 10: 587-598.

Wilson, D.O. and Trawatha, S.E. (1991). Physiological maturity and vigor in production of Florida Staysweet *shrunken-2* sweet corn. *Crop Science* 31: 1640-1647.

Wolfe, D.W., Topoleski, D.T., Gundersheim, N.A., and Ingall, B.A. (1995). Growth and yield sensitivity of four vegetable crops to soil compaction. *Journal of the American Society of Horticultural Science* 120: 956-963.

Wuebker, E.F., Mullen, R.E., and Koehler, K. (2001). Flooding and temperature effects on soybean germination. *Crop Science* 41: 1857-1861.

Wulff, R.D. (1995). Environmental maternal effects on seed quality and germination. In Kigel, J. and Galili, G. (Eds), *Seed Development and Germination* (pp. 491-505). New York: Marcel Dekker, Inc.

Yaklich, R.W., Vigil, E. L., and Wergin, W.P. (1986). Pore development and seed coat permeability in soybean. *Crop Science* 26: 616-624.

Yu, J. and Tuinstra, M.R. (2001). Genetic analysis of seedling growth under cold temperature stress in grain sorghum. *Crop Science* 41: 1438-1443.

Chapter 4

Methods to Improve Seed Performance in the Field

Peter Halmer

INTRODUCTION

A wide range of techniques is now used to help sow seeds and to improve or protect seedling establishment and growth under the changeable environments and seedbed constraints reviewed in Chapters 1 and 3. These techniques constitute the postharvest processing necessary to prepare seed for sowing and optional treatments that are generally described in the industry and scientific literature as "seed enhancements" or "seed treatments."

In the first comprehensive review of this subject, Heydecker and Coolbear (1977) distinguished the purposes of seed treatment as follows: to select, improve hygiene and mechanical properties, break dormancy, advance and synchronize germination, apply nutrients, and impart stress tolerance. Subsequent reviewers (e.g., Taylor et al., 1998) adopted similar schemes. Halmer (2000), for example, grouped practical seed treatment technologies into operational categories in the following way:

- *Conditioning or processing*—by cleaning, purification, and fractionation, using mainly mechanical techniques such as size and density grading, polishing, scarification, and color sorting
- *Protection*—by applying active ingredients, usually synthetic fungicides and insecticides (The agrochemical industry commonly calls this technology "seed treatment," using the term in a narrower sense.)
- *Physiological enhancement or "seed invigoration"*—by hydration techniques such as priming, or applying active substances such as plant growth regulators, to exploit the ability of most species to interrupt the germination process by drying, and to resume the process when seeds are reimbibed, without vital harm (Some authors restrict the expression "seed enhancement" specifically to describe these techniques.)

- *Coating*—by pelleting or encrusting, to alter handling or imbibitional characteristics or to carry pesticides, micronutrients, and beneficial microorganisms

The focus of this chapter is on the last two categories—in the main to review progress in seed coating and, especially, in physiological enhancement. These technologies, which some call "functional" seed treatments, are mainly used at present for high-value crops in intensive agriculture but in the future may have wider applications.

This review continues by considering the underlying mechanisms of physiological seed enhancement, previously evaluated by Bray (1995) and McDonald (2000). Recent research in the disciplines of molecular and cell biology, biophysics, and the modeling of germination is giving conceptual insights into these processes, which may provide convenient methods for further improvement of seed quality. In particular, attention has been addressed to identifying biochemical, biophysical, and morphological markers that can be used to dissect key germination processes. Study areas most directly relevant to an appreciation of physiological enhancement include (1) the mechanisms of cell and embryo expansion; (2) the preparation for cell division; (3) endosperm weakening by hydrolases; and (4) the mechanisms of desiccation tolerance, including protection of the state of cytoplasm and membranes and maintenance of DNA structure during drying, air-dry storage, reimbibition, and germination. A chapter such as this can extract only the main threads from the large quantity of detailed material on these topics, and attention will be directed to key reviews at appropriate points.

CHANGING SEED FORM AND LOT COMPOSITION

Sorting

Conventional processing technology includes sorting and grading seeds—exploiting superficial external seed features such as size, shape, color, surface texture, density, and buoyancy in air. Seed quality is purified and lots are "upgraded" by removing contaminants, and seed that is outside specifications is rejected. To supplement these well-known techniques, novel seed selection and sorting principles have been developed in recent years to remove fractions that have higher proportions of weak or dead seeds. Simak (1984) used water flotation to separate dead and viable forest plant seeds that had been previously imbibed and dried to amplify their density difference. Aqueous buoyancy sorting can also be effective after priming, e.g., by discarding low-density fractions before osmoprimed tomato and lettuce seeds are dried (Taylor et al., 1992). Taylor, McCarthy, and Chirco (1982) devised mixtures of polar organic solvents (chloroform and hexane) to separate dry seed batches by density. Equipment has been engineered to handle these solvents in a way that is safe for the seed and the operator, and this sorting technique is now used commercially for high-value horticultural and ornamental seeds. Jalink and colleagues (1998) recently developed an innovative variation on color sorting, using laser-induced fluorescence to detect the residual content of chlorophyll in seed coats, which in some cases is undetectable to the human eye. It is thought that the amount of the pigment is inversely related to the maturity of the seed. This technique appears to have practical value for upgrading tomato, pepper, leek, cucumber, and cabbage seed lots, and equipment is becoming commercially available to carry out this patented technique. Seeds are fed past a photoelectric cell, which triggers a jet of air to remove colored individuals one at a time. In the future, X-radiography might offer another real-time sorting principle, using decision logic to identify normal and anomalous embryo structures, e.g., in tomato seeds which develop an internal free space after priming and redrying (Downie, Gurusinghe, and Bradford, 1999).

Planting

Precision Sowing Systems

Many horticultural field root and salad crops and ornamental production systems are based on crop uniformity and must be precision sown in defined patterns to optimize yield and harvest quality. These crops are precision sown either directly where they are grown or are raised as seedlings in protected conditions—either in nursery beds or in soil blocks, paper pots, flat or plug trays in growing media—for later transplantation into pots or the field, or into phenolic foam cubes for hydroponic propagation. In contrast, plant spacing is not usually a critical factor in arable, grass, and cover crops, which are sown by broadcasting or drilling in rows in or onto bare soil, or are "direct seeded" directly into existing pastures, turf, or crop stubble.

The natural shape of many seed species is not ideal for precision seed drills, even after size grading. Also, although designed to cope with dry and dusty field conditions, most drills are vulnerable to blockage by misshapen seeds or dust, and seed flow can be impeded by sticky or rough seed surfaces. In these situations, coating and pelleting are valuable seed enhancements to improve the accuracy of mechanical singulation. Modern precision drills have three main designs. In the cell-feed system, seeds are collected in deep holes in the rim of a rotating metal wheel, into which they must fit completely, before being prised out by an ejector plate at the outlet point. In the belt-feed method, a small endless rubber belt with one to three rows of holes punched in it conveys seeds to the exit point from the seeder unit. In the vacuum-feed (or pneumatic or "air planter") system, suction is applied to one side of a rotating disc perforated with lines of regularly spaced holes, onto which the seeds are pulled and transported to the discharge point where a blanking plate cuts off the vacuum. Belt and vacuum seed drills are used not only for planting many types of horticultural species but, increasingly, for planting large-seeded crops such as maize, sunflower, cotton, soybeans, and beans. The vacuum-seeding principle is also frequently used to sow tray formats, e.g., using nozzle arrays or flat template plates drilled with holes to suit the layout. Nursery beds are sown using field drills or simpler perforated drum seeders. Grain drills have much simpler designs, e.g., with seed being carried by a fluted or studded feed roller to flexible tubes for delivery to the ground.

Coating: Pelleting, Encrusting, and Film Coating

The primary historical purposes of pelleting and encrusting is to build up seed to change its shape, weight, size, or surface structure, by applying variable amounts of filler materials and binders—typically to make seeds fit drills better. Pelleting is usually carried out to make irregularly shaped seed ovoid and smooth, or to make small seeds much larger. In comparison, seed coating ("minipelleting" or "encrusting") applies less material, so that the original seed shape is still more or less visible. Apart from improving drill performance, coating is also used to upgrade size ranges and to increase weight to prevent drift, e.g., for aerial seeding of range and amenity grasses. Pelleted and coated seed is commonly colored to make it easier to find seeds after drilling and check depth and spacing, as well as to identify companies, varieties, or treatments. Species pelleted in substantial commercial amounts include sugar beet (quantitatively by far the largest use), carrot, celery, chicory and endive, leek, lettuce, onion, pepper, tomato, and to a lesser extent some *Brassica* species and "super-sweet" corn varieties, and certain flower species, particularly those with tiny seeds. There are potential applications for seed coating in crops that do not need precision sowing, e.g., to reduce size variation of maize and sunflower kept in inventory, which are typically sold in up to six different size grades. Pelleting and coating can be used to carry nutrients and growth-stimulating materials, including plant growth regulators (PGRs).

Thin film coating is employed mainly to apply colorants and pesticide treatments onto seeds, in a firmer and more uniform way than can be achieved using conventional slurry application techniques. As well as improving treatment accuracy, film coating is used to minimize chemical dust-off losses during seed handling and drilling, and exposure of the operators who handle treated seed. It also presents seed for sale in a cosmetically attractive form. Characteristically, each seed is covered with a water-permeable polymer layer, which adds about 1 to 10 percent to the weight so that shape and size are little changed. Film-coating techniques are now well established for many high-value horticultural seed species and are being adopted for treating some higher-volume crops, such as maize, sunflower, canola, alfalfa, clover, and some grasses. Film coating is also widely used to apply insecticides and fungicides to the outside of pelleted seed: in some cases this is a preferred method of application to minimize phytotoxic effects, especially where these treatments have to be applied at very high loading rates.

The treatment of seeds with agrochemical formulations is now a market of huge worldwide value and importance that is steadily growing as alternatives to application by sprays or granules become available. Though most seed sown is treated in this way, this is not the place to cover the subject. Brandl (2001) has summarized recent progress, including the development in recent years of "active ingredients" with systemic modes of action that can protect plants for several months into the life of the crop. It is worth mentioning here in passing, however, that a number of active ingredients have moderate side effects on seed performance, by slowing germination or producing seedling abnormalities by imposing phytotoxic stresses; such defects are regarded as acceptable, commercially, considering the protection benefits delivered to the crop.

Commercial film coating, pelleting, and coating systems are often run as secret processes, and there are very few detailed investigations of this subject in the scientific literature, although patents give useful descriptions and insights into the technologies involved. Halmer (2000) has reviewed equipment and techniques and the general types of filler materials and binders in pelleting, encrusting, and film coating, and the processes for applying pesticide formulations onto seeds using these and other techniques.

Coating to Modify Imbibition and Germination

Usually and understandably, commercial pelleting, coating, and film-coating types are designed to impose minimal mechanical or physiological barriers on germination while reshaping and resizing seed strongly enough for drilling purposes and for the adhesion of protective treatments. However, the materials used can also be tailored to modify seed water availability and gaseous exchange, and so control the timing of germination and emergence.

Several studies have been published on ideas to use treatment or film-coating techniques to manipulate seed imbibition characteristics. Hydrophobic materials may be included within or around the pellet or coating fabric to allow seeds to germinate under wet conditions, for species such as onion where that can be erratic, or filler materials may be incorporated to give a more porous structure to the matrix. Some pellet types are designed to disintegrate rapidly or split after imbibition to expose the seed. Several studies over the years have investigated the promotion of emergence by including calcium or magnesium peroxide in the pellet to supply more oxygen in waterlogged environments (see Ollerenshaw, 1988).

Various film-coating polymers (e.g., Schmolka, 1988; Taylor et al., 1992; Ni, 2001) have been evaluated as potential barrier layers to alleviate seed imbibitional chilling injury leading to poor seedling establishment in vulnerable crops, such as certain cultivars of large-seeded grain legumes and super-sweet corn, especially when seed coat layers are abnormally thin or damaged. Damage can involve disruption of oil bodies and membranes, and the leakage of cell contents from the outermost embryo tissues, including the solutes measured in the electrical conductivity test for these species, and may lead to the death of cells on the cotyledon surface (see references in Chachalis and Smith, 2000). Seed coats of many species are not as vulnerable to rapid imbibition, due in part to the presence of semipermeable layers in the seed coat tissues, which restrict solute diffusion and leakage rates (see Taylor et al., 1998). Humidification is another approach to alleviate the imbibitional chilling injury problem—simply by raising seed moisture content in a damp atmosphere for several days (Taylor et al., 1992).

An extension of this concept is to delay imbibition with water-resistant polymers until climatic conditions become suitable for continued crop growth (e.g., Watts and Schreiber, 1974). In recent years this type of technology has attracted a great deal of commercial interest, though developments have mainly been relayed through the trade press and very little has been published yet in the scientific literature detailing mechanisms and field performance. Polymers with in vitro temperature-dependent water-permeability properties have been advocated to coat seeds for early planting so that they can imbibe only when favorable moisture and temperature conditions have developed (Stewart, 1992); among the applications being evaluated are the coordination of the flowering of parental lines planted at the same time for hybrid maize seed production. For a somewhat similar purpose, another water-resistant polymer coating has been evaluated as a tool to give a wider window of opportunity for sowing canola in the autumn in northern American latitudes, just before soils freeze over winter, for earlier emergence and crop maturation and greater yields compared to the normal spring-seeding time. These technologies are potentially powerful but will have to perform very reliably in changeable soil environments, particularly if they are to be used in space-planted crops that do not have a compensatory growth habit.

By contrast, the inclusion of water-attracting materials can aid imbibition and give more intimate seed-soil contact, or may retain moisture in the vicinity of the seed as soils dry. For example, some success has been reported using nonionic surfactants (Aksenova et al., 1994) and hydrophilic gels (Henderson and Hensley, 1987). The starch-based or polyacrylate/polyacrylamide polymers used commercially as soil amendments to retain water in agricultural and horticultural situations are also advocated to treat seeds. Such superabsorbent materials need to be applied and kept dry enough to prevent the seed batch congealing into an unusable mass.

Other Planting Systems

Hydroseeding and seed tapes. Some seed is sown using specialist techniques that do not involve conventional drills or coating. In hydroseeding aqueous slurries of seed and other materials are sprayed to enable fast and convenient seeding of amenity areas or steep slopes with grass, wildflowers, or other groundcover vegetation. The patent literature contains quite a few variations on the seed tape format, in which seed is stuck or embedded randomly or in patterns between layers of biodegradable paper or plastic, etc., in porous mats, grids, or narrow strips, some incorporating growing media, which are laid out dry in the ground (e.g., Holloway, 1999; Meikle and Smith, 2000). These sowing systems can help suppress weed growth and are suitable for placement of much higher doses of nutrients, moisture retention agents, and protective chemicals than could be directly coated onto seeds without encountering phytotoxic effects.

Pregermination. The slurry and tape-sowing concepts have each been developed for planting pregerminated hydrated seeds. Fluid drilling, the best-known technique of this type, is used in some places to establish small-seeded vegetables, e.g., celery and tomato. Seed is allowed to complete germination in an aerated medium of relatively high water potential and, after removal of ungerminated individuals by density separation, the sprouted seeds are suspended in a viscous gel and precision sown by extrusion into the soil (Pill, 1991; Far, Upadhyaya, and Shafii, 1994). Low water potentials or leachable plant growth inhibitors, such as abscisic acid, can be used to synchronize the germination process (Finch-Savage and McQuistan, 1989). A novel propagation concept proposes the use of seed tapes to raise and transplant germinated seedlings or more fully developed transplants in moist "paper pockets" containing hygroscopic material, etc. (Ahm, 2000). The success of such propagation systems relies in part on arranging timely seedling production and optimized seedbed conditions to ensure development of the young seedlings with minimal desiccation damage, and they are best suited to situations in which production follows a fixed plan and is not likely to be interrupted by bad weather.

In another type of pregermination treatment, development is suspended just after radicle emergence and seed is dried to produce high-viability lots for conventional sowing. In a patented technique that is marketed at present mainly for flower species, fully imbibed seeds are germinated to the point where radicles are just visible, sorted by machine vision, flotation, or other means to remove ungerminated seeds, and dried to induce desiccation tolerance. This can produce either damp pregerminated seed (30 to 55 percent moisture content) with a shelf life of a few weeks at ambient temperature, or dry seed that is viable for a few months (Bruggink and van der Toorn, 1995, 1996). McDonald (2000) reports that dehydration of pregerminated seeds using cool low relative humidity air (e.g., 11 percent and 35 percent RH, at 5°C) successfully imposes desiccation tolerance and extends storage life up to four weeks. When circumstances allow, it is also possible to use undried freshly primed seed. Alternatively, Sluis (1987) patented the idea of producing a moist pellet from primed seed incorporating materials such as osmotica or abscisic acid to slow germination; at refrigerated temperatures and/or under reduced atmospheric pressure the seed microenvironment would be sufficiently stabilized to allow several weeks of storage life.

PHYSIOLOGICAL ENHANCEMENT

Priming and Related Hydration Techniques

Germination enhancement technologies based on presowing seed hydration have attracted considerable interest in both seed physiological research and industry circles, where they have been extensively commercialized. Heydecker's work (Heydecker, Higgins, and Turner, 1975) is often taken as the starting point for modern research in this area, and a substantial literature has developed since. By manipulating water relations to exploit most seeds' natural ability to survive one or more cycles of imbibition and drying, subsequent germination is made faster and often more uniform—which Heydecker distinguished as the "advancement" and "priming" responses, respectively. In recent years, however, the meaning of the term *priming* has evolved from its original specific sense of increased germination synchrony and is now commonly used to describe seed presowing hydration methodologies without discrimination, where seeds are imbibed by whatever means (e.g., Khan, 1992; Parera and Cantliffe, 1994b; Taylor et al., 1998; McDonald, 2000).

The hydration treatments regulate the germination process by manipulating temperature, seed moisture content, and duration. Water is either made freely available to the seed (as in steeping or soaking) or restricted to a predetermined moisture content or a programmed sequence of moisture contents, typically using water potentials between –0.5 MPa and –2.0 MPa. Some positively photoblastic species benefit from treatment under light of appropriate wavelengths, and it is possible to include other materials such as nutrients and growth regulators with the water. Then, usually, seeds are dried back prior to further treatment as required, e.g., by coating or treatment with pesticides, for storage and sowing.

One practical drawback is that primed seeds often, but not always, deteriorate faster during storage and accelerated aging than untreated seeds. Symptoms include the onset of a reducing rate, uniformity, and final level of germination, and an increase in the proportion of abnormal seedlings— although the degree of the problem in susceptible species varies among seed lots and with the extent of priming performed and storage conditions. A related problem is the increasing injury seen as priming is allowed to proceed too far, reflecting the familiar fact that seeds' ability to survive drying and the dry state for extended periods of time is progressively lost as germination progresses. It is important to know how to optimize the priming for an individual seed lot. What applies to one lot need not necessarily apply to another; indeed, differences between lots can be more important than differences between cultivars.

Typical responses to priming are faster and closer spread of times to germination and emergence over all seedbed environments and wider temperature range for germination, leading to better crop stands, and hence improved yield and harvest quality, especially under suboptimal and stress growing conditions in the field, though responses can vary due to fluctuating water availability and temperature. Times to reach 50 percent of maximum emergence (T_{50}) can typically be decreased by up to one-third under environmental conditions in seedling production practice. Seed lots differ in the magnitude of their response to a standard priming treatment; in general, slower-germinating lots exhibit the greatest benefit (Brocklehurst and Dearman, 1983; Bradford, Steiner, and Trawatha, 1990).

Primed seeds are now used commercially in the production of many high-value crops where reliably uniform germination is important, such as the field seeding or plug production of leek, tomato, pepper, onion, and carrot, and the production of potted or bedding ornamental herbaceous plants, such as cyclamen, begonia, pansy (*Viola* sp.), *Polyanthus* sp., and primrose (*Primula* sp.), and several culinary herbs, as well as for some large-volume field crops, such as sugar beet and turf grasses. Due to its ability to raise the upper temperature limit for germination, priming is also very valuable for circumventing the secondary thermodormancy that results when imbibed seeds are likely to be exposed to supraoptimal temperatures for too long, e.g., in susceptible cultivars of lettuce, celery, and pansy (Hill, 1999; McDonald, 2000).

Technologies

Fundamentally, three strategies are used to deliver and restrict the amount of water and to supply air: submersion in solutions of osmotica in water, mixing with moist solid particulate materials, and hydration with water only. Though new descriptive names have proliferated in recent years, the basic principles of almost the entire array of priming technologies used in research and the seed industry today were mapped out in Heydecker and Coolbear's (1977) insightful review. Seed companies usually perform commercial priming treatments, using proprietary methodologies and systems that handle quantities ranging from tens of grams to several tonnes at a time, and that are often kept secret.

Priming protocols for a "new" species have been developed mainly on an experimental basis. There is a substantial research literature reporting and comparing priming approaches and conditions for many species: Welbaum, Shen, and colleagues (1998b) and McDonald (2000) provide selected bibliographies of priming techniques that have been used successfully on a wide range of crops. A rough rule of thumb is to start by using the temperatures considered optimal for germination of untreated seeds, water potentials equal to or less than the threshold water potential at which emergence of the embryonic axis (usually the radicle) is prevented, and durations from one to three weeks, but these conditions may not prove optimal for priming. Some seeds benefit from prewashing before priming to remove germination inhibitors, e.g., sugar beet and umbelliferous species.

Because of the variability in response between seed lots, optimum priming conditions—choosing the balance between the most rapid germination, and the longest storage life—often need in practice to be determined on a case-by-case basis for many species (Welbaum, Shen, et al., 1998). Conducting pilot priming runs on small samples achieves this empirically, i.e., by varying somewhat the final water potential and perhaps the stages taken to reach it, their duration and (less commonly) temperature, and testing germination responses.

A continuing goal for the seed industry is to find simple means to determine these parameters quickly and reliably, to complement or replace existing test procedures. Therefore, considerable research interest is directed toward identifying marker signals that correlate well with the degree of advancement and/or the loss of desiccation tolerance in individual seed lots. These indicators might provide means to assess the potential effectiveness of priming a seed lot, to help set operating parameters before the treatment is started, and to monitor its progress in real time prior to radicle emergence. They might also provide research tools to develop and distinguish new priming approaches and protocols (Job et al., 2000). The seed merchant and the grower would also value *post facto* tests that give a measure of whether, and how well, a seed lot has been physiologically enhanced, and to predict its storage life. Such tests should ideally give precise and reliable information across all varieties and seed lots and should also be fast and convenient to perform, in an industry in which many seed lots are processed on a just-in-time basis and decisions are needed quickly.

Osmopriming. Osmotic priming of seeds (also known as osmopriming or osmoconditioning) describes contacting seeds with aerated solutions of low water potential, usually by submersion, which are rinsed off afterward. This is still regarded by many researchers to be the standard priming technique, and treatment on the surface of paper or other fibers moistened with solution or immersed in small continuously aerated columns continues to be common study method in which only small quantities of seed are required.

Mannitol or inorganic salts [such as KH_2PO_4, $KH(PO_4)_2$, K_3PO_4, KCl, KNO_3, $Ca(NO_3)_2$, and various mixtures of these] have been used extensively as osmotica but, because of their low molecular size, these are capable of being absorbed by the seeds, which has been associated with toxic side effects in some cases. Na salts, however, which tend to be more toxic to some common agricultural seeds than K salts, have been proposed to induce tolerance to saline conditions, e.g., in tomato (Cano et al., 1991). Heydecker, Higgins, and Turner (1975) first suggested the alternative of using moderately high molecular weight fractions of polyethylene glycol (PEG, most commonly, 6 to 8 kDa), whose large size precludes it from entering the seed, and this is now widely used as a preferred osmoticum by many in research and the seed industry.

Care must be taken, particularly when using viscous PEG solutions, to ensure adequate gas exchange by constant vigorous aeration or by using stirred bioreactors (Bujalski et al., 1991). Some seeds, particularly onions, are reported to only osmoprime satisfactorily using air enriched with oxygen (Bujalski and Nienow, 1991). A patented process using a semipermeable membrane to separate seeds from an osmoticum contained in the outer jacket of a rotating tube has been devised for osmopriming small seed quantities, such as small-seeded flower species, and for seeds with mucilaginous coats that can cause difficulties in other priming methods (Rowse and McKee, 1999).

Matrix priming. Solid matrix priming (matripriming and the closely allied matriconditioning technique) mixes seed with solid insoluble matrix particles, such as exfoliated vermiculite, diatomaceous earth, or crosslinked highly water-absorbent polymers, and water in predetermined proportions (Taylor, Klein, and Whitlow, 1988; Eastin, 1990; Tsujimoto, Sato, and Matsushita, 1999). Seeds slowly imbibe to reach an equilibrium hydration level, determined by the reduced matrix potential of the water adsorbed on the particle surfaces, and after the incubation the moist solid material is removed by sieving. It is important in this approach to ensure adequate aeration and prevent the formation of temperature gradients in the seed mass, and that the surplus matrix material can be removed without mechanically damaging the seed or leaving too much dust on it.

Hydropriming. Hydropriming is currently used both in the sense of the continuous or staged addition of a limited amount of water (e.g., van Pijlen et al., 1996; McDonald, 2000) and also the sense of imbibition in water for a short period (e.g., Gurusinghe and Bradford, 2001) with or without subsequent incubation in humid air (Fujikura et al., 1993).

Slow imbibition is the basis of the patented drum priming and related experimental techniques (Rowse, 1996; Warren and Bennett, 1997), which evenly and slowly hydrate seeds up to a predetermined moisture content—typically about 25 to 30 percent on a fresh-weight basis—by misting, condensation, or dribbling. Tumbling in a rotating cylinder ensures that seed lots are evenly hydrated, aerated, and temperature controlled during the damp incubation stage. De Boer and Boukens (1999) have devised a priming system using direct hydration from a humid atmosphere (RH > 98 percent) to control the final stage of imbibition and maintain the moisture content in a static seed mass. These approaches have the practical economic advantages that the production of waste materials associated with osmopriming or matripriming is avoided, and that the relatively modest amounts of water involved are removed by drying.

Submerged aerated hydration, very akin to steeping, has been proposed as a treatment to enhance the germination of horticultural *Brassicas* (Thornton and Powell, 1992). Davidson (1981) proposed seed steeping in high oxygen atmospheres.

Steeping. Hydropriming for longer periods followed by drying back to the original seed moisture content is also commonly known as steeping. Steeping treatments, e.g., at up to 30°C for several hours (sometimes the duration needs to be adjusted for individual seed lots), are now widely performed to remove residual amounts of germination inhibitors and/or to infiltrate chemical fungicide treatments to control deep-seated seed-borne diseases, such as for sugar beet and umbelliferous species (Maude, 1996).

At its very simplest, on-farm steeping and sowing of wet seed has a long history. This was done where circumstances allowed in the days before the mechanisation of sowing, and similar overnight steeping is even now advocated as a pragmatic, low cost, and low risk agricultural method for enhancing crop establishment in developing countries, e.g., for groundnut, maize, upland rice, and chickpea crops (Massawe et al., 1999; Harris et al., 1999). Direct benefits are reported to include improved drought tolerance, earlier flowering, and higher seed/grain yield. Rice is also steeped in some mechanized farm situations, in part just to increase seed weight to aid in sowing from the air.

Other methods. The older research literature has a few reports (e.g., see Heydecker and Coolbear, 1977; Hegarty, 1978) of the benefits of seed hardening (two to three cycles of steeping and drying), particularly for drought tolerance, but the subject has not received much research attention of late and these approaches do not seem to be in wide commercial use, possibly because they are cumbersome to perform.

Changes in Endogenous Microflora

Seed-borne microflora, including pathogens, can increase during priming but cannot necessarily be fully controlled by conventional fungicides included in the osmopriming solutions alone (Maude et al., 1992; Nascimento and West, 1998), and seeds may need subsequent treatment after drying. However Petch and colleagues (1991) concluded that the presence of large number of microorganisms did not greatly affect seed performance when the same PEG osmoticum was used three times with leek and twice with carrot seed.

Wet Heat Treatments to Eradicate Seed-Borne Disease

Short hot water treatments are used to disinfect seeds, typically at temperatures of about 50° to 60°C for up to about 30 minutes for some small-seeded species such as flowers, and, for example, very brief exposure to steam is being evaluated as an organic treatment for cereal seed (Maude, 1996; Forsberg, 2001). Care needs to be taken in administering this type of heat treatment to avoid damaging seed quality. Klein and Hebbe (1994) found that short hot water treatments of tomato seeds produced plants that were 20 percent taller 30 days after sowing, but the beneficial results were not retained after three months of storage at 5°C.

Biopriming

Several researchers have investigated the use of so-called biopriming techniques, by including beneficial microorganisms in the priming processes as a crop delivery mechanism or to control disease proliferation during priming itself. For example, Warren and Bennett (2000) added *Pseudomonas aureofaciens* as a biological control organism in combination with an osmopriming treatment to control *Pythium ultimum* in tomato seedlings. Matrix priming and hydropriming are also suitable delivery mechanisms for beneficial microorganisms, akin to solid-state fermentation (see McQuilken, Halmer, and Rhodes, 1998).

Promotive and Retardant Substances

Many studies report the benefits of gibberellins, ethylene, and/or cytokinins such as benzyl adenine in combination with priming, e.g., for celery (Brocklehurst, Rankin, and Thomas, 1983) and for the O_2-enriched osmopriming of bedding plant species (Finch-Savage, 1991). Adding such plant growth regulators during priming can improve the germination performance of some seed species or lots compared to either treatment alone.

Alternatively, treatment with growth retardants has been advocated to dwarf the growth habit of transplants, such as bedding plants, which tend to develop an etiolated growth habit, especially if grown in low-light environments. For instance, Souza-Machado and colleagues (1996) reported that seed priming with a triazole (50 ppm paclobutrazol) in tomato cultivars produced seedlings that were shorter, greener, more uniform, with stronger thicker stems, and higher root:shoot weight ratios than nonprimed controls, though emergence itself was reduced: after five weeks in the field primed seedlings were taller than unprimed controls. Pill and Gunter (2001) found similar dwarfing responses in marigold seeds matrix-primed with paclobutrazol.

Drying

The technique and rate of drying after priming are also very important to subsequent seed performance. Slow drying at moderate temperatures is generally (e.g., Jum Soon, Young Whan, and Jeoung Lai, 1998) but not always (e.g., Parera and Cantliffe, 1994a) preferable. Various manipulations have been proposed to extend the storage life of primed seeds. Gurusinghe and Bradford (2001) found that a moisture reduction of 10 percent or more was effective in extending the longevity of hydroprimed tomato seeds. Heat shock is another tactic. For several species Bruggink, Ooms, and van der Toorn (1999) have found that greater longevity is obtained by keeping primed seeds under mild water and/or temperature stress for several hours (e.g., tomato) or days *(Impatiens)* before drying. These methods are very similar to those used to induce desiccation tolerance in just-germinated seeds, e.g., in cucumber radicles (Leprince et al., 2000).

Electromagnetic Treatments

Advantageous germination and seedling growth responses after treating seeds with continuous, intermittent, or rapidly pulsed exposure of seeds to stationary or alternating magnetic fields (typically up to about 0.25 to 1.0 Tesla) or electric fields (up to 100 kV/m or more) have been known for some time (Heydecker and Coolbear, 1977). These phenomena continue to receive research attention, though not prominently in the seed physiology literature (e.g., see Kornarzynski and Pietruszewski, 1999; de Souza Torres, Porras Leon, and Casate Fernandez, 1999; Carbonell, Martinez, and Amaya, 2000). The mechanisms underlying these intriguing responses, and the reproducibility of possible practical treatments based on them, remain to be investigated, and the subject will not be discussed further here.

PHYSIOLOGICAL RESPONSES TO ENHANCEMENT

From the previous section it can be appreciated that priming should be seen as taking the process of germination to various degrees, selecting from a continuum of water potentials, etc., and different durations and drying procedures. In practice therefore, priming is not a single treatment, but rather is the result of a choice between options. What constitutes optimal priming for a given seed lot is a compromise that will vary depending on research or commercial circumstances—including how quickly seed is to be sown. Because methods and conditions differ greatly, it is hard to generalize about responses, and care must be taken in drawing conclusions from the scientific literature about the mechanisms of priming.

Water Relations and Kinetics of Germination

Water Uptake

Seed germination in the strict sense is commonly defined as those events that begin with water uptake and end with the penetration by the embryonic axis (usually the radicle) through the structures surrounding the embryo. Seeds with permeable seed coats that are dry at maturity classically display a triphasic time course of water uptake: seed imbibition (Phase 1), reflecting the initial rapid absorption of water by the dry seed; a period of variable length in which seed water content is relatively constant or only slowly increasing (Phase 2), which may be greatly prolonged depending on the degree of dormancy; and a resumption of water uptake (Phase 3) associated with expansion and growth of the embryo after germination is complete (Bewley and Black, 1994).

Hydrothermal Time Models of Germination and Priming

Germination. Mathematical population models have been successfully developed in the past two decades to unify germination behavior in terms of what seem to be physiologically meaningful water potential (Ψ) and temperature (T) thresholds, and have been extended to cover dormancy, growth regulators, and—for our purposes in this chapter—priming. It might be hoped that tests based on a model which describes the key variables for a seed lot (some of which might turn out to be nearly constant at the species or cultivar level) could provide a simple guide to predict the operational parameters for optimal priming, which would be very valuable for the seed industry. These so-called hydrothermal time models will be outlined only briefly here since they are covered in depth by Finch-Savage in Chapter 2. In addition, the reader can see Bradford (1995), Cheng and Bradford (1999), and references therein for comprehensive reviews.

Most seed lots exhibit optimum temperatures within the thermal window in which seeds are capable of full germination, which is defined as the point at which time taken for germination of a certain fraction g is minimum. At supraoptimal temperatures, as T increases germination appears to decline linearly for each fraction and falls to zero at its upper limit $T_c(g)$, i.e., the time taken for germination becomes infinite. Similarly, there is a minimum temperature at which germination falls to zero. The germination time course of a seed lot sums the performance of individual seeds, which have inherently variable properties. According to hydrothermal-time thinking, an individual hydrated seed below its optimum temperature completes germination when it has accumulated the heat units characteristic of its rank (g) in the seed lot. (It is assumed that individual seeds in the population would germinate in the same order over the range of conditions covered.) Temperature affects germination rates (GR_n, the reciprocal of the time it takes for the nth radicle to emerge) on a *thermal* time basis: the T in excess of a base or minimum temperature (T_b) multiplied by the time to a given percent germination (t_g) is a thermal-time constant (θ_T), which differs for each seed fraction. Somewhat analogously, Gummerson (1986) suggested that reduced water potential delays germination on a *hydrotime* basis; the Ψ in excess of a threshold base water potential (Ψ_b, which differs for each seed fraction) multiplied by t_g has a constant value for *all* seed fractions, at a constant temperature. He further proposed that germination responses to T and Ψ could be combined into a single expression, using a hydrothermal time constant (θ_{HT}), defined for all fractions as $[(T - T_b)(\Psi - \Psi_b)] \cdot t_g$. By rearrangement of this equation, if the constants and variables are known, the germination rate of a fraction can be predicted: $GR = [(T - T_b)(\Psi - \Psi_b)]/(\theta_{HT})$. In theory, germination data obtained at just two water potentials would be sufficient to determine T_b and $\Psi_b(g)$, but more measurements may be needed in practice to give greater precision. Although it was developed for constant conditions, the model has been adapted to describe and predict behavior in irregularly changing temperature and water potential environments by integrating performance over a number of time intervals—as discussed in Chapter 2.

This hydrothermal time model has been found to match sets of germination time courses of nondormant seeds quite well, e.g., in tomato, lettuce, onion, mungbean, and melon, though more research is needed to cover a wider range of species and environmental conditions and to understand the genetic components (Dahal, Bradford, and Haigh, 1993; and see Welbaum, Bradford, et al., 1998; Cheng and Bradford, 1999). The general situation is as follows: (1) there is relatively little variation in T_b among individual seeds within a seed lot, or even within a species, although further work is needed; (2) the variations in Ψ_b between fractions in a seed lot are approximately normally distributed with σ_{Ψ_b}; but (3) agreement can be poor for some time courses—e.g., behavior can deviate close to T_b and Ψ_b, and the values of each can vary depending upon the environmental or hormonal conditions, including perhaps during test incubations.

In mathematical terms, then, the objectives of seed enhancement could be said to be one or more of the following: (1) to raise the optimum temperature and the upper limit $T_c(g)$; (2) to lower Ψ_b (which determines when and whether a given seed will germinate under specific environmental conditions); (3) to minimize θ_{HT}, θ_H, and $\theta_T(50)$ and the distribution of $\Psi_b(g)$ and

$\theta_T(g)$, which determine the rate and uniformity of germination. Three examples will illustrate responses that have been observed.

- Osmopriming of tomato cultivars, for instance, increased GR at all T > T_b and Ψ > $\Psi_b(g)$ without lowering T_b or Ψ_b, (i.e., it reduced θ_T and θ_H), but it also *increased* the variance in t_g (Dahal, Bradford, and Jones, 1990; Dahal and Bradford, 1990). Priming at -0.5 MPa > Ψ > -1.0 MPa appeared to shift distributions of $\Psi_b(g)$ to lower values, i.e., allowing subsequent germination to occur at a Ψ that initially would have blocked radicle emergence, but there was no shift in $\Psi_b(g)$ distributions after priming at Ψ > -0.5 MPa (Cheng and Bradford, 1999).
- In contrast, osmopriming of a mature muskmelon cultivar decreased T_b and Ψ_b but left θ_H unaffected (Welbaum and Bradford, 1991). Immature seeds were more responsive to priming than mature seeds, suggesting that the overall degree of response would be strongly influenced by the distribution of seed ages within a lot.
- The model has also been applied in studies of germination near the upper temperature limit $T_c(g)$. In lettuce seeds, as temperature was raised up to this point, germination became more and more sensitive to reduced water potentials; in hydrotime language, the $\Psi_b(g)$ distribution became progressively more positive. Osmopriming resulted in smaller increases in $\Psi_b(g)$ near the temperature limit, as well as reducing the hydrotime requirement for germination (Bradford and Somasco, 1994).

During priming. The fact that germination can be substantially advanced by priming seeds with water potentials <Ψ_b (as well as >Ψ_b) or at temperatures close to T_b, led to the proposal (see Bradford and Haigh, 1994) that the concept of *hydropriming time* or *hydrothermal priming time* may be applicable—invoking the idea that seeds can also accrue hydrotime and/or heat units in relation to a different base potential Ψ_{min} and temperature T_{min} (lower than Ψ_b and T_b); even though they cannot complete germination (unless they are being incubated for long enough above Ψ_b and T_b). The resultant median germination rate *(GR_{50})* after a priming duration of t_p is theoretically, $GR_{50} = GR_i + k'[(T-T_{min})(\Psi - \Psi_{min})] \cdot t_p$; where k' is the inverse of the hydrothermal priming time constant θ_{HTP}, and GR_i is the initial median rate for unprimed seeds, determined from the hydrothermal model expression. This model therefore provides a way of comparing a wide range of priming treatments on a common basis, across temperatures, water potentials, and durations. Theoretically, knowing the values of T_b, T_{min}, θ_{HT}, θ_{HTP} and, especially, the means and distributions of $\Psi_b(g)$ and Ψ_{min} might be practically useful before deciding to prime a seed lot.

In initial tests the hydrothermal priming time model proved capable of explaining a large part of the variance in tomato seed primed over a range of water potentials (Dahal, Bradford, and Haigh, 1993). However, in a recent in-depth critical study, Cheng and Bradford (1999) concluded that, despite providing a useful quantitative description of priming responses, the initial water relations characteristics of the tomato seed lots they studied did not predict the responses very precisely. They found that mean values of Ψ_{min} = –2.4 MPa and T_{min} = 9.1°C (both assumed for simplification to be the same for all seed lot fractions) fitted five out of six lots, and that $\Psi_b(g)$ values lay between –0.6 and –1.1 MPa and T_b between 12 and 13.5°C. But all parameters varied between seed lots, and in some of them T_{min} values were unreliable, suggesting that other factors remained to be accounted for. Seeds with a faster initial germination rate seemed to have lower Ψ_{min} values, meaning that a shorter duration of priming would be required to achieve a given degree of advancement.

It is apparent that, at least in the species and cases studied so far, these hydrothermal time models give only approximate predictions and for practical purposes do not appear by themselves to have the power to reliably prescribe how to prime individual seed lots. Refinements, or alternative models, may improve the outlook in future. For instance, Rowse, McKee, and Higgs (1999) have addressed one imperfection of the hydrothermal priming time model—its prediction that increase in resulting germination rate is proportional to priming time (t_p), whereas in reality at low water potentials the rate tends to reach a maximum that does not increase with further increases in duration—and have avoided the complications of the boundary condition at water potentials close to Ψ_b. They have proposed instead a new water relations model of seed germination using a different variable (the "virtual osmotic potential") to integrate the effect of constant or varying water potentials, without having to distinguish between priming and germinating potentials, and with certain assumptions this model seems to fit the performance of carrot and onion.

Despite their imperfections, the apparent general validity of the hydrotime and hydrothermal priming time models has given further impetus to research directed toward relating temperature and, in particular, water potential thresholds to the forces that drive and/or hold back radicle emergence from seeds, and to their biological determinants.

Completion of Germination

Seed Structural Constraints

Seeds (meaning both botanical true seeds and dry fruits) vary considerably in their internal morphology, as well as their conspicuous external form, which dictates how embryos enlarge and emerge from the seed, and is reflected in the physiological and biochemical germination mechanisms. In his classic comprehensive descriptive survey, Martin (1946) distinguished 12 gross anatomical types within gymnosperm and angiosperm genera, based on structure, organization, and compositional characteristics. Embryos may be (1) tiny, small, or dominant compared to the whole seed; (2) narrow, broad or spatulate, straight, curved, coiled, bent, or folded; and (3) located around the periphery, at the end or in the center, more or less surrounded by or sandwiched between the endosperm or perisperm tissues, which themselves may be living or wholly or partly dead, and have soft or hard textures.

In many seeds of agricultural importance, embryos are relatively unconstrained by surrounding tissues, such as in cereals, *Brassicas,* and many legumes and grasses (excepting cases of coat-imposed dormancy), and in these seeds germination involves only the onset of embryo growth. In many crop seeds, however, confining structures such as endosperm, testa, and pericarp contribute substantial mechanical barriers to embryo growth. In mature celery and carrot seeds, for instance, the embryo is underdeveloped and entirely embedded in the endosperm and must grow about two or three times at its expense by both cell expansion and cell division before visible radicle emergence occurs (Karssen et al., 1990; Gray, Steckel, and Hands, 1990). This embryo growth pattern, along with the variable presence of endogenous inhibitory materials, accounts for the typically slow germination in umbelliferous species. The endosperm also restrains the expansion of many seeds with relatively large embryos and is understood to be a major physiological determinant of the hydrotime threshold water potential, e.g., in tomato and pepper seeds in which the endosperm is a substantial tissue, and in lettuce where it is reduced to a thin but tough envelope layer—situations likened by Welbaum, Bradford, and colleagues (1998) to a rigid outer tire confining the pressurized inner tube of the embryo.

At the biochemical and cytological levels, then, germination and priming mechanisms may differ considerably between seed structural types due to the nature of the embryo and its enclosing tissues. Seed populations naturally conceal substantial differences in structure and physiological state, e.g., due to indeterminate flowering and seed developmental patterns, and often genetic variation as well, which determine the spread in time from one seed to another to complete germination.

Embryo Growth and Endosperm Cell-Wall Degradation

Plant cell growth is commonly accepted to be the result of the accumulation or generation of solutes within cells and osmotic water uptake, which generates sufficient turgor pressure to drive cell wall extension. The primary cell wall, whose main load-bearing component is a network of cellulose microfibrils tethered by hydrogen bonds to xyloglucan chains, is believed to yield in response to turgor by mechanisms mainly involving the activity of β-glucanases, xyloglucan endotransglycosylases (XETs), expansins, and hydroxyl radicals (see Cosgrove, 1999, for access to the large literature in this area). However, it is not yet known what determines the start of cell wall loosening in germinating seeds and the generation of sufficient turgor to drive radicle elongation to complete germination, as well as how these properties are distributed between individual extending cells. For seeds imbibed in water, direct measurements (e.g., in lettuce, melon, tomato) have indicated that embryo turgor per se would seldom be limiting for germination, as embryo osmotic potential values are generally quite negative (Ψ_π less than –2.0 MPa) (Welbaum, Bradford, et al., 1998). Either the radicle cell walls are too rigid or the structures surrounding the radicle prevent it from expanding.

One key mechanism regulating the timing of radicle emergence in seeds with enveloping endosperms is thought to be enzymatic weakening of the restraining tissues. Many species studied have endosperm cell walls rich in β-(1→4)mannan polysaccharides (thought to be galacto-mannans or galacto-glucomannans), and much recent research focus has been placed on the role of endo-ß-mannanase as the most prominent likely candidate for lowering the mechanical restraint by cleaving the polysaccharide backbone chain. Considerable evidence now suggests that β-mannanase activity is indeed involved in radicle emergence in these species, although it is doubtful that the enzyme is the sole determinant of the process. (The enzyme is also involved in mobilizing the endosperm as a food reserve, a process that mainly occurs after germination is completed.) The subject of enzymatic endosperm cell wall weakening during germination has been reviewed in detail by Black (1996), Bewley (1997), and Welbaum, Bradford, and colleages (1998) and will only be outlined here.

The relationship between endosperm cap weakening, the degree of β-mannanase activity and the time to germination after priming has been most thoroughly studied in tomato, which has become a well-studied experimental system due to advantageous features such as its size allowing easy dissection and the availability of PGR response mutants (see references in Toorop, van Aelst, and Hilhorst, 1998; Welbaum, Bradford, et al., 1998). In summary, there are few cases in tomato where germination is observed without at least some β-mannanase activity being present in the micropylar tip endosperm, but high β-mannanase activity does not ensure that germination will occur, or vice versa. Primed and redried seeds develop an internal free space between the embryo and endosperm, and the most rapid germinating seeds have the most extensive free space, observed nondestructively using X-radiography (Downie, Gurusinghe, and Bradford, 1999), and a germination-specific β-mannanase gene is expressed in the micropylar endosperm cap region (Nonogaki, Gee, and Bradford, 2001). However β-mannanase activity has been found to vary startlingly, by at least 1000-fold, even among individual homozygous inbred seeds (Still and Bradford, 1997), and it is necessary to work with individual seeds to get a clear picture of what is happening. A strong correlation has been found during osmopriming between the lowering of the mechanical restraint, the increase in β-mannanase activity, and the appearance of ice crystal-induced porosity, which was taken to reflect cell wall hydrolysis (Toorop, van Aelst, and Hilhorst, 1998). However, the magnitude of these changes in relation to germination advancement differed between priming conditions, even though germination was advanced by all of them. In endosperm caps measured singly, the restraint decreased and the enzyme activity increased during osmopriming at −0.4 MPa, but neither property changed during priming at −1.0 MPa, and two subpopulations could be distinguished at −0.7 MPa with and without changes. (To make matters worse, seed lots of the cultivar used in previous studies showed different absolute patterns of ß-mannanase activity across a similar range of water potentials.) The authors concluded that lowering of the endosperm restraint during priming positively affects the germination rate of primed seeds but is not a prerequisite for rapid germination. The finding that priming tomato seeds at more negative osmotic potentials decreases the base water potential Ψ_b suggests that a different mechanism becomes more prominent under these priming conditions than at higher potentials—the accumulation of solutes in the embryo, possibly, instead of requiring a substantial weakening of the force to puncture the endosperm cap.

Apparently contradictory conclusions have been reached in explaining how priming alleviates thermoinhibition in lettuce. Bradford and Somasco (1994) concluded from the modeling of water relations that the beneficial effects appeared to occur primarily in the embryo, by lowering the embryo yield threshold sufficiently to compensate for the increased endosperm resistance, rather than affecting the surrounding endosperm/pericarp envelope tissues. Sung Yu, Cantliffe, and Nagata (1998) found that the endosperm in thermotolerant cultivars had a lower resistance to puncture than thermosensitive ones, and furthermore priming reduced the initial force necessary to penetrate the seed and endosperm in several genotypes. These authors concluded that, for radicle protrusion to occur, there must first be a decrease in the resistance of the endosperm layer to the turgor pressure of the expanding embryo. Circumstantial support comes from the finding that seeds from thermotolerant lettuce genotypes had higher endo-β-mannanase activity before radicle protrusion at 35°C than thermosensitive ones. Enzyme activities increased during priming of thermosensitive varieties and therefore might be used as an indicator of priming (Nascimento, Cantliffe, and Huber, 2000).

Much remains to be understood about the mechanisms and physiological function of weakening the layers that surround embryos. Even in mannan-rich seeds, for instance, β-mannanase is by no means the only key enzyme or process likely to be involved. In germinating tomato seed, Bradford and colleagues have already identified polygalacturonase, arabinosidase, and expansin genes that are expressed predominately in the endosperm cap and radicle tip regions, and appear to have roles in cell wall modification or tissue weakening (Sitrit et al., 1999; Bradford et al., 2000) along with genes for β-1,3-glucanase and chitinase which are thought to have other functions (Wu et al., 2001). Similarly a β-1,3-glucanase appears to be important in rupturing the tobacco endosperm, which is the limiting step in seed germination (Leubner-Metzger and Meins, 2000). Interestingly, an extensin-like gene is specifically expressed during germination at the micropylar end of the single-cell-layer endosperm of *Arabidopsis,* though its function is not yet known (Dubreucq et al., 2000). The expression of some of these late-germination-stage genes may therefore prove of value as markers for priming.

Embryo Cell Division

As already mentioned, seeds such as carrot and celery have rudimentary immature embryos, which grow before the radicle emerges by a combination of cell division and cell expansion, and both processes also occur, to a lesser degree, during priming at lower water potentials (Karssen et al., 1990). By contrast, in seeds with proportionally large embryos the general case seems to be that cells do not complete mitosis, or even grow appreciably, before the embryo structures emerges from the seed. Priming seems to have no appreciable effect on either cell size or number in these cases, e.g., in leek and onion (Gray, Steckel, and Hands, 1990).

In dry tomato and pepper seeds most embryonic nuclei embryo are in the quiescent presynthetic G1 phase, with 2C amounts of DNA. De Castro and colleagues (2000) have demonstrated by elegant histochemistry that DNA synthesis during tomato germination in water starts in the radicle meristem tips before cell expansion begins, and the activation pattern then spreads toward the cotyledons as progressively more nuclei enter the G2 (4C) phase. At the same time, β-tubulin protein appears, accumulates, and is assembled into the microtubular cytoskeletal networks involved in the mitotic apparatus and establishing the plane of cell division. The situation is similar in sugar beet because in a substantial proportion of seeds a number of the radicle tip cells can enter the G2 phase, depending on environmental conditions during seed development, especially after heavy rainfall, and on seed maturity at time of harvest: on these grounds the G2:G1 ratio has been suggested as an indicator of the physiological status of a seed (Sliwinska, 2000). Here too, however, nuclear DNA contents increase one to two days after the start of imbibition, preceded by the accumulation of β-tubulin (Sliwinska et al., 1999), and DNA replication also occurs during priming (Redfearn and Osborne, 1997). Detailed studies of cell-cycle events in tomato, pepper, and sugar beet using flow cytometry and other techniques have revealed that the beneficial effects of priming are associated with the onset of replicative DNA synthetic processes in radicle meristem nuclei, leading to cells stably arrested in the G2 phase after drying. Priming-induced DNA replication and accumulation of β-tubulin have therefore been suggested as molecular markers for measuring the progression of events preceding radicle protrusion (Lanteri et al., 2000).

However, the relationship of cell cycle activity in radicle meristems prior to emergence and the degree of subsequent priming response has proved to be not at all straightforward. In tomato and pepper, the degree of change in DNA replication was found to vary considerably among similarly osmo-primed seed lots, even of the same cultivar, though all displayed more rapid radicle emergence; in some lots, the frequency of 4C nuclei increased in proportion to the accumulated hydrothermal priming time, while in other lots no increase was detected following priming (Lanteri et al., 1994; Gurusinghe, Cheng, and Bradford, 1999). In extensive studies on a single tomato seed lot, a positive linear relationship was found between the frequency of 4C nuclei (which could be as high as about 30 percent) and the improvement in median subsequent germination time when priming was performed at up to 25°C and above −1.5 MPa; but there was no correlation between the two after priming at higher temperatures or at −2.0 MPa, at which germination rates were improved without generating any increase in 4C signals at all (Özbingöl et al., 1999). Similarly in pepper, a range of osmotic treatments induced different frequencies of radicle tip nuclei to enter the synthetic phase despite producing very similar effects on germination rate (Lanteri et al., 1997). Also there was no consistent relationship between the frequencies and rates in cauliflower seeds after aerated hydration or osmopriming (Powell et al., 2000). In these three species at least, therefore, entry into G2 is not essential for germination advancement, especially in "suboptimal" priming conditions, and 4C:2C ratios are not a general measure of the efficiency of priming. Indeed, considerable increases in germination rates can be observed in the absence of any increases in 4C nuclei.

Lanteri and colleagues (2000) investigated the expression of β-tubulin in the root tips of pepper seeds as a complementary marker for priming. Concentrations of the protein, which increased and declined during seed development and were undetectable in mature dry seeds, accumulated prior to DNA replication after osmopriming for different durations at two water potentials, apparently by de novo synthesis. Approximately the same amount of β-tubulin was observed at the start of nuclear replication in each case, but there were no clear further increases after that point. This observation led the authors to suggest the possibility of using β-tubulin expression as an additional parameter to differentiate the effectiveness of priming treatments that do not induce nuclear replication.

Energy Metabolism

The drying of imbibed seeds may have profound and damaging effects on respiratory metabolism during water removal and after subsequent reimbibition. Using noninvasive photoacoustic techniques, Leprince and colleagues (2000) showed that dehydration induces imbalanced metabolism before membrane integrity is lost in desiccation-sensitive cucumber and pea radicles germinated in water. Compared to desiccation-tolerant organs, CO_2 production was much increased before and during dehydration; acetaldehyde and ethanol also appeared, and their emissions peaked well before the loss of membrane integrity, but these could be significantly reduced when dehydration occurred in 50 percent O_2 instead of air. Acetaldehyde was also found to disturb the phase behavior of phospholipid vesicles measured by infrared spectroscopy, suggesting that it may aggravate membrane damage induced by dehydration. Thus, desiccation tolerance appears to be associated with a balance between down-regulation of metabolism during drying and O_2 availability. Acetaldehyde and ethanol production therefore might prove to be sensitive hazard signals for the overpriming of seed.

In practice, though, priming may not often bring seeds to the point where such drastic changes occur. Very few studies have been conducted specifically relating to primed seeds. Corbineau and colleagues (2000) found that osmopriming raised both the energy charge (EC) and the ATP:ADP (adenosine triphosphate:adenosine diphosphate) ratio in tomato seeds, with the maximal effect obtained in osmopriming atmospheres containing more than 10 percent oxygen; these increases were partially retained after drying and were associated with much more intense respiratory metabolism during the first hours of subsequent imbibition in water.

Early Proteins Reserve Mobilization

A relationship has been found in sugar beet seeds between germination performance after priming and the first stage of mobilization of the 11-S globulin storage protein, which consists of A and B subunits attached by a disulfide bond. Both hydropriming and osmopriming substantially increased the solubilization of the B subunit, still linked to a fragment of the endoproteolytic cleaved A-chain (Job et al., 1997); this accounted for up to approximately 30 percent of the total B-subunit content in mature seeds, which in turn only was decreased by further proteolysis after radicle emergence. Similar behavior has been detected in *Arabidopsis,* in which degradation products of the 12S cruciferin B subunits accumulated during priming (Gallardo et al., 2001). Hydropriming sugar beet also reduced the range of soluble B-subunit content among individual seeds from 160-fold in untreated seeds to only fivefold (Bourgne, Job, and Job, 2000).

Job, Kersulec, and Job (1997) have therefore proposed that B-subunit solubilization can be used as a protein marker for the optimization of priming in sugar beet. Evidence supporting the robustness of this potential marker has come from the observation that, in two types of priming, the range of temperatures and oxygen concentrations that were effective in speeding germination were very similar to those which solubilized B subunits (Capron et al., 2000). A complication arises because soluble B subunits have been detected in "late mature" harvested seeds in some lots, suggesting again that events associated with germination can occur in some circumstances during the last stages of seed development, as already noted in the previous section for G2:G1 ratios (Sliwinska et al., 1999). Job and colleagues (2000) have also proposed that the disappearance of biotinylated protein can be used as marker of overextended osmopriming, which led to a substantial drop in germination under the conditions used.

Desiccation Tolerance and Storability

As a seed lot becomes overprimed, due to too long an exposure at the chosen Ψ and T, more and more individuals typically display damaged primary radicle meristems and, although they may complete germination in the strict sense that the radicles emerge, seedlings may not develop properly (e.g., more abnormal types are detected in the statutory germination test) and so may develop plants with a weakened root system or a damaged shoot tip or die before they emerge from the soil. Even though only a small percent of seeds in a lot may be affected, such losses are unacceptable for high-value seed in commercial practice. It is therefore valuable to determine safe limits for priming to minimize or preferably avoid these handicaps.

Desiccation tolerance in seeds is a fertile research area, with much focus on processes at the termination of seed development, including the physiology of recalcitrant seeds which cannot survive drying after development on the mother plant, recently reviewed by and Pammenter and Berjak (1999) and in Chapters 9 and 10 of this book. Vertucci and Farrant (1995) suggest that seeds might suffer different types of injury when water is withdrawn: mechanical damage, due to the reduction in cell volume; metabolic damage, due to failures in the coordinated regulation at intermediate water contents and the failure of protective antioxidant systems; and subcellular structural damage, due to removal of water intimately associated with the surface of macromolecular structures such as membranes. It is believed that desiccation-tolerant seed tissues require the interplay of a multifactorial suite of protective mechanisms to prevent these forms of damage and/or permit their repair on rehydration. Major roles are thought to be played by the composition of the cytoplasm (including soluble sugars), the presence of putatively protective molecules including late embryogenesis abundant (LEA) proteins, the efficient operation of antioxidant systems to protect against free radicals produced during dehydration that would otherwise lead to oxidative damage to membranes, and the presence and operation of repair mechanisms during rehydration (e.g., see Leprince, Hendry, and McKersie, 1993). The absence or ineffective expression of one or more of these mechanisms could determine desiccation sensitivity in primed seeds and could therefore serve as markers for safe priming.

Cell Stabilization: Sugars and LEA Proteins

It has been known for some years that sucrose and oligosaccharides, usually of the galactosyl-sucrose family (raffinose, etc.), occur in relatively large amounts in seeds and that their concentrations appear to correlate with the longevity of seeds and with the acquisition and loss of desiccation tolerance (Bernal-Lugo and Leopold, 1995; Koster and Leopold, 1988, Corbineau et al., 2000). During germination and priming the content and composition of intracellular soluble carbohydrates changes; for example, the oligosaccharide:sucrose ratio is reduced in primed pea and cauliflower (Hoekstra et al., 1994; Buitink, Hoekstra, and Hemminga, 1999). Soluble sugars are therefore widely understood to be important components in stabilizing cellular integrity during dehydration—though the mechanisms by which they do so are still a matter for research and debate—and are thus prominent candidates to consider as a contributing factor to the performance of primed seed after drying.

Cytoplasm. In air-dry storage conditions the cytoplasm of seeds and other anhydrobiotic organisms enter an aqueous glassy state; i.e., it becomes a solidlike liquid well below its normal freezing (crystallizing) point. The extremely high viscosity and low molecular mobility of the cytoplasm under these conditions are believed to be major factors in imparting longevity to anhydrobiotes, by restricting the rate of detrimental reactions associated with aging and stabilizing macromolecules, such as membranes, proteins, and DNA. It is thought, for instance, that direct measurements of molecular mobility might eventually lead to realistic predictions of longevity of seeds, such as those stored at low temperatures (Buitink, Hoekstra, and Hemminga, 2000). At one time it was suspected that supersaturated solutions of soluble sugars were the primary elements responsible for the glassy state in seeds, but recent physicochemical studies have revealed that the situation is more complex, and glycosides, larger carbohydrates (e.g., maltodextrin), and proteins also contribute to the property (Leopold, Sun, and Bernal-Lugo, 1994).

Membranes. Removal of water from membrane surfaces in desiccation-sensitive plant cells at low water contents causes the liquid crystalline lipid bilayer to convert into gel phase domains and, although they are readily reversed on rehydration, these transformations are associated with symptoms of injury and lethal effects, including extensive solute leakage and reorganization of the membrane protein complex. In desiccation-tolerant tissues, in contrast, according to the *water replacement hypothesis* (Crowe, Hoekstra, and Crowe, 1992), sugars and sugar alcohols replace the structural water that is normally hydrogen bonded to the membrane surfaces and macromolecules, and thereby maintain the correct of polar head group spacing of the membrane lipid and provide the hydrophilic interactions necessary to maintain the liquid-crystalline phase. However, a recent review (Hoekstra et al., 1997) has concluded that sugars may not be particularly effective in vivo in this way and proposes instead that the bilayer is stabilized by the migration of amphipathic molecules (e.g., flavinols) into membranes as water is lost, which substantially lowers the water content at which membrane lipids undergo the liquid-to-gel phase change. Bryant, Koster, and Wolfe (2001) have argued that the presence in the cytoplasm of small solutes that can form glasses, such as sugars, could limit the close approach of membranes and thereby diminish the physical stresses that could otherwise cause lipid fluid-to-gel phase transitions to occur during dehydration.

Whatever the stabilization mechanisms and the role of sugars in them, changed macroscopic physical properties of intracellular glasses do not seem to offer an explanation by themselves for the reduced longevity of primed seed. Differential scanning calorimetry revealed no changes in glass transition temperature (i.e., when the matrix "melts") associated with the osmopriming of pea, *Impatiens*, and pepper seeds. Nor was there any significant difference in molecular mobility, as determined by electron paramagnetic resonance spectroscopy of a spin probe introduced into the cytoplasm (Buitink, Hoekstra, and Hemminga, 1999; Buitink, Hemminga, and Hoekstra, 2000). Nevertheless, a circumstantial but perhaps relevant parallel can be drawn with the behavior of abscisic acid-pretreated carrot somatic embryos, which survive slow dehydration much better than fast, as primed seeds tend to do. Using in situ infrared microspectroscopy, Wolkers and colleagues (1999) found that fast drying resulted in much weaker average strength of hydrogen bonding at room temperature, a less clearly defined glassy matrix, apparently "less tight" molecular packing, and greater extent of protein denaturation than slow drying. LEA transcripts were also expressed after slow drying, suggesting a role in conferring stability within the glassy matrix.

Sugar composition is not consistently related to storage life performance of primed seeds. Gurusinghe and Bradford (2001) found that the short postpriming heat treatments that substantially restored longevity to hydroprimed tomato seeds only slightly changed sucrose and oligosaccharide (planteose) content. The suggestion here instead was that heat-shock proteins might be involved in the response, supported by the observation that the effectiveness of the heat treatment was correlated with expression of a constitutively expressed lumenal stress protein (BiP), a highly conserved member of the hsp 70 family associated with the endoplasmic reticulum (Gurusinghe, Powell, and Bradford, 2002).

Free Radicals and Oxidative Damage

It is widely recognized that toxic active oxygen species (AOS), e.g., superoxide radicals and H_2O_2, are produced as products of mitochondrial respiration and glyoxysomal lipid degradation, and are removed within plant cells by antioxidative enzymes, e.g., superoxide dismutase (SOD) and catalase (CAT), etc., which scavenge free radicals before they disrupt biomolecules.

One major harmful effect of free radical reactions is believed to be the accumulation of free fatty acids and other lipid-degradation products in membrane bilayers, which increase the lipid phase transition temperature and cause the irreversible formation of gel phase domains, which are lethal when the cell is rehydrated (McKersie, 1991). Bailly and colleagues (2000) found in sunflower seeds that malondialdehyde (MDA) content—a measure of the degree of lipid degradation—remained unchanged during osmopriming (–2.0 MPa), while activities of SOD and CAT increased strongly. Furthermore, although MDA concentrations increased markedly after drying, they declined again during six hours from the start of reimbibition, compared to an increase in control imbibed unprimed seed. This supports the idea that the enzymatic antioxidant defense system operates efficiently in sunflower seeds to scavenge AOS produced during osmopriming and that the system survives in an enhanced state in dried primed seeds to operate during the first hours after subsequent reimbibition, though it cannot cope with the effects of the intervening dehydration stage itself. The authors also suggest that the CAT isoenzyme pattern may be a good marker of whether sunflower seeds have been primed under their conditions.

Another possibility that might be investigated in primed seeds is whether reducing sugars, such as glucose and fructose, which decrease in embryos as seed matures and acquire desiccation tolerance, and increase following germination, will cause a reaction with metal ions such as Fe to generate oxidizing agents through the Maillard and Amadori rearrangement reactions of carbonyl groups with free amino groups in proteins, which can cause nonenzymatic modification during seed aging (Murthy and Sun, 2000).

DNA Damage and Repair

The ability to repair DNA damage after seeds are rehydrated so that a transcriptionally competent genome is assured has been proposed as an essential element of the suite of mechanisms contributing to survival of dehydrated orthodox seeds (Boubriak et al., 1997). Repair of lesions of the genome that occurs during drying and storage has been shown to be among the earliest events occurring when dry, orthodox seeds are rehydrated, and changes during priming might impair this ability. van Pijlen and colleagues (1996) speculated that the adverse storage performance of an osmoprimed (compared to a humidified or hydroprimed) tomato seed treatment could be explained by the fact that more nuclei in the osmoprimed embryonic root tips had replicated their DNA, and Lanteri and colleagues (1997) and Powell and colleagues (2000) have made similar observations in osmoprimed pepper and hydroprimed cauliflower. Perhaps the detrimental effects of overpriming are associated with a decreased ability to repair DNA as cells enter the S phase of the cell cycle and progress toward the G2 phase after subsequent rehydration.

In another hypothesis that remains to be tested, Boubriak and colleagues (2000) have pointed out the danger that, under certain low water potential regimes during the priming or subsequent drying of imbibed seeds, DNA might be subjected to irreparable enzymatic cleavage to nucleosome oligomers, as can occur during the accelerated aging of rye grains.

Conversely, during the process of the osmopriming itself, DNA and its synthesizing systems can be repaired and damaged rRNA replaced (Bray, 1995). McDonald, in Chapter 9, discusses the ability of priming to overcome such low vigor effects that result from seed deterioration in storage.

ECOLOGICAL ASPECTS OF SEED HYDRATION

Seed banks in the soil naturally experience a variable environment of temperature, water potential, oxygen, and other factors, both daily and seasonally. Seed burial of numerous species induces physiological changes that can improve the chances of establishment, allowing the seeds to respond to conditions favorable for germination and growth, through the development or breaking of specific types of dormancy, such as by alternate soil wetting and drying (Allen and Meyer, 1998; Baskin and Baskin, 1998). However, when a seed population is emerging from dormancy, those individuals that can germinate will have, in hydrotime language, so high a Ψ_b that their progress will be slow and unlikely to be completed during a short window of opportunity. Instead they will accumulate hydrothermal time and survive to germinate more rapidly at the next chance (Bradford, 1995).

The ecological significance of priming existing in nature to increase the chances of successful seedling establishment from the soil seed bank of different plant communities remains to be studied, but González-Zertuche and colleagues (2001) recently obtained circumstantial evidence that seeds of *Wigandia urens*, a Mexican shrub, do undergo changes while buried in their natural habitat that are similar to those seen after osmotic priming in the laboratory. Both treatments induced physiological changes that were primarily expressed in the heterogeneous burial environment, including the synthesis of heat-soluble proteins, of similar molecular size to LEA proteins. It is entirely plausible that the ability of seeds to survive interrupted germination and be primed is an expression of processes that have evolved in the soil seed bank to prepare some species for a rapid, uniform, and successful colonization of their environments.

CONCLUSIONS AND FUTURE DIRECTIONS

As seed becomes increasingly valuable, by the addition of input and output traits through genetic engineering and other breeding techniques, and due to increasing economic pressures for efficient and environmentally friendly crop production systems, the incentive to protect and ensure germination will increase. One trend already underway is that techniques of pelleting, coating, and priming are being considered commercially for large-volume crops, for which they were not previously cost-effective, but which demand new larger-scale process engineering approaches.

At the same time the need to understand and improve seed physiological quality continues to grow. Though the 1990s were a period of considerable advance in our understanding of the processes that transform germination into the start of growth and the loss of desiccation tolerance, our knowledge is still based on relatively few species—most extensively, tomato. As far as hydration treatments are concerned, it seems there may be no one explanation and no simple universal indicators of the processes operating, partly since in many respects germination patterns are species specific, as Bray (1995) observed. Bearing in mind too that priming of a given species can be conducted using a range of hydration and drying procedures and a spectrum of water potentials and durations, it is perhaps not surprising that metabolic and cellular events have been found to differ. Also, as Welbaum, Bradford, and colleagues (1998) pointed out, it has been valuable to recognize the problems posed by pooling large numbers of seeds for biochemical assays, which can obscure important seed-to-seed physiological variation, as noted, for example, by Still and Bradford (1997) and Bourgne, Job, and Job (2000). From the practical seed industry point of view, this understanding can provide experimental tools to help in the development of new hydration and drying procedures. What has been perhaps more disappointing—and setting aside its academic interest—in a production situation in which decisions have to be made reliably and often rapidly is the picture that often emerges of differences between seed lots in the magnitude or even the existence of changes during treatment (e.g., in cell cycle events or the appearance of cell wall degrading enzymes), at least in the much-studied tomato and pepper. As far as markers for priming and desiccation tolerance are concerned, this tends to reduce the hope that there might be tests with the precision required to be in a position to decide between alternative priming or drying conditions or procedures before treatment starts. There is a better prospect of tests which give qualitative indications that a particular dry seed lot has already been primed, but so far these tests by themselves probably cannot indicate the quantitative degree of priming without being combined with direct measurement of seed performance in comparison by some form of germination test.

Looking to the future, molecular tools are dramatically enhancing our knowledge of the biochemical and regulatory pathways underlying the complex physiological and developmental process of germination. Genomic and transgenic approaches can now establish the timing and identity of specifically activated genes and their tissue expression patterns, or the consequences of specific gene inactivation, and provide insights into their functions. The first fruits of this work are beginning to be seen in seed science, such as the identification in germinating tomato seeds of genes associated with cell wall weakening enzymes (mentioned earlier) and connected to the initiation of embryo growth and stress adaptation (see Bradford et al., 2000). Gallardo and colleagues (2001) conducted a broad proteomic analysis of changes occurring during germination and after drying of the model species, *Arabidopsis thaliana,* whose complete genome is now known. Using protein analysis in combination with sequence databases, these authors have identified, among many other changes, a total of eight unique proteins that accumulated during a hydropriming and/or an osmopriming treatment. One of the many stimulating outcomes from using the potential wealth of this type of information in classical physiological approaches could be the identification of diagnostic markers, which might be routinely employed to interrogate gene expression, e.g., using DNA array or ELISA technology, to characterize seed quality and to develop and perhaps optimize priming enhancement procedures.

REFERENCES

Ahm, H.P. (2000). A method of germinating seeds or the like growth-suited parts of a plant contained in germinating units, as well as a germinating box and a germinating assembly for use when carrying out the method. World Patent No. 0000006.

Aksenova, L.A., Dunaeva, M.V., Zak, E.A., Osipov, Y.F., and Klyachko, N.L. (1994). The effect of Tween 80 on seed germination in winter wheat cultivars differing in drought resistance. *Russian Journal of Plant Physiology* 41: 557-559.

Allen, P.S. and Meyer, S.E. (1998). Ecological aspects of seed dormancy loss. *Seed Science Research* 8: 183-191.

Bailly, C., Benamar, A., Corbineau, F., and Côme, D. (2000). Antioxidant systems in sunflower (*Helianthus annuus* L.) seeds as affected by priming. *Seed Science Research* 10: 35-42.

Baskin, C.C. and Baskin, J.M. (1998). *Seeds: Ecology, Biogeography, and Evolution of Dormancy and Germination.* San Diego, CA: Academic Press.

Bernal-Lugo, I. and Leopold, A.C. (1995). Seed stability during storage: Raffinose content and seed glassy state. *Seed Science Research* 5: 75-80.

Bewley, J.D. (1997). Breaking down the walls—A role for endo-β-mannanase in release from seed dormancy? *Trends in Plant Science* 2: 464-469.

Bewley, J.D. and Black, M. (1994). *Seeds: Physiology of Development and Germination.* New York: Plenum Press.

Black, M. (1996). Liberating the radicle: A case for softening-up. *Seed Science Research* 6: 39-42.

Boubriak, I., Kargiolaki, H., Lyne, L., and Osborne, D.J. (1997). The requirement for DNA repair in desiccation tolerance of germinating embryos. *Seed Science Research* 7: 97-105.

Boubriak, I., Naumenko, V., Lyne, L., and Osborne, D.J. (2000). Loss of viability in rye embryos at different levels of hydration: Senescence with apoptotic nucleosome cleavage or death with random DNA fragmentation. In Black, M., Bradford, K.J., and Vazquez-Ramos, J. (Eds.), *Seed Biology: Advances and Applications. Proceedings of the Sixth International Workshop on Seeds, Merida, Mexico, 1999* (pp. 205-214). Wallingford, UK: CABI Publishing.

Bourgne, S., Job, C., and Job, D. (2000). Sugarbeet seed priming: Solubilization of the basic subunit of 11-S globulin in individual seeds. *Seed Science Research* 10: 153-161.

Bradford, K.J. (1995). Water relations in seed germination. In Kigel, J. and Galili, G. (Eds.), *Seed Development and Germination* (pp. 351-396). New York: Marcel Dekker.

Bradford, K.J., Chen, F., Cooley, M.B., Dahal, P., Downie, B., Fukunaga, K.K., Gee, O.H., Gurusinghe, S., Mella, R.A., Nonogaki, H., et al. (2000). Gene expression prior to radicle emergence in imbibed tomato seeds. In Black, M., Bradford, K.J., and Vazquez-Ramos, J. (Eds.), *Seed Biology: Advances and Applications. Proceedings of the Sixth International Workshop on Seeds, Merida, Mexico, 1999* (pp. 231-251). Wallingford, UK: CABI Publishing.

Bradford, K.J. and Haigh, A.M. (1994). Relationship between accumulated hydrothermal time during seed priming and subsequent seed germination rates. *Seed Science Research* 4: 63-69.

Bradford, K.J. and Somasco, O.A. (1994). Water relations of lettuce seed thermoinhibition: I. Priming and endosperm effects on base water potential. *Seed Science Research* 4: 1-10.

Bradford, K.J., Steiner, J.J., and Trawatha, S.E. (1990). Seed priming influence on germination and emergence of pepper seed lots. *Crop Science* 30: 718-721.

Brandl, F. (2001). Seed treatment technologies: Evolving to achieve crop genetic potential. In Biddle, A. (Ed.), *Seed Treatment: Challenges and Opportunities Number 76* (pp. 3–18). Farnham, UK: British Crop Protection Council.

Bray, C.M. (1995). Biochemical processes during the osmoconditioning of seeds. In Kigel, J. and Galili, G. (Eds.), *Seed Development and Germination* (pp. 767-789). New York: Marcel Dekker.

Brocklehurst, P.A. and Dearman, J. (1983). Interactions between seed priming treatments and nine lots of carrot, celery and onion: I. Laboratory germination. *Annals of Applied Biology* 102: 577-584.

Brocklehurst, P.A., Rankin, W.E.F., and Thomas, T.H. (1983). Stimulation of celery seed germination and seedling growth with combined ethephon, gibberellin and polyethylene glycol seed treatments. *Plant Growth Regulation* 1: 195-202.

Bruggink, G.T., Ooms, J.J.J., and van der Toorn, P. (1999). Induction of longevity in primed seeds. *Seed Science Research* 9: 49-53.

Bruggink, G.T. and van der Toorn, P. (1995). Induction of desiccation tolerance in germinated seeds. *Seed Science Research* 5: 1-4.

Bruggink, G.T. and van der Toorn, P. (1996). Pregerminated seeds. U.S. Patent Nos. 5522907 and 5585536.

Bryant, G., Koster, K.L., and Wolfe, J. (2001). Membrane behavior in seeds and other systems at low water content: The various effects of solutes. *Seed Science Research* 11: 17-25.

Buitink, J., Hemminga, M.A., and Hoekstra, F.A. (2000). Is there a role for oligosaccharides in seed longevity? An assessment of intracellular glass stability. *Plant Physiology* 122: 1217-1224.

Buitink, J., Hoekstra, F.A., and Hemminga, M.A. (1999). A critical assessment of the role of oligosaccharides in intracellular glass stability. In Black, M., Bradford, K.J., and Vazquez-Ramos, J. (Eds.), *Seed Biology: Advances and Applications. Proceedings of the Sixth International Workshop on Seeds, Merida, Mexico, 1999* (pp. 461-466). Wallingford, UK: CABI Publishing.

Buitink, J., Hoekstra, F.A., and Hemminga, M.A. (2000). Molecular mobility in the cytoplasm of lettuce radicles correlates with longevity. *Seed Science Research* 10: 285-292.

Bujalski, W. and Nienow, A.W. (1991). Large-scale osmotic priming of onion seeds: A comparison of different strategies for oxygenation. *Scientia Horticulturae* 46: 13-24.

Bujalski, W., Nienow, A.W., Petch, G.M., Drew, R.L.K., and Maude, R.B. (1991). The process engineering of leek seeds: A feasibility study. *Seed Science and Technology* 20: 129-139.

Cano, E.A., Bolarín, M.C., Pérez Alfocea, F., and Caro, M. (1991). Effect of NaCl priming on increased salt tolerance in tomato. *Journal of Horticultural Science* 66: 621-628.

Capron, I., Corbineau, F., Dacher, F., Job, C., Côme, D., and Job, D. (2000). Sugarbeet seed priming: Effects of priming conditions on germination, solubilization of 11-S globulin and accumulation of LEA proteins. *Seed Science Research* 10: 243-254.

Carbonell, M.V. Martinez, E., and Amaya, J.M. (2000). Stimulation of germination in rice (*Oryza sativa* L.) by a static magnetic field. *Electro- and Magnetobiology* 19: 121-128.

Chachalis, D. and Smith, M.L. (2000). Imbibition behavior of soybean [*Glycine max* (L.) Merrill] accessions with different testa characteristics. *Seed Science and Technology* 28: 321-331.

Cheng, Z.Y. and Bradford, K.J. (1999). Hydrothermal time analysis of tomato seed germination responses to priming treatments. *Journal of Experimental Botany* 50: 89-99.

Corbineau, F., Picard, M.A., Fougereux, J.-A., Ladonne, F., and Côme, D. (2000). Effects of dehydration conditions on desiccation tolerance of developing pea seeds as related to oligosaccharide content and cell membrane properties. *Seed Science Research* 10: 329-339.

Cosgrove, D.J. (1999). Enzymes and other agents that enhance cell wall extensibility. *Annual Reviews in Plant Physiology and Plant Molecular Biology* 50: 391-417.

Crowe, J.H., Hoekstra, F.A., and Crowe, L.M. (1992). Anhydrobiosis. *Annual Review of Physiology* 54: 579-599.

Dahal, P. and Bradford, K.J. (1990). Effects of priming and endosperm integrity on seed germination rates of tomato genotypes: II. Germination at reduced water potential. *Journal of Experimental Botany* 41: 1441-1453.

Dahal, P., Bradford, K.J., and Haigh, A.M. (1993). The concept of hydrothermal time in seed germination and priming. In Côme, D. and Corbineau, F. (Eds.), *Basic and Applied Aspects of Seed Biology: Proceedings of the Fourth International Workshop on Seeds, Angers, France, 20-24 July 1992* (pp. 1009-1014). Paris: ASFIS.

Dahal, P., Bradford, K.J., and Jones, R.A. (1990). Effects of priming and endosperm integrity on seed germination rates of tomato genotypes: I. Germination at suboptimal temperature. *Journal of Experimental Botany* 41: 1431-1439.

Davidson, M.W. (1981). Improvements in or relating to grain germination processes and the like. U.K. Patent No. 1583148.

De Boer, J. and Boukens, M.A.N. (1999). Method and apparatus for priming seed. World Patent No. 9933331.

De Castro, R.D., van Lammeren, A.A.M., Groot, S.P.C., Bino, R.J., and Hilhorst, H.W.M. (2000). Cell division and subsequent radicle protrusion in tomato seeds are inhibited by osmotic stress but DNA synthesis and formation of microtubular cytoskeleton are not. *Plant Physiology* 122: 327-336.

De Souza Torres, A., Porras Leon, E., and Casate Fernandez, R. (1999). Efecto del tratamiento magnetico de semillas de tomate (*Lycopersicon esculentum* Mill) sobre la germinacion y el crecimiento de las plantulas. *Investigacion Agraria, Produccion y Proteccion Vegetales* 14: 437-444.

Downie, B., Gurusinghe, S., and Bradford, K.J. (1999). Internal anatomy of individual tomato seeds: Relationship to abscisic acid and germination physiology. *Seed Science Research* 9: 117-128.

Dubreucq, B., Berger, N., Vincent, E., Boisson, M., Pelletier, G., Caboche, M., and Lepiniec, L. (2000). The *Arabidopsis* AtEPR1 extensin-like gene is specifically expressed in endosperm during germination. *The Plant Journal* 23: 643-652.

Eastin, J.A. (1990). Solid matrix priming of seeds. U.S. Patent No. 4912874

Far, J.J., Upadhyaya, S.K., and Shafii, S. (1994). Development and field evaluation of a hydropneumatic planter for primed vegetable seeds. *Transactions of the American Society of Agricultural Engineers* 37: 1069-1075.

Finch-Savage, W.E. (1991). Development of bulk priming/plant growth regulator seed treatments and their effect on the seedling establishment of four bedding plant species. *Seed Science and Technology* 19: 477-485.

Finch-Savage, W.E. and McQuistan, C.I. (1989). The use of abscisic acid to synchronize carrot seed germination prior to fluid drilling. *Annals of Botany* 65: 195-199.

Forsberg, G. (2001). Heat sanitation of cereal seeds with a new, efficient, cheap and environmentally friendly method. In Biddle, A. (Ed.), *Seed Treatment: Challenges and Opportunities*, Number 76 (pp. 69-72). Farnham, UK: British Crop Protection Council.

Fujikura, Y., Kraak, H.L., Basra, A.S., and Karssen, C.M. (1993). Hydropriming, a simple and inexpensive priming method. *Seed Science and Technology* 21: 639-642.

Gallardo, K., Job, C., Groot, S.P.C., Puype, M., Demol, H., Vandekerckhove, J., and Job, D. (2001). Proteomic analysis of *Arabidopsis* seed germination and priming. *Plant Physiology* 126: 835-848.

González-Zertuche, L., Vázquez-Yanes, C., Gamboa, A., Sánchez-Coronado, M.E., Aguilera, P., and Orozco-Segovia, A. (2001). Natural priming of *Wigandia urens* seeds during burial: Effects on germination, growth and protein expression. *Seed Science Research* 11: 27-34.

Gray, D., Steckel, J.R.A., and Hands, L.J. (1990). Responses of vegetable seeds to controlled hydration. *Annals of Botany* 66: 227-235.

Gummerson, R.J. (1986). The effect of constant temperature and osmotic potential on the germination of sugar beet. *Journal of Experimental Botany* 37: 729-741.

Gurusinghe, S. and Bradford, K.J. (2001). Galactosyl-sucrose oligosaccharides and potential longevity of primed seeds. *Seed Science Research* 11: 121-133.

Gurusinghe S., Cheng, Z.Y., and Bradford, K.J. (1999). Cell cycle activity during seed priming is not essential for germination advancement in tomato. *Journal of Experimental Botany* 50: 101-106.

Gurusinghe, S., Powell, A.L.T., and Bradford, K.J. (2002). Enhanced expression of BiP is associated with treatments that extend longevity of primed tomato seeds. *Journal of the American Society for Horticultural Science* 127(4): 528-534.

Halmer, P. (2000). Commercial seed treatment technology. In Black, M. and Bewley, J.D. (Eds.), *Seed Technology and Its Biological Basis* (pp. 257-286). Sheffield, UK: Sheffield Academic Press.

Harris D., Joshi, A., Khan, P.A., Gothkar, P., and Sodhi, P.S. (1999). On-farm seed priming in semi-arid agriculture: Development and evaluation in maize, rice and chickpea in India using participatory methods. *Experimental Agriculture* 35: 15-29.

Hegarty, T.W. (1978). The physiology of seed hydration and dehydration, and the relation between water stress and the control of germination: A review. *Plant Cell and Environment* 1: 101-119.

Henderson, J.C. and Hensley, D.L. (1987). Effect of a hydrophilic gel on seed germination of three tree species. *HortScience* 22: 450-452.

Heydecker, W. and Coolbear, P. (1977). Seed treatments for improved performance—Survey and attempted prognosis. *Seed Science and Technology* 3: 353-425.

Heydecker, W., Higgins, J., and Turner, Y.J. (1975). Invigoration of seeds? *Seed Science and Technology* 3: 881-888.

Hill, H.J. (1999). Recent developments in seed technology. *Journal of New Seeds* 1: 105-112.

Hoekstra, F.A., Haigh, A.M., Tetteroo, F.A.A., and van Roekel, T. (1994). Changes in soluble sugars in relation to desiccation tolerance in cauliflower seeds. *Seed Science Research* 4: 143-147.

Hoekstra, F.A., Wolkers, W.F., Buitink, J., Golovina, E.A., Crowe, J.H., and Crowe, L.M. (1997). Membrane stabilization in the dry state. *Comparative Biochemistry and Physiology* 117A: 335–341.

Holloway, D.H. (1999). Seed germination medium. U.K. Patent No. 2330998.

Jalink, H., van der Schoor, R., Frandas, A., van Pijlen, J.G., and Bino, R.J. (1998). Chlorophyll fluorescence of *Brassica oleracea* seeds as a non-destructive marker for seed maturity and seed performance. *Seed Science Research* 8: 437-443.

Job, C., Kersulec, A., Ravasio, L., Chareyre, S., Pépin, R., and Job, D. (1997). The solubilization of the basic subunit of sugarbeet seed 11-S globulin during priming and early germination. *Seed Science Research* 7: 225-243.

Job, D., Capron, I., Job, C., Dacher, F., Corbineau, F., and Côme, D. (2000). Identification of germination-specific protein markers and their use in seed priming technology. In Black, M., Bradford, K.J., and Vazquez-Ramos, J. (Eds.), *Seed Biology: Advances and Applications. Proceedings of the Sixth International Workshop on Seeds, Merida, Mexico, 1999* (pp. 449-459). Wallingford, UK: CABI Publishing.

Job, D., Kersulec, A., and Job, C. (1997). Globulin protein 11S, usable as a seed impregnation marker during plant seed germination. World Patent No. 9743418.

Jum Soon, K., Young Whan, C., and Jeoung Lai, C. (1998). Effect of dehydration conditions on the germination and membrane integrity of tomato seeds after priming. *Journal of the Korean Society for Horticultural Science* 39: 250-255.

Karssen, C.M., Haigh, A., Toorn, P., and Weges, R. (1990). Physiological mechanisms involved in seed priming. In Taylorson, R.B. (Ed.), *Recent Advances in the Development and Germination of Seeds* (pp. 269-280). New York and London: Plenum Press.

Khan, A.A. (1992). Preplant physiological seed conditioning. *Horticultural Reviews* 13: 131-181.

Klein, J.D. and Hebbe, Y. (1994). Growth of tomato plants following short-term high-temperature seed priming with calcium chloride. *Seed Science and Technology* 22: 223-230.

Kornarzynski, K. and Pietruszewski, S. (1999). Effect of the stationary magnetic field on the germination of wheat grain. *International Agrophysics* 13: 457-461.

Koster, K.L. and Leopold, A.C. (1988). Sugars and desiccation tolerance in seeds. *Plant Physiology* 88: 829-832.

Lanteri, S., Belletti, P., Marzach, C., Nada, E., Quagliotti, L., and Bino, R.J. (1997). Priming-induced nuclear replication activity in pepper (*Capsicum annuum* L.) seeds: Effect on germination and storability. In Ellis, R.H., Black, M., Murdoch, A.J., and Hong, T.D. (Eds.), *Basic and Applied Aspects of Seed Biology: Proceedings of the Fifth International Workshop on Seeds, Reading* (pp. 451-459). Dordrecht: the Netherlands, Kluwer Academic Publishers.

Lanteri, S., Portis, E., Bergervoet, H.W., and Groot, S.P.C. (2000). Molecular markers for the priming of pepper seeds (*Capsicum annuum* L.). *Journal of Horticultural Science and Biotechnology* 75: 607-611.

Lanteri, S., Saracco, F., Kraak, H.L., and Bino, R.J. (1994). The effects of priming on nuclear replication activity and germination of pepper (*Capsicum annuum*) and tomato (*Lycopersicon esculentum*) seeds. *Seed Science Research* 4: 81-87.

Leopold, A.C., Sun, W.Q., and Bernal-Lugo, I. (1994). The glassy state in seeds: Analysis and function. *Seed Science Research* 4: 267-274.

Leprince, O., Harren, F.J.M., Buitink, J., Alberda, M., and Hoekstra, F.A. (2000). Metabolic dysfunction and unabated respiration precede the loss of membrane integrity during dehydration of germinating radicles. *Plant Physiology* 122: 597-608.

Leprince, O., Hendry, G.A.F., and McKersie, B.D. (1993). The mechanisms of desiccation tolerance in developing seeds. *Seed Science Research* 3: 231-246.

Leubner-Metzger, G. and Meins, F., Jr (2000). Sense transformation reveals a novel role for class I β-1,3-glucanase in tobacco seed germination. *The Plant Journal* 23: 215-221.

Martin, A.C. (1946). The comparative internal morphology of seeds. *The American Midland Naturalist* 36: 513-660.

Massawe, F.J., Collinson, S.T., Roberts, J.A., and Azam Ali, S.N. (1999). Effect of pre-sowing hydration on germination, emergence and early seedling growth of bambara groundnut (*Vigna subterranea* L. Verdc). *Seed Science and Technology* 27: 893-905.

Maude, R.B. (1996). *Seedborne Diseases and Their Control: Principles and Practices*. Wallingford, UK: CABI Publishing.

Maude, R.B., Drew, R.L.K., Gray, D., Petch, G.M., Bujalski, W., and Nienow, A.W. (1992). Strategies for control of seedborne *Alternaria dauci* (leaf blight) of carrots in priming and process engineering system. *Plant Pathology* 41: 204-214.

McDonald, M. (2000). Seed priming. In Black, M. and Bewley, J.D. (Eds.), *Seed Technology and Its Biological Basis* (pp 287-325). Sheffield, UK: Sheffield Academic Press.

McKersie, B.D. (1991). The role of oxygen free radicals in mediating freezing and desiccation stress in plants. In Pell, E. and Steffer, K. (Eds.), *Active Oxygen/Oxidative Stress and Plant Metabolism* (pp. 107-118). Rockville, MD: American Society of Plant Physiologists.

McQuilken, M.P., Halmer, P., and Rhodes, D.J. (1998). Application of microorganisms to seeds. In Burges, H.D. (Ed.), *Formulation of Microbial Biopesticides, Beneficial Microorganisms and Nematodes* (pp. 255-285). Dordrecht, the Netherlands: Kluwer Academic Publishers.

Meikle, R.A.R. and Smith, D. (2000). Seed germination system. U.S. Patent No. 6070358.

Murthy, U.M.N. and Sun, W.Q. (2000). Protein modification by Amadori and Maillard reactions during seed storage: Roles of sugar hydrolysis and lipid peroxidation. *Journal of Experimental Botany* 51: 1221-1228.

Nascimento, W.M., Cantliffe, D.J., and Huber, D.J. (2000). Endo-beta-mannanase activity during lettuce seed germination at high temperature conditions. *Acta Horticulturae* 517: 107-112.

Nascimento, W.M. and West, S.H. (1998). Microorganism growth during muskmelon seed priming. *Seed Science and Technology* 26: 531-534.

Ni, B.-R. (2001). Alleviation of seed imbibitional chilling injury using polymer film coating. In Biddle, A. (Ed.), *Seed Treatment: Challenges and Opportunities, Number 76* (pp. 73–80). Farnham, UK: British Crop Protection Council.

Nonogaki, H., Gee, O.H., and Bradford, K.J. (2001). A germination-specific endo-β-mannanase gene is expressed in the micropylar endosperm cap of tomato seeds. *Plant Physiology* 123: 1235-1246.

Ollerenshaw, J.H. (1988). Calcium peroxide as a seed coating to alleviate stresses on crop plants. In Martin, T.J. (Ed.), *Application to Seeds and Soil* (pp. 285-292). British Crop Protection Council Monograph No. 39. Farnham, UK: British Crop Protection Council.

Özbingöl, N., Corbineau, F., Groot, S.P.C., Bino, R.J., and Côme, D. (1999). Activation of the cell cycle in tomato (*Lycopersicon esculentum* Mill.) seeds during osmoconditioning as related to temperature and oxygen. *Annals of Botany* 84: 245-251.

Pammenter, N.W. and Berjak, P. (1999). A review of recalcitrant seed physiology in relation to desiccation-tolerance mechanisms. *Seed Science Research* 9: 13-37.

Parera, C.A. and Cantliffe, D.J. (1994a). Dehydration rate after solid matrix priming alters seed performance of shrunken-2 corn. *Journal of the American Society for Horticultural Science* 119: 629-635.

Parera, C.A. and Cantliffe, D.J. (1994b). Presowing seed priming. *Horticultural Reviews* 16: 109-141.

Petch, G.M., Maude, R.B., Bujalski, W., and Nienow, A.W. (1991). The effects of re-use of polyethylene glycol priming osmotica upon the development of microbial populations and germination of leeks and carrots. *Annals of Applied Biology* 119: 365-372.

Pill, W.G. (1991). Advances in fluid drilling. *HortTechnology* 1: 59-65.

Pill, W.G. and Gunter, J.A., Jr. (2001). Emergence and shoot growth of cosmos and marigold from paclobutrazol-treated seed. *Journal of Environmental Horticulture* 19: 11-14.

Powell, A.A., Yule, L.J., Jing, H.-C., Groot, S.P.C., Bino, R.J., and Pritchard, H.W. (2000). The influence of aerated hydration seed treatment on seed longevity as assessed by the viability equations. *Journal of Experimental Botany* 51: 2031-2043.

Redfearn, M. and Osborne, D.J. (1997). Effects of advancement on nucleic acids in sugarbeet. *Seed Science Research* 7: 261-267.

Rowse, H.R. (1996). Drum-priming—A non-osmotic method of priming seeds. *Seed Science and Technology* 24: 281-294.

Rowse H.R. and McKee, J.M.T. (1999). Seed priming. World Patent No. 9608132.

Rowse, H.R., McKee, J.M.T., and Higgs, E.C. (1999). A model of the effects of water stress on seed advancement and germination. *New Phytologist* 143: 273-279.

Schmolka, I.R. (1988). Seed protective coating. U.S. Patent No. 4735015.

Simak, M. (1984). Method for sorting of seed. U.S. Patent No. 4467560.

Sitrit, Y., Hadfield, K.A., Bennett, A.B., Bradford, K.J., and Downie, A.B. (1999). Expression of a polygalacturonase associated with tomato seed germination. *Plant Physiology* 121: 419-428.

Sliwinska, E. (2000). Analysis of the cell cycle in sugarbeet seed during development, maturation and germination. In Black, M., Bradford, K.J., and Vazquez-Ramos, J. (Eds.), *Seed Biology: Advances and Applications. Proceedings of the Sixth International Workshop on Seeds, Merida, Mexico, 1999* (pp. 133-139). Wallingford, UK: CABI Publishing.

Sliwinska, E., Jing, H.-C., Job, C., Job, D., Bergervoet, J.H.W., Bino, R.J., and Groot, S.P.C. (1999). Effect of harvest time and soaking treatment on cell cycle activity in sugarbeet seeds. *Seed Science Research* 9: 91-99.

Sluis, S.J. (1987). Process for bringing pregerminated seed in a sowable and for some time storable form, as well as pilled pregerminated seeds. U.S. Patent No. 4658539.

Souza-Machado, V., Ali, A., Pitblado, R., and May, P. (1996). Enhancement of tomato seedling quality involving triazole seed priming and seedling nutrient loading. In *Proceedings of the First International Conference on the Processing Tomato* (Recife, Brazil, November 1996) (pp. 71-72). Alexandria, VA: ASHS Press.

Stewart, R.F. (1992). Temperature sensitive seed germination control. U.S. Patent No. 5129180

Still, D.W. and Bradford, K.J. (1997). Endo-β-mannanase activity from individual tomato endosperm caps and radicle tips in relation to germination rates. *Plant Physiology* 113: 21-29.

Sung Y., Cantliffe, D.J., and Nagata, R. (1998). Using a puncture test to identify the role of seed coverings on thermotolerant lettuce seed germination. *Journal of the American Society for Horticultural Science* 123: 1102-1110.

Taylor, A.G., Allen, P.S., Bennett, M.A., Bradford, K.J., Burris, J.S., and Misra, M.K. (1998). Seed enhancements. *Seed Science Research* 8: 245-256.

Taylor, A.G., Klein, D.E., and Whitlow, T.H. (1988). SMP: Solid matrix priming of seeds. *Scientia Horticulturae* 37: 1-11.

Taylor, A.G., McCarthy, A.M., and Chirco, E.M. (1982). Density separation of seeds with hexane and chloroform. *Journal of Seed Technology* 7: 78-83.

Taylor, A.G., Prusinski, J., Hill, H.J., and Dickson, M.D. (1992). Influence of seed hydration on seedling performance. *HortTechnology* 2: 336-344.

Thornton, J.M. and Powell, A.A. (1992). Short-term aerated hydration for the improvement of seed quality in *Brassica oleracea* L. *Seed Science Research* 2: 41-49.

Toorop, P.E., van Aelst, A.C., and Hilhorst, H.W M. (1998). Endosperm cap weakening and endo-β-mannanase activity during priming of tomato (*Lycopersicon esculentum* cv. Moneymaker) seeds are initiated upon crossing a threshold water potential. *Seed Science Research* 8: 483-491.

Tsujimoto, T., Sato, H., and Matsushita, S. (1999). Hydration of seeds with partially hydrated super absorbent polymer particles. U.S. Patent No. 5930949.

van Pijlen, J.G., Groot, S.P.C., Kraak, H.L., Bergervoet, J.H.W., and Bino, R.J. (1996). Effects of pre-storage hydration treatments on germination performance, moisture content, DNA synthesis and controlled deterioration tolerance of tomato (*Lycopersicon esculentum* Mill.) seeds. *Seed Science Research* 6: 57-63.

Vertucci, C.W. and Farrant, J.M. (1995). Acquisition and loss of desiccation tolerance. In Kigel, J. and Galili, G. (Eds.), *Seed Development and Germination* (pp. 237–271). New York: Marcel Dekker.

Warren, J.E. and Bennett, M.A. (1997). Seed hydration using the drum priming system. *HortScience* 32: 1220-1221.

Warren, J.E. and Bennett, M.A. (2000). Bio-osmopriming tomato (*Lycopersicon esculentum* Mill.) seeds for improved stand establishment. In Black, M., Bradford, K.J., and Vazquez-Ramos, J. (Eds.), *Seed Biology: Advances and Applications. Proceedings of the Sixth International Workshop on Seeds, Merida, Mexico, 1999* (pp. 477-487). Wallingford, UK: CABI Publishing.

Watts, H. and Schreiber, K. (1974). Manufacture of dormant pellet seed by coating with non-elastomeric polymer. U.S. Patent No. 3803761.

Welbaum, G.E. and Bradford, K.J. (1991). Water relations of seed development and germination in muskmelon (*Cucumis melo* L.): VI. Influence of priming on germination responses to temperature and water potential during seed development. *Journal of Experimental Botany* 42: 393-399.

Welbaum, G.E., Bradford, K.J., Yim, K.O., Booth, D.T., and Oluoch, M.O. (1998). Biophysical, physiological and biochemical processes regulating seed germination. *Seed Science Research* 8: 161-172.

Welbaum, G.E., Shen, Z., Oluoch, M.O., and Jett, L.W. (1998). The evolution and effects of priming vegetable seeds. *Seed Technology* 20: 209-235.

Wolkers, W.F., Tetteroo, F.A.A., Alberda, M., and Hoekstra, F.A. (1999). Changed properties of the cytoplasmic matrix associated with desiccation tolerance of dried carrot somatic embryos: An in situ Fourier transform infrared spectroscopic study. *Plant Physiology* 120: 153-163.

Wu, C.T., Leubner-Metzger, G., Meins, F., Jr., and Bradford, K.J. (2001). Class I β-1,3-glucanase and chitinase are expressed specifically in the micropylar endosperm of tomato seeds prior to radicle emergence. *Plant Physiology* 126: 1299-1313.

SECTION II:
DORMANCY AND THE BEHAVIOR
OF CROP AND WEED SEEDS

Chapter 5

Inception, Maintenance, and Termination of Dormancy in Grain Crops: Physiology, Genetics, and Environmental Control

Roberto L. Benech-Arnold

INTRODUCTION

Dormancy is the failure to germinate because of some internal block that prevents the completion of the germination process (Black, Butler, and Hughes, 1987). For completeness it should added that dormant seeds cannot germinate in the same conditions (e.g., water, air, temperature) under which nondormant seeds do so. Although the adaptive significance of dormancy is quite evident for plants living in the "wild" (see Chapter 8), it has always been a complication in seeds from plants that are grown as crops. Indeed, a persistent dormancy would prevent the utilization of a seed lot either for the generation of a new crop or for industrial purposes (i.e., malting). On the other hand, most crops that originally must have had dormancy have been selected so heavily against dormancy throughout their domestication process that seeds are germinable even prior to crop harvest; this frequently leads to preharvest sprouting, a phenomenon whose consequences are widely described in Chapter 6.

Due to the paucity of our knowledge on the genetic, physiological, and environmental control of dormancy, it is very difficult to adjust the timing of dormancy loss to a precise and narrow time window (i.e., neither as early as to expose the crop to the risk of preharvest sprouting, nor as late as to have a dormant seed lot at the time of the next sowing or industrial utilization). Among the cereals, malting barley is possibly the most problematic crop. The malting process itself requires grain germination, so a low dormancy level at harvest is a desirable characteristic because the grain can be malted immediately after crop harvest, thus avoiding costs and deterioration resulting from grain storage until dormancy is terminated. Therefore, breeders are compelled to work within a narrow margin. In this case, the possibility of solving the conflict between obtaining genotypes with low dormancy at harvest, but not with such an anticipated termination of dormancy that leads to sprouting risks, requires a thorough knowledge of the mechanisms determining dormancy release in the maturing grain. Moreover, it is essential to understand how those mechanisms are genetically and environmentally controlled.

Problems derived from either a short or a persistent dormancy are less frequent in oil crops, though they do exist. Sprouting, for example, has not been reported to occur in the most important oil crops (i.e., soybean, sunflower, canola); this is in spite of the fact that both soybean and canola seeds are germinable as soon as the grain has undergone desiccation in the mother plant. Soybean and canola seeds develop within legumes and siliques, respectively, which must impede direct contact between the grain and rain water in the field. Sunflower, however, does not sprout because its seeds are highly dormant at harvest time, and this deep dormancy may persist for several months. Indeed, having a dormant lot by the time sunflower seeds are sold for sowing is a significant problem that most seed companies face frequently.

The aim of this chapter is to discuss the physiology, genetics, and environmental control of dormancy inception, maintenance, and loss in some grain crops, namely, cereals and sunflower. It is intended, also, to analyze the perspectives of controlling the timing of occurrence of these processes through manipulation of the genes that regulate the physiological mechanisms involved.

PHYSIOLOGY OF DORMANCY IN THE CEREAL GRAIN

Where Is Dormancy Located in Cereal Grains?

Dormancy inception occurs very early in cereals. Embryos are usually fully germinable from early stages of development (i.e., 15 to 20 days after pollination [DAP]) if isolated from the entire grain and incubated in water (Walker-Simmons, 1987; Benech-Arnold, Fenner, and Edwards, 1991; Benech-Arnold et al., 1999). The entire grain, however, reaches full capacity of germination well after it has been acquired by the embryo. This coat (endosperm plus pericarp)-imposed dormancy is the barrier preventing untimely germination, and its duration depends on the genotype and on the environment experienced during maturation and beyond. In summary, though cases of embryo dormancy have been reported for grains of some cereal crops (Norstog and Klein, 1972; Black, Butler, and Hughes, 1987), the duration of coat-imposed dormancy determines the timing of acquisition of grain germinability. For example, sprouting-susceptible cultivars are those whose coat-imposed dormancy is terminated well before harvest maturity.

In some cereals (i.e., barley, rice), glumellae (the hull) adhering to the caryopsis represents another constraint for embryo germination in addition to that already imposed by endosperm plus pericarp (Corbineau and Come, 1980). Benech-Arnold and colleagues (1999) followed the dynamics of the release from dormancy imposed by the different structures surrounding the embryo in grains from cultivars with short (cv. B1215) and longer-lasting dormancy (cv. Quilmes Palomar). As expected, embryos from both cultivars germinated precociously from early stages of development if excised from the entire grain (Figure 5.1a). In both cultivars dormancy imposed by endosperm plus pericarp was steadily overcome at a similar rate throughout development (Figure 5.1b). However, although caryopses presented low dormancy from well before physiological maturity (PM, defined as the moment when the grain has attained maximum dry weight), the presence of the hull prevented grain germination prior to that stage. Hull-imposed dormancy started to be removed from PM onward, with a rate that was different depending on the cultivar: in 'B1215' grains this restriction was removed abruptly, while in 'Q. Palomar' grains, the removal occurred at a lower rate (Figure 5.1c).

Hormonal Regulation of Dormancy in Cereal Grains

The Role of Abscisic Acid

Research on the mechanisms of dormancy in the developing seeds of many species suggests a strong involvement of the phytohormone abscisic acid (ABA) (King, 1982; Fong, Smith, and Koehler, 1983; Karssen et al., 1983; Walker-Simmons, 1987; Black, 1991; Benech-Arnold, Fenner, and Edwards, 1991; Benech-Arnold et al., 1995; Steinbach et al., 1995; Steinbach, Benech-Arnold, and Sánchez, 1997). ABA-deficient or -insensitive mutants of *Arabidopsis* and maize precociously germinate (Robichaud, Wong, and Sussex, 1980; Karssen et al., 1983), and application of the ABA-synthesis inhibitor fluridone has been shown to anticipate the release from dormancy in developing seeds of some species (Fong, Smith, and Koehler, 1983; Xu, Coulter, and Bewley, 1991; Steinbach, Benech-Arnold, and Sánchez, 1997). In cereals, the imposition of dormancy to the embryo by the structures that surround it might be mediated by the high levels of endogenous ABA existing in the embryos during grain development (Walker-Simmons, 1987; Steinbach et al., 1995). ABA content in embryos is usually low until 15 DAP (Walker-Simmons, 1987; Steinbach et al., 1995; Benech-

(a)

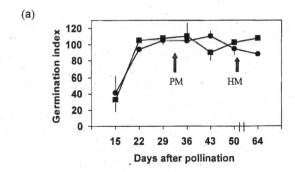

(b)

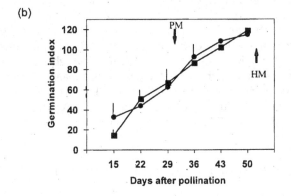

(c)

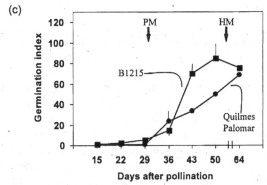

FIGURE 5.1. Germination indexes of (a) embryos, (b) dehulled caryopses, and (c) grains from a sprouting-susceptible ('B1215', squares) and a sprouting-resistant ('Quilmes Palomar', circles) cultivar, harvested at different days after pollination and incubated at 20°C for 12 days. PM and HM are the moments the crops reached physiological and harvest maturity, respectively. (*Source:* Adapted from figures in Benech-Arnold et al., 1999.)

Arnold et al., 1999). From that moment onward, ABA content goes up coinciding with the acquisition of the capacity of the embryo to germinate if isolated from the rest of the grain; hence, one possibility is that precocious germination would be prevented by the surrounding structures by impeding ABA from leaching outside the embryo (Bewley and Black, 1994).

ABA content has been reported to peak at around PM and to decline afterward when the grain undergoes desiccation (Goldbach and Michael, 1976; Walker-Simmons, 1987; Quarrie, Tuberosa, and Lister, 1988; Morris, Jewer, and Bowles, 1991; Steinbach et al., 1995; Benech-Arnold et al., 1999). However, and in contrast to what might have been expected, no correlations have been found between ABA embryonic content during seed development and timing of exit from dormancy. In other words, although inhibiting ABA synthesis (either genetically or through chemicals) has been shown to accelerate the termination of dormancy, genotypes with a short dormancy usually do not have lower ABA content during grain development than those with long-lasting dormancy. One exception for this lack of correlation, however, has been reported for barley. In barley cultivars with contrasting timing of exit from dormancy, ABA embryonic content is usually similar until PM, and maximum ABA content also occurs prior to PM (Figure 5.2). However, immediately after PM, a dramatic reduction in embryonic ABA content takes place in embryos from the sprouting-susceptible 'B1215', coinciding with the abrupt termination of hull-imposed dormancy that takes place in these grains after PM (Figure 5.1c); in 'Q. Palomar' (a cultivar with longer-lasting dormancy) embryos, in contrast, ABA content is kept at high levels for longer (i.e., until 43 DAP).

It has been suggested that dormancy imposed by the hull is mediated by high polyphenol-oxidase activity existing in the barley glumellae which results in oxygen deprivation for the embryo (Lenoir, Corbineau, and Côme, 1986). The way in which oxygen influences germination of dormant seeds is largely unknown, but it has been hypothesized that oxygen concentration might determine the rate with which germination inhibitors are catabolized (Neil and Horgan, 1987). This proposition is strongly supported by results presented by Wang and colleagues (1998) showing that the dormancy breaking effect of a strong oxidant such as hydrogen peroxide is through a reduction in the endogenous level of the germination inhibitor abscisic acid. The question arising is, How can this mechanism operate differentially throughout development and between genotypes presenting different timing of exit from dormancy? In the light of these results and within the frame of the proposition that the hull impedes embryo germination because it interferes with ABA oxidation (or metabolization) through oxygen deprivation, it could be argued that release from hull-imposed dormancy occurs because oxygen in high concentrations is not necessary when germination

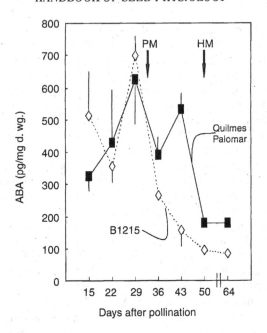

FIGURE 5.2. Abscisic acid content (expressed as picograms of ABA per milligram of dry weight) in embryos from a sprouting-susceptible ('B1215', white diamond) and sprouting-resistant ('Quilmes Palomar', black squares) cultivar, harvested at different days after pollination. PM and HM are the moments the crops reached physiological and harvest maturity, respectively. (*Source:* Adapted from figures in Benech-Arnold et al., 1999.)

inhibitors (i.e., ABA) are no longer present. These results explain the different timing of exit from dormancy between cultivars whose grains acquire germinability immediately after PM or few days after harvest. However, in most barley cultivars dormancy may last several months; in such cases, the correlation between ABA and germinability does not hold. Indeed, although inhibiting ABA synthesis with fluridone can anticipate exit from dormancy, these cultivars do not present higher ABA content during grain development and, on the other hand, ABA levels are barely detectable after harvest maturity. Some authors have proposed that the maintenance of dormancy in those cultivars is mediated by de novo synthesis of ABA upon grain incubation, which would not occur in grains without dormancy (Wang et al., 1998). However, this possibility is still a subject of debate.

The role of changes in embryo responsiveness to ABA has been suggested to be a key one for controlling release from dormancy in cereals and other species (Robichaud, Wong, and Sussex, 1980; Walker-Simmons,

1987; Corbineau, Poljakoff-Mayber, and Côme, 1991; Steinbach et al., 1995; Benech-Arnold et al., 2000; Van Beckum, Libbenga, and Wang, 1993; Wang et al., 1994; Wang, Heimovaara-Dijkstra, and Van Duijn, 1995; Visser et al., 1996). Embryo sensitivity to ABA is measured as the embryo capacity to overcome the inhibitory action of a certain concentration of the hormone. In the system 'B1215'-'Q. Palomar' termination of hull-imposed dormancy is also correlated with changes in embryo sensitivity to ABA (Figure 5.3); release of 'B1215' grains from dormancy coincides with an abrupt loss of embryo sensitivity to ABA, while high responsiveness to ABA is maintained for longer in 'Q. Palomar' embryos. Cultivars that have a lower embryo sensitivity to ABA during seed development usually present a faster release from dormancy. For example, a tenfold higher concentration of ABA is required to inhibit germination of embryos from a sorghum variety whose grains are released from dormancy prior to PM than is necessary to inhibit germination of embryos from a variety with a long-lasting dormancy (Steinbach et al., 1995). The nature of the low sensitivity to ABA observed in embryos from genotypes with short dormancy remains unclear, though some possibilities have been proposed. In an interesting paper, Visser and colleagues (1996) showed results suggesting that the low embryo sensitivity to ABA exhibited by a barley cultivar with no dormancy was not related to alterations in the ABA transduction pathway but to a high rate of degradation of the hormone in the outside walls of the embryo.

The Role of Gibberellins

The central role of gibberellins (GAs) in promoting seed germination was suggested decades ago and confirmed clearly since the identification of GA-deficient mutants of *Arabidopsis* and tomato seeds that will not germinate unless exogenously supplied with GAs (Lona, 1956; Karssen et al., 1989; Hilhorst, 1995; Karssen, 1995). Similarly, dormant developing sorghum caryopses can be induced to germinate if incubated in the presence of GAs (Steinbach, Benech-Arnold, and Sánchez, 1997). This role should not be confounded with the postgerminative one referred to in Chapters 6 and 13 when describing production of α-amylase in barley and other germinating grains. It has been proposed that endogenous GAs control germination through two processes: a decrease in the mechanical resistance of the tissues surrounding the embryo and promotion of the growth potential of the embryo (see Chapter 7 for details; Bradford et al., 2000), thus antagonizing the effect of ABA (Schopfer and Plachy, 1985). In cereals in which the tissues covering the embryo are weak or are split during imbibition, GA action must be restricted to promote embryo growth potential. Benech-Arnold and

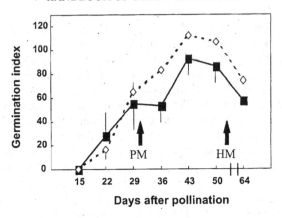

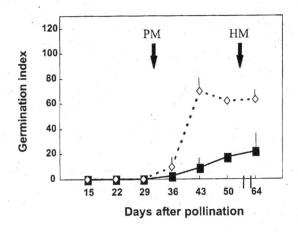

FIGURE 5.3. Germination indexes (GI) of embryos from a barley cultivar with a short dormancy ('B1215', ◊) and one with a longer lasting dormancy ('Q. Palomar', ■), harvested at different days after pollination, after 12 days of incubation at 20°C, in the presence of 5 µM ABA (upper panel) or 50 µM ABA (lower panel). PM and HM are the moments the crops reached physiological and harvest maturity, respectively. (*Source:* Adapted from figures in Benech-Arnold et al., 1999.)

colleagues (2000) hypothesized that the low dormancy presented by developing sorghum caryopses from sprouting-susceptible genotypes should be expressed as a high capacity of the embryo to produce GA de novo synthesis upon grain imbibition; these GAs would be necessary to counterbalance the inhibitory effect imposed by the high ABA content existing during grain development.

In addition to its role as germination promoter, it has been demonstrated that the pattern of exit from dormancy in developing cereal grains can be altered by inhibiting GA synthesis, suggesting that this pattern depends on the extent to which ABA action as a dormancy imposer is counterbalanced by the effect of GAs (Steinbach, Benech-Arnold, and Sánchez, 1997; Benech-Arnold et al., 1999). Applications of the GA synthesis inhibitor paclobutrazol almost immediately after anthesis of barley and sorghum varieties with short dormancy results in a pattern of exit from dormancy that resembles the characteristic pattern of varieties with a long-lasting dormancy, even though genotypes with a short dormancy have not been found to present a lower GA content during development (Benech-Arnold et al., 2000). However, it could be predicted from experiments with paclobutrazol that lowering GA content through genetic means should result in genotypes with extended dormancy.

PHYSIOLOGY OF DORMANCY IN THE SUNFLOWER SEED

At harvest time sunflower seeds are dormant and germinate poorly (Corbineau, Bagniol, and Côme, 1990; Corbineau and Côme, 1987; Cseresnyes, 1979). This dormancy is the result of true embryo dormancy (Corbineau, 1987; Corbineau, Bagniol, and Côme, 1990) and the inhibitory action of the envelopes (Corbineau, 1987; Corbineau, Bagniol, and Côme, 1990; Corbineau and Côme, 1987) including the seed coat and the pericarp since sunflower "seeds" are achenes.

The inception of embryo dormancy occurs relatively early throughout seed development. Sunflower embryos are germinable if isolated from the entire seed from as early as 7 DAP and until approximately 12 DAP; the entire seed, however, germinates very poorly during this period, showing the existence of coat (seed coat plus pericarp)-imposed dormancy (Figure 5.4) (LePage-Degivry and Garello, 1992; Corbineau, Bagniol, and Côme, 1990). From 12 DAP onward, embryo dormancy progressively develops, and at 20 to 22 DAP, embryos are fully dormant (Figure 5.4). This embryo dormancy is not eliminated if the axis is separated from the cotyledons, indicating that the axis itself is dormant (LePage-Degivry and Garello, 1992). While the seed progresses toward maturation, embryos are slowly released from dormancy; by the time the grain has attained harvest maturity, some embryo dormancy still persists (Figure 5.4). Therefore, the deep dormancy that sunflower grains present at harvest results from the coexistence of coat-imposed dormancy and some remnant embryo dormancy (Corbineau, Bagniol, and

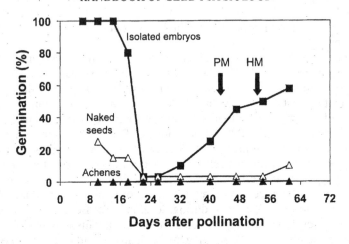

Days after pollination

FIGURE 5.4. Germination percentage of sunflower achenes (solid triangles), naked seeds (open triangles), and isolated embryos (solid squares) harvested at different days after pollination and incubated at 25°C. Whole achenes were totally unable to germinate when incubated at any of the DAP displayed in the graph. PM and HM are, approximately, the moments the crop reached physiological and harvest maturity, respectively. (*Source:* Redrawn with data from LePage-Degivry and Garello, 1992, and Corbineau, Bagniol, and Côme, 1990.)

Côme, 1990). Embryo dormancy is lost shortly after harvest if the seed is subjected to dry after-ripening, but coat-imposed dormancy persists for longer and may require several weeks of dry after-ripening to be overcome.

The plant growth regulator ABA appears to be involved in the imposition of embryo dormancy. The inclusion of fluridone in culture media for sunflower embryo development prevents the induction of embryo dormancy (LePage-Degivry, Barthe, and Garello, 1990; LePage-Degivry and Garello, 1992). Nevertheless, the pattern of accumulation of ABA in the developing embryo does not coincide with the embryo physiological behavior. During seed development, embryos germinate well at the time when the endogenous ABA level is at its highest (7 to 12 DAP); thereafter, ABA decreases to a low value when embryo dormancy becomes established (LePage-Degivry, Barthe, and Garello, 1990). It seems, then, that the ABA peak at early stages is responsible for the imposition of the dormant state that is established immediately after that peak has taken place.

Moreover, it appears that ABA needs to be present during a critical time period to induce dormancy. In an interesting study, LePage-Degivry and Garello (1992) showed that when young (7 DAP), nondormant embryos were cultured in the presence of ABA, the hormone produced a temporary inhibition of germination but did not induce dormancy (i.e., embryos were able to germinate when transferred to a basal medium). In contrast, exogenous ABA became effective if applied immediately prior to the natural induction of dormancy. For example, five days culture on a medium containing 5×10^{-5} M ABA resulted in partial dormancy in 13 DAP embryos, while total induction of dormancy occurred in 17 DAP embryos. The authors concluded that either a change in sensitivity to ABA occurs during development, or the existence of a second factor is necessary along with ABA to induce dormancy. Regarding this second possibility, the authors speculate about the existence of a regulatory protein called VP that binds to ABA for the induction of dormancy; ABA would not be able to induce dormancy in 7 to 10 DAP embryos due to the absence of this protein. LePage-Degivry and Garello (1992) suggested that the capacity of the embryo to produce in situ ABA synthesis, which appears during seed development along with the onset of dormancy, is necessary not only to induce, but also to maintain dormancy.

As mentioned previously, embryo dormancy can be terminated by dry storage. Bianco, Garello, and LePage-Degivry (1994) attempted to elucidate the mechanism through which dry storage terminates embryo dormancy by drying artificially dormant 17 to 26 DAP embryos and testing for germinability either immediately after drying or after leaving the embryos for six weeks in a desiccator (dry storage). They observed a decrease in ABA content immediately after the drying process that was not accompanied by a complete release from dormancy. On the other hand, additional dry storage did not affect the ABA content but instead promoted germination. In addition, the authors found that the drying treatment also stimulated immature sunflower embryos and axes to respond to gibberellins upon rehydration. The authors concluded from these results that although the drying treatment induced both a decline in ABA and an increase in sensitivity to GA, additional dry storage is necessary to obtain germination. They propose the suppression induced during this dry storage of the aforementioned capacity to produce in situ ABA synthesis in the embryo, as the mechanism behind the response, though they did not show the extent to which the drying treatment by itself could also result in such suppression.

The inception of seed coat plus pericarp-imposed dormancy occurs early throughout seed development: by the stage at which young (7 to 13 DAP), nondormant embryos can germinate readily if isolated from the entire seed, germination of the whole grain is prevented by the presence of the envelopes (Figure 5.4). Coat-imposed dormancy possibly continues during the rest of the developmental period, but its existence is difficult to corroborate because the embryo itself is dormant during most of this period. Once the seed has completed maturation and while the embryo gradually loses its dormancy, coat-imposed dormancy persists for longer, in some cases for several months.

The nature of this inhibition imposed on embryo germination is highly unknown, though it has been suggested that both pericarp and seed coat interfere with oxygen difussion toward the embryo (Gay, Corbineau, and Côme, 1991). As in the case of hull-imposed dormancy in barley, it could be speculated that the envelope impedes embryo germination because it interferes with ABA and/or other inhibitor oxidation (or metabolization) through oxygen deprivation. Similarly, the mechanism through which dry after-ripening alleviates coat-imposed dormancy has not been explored to the best of our knowledge. It could be that, even after the embryo has been released from dormancy, it retains the capacity to produce ABA synthesis upon imbibition, which might be necessary to maintain (coat-imposed) dormancy; indeed, oxygen deprivation caused by the presence of the embryo would prevent ABA oxidation. If, as mentioned before, dry storage suppresses the capacity of the embryo to produce ABA synthesis (Bianco, Garello, and LePage-Degivry, 1994), then coat-imposed dormancy would be terminated because oxygen in high concentrations should not be necessary when ABA is no longer present. This hypothesis should be thoroughly tested. Unfortunately, we are not aware of any study in which the physiology of dormancy in sunflower has been comparatively investigated in genotypes with contrasting duration of dormancy.

THE EXPRESSION OF DORMANCY IN GRAIN CROPS

Except for the case of seeds that present full dormancy and consequently do not germinate at either temperature, it is a common feature that dormancy is expressed at certain temperatures. Vegis (1964) introduced the concept of degrees of relative dormancy from the observation that, as dormancy is released, the temperature range permissive for germination widens, until germination is maximal under a wide thermal range. This is also the case for dormant cereal grains: in summer cereals such as sorghum, dormancy is not expressed at high temperatures (i.e., 30°C) (Benech-Arnold et al., 1995; Benech-Arnold, Enciso, and Sánchez, 1999), and in winter cereals such as wheat and barley it is not expressed at low temperatures (i.e., 10°C or lower) (Bewley and Black, 1994; Gosling et al., 1981; Mares, 1984; Black, Butler, and Hughes, 1987; Walker-Simmons, 1988). It should be emphasized that the depressed germination which occurs as temperatures exceed (in the case of winter cereals) or are below (in the case of summer cereals) certain values is truly an expression of dormancy and not an inevitable effect of temperature on germination, for it does not take place in isolated embryos or in grains which have after-ripened (Mares, 1984). Moreover, it has been shown in wheat that isolated embryos incubated at high temperatures (i.e., 25 to 30°C) are more effectively inhibited by ABA than embryos incubated at lower temperatures (Walker-Simmons, 1988). This thermal range permissive for germination widens with after-ripening so grains become able to germinate at most temperatures. Similarly, it was observed for barley grains that, so long they are released from dormancy throughout development and maturation, they are able to germinate at higher temperatures (Benech-Arnold, Enciso, and Sánchez, 1999). This differential expression of dormancy which depends on the incubation temperature also has implications for crop behavior in the field. For example, the lack of expression of dormancy at low temperatures, characteristic of winter cereals, implies that in years when damp conditions are combined with low air temperatures around harvest time, both resistant (high dormancy) and susceptible (low dormancy) cultivars might be expected to sprout. The inverse could be said about summer cereals such as sorghum; high temperatures combined with damp conditions around harvest permit the germination *in planta* of both dormant and nondormant cultivars.

The amount of water in the incubation medium also allows differential expression of dormancy in barley grains. Indeed, most barley cultivars which present some dormancy at harvest will not germinate if the grains are incubated in a petri dish at favorable temperatures but with 8 or even 6 ml instead of 4 ml of distilled water (Pollock, 1962); the same does not occur in grains from cultivars with low dormancy, or in those that have after-ripened, showing that it is truly an expression of dormancy (Figure 5.5). This phenomenon is known by the malting industry as "sensitivity to water" and is one of the quality parameters assessed upon reception of a grain lot. This sensitivity to water must be related to the oxygen deprivation imposed by the presence of the hull, described previously, which might be enhanced by the hypoxia that results from an excess of water in the incubation media.

In the case of freshly harvested sunflower seeds, dormancy is expressed at temperatures lower and higher than 25°C (Corbineau, Bagniol, and Côme, 1990). Dormancy expression at low temperatures is attributed to embryo dormancy which is not expressed at high temperatures (Corbineau, Bagniol, and Côme, 1990); conversely, dormancy expressed at high temperatures results from coat-imposed dormancy (Corbineau, Bagniol, and Côme, 1990). Consequently, a few weeks of dry after-ripening allows seed germination at low temperatures due to termination of embryo dormancy; the acquisition of the capacity to germinate at high temperatures, in contrast, may take several weeks of dry after-ripening (Corbineau, Bagniol, and Côme, 1990).

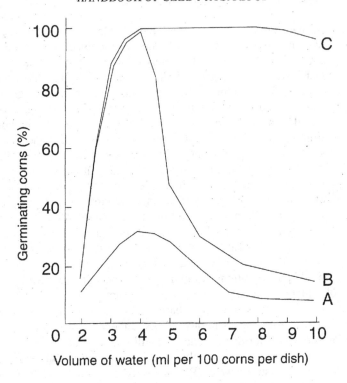

FIGURE 5.5. Response of barley germination to the amount of water present: (A) freshly harvested dormant barley; (B) the same barley, still water sensitive though not fully dormant; (C) the same barley after some time of after-ripening in dry storage. (*Source:* Adapted from a figure in Pollock, J.R.A., 1962.)

REMOVING DORMANCY AT AN INDUSTRIAL SCALE

In some cases it is not possible to wait for the effect of dry after-ripening to take place and termination of grain dormancy must be anticipated. This is frequently the case with malting barley, whose germination is required for industrial utilization (see Chapter 13) and also with sunflower whose grains are usually dormant by the time they are needed for generating a new crop.

One of the most popular methods used by the malting industry, whenever allowed by the customer, is the addition of gibberellic acid (GA_3) to the incubation medium to promote the germination of dormant barley grains. Indeed, it is well known that gibberellic acid at low concentrations (0.1 to 0.2 ppm) stimulates germination in these grains (Brookes, Lovett, and MacWilliam, 1976). Studies on the most appropriate point in the malting process at which to add gibberellic acid have concluded that it should be sprayed on soon after the grain is removed from the steep (Brookes, Lovett, and MacWilliam, 1976). Other methods to remove dormancy in barley include the use of dilute solutions of hydrogen sulfide and keeping the grains for three days at 40°C, either in the open air when their moisture content fell to about 8 percent, or in closed vessels, when moisture contents were unchanged at between 17 and 20 percent (Pollock, 1962).

As with other cultivated species such as *Lactuca sativa* (Abeles, 1986) and *Arachis hypogaea* (Ketring, 1977), ethylene (C_2H_4) and etephon strongly stimulate the germination of dormant sunflower seeds (Srivastasa and Dey, 1982; Corbineau and Côme, 1987; Corbineau, Bagniol, and Côme, 1990). In contrast, gibberellic acid and cold stratification do not overcome dormancy in this species (Bagniol, 1987) though it was shown that 1 mM GA_3 is effective for overcoming dormancy in some wild sunflowers (Seiler, 1998). Corbineau, Bagniol, and Côme (1990) showed that ethylene and its immediate precursor (1-aminocyclopropane-1-carboxylic acid) strongly stimulated germination of primary dormant sunflower seeds; on the contrary, inhibitors of ethylene (i.e., amino-oxyacetic acid and $CoCl_2$) or ethylene action (silver thiosulfate and 2.5 norbomadiene) inhibited germination of nondormant seeds. Beyond the evident practical implications of these results, they also indicate that ethylene synthesized by the seeds themselves is involved in the regulation of sunflower seed germination. The use of ethylene or its precursors appears as a promising technology to stimulate the germination of dormant sunflower lots. Possibly, seed companies have not adopted it yet, due to the inexistence of adequate devices to treat large amounts of seeds.

GENETICS AND MOLECULAR BIOLOGY
OF DORMANCY IN GRAIN CROPS

Although some investigations indicate that dormancy is controlled by one or two recessive genes (Bhatt, Ellison, and Mares, 1983), in several studies, seed dormancy has been revealed as a quantitative trait (i.e., a trait with continuous phenotypic variation). Consequently, modern approaches for determining the genetic bases of seed dormancy include the use of molecular markers (AFLP [amplified fragment length polymorphism] or RFLP [restriction fragment length polymorphism]) to identify QTLs (quantitative trait loci) or, in other words, loci controlling the quantitative trait "dormancy." Wheat had three QTLs that explained more than 80 percent of the total phenotypic variance in seed dormancy (Kato et al., 2001). A major QTL was located on the long arm of chromosome 4A, and two minor QTLs were on chromosomes 4B and 4D. In sorghum, two unlinked QTL, *phsE* and *phsF*, were found to influence dormancy in an F_2 population generated by the cross of a sprouting-susceptible variety with a sprouting-resistant one. These two QTLs accounted together for 53 percent of the phenotypic variance for preharvest sprouting (Lijavetzky et al., 2000).

Early genetic investigations (Buraas and Skinnes, 1984) revealed that seed dormancy in Scandinavian barleys was governed by several recessive, nucleoplasmic loci with high heritability. Genetic control of barley seed dormancy has also been studied by means of QTL mapping (Ullrich et al., 1993; Takeda, 1996). A saturated molecular marker linkage map based on the six-row Steptoe/Morex (S/M) mapping population has been developed (Kleinhofs et al., 1993) and extensively used for QTL analysis by the North American Barley Genome Mapping Project (Hayes et al., 1993; Han et al., 1996; Romagosa et al., 1996). Steptoe is a six-row feed barley with high levels of dormancy (Muir and Nilan, 1973). Morex is a six-row malting type that does not express dormancy (Rasmusson and Wilcoxson, 1979). Four regions of the barley genome on chromosomes 1 (7H), 4 (4H), and 7 (5H) were associated with most of the differential genotypic expression for dormancy in the S/M cross (Ullrich et al., 1996; Oberthur et al., 1995; Han et al., 1996; Larson et al., 1996). They were designated SD1 to SD4 by Han and colleagues (1996) and accounted for approximately 50, 15, 5, and 5 percent of the phenotypic differences, respectively, in germination following several post-harvest periods. In an early study, Livers (1957; cited by Romagosa et al., 1999) found some evidence that one or more postharvest dormancy *(phd)* genes may be located on chromosome 7, which is where two of the S/M QTLs are located. Takeda (1996), using QTL analysis with the Harrington/TR306 (H/T), population identified one region each on chromosomes 5 (1H) and 7 (5H) that controlled dormancy. The chromosome 7 (5H) H/T QTL coincides with the S/M QTL SD2 on the end of the long arm and was suggested to be allelic. In a recent study, Romagosa and colleagues (1999) investigated the individual effects on the S/M SD QTL on dormancy during seed development and after-ripening. With this aim, three pairs each of doubled haploid lines (DHLs) derived from Steptoe/Morex F_1s with the MM SS, SS MM, and SS SS genotypes at the SD1 and SD2 QTL and fixed M genotypes (MM MM) at the SD3 and SD4 QTL were identified by RFLP analysis. Morex and genotype MM SS MM MM were the first to start losing dormancy throughout development; the other genotypes remained dormant until the end of seed development (Figure 5.6a). Similarly, Morex and genotypes MM SS MM MM and MM SS SS SS had completely lost dormancy after 30 days of after-ripening, while other genotypes presented a pattern of exit from dormancy which progressively resembled that observed for the highly dormant Steptoe (Figure 5.6b). Since the

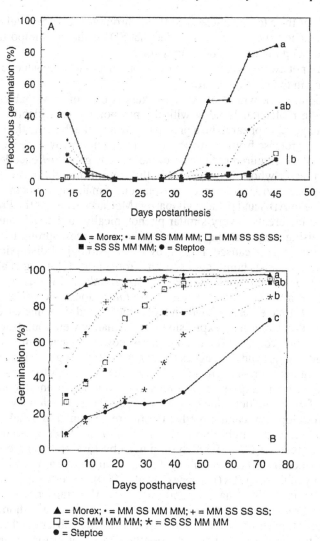

FIGURE 5.6. Germination percentage of various barley genotypes during seed development (A) and during after ripening after crop harvest (B). Genotypic means followed by the same letter are not statistically significant according to Duncan test ($\alpha < 0.05$) (a) or LSD test ($\alpha < 0.05$) (b) MM (Morex) and SS (Steptoe) designation refer to the genotypes of the pair of flanking markers for the four seed dormancy (SD) QTLs in order: SD1, SD2, SD3, SD4, e.g., MM SS MM MM. (*Source:* From Romagosa et al., 1999. Reproduced with permission.)

presence of the Steptoe allele at SD1 on chromosome 7 (5H) delayed exit from dormancy, the authors concluded that SD1 is the most important QTL in determining the time of dormancy release.

We are not aware of any study carried out to identify the genetic basis of dormancy in the sunflower crop.

Although work with molecular markers is extremely valuable, studies linking the molecular biology with the physiology (i.e., identification of candidate genes) appear to be a promising means of achieving the objective of adjusting release from dormancy to a precise and narrow time window as, for example, is required in the case of barley. The gene *Vp1* encodes a transcription factor whose involvement in the control of embryo sensitivity to ABA has been evidenced since the isolation of maize *vp1* mutants that are insensitive to ABA and present viviparity (McCarty et al., 1991). Preharvest sprouting in cereals is very similar phenotypically to the *vp1* mutation in maize, raising the interesting possibility that preharvest sprouting in barley and other cereals is caused, in part, by the physiological disruption of the *Vp1* function. Genes homologous to *Vp1* from barley (Hollung et al., 1997) and other Gramineae such as rice (Hattori, Terada and Hamasuna, 1994), sorghum (Carrari et al., 2001), and *Avena fatua* (Jones, Peters, and Holdsworth, 1997) have been cloned and sequenced, and, in some cases, close correlations between *Vp1* expression and dormancy were found (Jones, Peters, and Holdsworth, 1997). In other cases, however, such a correlation was not found. Carrari and colleagues (2001), using two sorghum varieties with contrasting duration of dormancy, did not see any straightforward relationship between *Vp1* expression during seed development and the particular pattern of exit from dormancy. In other words, the expression levels of *Vp1* during development cannot predict the future germination behavior of the immature seed upon imbibition. Moreover, *Vp1* has recently been mapped using the Redland B2/IS 9530 system used by Lijavetzky and colleagues (2000) to identify QTLs controlling dormancy in sorghum. *Vp1* did not map within any of these QTLs (Lijavetzky et al., 2000). Nevertheless, McKibbin and colleagues (2002) have recently analyzed *Vp1*-transcript structure in wheat embryos during grain development and found that a homeologue produces cytoplasmic mRNAs of different size. They observed that the majority of transcripts are spliced incorrectly, contain insertions of intron sequences or deletions of coding region, and do not have the capacity to encode full-length proteins. These authors suggest that missplicing of wheat *Vp1* genes contributes to an early release of dormancy of the grains which frequently results in preharvest sprouting (McKibbin et al., 2002). In contrast, *Avena fatua Vp1* genes do not show the same missplicing and, in agreement with the idea that *Vp1* gene exerts control on dormancy, *A. fatua* grains present a persistent dormancy. Interestingly, developing embryos from transgenic wheat grains expressing the *Avena fatua Vp1* showed enhanced responsiveness to applied ABA, and ripening ears were less susceptible to preharvest sprouting (McKibbin et al., 2002). These results, then, identify a possible route to manipulate dormancy duration in wheat.

Protein kinases often act in the transduction of external signals and could have a role in the effects of environmental conditions on expression of dormancy. For that reason a protein kinase mRNA (PKABA1) that accumulates in mature wheat seed embryos and that is responsive to applied ABA was cloned and its expression analyzed during imbibition of dormant and nondormant wheat seeds. When dormant seeds are imbibed, embryonic PKABA1 mRNA levels remain high for as long as the seeds are dormant, while they decline and disappear in embryos of germinating seeds (Anderberg and Walker-Simmons, 1991; Holappa and Walker-Simmons, 1995). The role of this kinase in dormant seeds is currently under investigation, but a potential role of phosphorylation-dependent responses in maintenance of seed dormancy is also supported by characterization of the *abi1* mutant of *Arabidopsis* (an ABA-insensitive with no dormancy) (Meyer, Leube, and Grill, 1994). The participation of this protein kinase in maintaining dormancy in grains from other crops remains to be investigated. G-protein-coupled receptors can participate in hormone-dependent signaling cascades affecting germination-related genes. Recently it has been shown that over-expression of GCR1, a G-protein-coupled receptor gene, decreases seed dormancy in *Arabidopsis* (Colucci et al., 2002). Whether expression of this type of gene is related to dormancy depth in cereal seeds is not yet known, although it is an interesting possibility.

Differences in gene expression in imbibed dormant and nondormant caryopses of *Avena fatua* (wild oats) have been determined through the technique of differential display (Li and Foley, 1994, 1995; Johnson et al., 1995). Monitoring gene expression in dormant and nondormant caryopses of barley through differential display could eventually evince yet unknown physiological and biochemical mechanisms controlling dormancy, provided the function of genes that are differentially expressed is finally elucidated.

In summary, both genetics and molecular biology studies could aid in the search for cultivars that, without having a long-lasting dormancy, could present resistance to preharvest sprouting. However, the complementarity with physiological studies is essential if such a goal is to be attained. The most profitable genetic investigations would be those that, for example, through QTL analysis, demonstrate the participation of genes with known physiological function. If in the end the phenotype happens to correlate well with some characteristic of that gene (i.e., differences between phenotypes in terms of gene expression timing, sequence, regulation, etc.) then the possibilities for manipulating the system are high.

ENVIRONMENTAL CONTROL OF DORMANCY IN GRAIN CROPS

The duration of dormancy is determined mainly by the genotype, but, as in many other species, it is known that dormancy in grain crops can also be influenced by the environment experienced by the mother plant (Kahn and Laude, 1969; Nicholls, 1982; Reiner and Loch, 1976; Schuurink, Van Beckum, and Heidekamp, 1992; Cochrane, 1993; see also Auranen, 1995, for references). Indeed, the effects of the parental environment on seed dormancy have been reported for a wide range of species (for reviews, see Fenner, 1991; Wulff, 1995). Some well-defined patterns emerge, however, with certain environmental factors tending to have similar effects in different species. For example, low dormancy is generally associated with high temperatures, short days, drought and nutrient availability during seed development (Walker-Simmons and Sesing, 1990; Fenner 1991; Benech-Arnold, Fenner, and Edwards, 1991, 1995; Gate, 1995). The assessment and quantification of these effects might lead to the development of predictive models that could be of great help for reducing the incidence of problems derived from either a short or a persistent dormancy.

Among the different factors acting on the mother plant, temperature appears to be primarily responsible for year-to-year variation in grain dormancy within a genotype. Evidence suggests that temperature might be effective only within a sensitivity period during grain filling (Kivi, 1966; Reiner and Loch, 1976; Buraas and Skinnes, 1985). Reiner and Loch (1976) determined that low temperatures during the first half of grain filling, combined with high temperatures during the second half, result in a low dormancy level of the barley grain and, presumably, in preharvest sprouting susceptibility. Conversely, high temperatures during the first half combined with low temperatures during the second produced the highest dormancy levels. The authors established a linear relationship between the ratio of the temperatures prevailing at both halves of the filling period and the dormancy level of the grains three weeks after harvest. This model has since been used by the German malting industry to predict dormancy level in the malting barley harvest lots.

In a recent work, Rodriguez and colleagues (2001) identified a time window within the grain-filling period of cultivar Quilmes Palomar with sensitivity to temperature for the determination of dormancy. This time window was found to occur a few days before physiological maturity (PM). Specifically, the duration of the phase heading PM for this cultivar was determined to last, on a thermal time scale, 420°C days (accumulated over a base temperature of 5.5°C). The sensitivity window was found to start at 300°C day after heading and to finish at 350°C day after heading. A positive linear relationship was established between the average temperature perceived by the crop during this time window and the germination index of the grains 12 days after PM (Figure 5.7).

Twelve days after PM is approximately halfway between physiological and harvest maturity; grain germination index measured at this stage is a good estimate of the rate at which the grains are being released from dormancy after PM. According to this model, the higher the temperature experienced during the sensitivity time window, the faster the rate with which grains will be released from dormancy after PM and, consequently, the lower the dormancy level prior to crop harvest. Such a situation, combined with a forecast of heavy rains for the forthcoming days, implies a risk for the crop and the farmer could decide to harvest before the crop has reached full maturity. Conversely, low temperatures experienced by the crop during the sensitivity window would result in a high dormancy level prior to harvest, making the crop resistant to sprouting. This model was successfully validated with data collected from commercial plots 700 km away from the

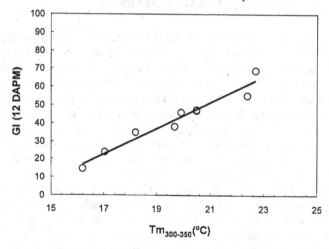

FIGURE 5.7. The relationship between temperature experienced by the crop in the sensitivity window going from 300 to 350°C days after heading ($Tm_{300-350}$), and the germination index of grains harvested 12 days after physiological maturity (GI [12 DAPM]) and incubated at 20°C. (*Source:* Redrawn with data from Rodriguez et al., 2001.)

site where the model was produced (Rodriguez et al., 2001). However, it was also noted that temperature explains only one dimension of the observed variability in dormancy. Indeed, some other unknown factors were responsible for influencing the relationship between temperature and dormancy (Rodriguez et al., 2001). Current efforts are directed toward identifying these factors and quantifying their effects.

As an exception to the general rule stating that low dormancy is generally associated with high temperatures experienced during grain filling, it has been found for sunflower that high temperatures during grain development result in an extended period of dormancy (Fonseca and Sánchez, 2000). In this case, germination was tested at low incubation temperatures (i.e., grains that had developed at high temperatures required longer time of dry after-ripening to acquire the capacity to germinate at low temperatures). Since embryo dormancy is expressed at low temperatures, it might be, then, that high temperatures during grain filling extended the duration of embryo dormancy. Germination at high temperatures, however, was not tested, and therefore it is not possible to say whether the duration of seed coat dormancy was also extended by high temperatures during grain development.

CONCLUDING REMARKS

The task of adjusting the timing of exit from dormancy of grain crops to the needs of both farmers and industry does not seem to be an easy one. However, an adequate knowledge of the physiology and the genetics of dormancy in grain crops should help to solve the conflict between obtaining cultivars with low dormancy at harvest but not with such an anticipated termination of dormancy that leads to sprouting. Although much progress has been made in recent years, we are still far away from having detailed knowledge on the physiology and genetics of dormancy. It is worth emphasizing that studies linking the genetics (and the molecular biology) with the physiology appear to be the most promising ones. For example, if genes controlling sensitivity to hormones (either ABA or GAs) are finally identified and their participation in the control of dormancy is eventually evidenced, then efforts should be directed to understand the regulation of those genes. It would not be surprising to find out that the action of genes controlling, for example, sensitivity to ABA, is cancelled after the grain has undergone desiccation (Kermode, 1995). If the transduction pathway is finally understood, then it should not be very difficult to manipulate the timing of such cancellation. This is just one example to illustrate how molecular studies oriented by physiological studies could yield tools for the production of genotypes with a precise timing of dormancy release.

Our knowledge on how the environment modulates the timing of exit from dormancy in grain crops could also help to make management decisions to reduce the incidence of problems derived from dormancy. Throughout this chapter it was described how a comprehensive assessment of the effects of temperature on dormancy during seed development can be used for deciding management practices. It is quite evident, however, that other factors in addition to temperature modulate the timing of exit from dormancy. When these factors are identified and their effects quantified, decisions on management practices will be made on an even more solid basis.

REFERENCES

Abeles, F.B. (1986). Role of ethylene in *Lactuca sativa* cv. Grand Rapids seed germination. *Plant Physiology* 81: 780-787.

Anderberg, R.J. and Walker-Simmons, M.K. (1991). Isolation of a wheat cDNA clone for an abscisic acid-inducible transcript with homology to protein kinases. *Proceedings of the National Academy of Sciences, USA* 89: 10183-10187.

Auranen, M. (1995). Pre-harvest sprouting and dormancy in malting barley in northern climatic conditions. *Acta Agriculturae Scandinavica* 45: 89-95.

Bagniol, S. (1987). Mise en évidence de l'intervention de l'ethylene dans la germination et la dormance des semences de tournesol (*Helianthus annuus* L.). Diplôme d'Ëtudes Approfondies. Université Pierre et Marie Curie, Paris.

Benech-Arnold, R.L., Enciso, S., and Sánchez, R.A. (1999). Fluridone stimulus of dormant sorghum seeds germination at low temperatures is not accompanied by changes in ABA content. In Weipert, D. (Ed.), *Pre-Harvest Sprouting in Cereals 1998* (Part II) (pp. 76-80). Detmold, Germany: Association of Cereal Research.

Benech-Arnold, R.L., Enciso, S., Sánchez, R.A., Carrari, F., Perez-Flores, L., Iusem, N., Steinbach, H.S., Lijavetzky, D., and Bottini, R. (2000). Involvement of ABA and GAs in the regulation of dormancy in developing sorghum seeds. In Black, M., Bradford, K.J., and Vázquez Ramos, J. (Eds.), *Seed Biology: Advances and Applications* (pp. 101-111). Oxon, UK: CAB International.

Benech-Arnold, R.L., Fenner, M., and Edwards, P.J. (1991). Changes in germinability, ABA levels and ABA embryonic sensitivity in developing seeds of *Sorghum bicolor* induced by water stress during grain filling. *New Phytologist* 118: 339-347.

Benech-Arnold, R.L., Fenner, M., and Edwards, P.J. (1995). Influence of potassium nutrition on germinability, ABA content and embryonic sensitivity to ABA of developing seeds of *Sorghum bicolor* (L.) Moench. *New Phytologist* 130: 207-216.

Benech-Arnold, R.L., Giallorenzi, M.C., Frank, J., and Rodriguez, V. (1999). Termination of hull-imposed dormancy in barley is correlated with changes in embryonic ABA content and sensitivity. *Seed Science Research* 9: 39-47.

Benech-Arnold, R.L., Kristof, G., Steinbach, H.S., and Sánchez, R.A. (1995). Fluctuating temperatures have different effects on embryonic sensitivity to ABA in

Sorghum varieties with contrasting pre-harvest sprouting susceptibility. *Journal of Experimental Botany* 46: 711-717.

Bewley, J.D. and Black, M. (1994). *Seeds: Physiology of Development and Germination*, Second Edition. New York: Plenum Press.

Bhatt, G.M., Ellison, F.W., and Mares, D.J. (1983). Inheritance studies on dormancy in three wheat crosses. In Kruger, J.E. and LaBarge, D.E. (Eds.), *Third International Symposium on Preharvest Sprouting in Cereals* (pp. 274-278). Boulder, CO: Westview Press.

Bianco, J., Garello, G., and Le Page-Degivry, M.T. (1994). Release of dormancy in sunflower embryos by dry storage: Involvement of gibbrellins and abscisic acid. *Seed Science Research* 4: 57-62.

Black, M. (1991). Involvement of ABA in the physiology of developing and mature seeds. In Davies, W.J. (Ed.), *Abscisic Acid Physiology and Biochemistry* (pp. 99-124). Oxford: Bios Scientific Publishers Limited.

Black, M., Butler, J., and Hughes, M. (1987). Control and development of dormancy in cereals. In Mares, D. (Ed.), *Proceedings of the Fourth Symposium on Preharvest Sprouting in Cereals* (pp. 379-392). Boulder, CO: Westview Press.

Bradford, K.J., Chen, F., Cooley, M.B., Dahal, P., Downie, B., Fukunaga, K.K., Gee, O.H., Gurusinghe, S., Mella, R.A., Nonogaki, H., et al. (2000). Gene expression prior to radicle emergence in imbibed tomato seeds. In Black, M., Bradford, K.J., and Vázquez-Ramos, J. (Eds.), *Seed Biology Advances and Applications* (pp.231-251). Oxon, UK: CAB International.

Brookes, P.A., Lovett, D.A., and MacWilliam, I.C. (1976). The steeping of barley: A review of the metabolic consequences of water uptake, and their practical implications. *Journal of the Institute of Brewing* 82: 14-26.

Buraas, T. and Skinnes, H. (1984). Genetic investigations on seed dormancy in barley. *Hereditas* 101: 235-244.

Buraas, T. and Skinnes, H. (1985). Development of seed dormancy in barley, wheat and triticale under controlled conditions. *Acta Agriculturae Scandinavica* 35: 233-244.

Carrari, L., Perez-Flores, J., Lijavetzky, D., Enciso, S., Sanchez, R.A., Benech-Arnold, R.L., and Iusem, N. (2001). Cloning and expression of a sorghum gene with homology to maize *vp1:* Its potential involvement in pre-harvest sprouting resistance. *Plant Molecular Biology* 45: 631-640.

Cochrane, M.P. (1993). Effects of temperature during grain development on the germinability of barley grains. *Aspects of Applied Biology* 36: 103-113.

Colucci, G., Apone, F., Alyeshmerni, N., Chalmers, D., and Chrispeels, M.J. (2002). GCR1, the putative *Arabidopsis* G protein-coupled receptor gene is cell cycle-regulated, and its overexpression abolishes seed dormancy and shortens time to flowering. *Proceedings of the National Academy of Sciences* 99: 4736-4741.

Corbineau, F. (1987). La germination des semences de tournesol et sa regulation par l'éthylene. *Comptes Rendus de l'Academie des Sciences de Paris, Série D.* 266: 477-479.

Corbineau, F., Bagniol, S., and Côme, D. (1990) Sunflower (*Helianthus annuus* L.) seed dormancy and its regulation by ethylene. *Israel Journal of Botany* 39: 313-325.

Corbineau, F. and Côme, D. (1980). Quelques caractéristiques de la dormance du caryopse d'Orge (*Hordeum vulgare* variété Sonja). *Comptes Rendus de l'Academie des Sciences de Paris, Série D.* 280: 547-550.

Corbineau, F. and Côme, D. (1987). Regulation de las semences de tournesol par l'éthylene. In *Annales ANPP, 2ème Colloque sur les substances de croissance et leurs utilisations en agriculture,* Volume 1 (pp. 271-282). Paris: Association Nationale de Protection des Plantes.

Corbineau, F., Poljakoff-Mayber, A., and Côme, D. (1991). Responsiveness to abscisic acid of embryos of dormant oat *(Avena sativa)* seeds: Involvement of ABA-inducible proteins. *Physiologia Plantarum* 83: 1-6.

Cseresnyes, Z. (1979). Studies on the duration of dormancy and methods of determining the germination of dormant seeds of *Helianthus annuus. Seed Science and Technology* 7: 179-188.

Fenner, M. (1991). The effects of the parent environment on seed germinability. *Seed Science Research* 1: 75-84.

Fong, F., Smith, J.D., and Koehler, D.E. (1983). Early events in maize seed development. *Plant Physiology* 73: 899-901.

Fonseca, A. and Sánchez, R.A. (2000). Efecto de la temperatura durante el llenado de grano sobre la germinación de semillas de girasol (*Helianthus annuus* L.). In Abstracts from the XXIII Reunión Argentina de Fisiología Vegetal (pp. 216-217). Córdoba, Argentina, November 27-30.

Gate, P. (1995). Ecophysiologie de la germination sur pied. *Perspectives Agricoles* 204: 22-29.

Gay, G., Corbineau, F., and Côme, D. (1991). Effects of temperature and oxygen on seed germination and seedling growth in sunflower (*Helianthus annus* L.) *Environmental and Experimental Botany* 31: 193-200.

Goldbach, H. and Michael, G. (1976). Abscisic acid content of barley grains during ripening as affected by temperature and variety. *Crop Science* 16: 797-799.

Gosling, P.G., Butler, R.A., Black, M., and Chapman, J.M. (1981). The onset of germination ability in developing wheat. *Journal of Experimental Botany* 32: 621-627.

Han, F., Ullrich, S.E., Claney, J.A., Jitkov, V., Kilian, A., and Romagosa, I. (1996). Verification of barley seed dormancy loci via linked molecular markers. *Theoretical and Applied Genetics* 92: 87-91.

Hattori, T., Terada, T., Hamasuna, S.T. (1994). Sequence and functional analyses of the rice gene homologous to the maize *Vp1. Plant Molecular Biology* 24: 805-810.

Hayes, P.M., Liu, B.H., Knapp, S.J., Chen, F., Jones, B., Blake, T., Franckowiak, J., Rasmusson, D., Sorrells, M., Ullrich, S.E., et al. (1993). Quantitative trait locus effects and environmental interaction in a sample of North American barley germplasm. *Theoretical and Applied Genetics* 87: 392-401.

Hilhorst, H.W.M. (1995). A critical update on seed dormancy: I. Primary dormancy. *Seed Science Research* 5: 61-73.

Holappa, L.D. and Walker-Simmons, M.K. (1995). The wheat abscisic acid-respon-sive protein kinase mRNA, PKABA1, is up-regulated by dehydration, cold temperature and osmotic stress. *Plant Physiology* 108: 1203-1210.

Hollung, K., Espelund, M., Schou, K., and Jakobsen, K.S. (1997). Developmental, stress and ABA modulation of mRNA levels for bZip transcription factors and *Vp1* in barley embryos and embryo-derived suspension cultures. *Plant Molecular Biology* 35: 561-571.

Johnson, R.R., Cranston, H.J., Chaverra, M.E., and Dyer, W.E. (1995). Characterization of cDNA clones for differently expressed genes in embryos of dormant and nondormant *Avena fatua* L. caryopses. *Plant Molecular Biology* 28: 113-122.

Jones, H.D., Peters, N.C.B., Holdsworth, M.J. (1997). Genotype and environment interact to control dormancy and differential expression of the *VIVIPAROUS-1* homologue in embryos of *Avena fatua*. *Plant Journal* 12: 911-920.

Karssen, C.M. (1995). Hormonal regulation of seed development, dormancy, and germination studied by genetic control. In Kigel, J. and Galili, G. (Eds.), *Seed Development and Germination* (pp. 333-350). New York: Marcel Dekker, Inc.

Karssen, C.M., Brinkhorst-Van der Swan, D.L.C., Breekland, A.E., and Koorneef, M. (1983). Induction of dormancy during seed development by endogenous abscisic acid: Studies on abscisic acid deficient genotypes of *Arabidopsis thaliana* (L.). *Planta* 157: 158-165.

Karssen, C.M., Zagorski, S., Kepczynski, J., and Groot, S.P.C. (1989). Key role for endogenous gibberellins in the control of seed germination. *Annals of Botany* 63: 71-80.

Kato, K., Nakamura, W., Tabiki, T., Mura, H., and Sawada, S. (2001). Detection of loci controlling seed dormancy on group 4 chromosomes of wheat and comparative mapping with rice and barley genomes. *Theoretical and Applied Genetics* 102: 980-985.

Kermode, A.R. (1995). Regulatory mechanisms in the transition from seed development to germination: Interactions between the embryo and the seed environment. In Kigel, J. and Galili, G. (Eds.), *Seed Development and Germination* (pp. 273-332). New York: Marcel Dekker, Inc.

Ketring, D.L. (1977). Ethylene and seed germination. In Khan, A.A. (Ed.), *The Physiology and Biochemistry of Seed Dormancy and Germination* (pp. 157-178). Amsterdam: Elsevier, North Holland Biomedical Press.

Khan, R.A. and Laude, H.M. (1969). Influence of heat stress during seed maturation on germinability of barley seed at harvest. *Crop Science* 9: 55-58.

King, R.W. (1982). Abscisic acid in seed development. In Khan, A.A. (Ed.), *The Physiology and Biochemistry of Seed Dormancy and Germination* (pp. 157-181). Amsterdam: Elsevier, North Holland Biomedical Press.

Kivi, E. (1966). The response of certain pre-harvest climatic factors on sensitivity to sprouting in the ear of two-row barley. *Acta Agriculturae Fennica* 107: 228-246.

Kleinhofs, A., Kilian, A., Saghai Maroof, M.A., Biyashev, R.M., Hayes, P.M., Chen, F.Q., Lapitan, N., Fenwich, A., Blake, T.K., Kanazin, V., et al. (1993). A molecular, isozyme and morphological map of the barley *(Hordeum vulgare)* genome. *Theoretical and Applied Genetics* 86: 705-712.

Larson, S., Bryan, G., Dyer, W., and Blake, T. (1996). Evaluating gene effects of a major barley seed dormancy QTL in reciprocal backcross populations. *Journal of Agricultural Genomics* 2: Article 4. Available at <http://www.cabi-publishing. org/gateways/jag/index.html>.

Le Page-Degivry, M.T., Barthe, P., and Garello, G. (1990). Involvement of endogenous abscisic acid in onset and release of *Helianthus annuus* embryo dormancy. *Plant Physiology* 92: 1164-1168.

Le Page-Degivry, M.T. and Garello, G. (1992). In situ abscisic acid synthesis: A requirement for induction of embryo dormancy in *Helianthus annuus*. *Plant Physiology* 98: 1386-1390.

Lenoir, C., Corbineau, F., and Côme, D. (1986). Barley *(Hordeum vulgare)* seed dormancy as related to glumella characteristics. *Physiologia Plantarum* 68: 301-307.

Li, B. and Foley, M.E. (1994). Differential polypeptide patterns in imbibed dormant and after-ripened *Avena fatua* embryos. *Journal of Experimental Botany* 45: 275-279.

Li, B. and Foley, M.E. (1995). Cloning and characterization of differentially expressed genes in imbibed dormant and after-ripened *Avena fatua* embryos. *Plant Molecular Biology* 29: 823-831.

Lijavetzky, D., Martinez, M.C., Carrari, F., and Hopp, H.E. (2000). QTL analysis and mapping of pre-harvest sprouting resistance in *Sorghum*. *Euphytica* 112: 125-135.

Livers, R.W. (1957). Linkage studies with chromosomal translocation stocks in barley. PhD Thesis (Diss. Abstr. AAT 5801125). St. Paul: University of Minnesota.

Lona, F. (1956). L'acido gibberéllico determina la germinazione del semi di *Lactuca scariola* in fase di scotoinhibizione. *Ateneo Pamense* 27: 641-644.

Mares, D. (1984). Temperature dependence of germinability of wheat *(Triticum aestivum)* grain in relation to pre-harvest sprouting. *Australian Journal of Agricultural Research* 35: 115-128.

McCarty, D.R., Hattori, T., Carson, C.B., Vasil, V., Lazar, M., and Vasil, I.K. (1991). The *viviparous-1* developmental gene of maize encodes a novel transcriptional activator. *Cell* 66: 895-905.

McKibbin, R.S., Wilkinson, M.D., Bailey, P.C., Flintham, J.E., Andrew, L.M., Lazzeri, P.A., Gale, M.D., Lenton, J.R., and Holdsworth, M.J. (2002). Transcripts of *Vp-1* homeologues are misspliced in modern wheat and ancestral species. *Proceedings of the National Academy of Sciences* 99: 10203-10208.

Meyer, K., Leube, M.P., and Grill, E. (1994). A protein phosphatase 2C involved in ABA signal transduction in *Arabidopsis thaliana*. *Science* 264: 1452-1455.

Morris, P.C., Jewer, P.C., and Bowles, D.J. (1991). Changes in water relations and endogenous abscisic acid content of wheat and barley grains and embryos during development. *Plant, Cell and Environment* 14: 443-446.

Muir, C.E. and Nilan, R.A. (1973). Registration of Steptoe barley. *Crop Science* 13: 770.

Neil, S.J. and Horgan, R. (1987). Abscisic acid and related compounds. In Rivier, L. and Crozier, A. (Eds.), *The Principles and Practice of Plant Hormone Analysis* (pp. 111-167). London: Academic Press.

Nicholls, P.B. (1982). Influence of temperature during grain growth and ripening of barley on the subsequent response to exogenous gibberellic acid. *Australian Journal of Plant Physiology* 9: 373-383.

Norstog, K. and Klein, R.M. (1972). Development of cultured barley embryos: II. Precocious germination and dormancy. *Canadian Journal of Botany* 50: 1887-1894.

Oberthur, L., Dyer, W., Ullrich, S.E., and Blake, T.K. (1995). Genetic analysis of seed dormancy in barley (*Hordeum vulgare* L.) *Journal of Agricultural Genomics* 1: Article 5. Available at <http://www.cabi-publishing.org/gateways/jag/index.html>.

Pollock, J.R.A. (1962). The nature of the malting process. In Cook, A.M. (Ed.), *Barley and Malt: Biology Biochemistry, Technology* (pp. 303-398). New York: Academic Press.

Quarrie, S.A., Tuberosa, R., and Lister, P.G. (1988). Abscisic acid in developing grains of wheat and barley genotypes differing in grain weight. *Plant Growth Regulation* 7: 3-17.

Rasmusson, D.C. and Wilcoxson, R.D. (1979). Registration of Morex barley. *Crop Science* 19: 293.

Reiner, L. and Loch, V. (1976). Forecasting dormancy in barley—Ten years experience. *Cereal Research Communication* 4: 107-110.

Robichaud, C.S., Wong, J., and Sussex, I.M. (1980). Control of in vitro growth of viviparous embryo mutants of maize by abscisic acid. *Developmental Genetics* 1: 325-330.

Rodriguez, V., González Martín, J., Insausti, P., Margineda, J.M., and Benech-Arnold, R.L. (2001). Predicting pre-harvest sprouting susceptibility in barley: A model based on temperature during grain filling. *Agronomy Journal* 93: 1071-1079.

Romagosa, I., Han, F., Clancy, J.A., and Ullrich, S.E. (1999). Individual locus effects on dormancy during seed development and after ripening in barley. *Crop Science* 39: 74-79.

Romagosa, I., Ullrich, S.E., Han, F., and Hayes, P.M. (1996). Use of additive main effects and multiplicative interaction model in QTL mapping for adaptation in barley. *Theoretical and Applied Genetics* 93: 30-37.

Schopfer, P. and Plachy, C. (1985). Control of seed germination by abscisic acid: III. Effect on embryo growth potential (minimum turgor pressure) and growth coefficient (cell wall extensibility) in *Brassica napus* L. *Plant Physiology* 77: 676-686

Schuurink, R.C., Van Beckum, J.M.M., and Heidekamp, F. (1992). Modulation of grain dormancy in barley by variation of plant growth conditions. *Hereditas* 117: 137-143.

Seiler, G.J. (1998). Seed maturity, storage time and temperature, and media treatment effects on germination of two wild sunflowers. *Agronomy Journal* 90: 221-226.

Srivastava, A.K. and Dey, S.C. (1982). Physiology of seed dormancy in sunflower. *Acta Agronomica Academiae Scientarum Hungaricae* 31: 70-80.

Steinbach, H.S., Benech-Arnold, R.L., Kristof, G., Sánchez, R.A., and Marcucci Poltri, S. (1995). Physiological basis of pre-harvest sprouting resistance in *Sorghum bicolor* (L.) Moench. ABA levels and sensitivity in developing embryos of sprouting resistant and susceptible varieties. *Journal of Experimental Botany* 45: 701-709.

Steinbach, H.S., Benech-Arnold, R.L., and Sánchez, R.A. (1997). Hormonal regulation of dormancy in developing *Sorghum* seeds. *Plant Physiology* 113: 149-154.

Takeda, K. (1996). Varietal variation and inheritance of seed dormancy in barley. In Noda, K. and Mares, D. (Eds.), *Pre-Harvest Sprouting in Cereals 1995* (pp. 205-212). Osaka, Japan: Center for Academic Societies.

Ullrich, S.E., Han, F., Blake, T.K., Oberthur, L.E., Dyer, W.E., and Clancy, J.A. (1996). Seed dormancy in barley: Genetic resolution and relationship to other traits. In Noda, K. and Mares, D. (Eds.), *Pre-Harvest Sprouting in Cereals 1995* (pp. 157-163). Osaka, Japan: Center for Academic Societies.

Ullrich, S.E., Hayes, P.M., Dyer, W.E., Blake, T.K., and Clancy, J.A. (1993). Quantitative trait locus analysis of seed dormancy in 'Steptoe' barley. In Walker-Simmons, M.K. and Ried, J.L. (Eds.), *Pre-harvest sprouting in cereals 1992* (pp. 136-145). St. Paul, MN: American Society of Cereal Chemistry.

Van Beckum, J.M.M., Libbenga, K.R., and Wang, M. (1993). Abscisic acid and gibberellic acid-regulated responses of embryos and aleurone layers isolated from dormant and nondormant barley grains. *Physiologia Plantarum* 89: 483-489.

Vegis, A. (1964). Dormancy in higher plants. *Annual Review of Plant Physiology* 15: 185-224.

Visser K., Visser, A.P.A., Cagirgan, M.A., Kijne, J.W., and Wang, M. (1996). Rapid germination of a barley mutant is correlated with a rapid turnover of abscisic acid outside the embryo. *Plant Physiology* 111: 1127-1133.

Walker-Simmons, M.K. (1987). ABA levels and sensitivity in developing wheat embryos of sprouting resistant and susceptible cultivars. *Plant Physiology* 84: 61-66.

Walker-Simmons, M.K. (1988). Enhancement of ABA responsiveness in wheat embryos by high temperature. *Plant, Cell and Environment* 11: 769-775.

Walker-Simmons, M.K. and Sesing, J. (1990). Temperature effects on embryonic abscisic acid levels during development of wheat grain dormancy. *Journal of Plant Growth Regulation* 9: 51-56.

Wang, M., Bakhuizen, R., Heimovaara-Dijkstra, S., Zeijl, M.J., De Vries, M.A., Van Beckum, J.M., and Sinjorgo, K.M.C. (1994). The role of ABA and GA in barley grain dormancy: A comparative study between embryo dormancy and aleurone dormancy. *Russian Journal of Plant Physiology* 41: 577-584.

Wang, M., Heimovaara-Dijkstra, S., and Van Duijn, B. (1995). Modulation of germination of embryos isolated from dormant and nondormant grains by manipulation of endogenous abscisic acid. *Planta* 195: 586-592.

Wang, M., van der Meulen, R.M., Visser, K., Van Schaik, H.-P., Van Duijn, B., and de Boer, A.H. (1998). Effects of dormancy-breaking chemicals on ABA levels in barley grain embryos. *Seed Science Research* 8: 129-137.

Wulff, R.D. (1995). Environmental maternal effects on seed quality and germination. In Kigel, J. and Galili, G. (Eds.), *Seed Development and Germination* (pp. 491-505). New York: Marcel Dekker, Inc.

Xu, N., Coulter, K.M., and Bewley, J.D. (1991). Abscisic acid and osmoticum prevent germination of developing alfalfa embryos, but only osmoticum maintains the synthesis of developmental proteins. *Planta* 182: 382-390.

Chapter 6

Preharvest Sprouting of Cereals

Gary M. Paulsen
Andrew S. Auld

INTRODUCTION

Preharvest sprouting of cereals is defined as germination of physiologically ripe kernels before harvest (Derera, 1989b). This simple definition encompasses numerous factors: maturation, ripening, and after-ripening of grain; innate dormancy; the presence of conditions to initiate germination; induction of enzymatic activities; involvement of plant hormones; and suitability of the grain for its intended use. The classical definition of germination as the sum total of processes preceding and including protrusion of the radicle/coleorhiza through the surrounding structures (Hilhorst and Toorop, 1997) may not be entirely appropriate to the study of preharvest sprouting. Changes that occur early in the endosperm before new seedling tissues may be so deleterious as to make sprouted grain unfit for many purposes.

Preharvest sprouting is usually associated with prolonged or repeated rain, heavy dew, high humidity, and low temperature following ripening of the grain (Nielsen et al., 1984). The conditions that favor sprouting often compound the problem by delaying harvest. Such conditions occur throughout the world: northern and western Europe; parts of Africa; tropical and semitropical Asia, including southeastern China; northern Australia; northern and northwestern areas of the United States and adjacent areas in Canada; and a broad band across South America (Derera, 1989). The problem may be exacerbated by cultivation of susceptible crops in those areas. An example is production of white wheat (*Triticum aestivum* L.), which has little resistance to preharvest sprouting in the northern wheatbelt of Australia. However, damage also occurs with some frequency even in areas where conditions do not normally favor sprouting. For instance, in Kansas, the major wheat state in the United States, hot, dry weather following ripening

This chapter is contribution number 02-316-B from the Kansas Agricultural Experiment Station.

of resistant hard red wheat usually results in little preharvest sprouting. Still, significant damage occurred in parts of the state during 1979, 1989, 1993, and 1999, when conditions were particularly favorable.

Instances of preharvest sprouting have been reported for all cereals. Most damage occurs to common wheat because it is the most widely grown of all cereals, including cultivation in areas where sprouting is likely to occur. Many cultivars are susceptible, and sprouting is highly deleterious to some of the products.

Rye (*Secale cereale* L.) is particularly susceptible to preharvest sprouting because the flowers are cross-pollinated, and the open structures of the glumes allow water to reach the grain (Derera, 1989b). Preharvest sprouting of barley (*Hordeum vulgare* L.) damages the quality of the grain for baking and viability of the kernels for malting. However, germination may increase digestibility of both crops and enhance their feeding value for livestock. Sprouting of oat (*Avena sativa* L.) occurs episodically in some areas, particularly northern regions, but has little effect on the quality of the grain for feed. Triticale (×*Triticosecale* Wittmack), like rye, is extremely liable to preharvest sprouting. Sprouting has the same deleterious effect on baking quality of triticale as on wheat, but the bulk of the crop is used for livestock and its value is affected little. Preharvest sprouting of maize (*Zea mays* L.) is usually associated with vivipary (Smith and Fong, 1993) because the grain is protected by the husk from the moist conditions that promote germination in other cereals. Sorghum [*Sorghum bicolor* (L.) Moench] and pearl millet [*Pennisitum glaucum* (L.) R. Br.] are rarely subject to preharvest sprouting because of the semiarid nature of the regions where they are grown. However, grain of both species sprouts when conditions are appropriate. Japonica rice (*Oryza sativa* L.), which is usually grown in more temperate regions, sprouts more easily than Indica rice of the tropics, which is highly resistant (Yamaguchi et al., 1998).

The most extensive survey of direct losses to producers from preharvest sprouting of cereals was reported by Derera (1989c). Average annual losses in 37 countries totaled over US$450 million, mostly to wheat, from 1978 to 1988. However, the major cereal-producing countries of China, India, USSR, and Argentina were not included in the survey, and estimates were not available from the United States and several other countries. Sprouting of durum (*Triticum durum* Desf.) wheat in the northern United States alone caused several hundred million dollars of damage over a decade (Dick et al., 1989). It is likely that total worldwide direct annual losses currently approach US$1 billion.

Direct economic losses to producers from preharvest sprouting occur in several ways. The yield may be reduced by loss of dry matter and shattering of the grain, the volume density (test weight) may decrease from loss of dry matter and irreversible swelling of the kernels, and suitability of the grain for many food products may be diminished. Because payments to producers in the United States and many other countries are determined by the yield, volume density, and grade of the grain, any of the effects of sprouting reduce their income. In the United States, for instance, more than 4 percent damaged kernels—including sprouted kernels—causes hard wheat to be rated Grade 3 or lower and unacceptable for bread making. The loss in value from diminished quality, however, is usually offset by use of sprouted grain for livestock feed.

Indirect losses should be added to the total economic losses from preharvest sprouting of cereals. Traditional markets are lost when exporters cannot supply sound grain to customers (Briggle, 1979). Producers in parts of China would benefit from growing white wheat because of higher payments from the government but must grow red wheat because of the hazard of sprouting damage (Paulsen, 1985). Production of some cereals in the humid tropics is limited, in part, by the possibility of preharvest sprouting.

THE PREHARVEST SPROUTING PROCESS

Absorption of moisture by kernels is influenced by morphology of the inflorescence, characteristics of the seedcoat, turgor of the embryo, and chemical properties of the caryopsis (King, 1989). Environmental factors, particularly temperature, also affect imbibition by influencing the properties of water (Murphy and Noland, 1982). Dry grain (9 to 12 percent moisture) has an extremely low water potential, -400 MPa in the case of wheat, and so readily imbibes water (Shakeywich, 1973). Fifty percent germination occurs at a threshold water potential of 0.8 to 1.0 MPa or about 45 percent seed moisture content (King, 1989). Cereals do not have impermeable seedcoats as do legumes, and the critical moisture content for germination in freely available water is reached in about 3 h. However, cultivars differ substantially in the rate of imbibition. Many factors have been implicated in controlling imbibition, but no single factor has been identified. Conditions that influence imbibition by wheat and other cereals were reviewed by King (1989). Imbibition is increased by features associated with awns in wheat and is affected by waxiness, pubescence, and angle of the inflorescence in barley. Grain hardness, color, restriction by the seedcoat, thickness of the testa and other layers, size, and surface:volume ratio are implicated in some studies but not others (King, 1989). The rate of drying of the spike and grain after moisture becomes unavailable is also likely to affect sprouting, but appears to be determined solely by evaporation and does not differ among cultivars. Temperature affects imbibition by influencing the viscosity of water and probably the wetability of tissues (Vertucci and Leopold, 1986), as well as the rate of drying by evaporation.

Water enters the grain most rapidly via tissues that overlay the embryo (Evers, 1989). King (1989) concluded that the main path of water to the embryo must be laterally through the pericarp. Starch in the endosperm of cereals is much more hydrophobic than the contents of the embryo (Chung and Pfost, 1967).

Movement of the water front through the kernel initiates numerous processes in the embryo, endosperm, and associated tissues. Absorbed gasses are released, membranes are reorganized, mitochondria develop, endogenous enzymes are activated, and new enzymes appear by de novo synthesis (McDonald, 1994). Most of the deleterious changes during sprouting occur from mobilization of reserves in the endosperm. Most attention has been given to hydrolysis of starch, but many other substrates in the endosperm—proteins, lipids, phytin, etc.—are degraded to provide substance for the embryo and developing seedling. Although the changes in the endosperm are most prominent, they are mostly controlled by the embryo/scutellum (King, 1989).

Investigations of biochemical and physiological changes in cereals during preharvest sprouting have emphasized the enzymes involved. Enzymes catalyze the biochemical processes, and changes in their activities are among the most pronounced effects of preharvest sprouting. They are also responsible for most of the deleterious changes that occur. Several measures of preharvest sprouting are based on changes in enzyme activity.

Hydrolysis of starch in the endosperm to simple sugars for use by the embryo involves numerous enzymes: endoamylases (α-amylases), debranching enzymes (R-enzyme, pullulanase), isoamylase, exoamylase (β-amylase), and α-glucosidase (maltase) (Beck and Ziegler, 1989). α-Amylase is the only enzyme that can hydrolyze raw starch. It cleaves 2(164) glucosidic bonds in amylose and amylopectin and is often considered to be the key to the problem of preharvest sprouting (Duffus, 1989).

α-Amylase in cereals is commonly divided into two types, an endogenous late maturity, green, or low-pI group and a germination or high-pI group that is associated with sprouting (Kruger, 1989). Numerous isozymes occur within both groups, and differences between groups are not distinct. Late-maturity α-amylase also occurs during germination, and some isozymes that form during maturation have pIs that are typical of isozymes that appear during germination (Mares and Mrva, 1993). Production of late-maturity α-amylase is controlled by the pericarp or embryo, whereas α-amylases that form during germination are associated with the aleurone and/or the scutellum (Kruger, 1989).

The relative importance of α-amylase activity that was retained in the pericarp or formed during late maturity, during germination before maturation, or during germination after maturation of wheat was assessed by Lunn and colleagues (2001). Late-maturity α-amylase was most widespread, occurring in 25 of 32 cultivar × location × year instances where sprouting was identified. However, α-amylase that formed during sprouting occurred in 21 of the 32 instances and was primarily responsible for damage to the grain from preharvest sprouting. Late-maturity, low-pI α-amylase in the wheat cultivar Chinese Spring was controlled by a single recessive gene on the long arm of chromosome 6BL, and synthesis of high-pI α-amylase in the aleurone also involved a gene on the long arm of chromosome 6BL (Mrva and Mares, 1998).

Debranching enzymes, including isoamylase, as their name suggests, hydrolyze α-(1→6)-glucosidic bonds in amylopectin to accelerate breakdown of the starch by α-amylase (Kruger, 1989). The enzyme is formed during early stages of maturation, and at least two isozymes occur in wheat. In some species, the enzyme accumulates in an inactive form that is liberated by proteolysis during germination (Beck and Ziegler, 1989).

β-Amylases hydrolyze alternate α (164) bonds of starch to form maltose. They develop during maturation and occur in free and bound forms in ripe grain (Beck and Ziegler, 1989). The enzyme is present in both the pericarp and the endosperm, although the former disappears during maturation and only the latter structure contains the enzyme when the grain ripens (Kruger, 1989). β-Amylases may be important in preharvest sprouting of wheat if insufficient activity relative to α-amylases leads to an accumulation of dextrins that make bread crumbs sticky (Duffus, 1989).

Maltase is generally present at low levels in mature grain and increases by de novo synthesis during germination. It is mostly produced in the scutellum and secreted into the endosperm (Beck and Ziegler, 1989).

Proteolytic enzymes function during sprouting to mobilize N for the embryo and seedling and to release bound or inactive enzymes. β-Amylase, debranching enzymes, and probably other enzymes are complexed with proteins during maturation and then freed by proteolysis during germination (Beck and Ziegler, 1989).

Many types of proteolytic enzymes—endopeptidases, carboxypeptidases, aminopeptidases, etc.—are associated with sprouting. Knowledge of their roles is complicated by their complexity, difficulty of extraction and purification, and differing reactions to assay conditions and substrates (Kruger, 1989). An endopeptidase that attacks modified gluten occurs in nongerminated wheat, and it and another endopeptidase increase rapidly during germination (McMaster et al., 1989). The first enzyme was apparently distributed throughout the endosperm, whereas the enzyme that was activated by germination was derived from the aleurone and scutellum. Activity of the enzymes in damaged grain affected quality of the dough for bread, and activity during processing affected the quality for alkaline noodles. Nongerminated barley, in contrast, contained endopeptidase that hydrolyzed edistin but not gelatin or hordein (Jones and Wrobel, 1993). Germination induced a number of proteinases, most of which were associated with aleurone, scutellar, and endosperm tissues.

A carboxypeptidase in the endosperm of wheat increases throughout maturation, whereas one in the outer layers of the kernel disappears (Kruger, 1989). During germination, activity of the enzyme in the endosperm near the scutellar epithelium increases, apparently due to dissipation of inhibitors.

Lipases have little or no activity in nongerminated cereals (Jensen and Heltved, 1982). During germination, activity appears first in the scutellum, followed by the scutellum-endosperm interface, and then gradually progresses throughout the endosperm. Enhanced activity of lipases may affect the viability of sprouted kernels during storage (Kruger, 1989). However, the increase in lipases is usually much smaller than the change in α-amylase (Fretzdorff, 1993).

Activity of many other enzymes increases markedly during sprouting of cereals. Phytases increase to release phosphorus for the new seedling. Monophenol oxidase and polyphenol oxidase may increase up to 33-fold and cause the gray crumb discoloration of bread and the off color of noodles made from sprouted wheat (Kruger, 1989). However, the increase in polyphenol oxidase is typically much less than α-amylase, and little of the enzyme occurs in flour after milling (Kruger and Hatcher, 1993). Catalases and peroxidases catalyze oxidative reactions that may affect the rheological properties of dough. Other enzymes that increase during sprouting, such as ribonucleases, are important for the developing seedling but have no known effects on cereal products.

PHYSIOLOGICAL CONTROL
OF PREHARVEST SPROUTING

Preharvest sprouting is affected at numerous levels by factors ranging from inhibitors in awns (bracts) to control of α-amylase synthesis by gibberellic acid (GA) (Gale, 1989). As discussed in Chapter 5, many of these factors involve dormancy, which is the inability of a viable, mature seed to germinate even under favorable conditions (Hilhorst and Toorop, 1997). Of the two types of dormancy— coat-imposed dormancy derived from the presence of endosperm plus pericarp (plus glumellae in the case of barley) and true embryo dormancy—only the former occurs in cereals. Immature embryos of barley, rice, and wheat, for example, rapidly germinate when they are removed from developing kernels and placed in water or other media (Kermode, 1990).

All of the factors discussed earlier that affect imbibition also influence sprouting. Resistance to expansion of the germinating kernel and its embryo by the pericarp and testa may also inhibit germination (Wellington, 1956). Other unknown factors may cause differences in the rate of germination among genotypes even when the rate of imbibition and other traits are similar (Gale, 1989).

Inhibitors of various types play central roles in preharvest sprouting. The glumes (bracts) of wheat, for instance, contain an unknown inhibitor that delays sprouting and is simply inherited (Derera and Bhatt, 1980; Wu and Carver, 1999). Similarly, the well-known resistance to preharvest sprouting of red wheats relative to white wheats has been attributed to precursors of the pigment phlobaphene in the testa layer of the former (Miyamoto, Tolbert, and Everson, 1961). These compounds, catechin and tanninlike materials, occurred in lower amounts in white wheats than in red wheats, in which they declined during after-ripening to permit germination. Pigments in the seedcoat may be part of a two-factor system that inhibits germination directly or by interfering with gaseous exchange (Mares, 1998).

Numerous plant growth substances are directly implicated in preharvest sprouting. In addition to the pregerminative action of gibberellic acid discussed in Chapter 5, the postgerminative role of GA from the embryo and scutellum in inducing synthesis of α-amylase in the aleurone is well known (Beck and Ziegler, 1989). Debranching enzymes, maltase, some proteinases, phytase, ribonuclease, and others are also induced by GA, but the mechanisms may differ. Whereas de novo synthesis and secretion of α-amylase are stimulated by GA, other enzymes may be activated by GA or would increase even without GA (King, 1989). In barley and presumably other cereals, GA promotes accumulation of a-amylase mRNA in the aleurone and involves synthesis of a protein factor for efficient expression (Muthukrishnan, Chaudra, and Maxwell, 1983). The GA acts as a positive regulator of expression of α-amylase genes in vivo in barley (Chandler and Mosleth, 1989).

Abscisic acid has many imputed functions in addition to regulating developmental changes from maturation to germination (Kermode, 1990). Dormancy of cereals is roughly proportional to their abscisic acid (ABA) content, suggesting that the growth substance is involved in both the initiation and maintenance of nongerminability (King, 1989). Other work suggests that wheat cultivars that differ in dormancy have similar contents of ABA but vary in sensitivity to the compound (Walker-Simmons, 1987). The role of this hormone in controlling the timing of dormancy release in cereal grains is thoroughly treated in Chapter 5. In addition, and from a post-germinative standpoint, it has been found that an ABA-responsive protein kinase mRNA mediated the suppression of GA-inducible genes in the aleurone of wheat (Walker-Simmons et al., 1998).

Other growth substances—jasmonic acid, ethylene, and cytokinins—modify germination of many species but have not been studied extensively in cereals (Hilhorst and Toorop, 1997). Liu and colleagues (1998) con-cluded that indole-3-acetic acid (IAA) inhibited germination and acted in concert with GA and cytokinins to regulate the process. The observation that tryptophan, a purported precursor of IAA, inhibited sprouting of resis-tant wheat cultivars supports a role for the auxin in controlling germination (Morris et al., 1988).

Several proteins, mostly albumins, that inhibit endogenous α-amylase occur in wheat and barley (Gale, 1989). The proteins increase during ger-mination and may control α-amylase activity. Only proteins from sprout-ing-resistant genotypes inhibited α-amylase from a sprouting-susceptible . genotype of wheat (Abdul-Hussain and Paulsen, 1989). However, adding ethylenediaminetetraacetic acid (EDTA) to chelate calcium caused inhibi-tion by all genotypes, suggesting that the proteins interacted with the metal. Phytic acid from the bran also inhibited α-amylase activity in wheat, again by lowering the level of the calcium cofactor (Cawley and Mitchell, 1968).

QUALITY OF PRODUCTS FROM SPROUTED CEREALS

The consequences of preharvest sprouting directly depend on the types of products for which the cereal is intended and on the processing methods used. Severely sprouted grain might be blended with sound grain in some cases and almost always has considerable residual value as livestock feed. Sprouting might even increase the value of cereals for feed by making them more palatable and digestible.

Breads

Breads baked from hard wheats are affected more than most products by preharvest sprouting of the grain. Production of the bread is complicated by extreme stickiness of the dough, which necessitates special handling in small bakeries and can disrupt operations of large bakeries. Even slicing bread made from sprouted wheat can be difficult, and the resulting loaves are often cavitated and grayish.

Extensive studies with field-sprouted wheats by Kulp, Roewe-Smith, and Lorenz (1983) and Lorenz and colleagues (1983) illustrate the problem. Sprouting weakened the dough strength, decreased the amylograph peak viscosities, and caused poor handling and machining properties. Loaf volume increased, but the internal quality was poor. Thickening ability of starch from sprouted wheat was adversely affected.

Stickiness of dough is usually attributed to extensive enzymic hydrolysis of damaged starch and altered rheological properties to proteolytic enzymes (Kruger, 1989). Kulp, Roewe-Smith, and Lorenz (1983) and Lorenz and colleagues (1983) also concluded that elevated α-amylase and proteinase activities were responsible. However, electron microscopy and X-ray diffraction found no changes in starch that were attributable to sprouting. Sticky dough might also be caused by the limit dextrins that result from excessive α-amylase relative to β-amylase activity and possibly α-amylase to debranching activity (Duffus, 1989). Proteinases and lipases that open the starch granule to amylosis might also be involved.

Hearth breads appear to be degraded less than Western-style pan breads by sprouted grain. Seven of nine international leavened and unleavened breads from flour from soft white wheat that sprouted in the field had acceptable quality (Finney et al., 1980).

Cakes and Cookies

Field sprouting of soft wheats had little effect on crumb properties of sponge cake but increased the cake volume at low levels and decreased it at high levels of sprouting (Finney et al., 1981). Sprouting of hard wheats also increased the volume and coarsened the grain of yellow cake (Lorenz et al., 1983); however, the cake texture was smoother and softer. In other studies, sprouting caused poor baking quality, a depressed center, coarse grain, and a firm texture in cakes (Lorenz and Valvano, 1981). When flour from sprouted grain was used for cookies, the spread increased and the top grain score improved, but the crust darkened.

Speciality Batters

High α-amylase in sprouted wheat generally reduces the quality of batters for many uses (Nagao, 1995). Batter for coating fish and vegetables as tempura loses its light and viscous character and coats poorly. Batter for *takoyaki*, coated octopus tentacles, loses its shape. When used for Japanese muffins with a sweet bean filling, the batter may not be viscous enough to cover the contents.

Pasta

Sprouting affects both the processing and quality of the many kinds of noodles that are made from wheat. In dry noodles, high α-amylase weakens the dough so that the noodles cannot support their own weight and break during the dehydration process (Nagao, 1995). For wet noodles, where color, brightness, and texture are of major importance, the enzymes that increase during sprouting seriously affect product quality (Kruger, Hatcher, and Dexter, 1995). Cantonese noodles were slightly less bright when they were prepared from sprouted wheat but had similar textural properties as noodles from sound wheat. Raw noodles differed only slightly in firmness and resistance to compressibility.

Changes in noodle quality are usually attributed to α-amylase, proteinase, and polyphenol oxidase enzymes that increase during sprouting (Kruger, 1989). α-Amylase might be most problematic, since it typically increases several-thousand-fold during sprouting, and over 75 percent of the activity in whole meal occurs in the flour (Kruger and Hatcher, 1993). Effects of increased proteinase activity may be overshadowed by the α-amylase (Kruger, 1989). Polyphenol oxidase, in contrast to α-amylase, increases only about 2.5-fold during sprouting and is localized in the bran so that only about 1 percent of the activity occurs in the flour. Low water absorption of flours may limit the mobility between the enzymes and their substrates, and the brief processing time may limit the period for deterioration to occur during preparation of noodles compared with bread (Kruger, Hatcher, and Dexter, 1995).

Alcoholic Beverages and Glucose Syrups

Amylosis is a primary step in the processing of cereals for beverages and syrups. Starch is hydrolyzed to dextrins for beer and glucose syrup but must be completely converted to fermentable sugars for production of alcohol and spirits (Peiper, 1998). Malt from barley is generally used for controlled, uniform amylosis in most fermentation processes. Cultivars that have low or no dormancy are greatly preferred for malting (Aastrup, Riis, and Munck, 1989). However, preharvest sprouting of the barley may shorten its viability during storage, lower conversion of the malt and extractability of fermentable material, and increase the growth of molds (Kruger, 1989).

Seed Quality

The quality of sprouted grain for seed concerns seedsmen and farmers. Wheat seed that is severely sprouted loses viability rapidly, is easily damaged, and deteriorates quickly during storage (Elias and Copeland, 1991; Barnard and Purchase, 1998). The emergence percentage in the field may be substantially lower than the germination percentage in the laboratory, and stands may be reduced further by treatment of the seed with fungicides (Barnard and Purchase, 1998). Adverse storage conditions may accelerate the decline in seed germination and seedling vigor (Stahl and Steiner, 1998). However, it is doubtful that severely sprouted seed appears often in commercial channels.

Wheat seed that has low or incipient levels of sprouting (mean Hagberg falling number of 107) but is otherwise sound may be used for planting (Foster, Burchett, and Paulsen, 1998). Germination and emergence from deep planting and field establishment declined in some cultivars, but grain yields were not affected even after storing the seed for 27 months.

MEASUREMENT OF PREHARVEST SPROUTING

Evaluation of damage to grain from preharvest sprouting and of resistance to the problem involves numerous considerations (Mares, 1989; Wu and Carver, 1999). Routine assays of sprouting mostly involve proper storage and preparation of the samples and measurement of sprouting damage. Experimental evaluation of sprouting resistance has similar requirements plus the inclusion of a suitable wetting treatment to induce sprouting.

Sampling

Grain should be sampled at uniform stages of development to avoid differences in dormancy during maturation. This might be at physiological maturity, when the grain contains 25 to 35 percent moisture, or at harvest ripeness, when the grain moisture is at 12 to 13 percent. Samples may be either assayed immediately or dried to 15 percent or less moisture and stored for future use. Drying with ambient air is often satisfactory and forced air, if used, should not exceed 30°C and only for the briefest duration. Lyophilization is not recommended in most cases because of the possibility of freeze damage to the embryo and other parts of the grain. Once the sample is dry, it can be held at room temperature, where it will after-ripen naturally but incur little change in α-amylase activity and most constituents. Alternatively, dried samples may be held at –20°C to arrest the after-ripening process and preserve dormancy. If the intact spike is used, 5 to 10 cm of culm should be left for handling and if the grain is threshed, hand rubbing or dissection should be used instead of mechanical methods to avoid damage to the kernels. Detailed discussions of these considerations are given by Mares (1989).

Controlled Sprouting

Assaying sprouting in intact spikes is often preferred over other methods because it incorporates differences in wetting, water movement, inhibitors in the glumes, and other factors. Rain simulators of various types, misting chambers, immersion in water, burial in moist sand, and other methods are used. Rain simulators do not duplicate the kinetic energy, velocity, and random size of natural rain (King, 1989), but consistent results among various methods suggest that the effect is small (Mares, 1989). Sprouting of kernels in the spike can be assessed visually, by dissecting the grain, and by analyzing α-amylase activity.

Sprouting of kernels on commercial germination paper, filter paper, or other media, often in petri dishes, is practiced routinely. The technique is criticized for its lack of physiological integrity, particularly in water content, but it does measure relative dormancy under standard conditions (Mares, 1989). Results may be expressed as percentage of kernels germinating, time to 50 percent germination, or other expressions. A germination promptness index,

$$GPI = \sum_{i=1}^{n} \frac{d_i}{i}, \qquad (6.1)$$

where d_i = number of kernels germinating on day i, incorporates both the rate and magnitude of germination.

Excised embryos are germinated on media with various adjuncts because of the importance of their response to preharvest sprouting. Sensitivity of the embryos to ABA, catechins, or other substances is associated with resistance to sprouting and controlled simply by a few genes in wheat and barley (Gale, 1989). Embryos are easily dissected from kernels and germinate rapidly in water or basal media.

Measuring Sprouting

Various methods are available for measuring sprouting, and their choice depends on the requirements for the test. Visual counting of sprouted kernels and the Hagberg falling number are generally used for commercial samples, whereas these and other methods, many of which measure activities of enzymes, are used experimentally.

Visual counting of sprouted kernels is usually the first measure of damage to grain at local elevators and terminal elevators. In the United States, sprouted wheat is rated by the Federal Grain Inspection Service (1997) as "kernels with the germ end broken open from germination and showing sprouts or from which the sprouts have broken off" (p. 32). The method, which judges kernels to be sprouted or not sprouted, lacks precision and typically has extremely high coefficients of variation for experimental use. In addition, considerable damage from elevated α-amylase activity may occur well before any seedling structures are evident.

The Hagberg falling number method (Falling Number Corp., Huddinge, Sweden) measures the time in seconds for a plunger to fall through the gelatinized starch in a slurry of ground grain. Values range from 60 for highly sprouted grain to 500 or higher for sound grain; a minimum of 250 to 300 is generally required for wheat grain for bread. The procedure is affected by a number of factors but is rapid, simple, and gives good precision. It has been adopted as the official method by several associations and much of the grain industry. The relatively large sample sizes, 300 g for the sample and 7 g for the assay, are barriers for many experimental uses.

The stirring number as measured by the Rapid Visco Analyzer (Foss Technology Corp., Eden Prairie, Minnesota) is another viscometric method that uses a slightly smaller sample (4 g) than the falling number procedure and is highly reproducible. The Brabender amylograph (C.W. Brabender Co., South Hackensack, New Jersey) measures changes in the viscosity of a flour-water paste with increased temperature. The method detects low levels of sprouting and other properties. It has the disadvantages of requiring a large sample (60 g) and long running time (ca. 60 min).

Falling number, stirring number, and amylograph peak viscosity values are affected by sample size, temperature, and often by barometric pressure (Koeltzow and Johnson, 1993). The values are usually highly correlated, but results cannot be easily converted from one method to another.

A number of methods are available for measuring α-amylase. Older methods that determine gas production or reducing power are used infrequently because they are time-consuming and often require specialized equipment. Several procedures measure α-amylase activity with a dye-labeled starch substrate. The action of the enzyme on the substrate liberates the soluble dye, which is usually measured spectrophotometrically. The method requires a constant temperature water bath, shaker, centrifuge, spectrophotometer, and other equipment and is not suited for many uses. However, it provides a direct measure of α-amylase activity, which is often of interest for experimental purposes.

Nephelometry, which measures scattering of light by suspended particles, also directly measures α-amylase activity. As the β-limit dextrin substrate is hydrolyzed by the enzyme, the nephelor decreases linearly. The method is highly sensitive to low levels of α-amylase but requires substantial expertise for its use (Kruger and Hatcher, 1993). A kinetic microplate modification of the method has been described (Kruger and Hatcher, 1993).

Several procedures measure α-amylase that diffuses from sectioned grain into an agar-starch substrate. Diffusion of the enzyme is deleted with iodine-iodide or by using a dye-labeled starch; the logarithm of activity is measured by the diameter of the digested area. The procedure takes about one day to complete but requires little equipment and expense.

Other enzymes, including proteinases, oxalate oxidase, and lipases, have been suggested as measures of preharvest sprouting. Proteinases offer little advantage over α-amylase, which causes most of the damage during sprouting and can be assayed by a variety of methods. Oxalate oxidase might be useful for detecting early sprouting, but it may not be applicable to severely sprouted samples (Fretzdorff and Betsche, 1998). The increase in lipase activity during sprouting can be visualized by hydrolysis of nonfluorescent fluorescein dibutyrate to fluorescent fluorescein (Jensen and Heltved, 1982).

Monoclonal antibodies are also used to measure sprouting by detecting specific enzymes or other constituents. The antibodies are typically labeled with fluorescein to visualize the reaction. Several commercial systems that employ the method for routine sampling have been developed, and the procedure is extremely useful for determining specific changes during sprouting.

CONTROLLING SPROUTING BY BREEDING

Similar to many other plant adversities, preharvest sprouting is controlled most effectively and economically by genetic resistance. However, as noted by Derera (1989b), breeding practices by scientists and production practices by farmers often work against dormancy of cereals. Breeders, for instance, often shuttle their experimental lines between the field and the greenhouse or between summer and winter nurseries to raise as many generations as possible each year. Farmers in northern latitudes may plant winter cereals from seed that was recently harvested. In some cases, industry may require cereals that have little or no dormancy, as with barley for malting (Aastrup, Riis, and Munck, 1989).

Breeding for resistance to preharvest sprouting can take many approaches because of the numerous morphological, physiological, and biochemical factors that influence the trait. Multiple sources of genetic resistance are available in wheat, barley, and rye (Derera, 1989a), and their dissimilar genetic mechanisms suggest that different genes can be "pyramided" into single genotypes to create highly resistant genotypes (Allan, 1993). Genetic engineering may likewise hold considerable promise for increasing resistance of cereals to preharvest sprouting (Anderson, Sorrels, and Tanksley, 1993).

Resistance to preharvest sprouting is strongly associated with red grain color in wheat, and much of the effort to improve the trait has focused on white wheat (Gale, 1989). Most studies find that resistance of white wheat to preharvest sprouting is a quantitative trait (e.g., Upadhyay and Paulsen, 1988; Paterson and Sorrells, 1990; Allan, 1993). This result may be associated with the different dormancy mechanisms that occur (Paterson and Sorrells, 1990). Other studies indicate that dormancy is controlled by one or two recessive genes (Bhatt, Ellison, and Mares, 1983).

Four quantitative trait loci (QTLs) for dormancy were detected in barley (Han et al., 1996), and five QTLs for dormancy were detected in rice (Lin, Sasaki, and Yano, 1998). Wheat had three QTLs that explained more than 80 percent of the total phenotypic variance in seed dormancy (Kato et al., 2001). A major QTL was located on the long arm of chromosome 4A, and two minor QTLs were on chromosomes 4B and 4D. Comparative maps suggested a homologous relationship between the major QTL and barley gene *SD4*.

'Clark's Cream' white wheat illustrates the use of resistance in an unadapted cultivar to improve resistance to preharvest sprouting in a modern cultivar. One of the present authors (GMP) received a bushel of 'Clark's Cream' seed from Mr. Earl Clark, a breeder-farmer who developed many important early cultivars for the Great Plains, for agronomic studies in 1976. The cultivar had a low level of sprouting after persistent rains in 1979, and subsequent studies showed that it expressed both high dormancy and low α-amylase production (McCrate et al., 1981). The traits were later associated with extreme embryo sensitivity to an endogenous inhibitor (Morris et al., 1989). An earlier report by Heyne (1956) stated that Clark normally left his experimental lines in the field for eight weeks after they ripened, suggesting that modifiers accumulated for resistance to sprouting. Broad-sense heritability estimates for sprouting resistance were moderate, and the trait was quantitatively inherited (Upadhyay and Paulsen, 1988). 'Clark's Cream' was used to develop the sprouting-resistant cultivar Cayuga and identify four genetic markers for the trait (Anderson, Sorrels, and Tanksley, 1993).

CONTROLLING SPROUTING IN THE FIELD

Few remedies are available for preventing preharvest sprouting when weather conditions promote germination. Planting of resistant species and cultivars, if they are available, is obvious. In some cases, cultivars that have the appropriate maturity to ripen before or after seasonal rains that cause sprouting may be selected.

Prompt harvest of the grain is usually the best means to prevent preharvest sprouting. In some cases, this involves harvesting immediately after the grain ripens, i.e., contains 12 to 13 percent moisture and can safely be stored. In other cases, the grain must be harvested earlier at a moisture level of 16 to 20 percent and artificially dried at extra expense. Grain that is swathed before it is ripe may be more dormant and resistant to sprouting than grain that is allowed to dry while standing. However, grain that is swathed dries slowly, and rain may increase the level of preharvest sprouting.

REFERENCES

Aastrup, S., Riis, P., and Munck, L. (1989). Controlled removal of dormancy renders prolonged storage of sprouting resistant malting barley superfluous. In Ringlund, K., Mosleth, E., and Mares, D.J. (Eds.), *Fifth International Symposium on Preharvest Sprouting in Cereals* (pp. 329-337). Boulder, CO: Westview Press.

Abdul-Hussain, S. and Paulsen, G.M. (1989). Role of proteinaceous α-amylase enzyme inhibitors in preharvest sprouting of wheat grain. *Journal of Agriculture and Food Chemistry* 37: 295-299.

Allan, R.E. (1993). Genetic expression of grain dormancy in a white-grain wheat cross. In Walker-Simmons, M.K. and Reid, J.L. (Eds.), *Pre-Harvest Sprouting in Cereals 1992* (pp. 37-46). St. Paul, MN: American Association of Cereal Chemists.

Anderson, J.A., Sorrels, M.E., and Tanksley, S.D. (1993). RFLP analysis of genomic regions associated with resistance to preharvest sprouting in wheat. *Crop Science* 33: 453-459.

Barnard, A. and Purchase, J. (1998). The effect of seed treatment and preharvest sprouted seed on the emergence and yield of winter wheat in South Africa. In Weipert, D. (Ed.), *Eighth International Symposium on Pre-Harvest Sprouting in Cereals* (pp. 26-35). Detmold, Germany: Association of Cereal Research.

Beck, E. and Ziegler, P. (1989). Biosynthesis and degradation of starch in higher plants. *Annual Reviews of Plant Physiology and Plant Molecular Biology* 40: 95-117.

Bhatt, G.M., Ellison, F.W., and Mares, D.J. (1983). Inheritance studies on dormancy in three wheat crosses. In Kruger, J.E. and LaBarge, D.E. (Eds.), *Third*

International Symposium on Preharvest Sprouting in Cereals (pp. 274-278). Boulder, CO: Westview Press.

Briggle, L.W. (1979). Pre-harvest sprout damage in wheat in the U.S. *Cereal Research Communications* 8: 245-250.

Cawley, R.W. and Mitchell, T.A. (1968). Inhibition of wheat α-amylase by bran phytic acid. *Journal of the Science of Food and Agriculture* 19: 106-108.

Chandler, P.M. and Mosleth, E. (1989). Do gibberellins play an in vivo role in controlling alpha-amylase gene expression? In Ringlund, K., Mosleth, E., and Mares, D.J. (Eds.), *Fifth International Symposium on Preharvest Sprouting in Cereals* (pp. 100-109). Boulder, CO: Westview Press.

Chung, D.S. and Pfost, H.B. (1967). Adsorption and desorption of water vapor by cereal grains and their products. *Transactions of the ASAE* 10: 549-555.

Derera, N.F. (1989a). Breeding for preharvest sprouting tolerance. In Derera, N.F (Ed.), *Preharvest Field Sprouting in Cereals* (pp. 111-128). Boca Raton, FL: CRC Press.

Derera, N.F. (1989b). The effects of preharvest rain. In Derera, N.F (Ed.), *Preharvest Field Sprouting in Cereals* (pp. 1-25). Boca Raton, FL: CRC Press.

Derera, N.F. (1989c). A perspective of sprouting research. In Ringlund, K., Mosleth, E., and Mares, D.J. (Eds.), *Fifth International Symposium on Preharvest Sprouting in Cereals* (pp. 3-11). Boulder, CO: Westview Press.

Derera, N.F. and Bhatt, G.M. (1980). Germination inhibition of the bracts in relation to pre-harvest sprouting tolerance in wheat. *Cereal Research Communications* 8: 199-201.

Dick, J.W., Hansen, D., Holm, Y.F., and Cantrell, R.G. (1989). Pearling and milling of sprouted durum wheat. In Ringlund, K., Mosleth, E., and Mares, D.J. (Eds.), *Fifth International Symposium on Preharvest Sprouting in Cereals* (pp. 344-350). Boulder, CO: Westview Press.

Duffus, C.M. (1989). Recent advances in the physiology and biochemistry of cereal grains in relation to pre-harvest sprouting. In Ringlund, K., Mosleth, E., and Mares, D.J. (Eds.), *Fifth International Symposium on Preharvest Sprouting in Cereals* (pp. 47-56). Boulder, CO: Westview Press.

Elias, S. and Copeland, L.O. (1991). Effect of preharvest sprouting on germination, storability, and field performance of red and white wheat seed. *Journal of Seed Technology* 15: 67-78.

Evers, A.D. (1989). Grain morphology in the sprouting context. In Ringlund, K., Mosleth, E., and Mares, D.J. (Eds.), *Fifth International Symposium on Preharvest Sprouting in Cereals* (pp. 57-64). Boulder, CO: Westview Press.

Federal Grain Inspection Service (1997). Wheat. In *Grain Inspection Handbook*, Book II (Chapter 13). Washington, DC: Grain Inspection, Packers and Stockyards Administration, U.S. Department of Agriculture.

Finney, K.F., Natsuaki, O., Bolte, L.C., Mathewson, P.R., and Pomeranz, Y. (1981). Alpha-amylase in field-sprouted wheats: Its distribution and effect on Japanese-type sponge cake and related physical and chemical tests. *Cereal Chemistry* 58: 355-359.

Finney, P.L., Morad, M.M., Patel, K., Chaudhury, S.M., Ghiasi, K., Ranhotra, G., Seitz, L.M., and Sebti, S. (1980). Nine international breads from sound and highly-field-sprouted Pacific Northwest soft white wheat. *Baker's Digest* 54: 22-27.

Foster, N.R., Burchett, L.A., and Paulsen, G.M. (1998). Seed quality of hard red wheat after incipient preharvest sprouting. *Journal of Applied Seed Production* 16: 87-91.

Fretzdorff, B. (1993). Lipolytic enzyme activities in germinating wheat, rye and triticale. In Walker-Simmons, M.K. and Reid, J.L. (Eds.), *Pre-Harvest Sprouting in Cereals 1992* (pp. 270-277). St. Paul, MN: American Association of Cereal Chemists.

Fretzdorff, B. and Betsche, T. (1998). Is oxalate oxidase indicative of pre-harvest sprouting related deterioration in cereal grains? In Weipert, D. (Ed.), *Eighth International Symposium on Pre-Harvest Sprouting in Cereals* (pp. 119-122). Detmold, Germany: Association of Cereal Research.

Gale, M.D. (1989). The genetics of preharvest sprouting in cereals, particularly in wheat. In Derera, N.F (Ed.), *Preharvest Field Sprouting in Cereals* (pp. 85-110). Boca Raton, FL: CRC Press.

Han, F., Ullrich, S.E., Clancy, J.A., Jitkow, V., Kilian, A., and Romagosa, I. (1996). Verification of barley seed dormancy loci via linked molecular markers. *Theoretical and Applied Genetics* 92: 87-91.

Heyne, E.G. (1956). Earl G. Clark, Kansas farmer wheat breeder. *Transactions of the Kansas Academy of Science* 59: 391-404.

Hilhorst, H.W.M. and Toorop, P.E. (1997). Review on dormancy, germinability, and germination in crop and weed seeds. *Advances in Agronomy* 61: 111-165.

Jensen, S.A. and Heltved, F. (1982). Visualization of enzyme activity in germinating cereal seeds using a lipase sensitive fluorochrome. *Carlsberg Research Communications* 47: 297-303.

Jones, B.L. and Wrobel, R. (1993). The endoproteinases of germinating barley. In Walker-Simmons, M.K. and Reid, J.L. (Eds.), *Pre-Harvest Sprouting in Cereals 1992* (pp. 262-269). St. Paul, MN: American Association of Cereal Chemists.

Kato, K., Nakamura, W., Tabiki, T., Mura, H., and Sawada, S. (2001). Detection of loci controlling seed dormancy on group 4 chromosomes of wheat and comparative mapping with rice and barley genomes. *Theoretical and Applied Genetics* 102: 980-985.

Kermode, A.R. (1990). Regulatory mechanisms involved in the transition from seed development to germination. *Critical Reviews in Plant Sciences* 9: 155-195.

King, R.W. (1989). Physiology of sprouting resistance. In Derera, N.F (Ed.), *Preharvest Field Sprouting in Cereals* (pp. 27-60). Boca Raton, FL: CRC Press.

Koeltzow, D.E. and Johnson, A.C. (1993). Comparison of sprout damage analysis techniques. In Walker-Simmons, M.K. and Reid, J.L. (Eds.), *Pre-Harvest Sprouting in Cereals 1992* (pp. 391-399). St. Paul, MN: American Association of Cereal Chemists.

Kruger, J.E. (1989). Biochemistry of preharvest sprouting in cereals. In Derera, N.F (Ed.), *Preharvest Field Sprouting in Cereals* (pp. 61-84). Boca Raton, FL: CRC Press.

Kruger, J.E. and Hatcher, D.W. (1993). Comparison of newer methods for the determination of α-amylase in wheat or wheat flour. In Walker-Simmons, M.K. and Reid, J.L. (Eds.), *Pre-Harvest Sprouting in Cereals 1992* (pp. 400-408). St. Paul, MN: American Association of Cereal Chemists.

Kruger, J.E., Hatcher, D.W., and Dexter, J.E. (1995). Influence of sprout damage on Oriental noodle quality. In Noda, K. and Mares, D.J. (Eds.), *Seventh International Symposium on Pre-Harvest Sprouting in Cereals 1995* (pp. 9-18). Osaka, Japan: Center for Academic Societies.

Kulp, K., Roewe-Smith, P., and Lorenz, K. (1983). Preharvest sprouting of winter wheat: I. Rheological properties of flours and physicochemical characteristics of starches. *Cereal Chemistry* 60: 355-359

Lin, S.Y., Sasaki, T., and Yano, M. (1998). Mapping quantitative trait loci controlling seed dormancy and heading date in rice, *Oryza sativa* L., using backcross inbred lines. *Theoretical and Applied Genetics* 96: 997-1003.

Liu, X., Wang, G., Jin, Y., Yang, S., and Li, Y. (1998). Endogenous hormone activity during grain filling of wheat genotypes differing in pre-harvest sprouting. In Weipert, D. (Ed.), *Eighth International Symposium on Pre-Harvest Sprouting in Cereals 1998* (pp. 99-101). Detmold, Germany: Association of Cereal Research.

Lorenz, K., Roewe-Smith, P., Kulp, K., and Bates, L. (1983). Preharvest sprouting of winter wheat: II. Amino acid composition and functionality of flour and flour fractions. *Cereal Chemistry* 60: 360-366.

Lorenz, K. and Valvano, R. (1981). Functional characteristics of sprout-damaged soft white wheat flours. *Journal of Food Science* 46: 1018-1020.

Lunn, G.D., Major, B.J., Kettlewell, P.S., and Scott, R.K. (2001). Mechanisms leading to excess alpha-amylase activity in wheat (*Triticum aestivum*, L.) grain in the U.K. *Journal of Cereal Science* 33: 313-329.

Mares, D.J. (1989). Preharvest sprouting damage and sprouting tolerance: Assay methods and instrumentation. In Derera, N.F (Ed.), *Preharvest Field Sprouting in Cereals* (pp. 129-170). Boca Raton, FL: CRC Press.

Mares, D.J. (1998). The seed coat and dormancy in wheat grains. In Weipert, D. (Ed.), *Eighth International Symposium on Pre-Harvest Sprouting in Cereals 1998* (pp. 77-81). Detmold, Germany: Association of Cereal Research.

Mares, D.J. and Mrva, K. (1993). Late maturity α-amylase in wheat. In Walker-Simmons, M.K. and Reid, J.L. (Eds.), *Pre-Harvest Sprouting in Cereals 1992* (pp. 178-184). St. Paul, MN: American Association of Cereal Chemists.

McCrate, A.J., Nielson, M.T., Paulsen, G.M., and Heyne, E.G., (1981). Preharvest sprouting and α-amylase activity in hard red and hard white winter wheat cultivars. *Cereal Chemistry* 58: 424-428.

McDonald, M.B. (1994). Seed germination and seedling establishment. In Boote, K.J., Bennett, J.M., Sinclair, T.R., and Paulsen, G.M. (Eds.), *Physiology and Determination of Crop Yield* (pp. 37-60). Madison, WI: ASA-CSSA-SSSA.

McMaster, G.J., Tomlinson, D., Edwards, R., Ross, A., and Moss, H.J. (1989). Endoprotease activity in rain-damaged Australian wheats. In Ringlund, K., Mosleth, E., and Mares, D.J. (Eds.), *Fifth International Symposium on Preharvest Sprouting in Cereals* (pp. 65-74). Boulder, CO: Westview Press.

Miyamoto, T., Tolbert, N.E., and Everson, E.H. (1961). Germination inhibitors related to dormancy in wheat seeds. *Plant Physiology* 36: 739-746.

Morris, C.F., Moffatt, J.M., Sears, R.G., and Paulsen, G.M. (1989). Seed dormancy and responses of caryopses, embryos, and calli to abscisic acid in wheat. *Plant Physiology* 90: 643-647.

Morris, C.F., Mueller, D.D., Faubion, J.M., and Paulsen, G.M. (1988). Identification of L-tryptophan as an endogenous inhibitor of embryo germination in white wheat. *Plant Physiology* 88: 435-440.

Mrva, K. and Mares, D. (1998). Co-ordinated synthesis of high pI alpha-amylase isozymes in the aleurone of wheat grains. In Weipert, D. (Ed.), *Eighth International Symposium on Pre-Harvest Sprouting in Cereals 1998* (pp. 131-135). Detmold, Germany: Association of Cereal Research.

Murphy, J.B. and Noland, T.L. (1982). Temperature effects on seed imbibition and leakage mediated by viscosity and membranes. *Plant Physiology* 69: 428-431.

Muthukrishnan, S., Chaudra, G.R., and Maxwell, E.S. (1983). Hormonal control of α-amylase gene expression in barley. *Journal of Biological Chemistry* 258: 2370-2375.

Nagao, S. (1995). Detrimental effect of sprout damage on wheat flour products. In Noda, K. and Mares, D.J. (Eds.), *Seventh International Symposium on Pre-Harvest Sprouting in Cereals* (p. 3-8). Osaka, Japan: Center for Academic Societies.

Nielsen, M.T., McCrate, A.J., Heyne, E.G., and Paulsen, G.M. (1984). Effect of weather variables during maturation on preharvest sprouting of hard white winter wheat. *Crop Science* 24: 779-782.

Paterson, A.H. and Sorrells, M.E. (1990). Inheritance of grain dormancy in white-kerneled wheat. *Crop Science* 30: 25-30.

Paulsen, G.M. (1985). Technology for improvement and production of wheat in China. *Journal of Agronomic Education* 14: 63-68.

Peiper, H.J. (1998). Enzyme potential of cereals suitable for the production of ethanol and glucose syrup. In Weipert, D. (Ed.), *Eighth International Symposium on Pre-Harvest Sprouting in Cereals 1998* (pp. 56-66). Detmold, Germany: Association of Cereal Research.

Shakeywich, C.F. (1973). Proposed method of measuring swelling pressure of seeds prior to germination. *Journal of Experimental Botany* 24: 1056-1061.

Smith, J.D. and Fong, F. (1993). Classification and characterization of the viviparous mutants of maize (*Zea mays* L.). In Walker-Simmons, M.K. and Reid, J.L. (Eds.), *Pre-Harvest Sprouting in Cereals 1992* (pp. 295-302). St. Paul, MN: American Association of Cereal Chemists.

Stahl, M. and Steiner, A. (1998). Viability loss of sprouted wheat seeds during storage. In Weipert, D. (Ed.), *Eighth International Symposium on Pre-Harvest Sprouting in Cereals 1998* (pp. 123-130). Detmold, Germany: Association of Cereal Research.

Upadhyay, M. and Paulsen, G.M. (1988). Heritabilities and genetic variation for preharvest sprouting in progenies of Clark's Cream white winter wheat. *Euphytica* 38: 93-100.

Vertucci, C.W. and Leopold, A.C. (1986). Physiological activities associated with hydration levels in seeds. In Leopold, A.C. (Ed.), *Membranes, Metabolism, and Dry Organisms* (pp. 35-49). Ithaca, NY: Comstock Publishing Association.

Walker-Simmons, M.K. (1987). ABA levels and sensitivity in developing wheat embryos of sprouting resistant and susceptible cultivars. *Plant Physiology* 84: 61-66.

Walker-Simmons, M.K., Aurelio-Cadenas, A., Verhey, S.D., Zhang, P., Holappa, L.D., and David Ho, T.-H. (1998). Involvement of PKABA/protein kinase in molecular control of α-amylase and protease gene expression in barley aleurone cells. In Weipert, D. (Ed.), *Eighth International Symposium on Pre-Harvest Sprouting in Cereals 1998* (pp. 203-217). Detmold, Germany: Association of Cereal Research.

Wellington, P.S. (1956). Studies on the germination of cereals: 2. Factors determining the germination behaviour of wheat grains during maturation. *Annals of Botany* 20: 481-486.

Wu, J. and Carver, B.F. (1999). Sprout damage and preharvest sprout resistance in hard white winter wheat. *Crop Science* 39: 441-447.

Yamaguchi, J., Toyofuku, K., Morita, A., Ikeda, A., Matsukura, C., and Perata, P. (1998). Sugar repression of alpha-amylase genes in rice embryos. In Weipert, D. (Ed.), *Eighth International Symposium on Pre-Harvest Sprouting in Cereals 1998* (pp. 136-145). Detmold, Germany: Association of Cereal Research.

Chapter 7

The Exit from Dormancy
and the Induction of Germination:
Physiological and Molecular Aspects

Rodolfo A. Sánchez
R. Alejandra Mella

INTRODUCTION

Timing and location of germination are crucial for the chances of success of the newly produced plant, and, accordingly, the temporal and spatial patterns of germination of the seeds of many species are finely tuned to the environmental scenario. Dormancy plays a central role in the adjustment of the behavior of seed populations to the restrictions and opportunities of a given environment (Chapter 8).

Dormancy is a physiological condition that prevents germination in an otherwise favorable set of external conditions (for a more detailed discussion of the definition of dormancy see Chapter 8). At first sight it may seem paradoxical that a sophisticated mechanism evolved in seeds to block their only function. However, it takes only a very brief inspection of the consequences of the lack of dormancy to appraise its importance. Were it not for dormancy, the seeds of many species would germinate when still attached to the parent plant, not surviving to be established in the soil, or the seedlings of an annual summer weed would be produced at the end of fall and so be condemned to die because of the cold of winter. The varied aspects of the relationships between dormancy and population dynamics of many species have been treated in excellent books such as those by Bewley and Black (1994) and Baskin and Baskin (1998).

The level of dormancy varies with time, provoking changes in the sensitivity of germination to various environmental factors (Chapter 8). At certain times dormancy is low enough and allows that a certain environmental factor (i.e., light, temperature, nitrates, or combinations of them) can terminate dormancy and induce germination. For seeds in soil, these factors represent important signals carrying essential information, cueing germination to the proper environmental situation.

Once the proper signal is perceived, a series of processes are set in motion that finally result in reactivation of embryo growth and radicle protrusion through the covering structures. Expansion of the embryonic axis requires changes in the embryo as well as in the surrounding tissues. In this chapter we will consider the physiological and molecular aspects of these processes. A significant proportion of the physiological research has been carried out with seeds of lettuce (Bewley and Halmer, 1980/1981; Borthwick and Robbins, 1928; Carpita, Ross, and Nabors, 1979; Carpita et al., 1979; Psaras and Georghiu, 1983; Psaras, Georghiu, and Mitrakos, 1981) and some Solanaceae species (tomato, pepper, Datura) (Dahal, Nevins, and Bradford, 1996; de Miguel et al., 2000; de Miguel and Sánchez, 1992; Groot and Karssen, 1987, 1992; Groot et al., 1988; Mella, Maldonado, and Sánchez, 1995; Ni and Bradford, 1993; Nonogaki and Morohashi, 1996, 1999; Nonogaki, Nomaguchi, and Morohashi, 1995, 1998; Sánchez and de Miguel, 1985, 1997; Sánchez et al., 1990; Watkins and Cantliffe, 1983; Watkins et al., 1985), although most of our knowledge of the molecular and genetic aspects of germination has been derived mainly from work with seeds of Arabidopsis, tomato, and tobacco (Bradford et al., 2000; Grappin et al., 2000; Karssen and Lacka, 1986; Koornneef and van der Veen, 1980). Several of these aspects have been covered by excellent reviews (Bewley, 1997; Hilhorst, 1995; Koornneef, Bentsink, and Hilhorst, 2002; Peng and Harberd, 2002). Since in almost all of these species germination is controlled by light, this chapter will focus in the light control of germination in species with coat-imposed dormancy. In seeds with this type of dormancy, germination depends on the balance between the expansive capacity of the embryo and the restrictions imposed by the surrounding tissues. Light perception by the photoreceptors may affect processes in components of the balance, initiating or blocking germination. In the following sections we will describe, after a brief description of the known photoreceptors, the changes in the embryo and the surrounding tissues (mainly the endosperm) related to germination.

THE EFFECTS OF LIGHT PHOTORECEPTORS

Light can promote or inhibit germination depending on its spectral composition and irradiance, the physiological status of the seeds, and the conditions of the other environmental factors, particularly temperature and water potential (Bewley and Black, 1994; Casal and Sánchez, 1998).

Promotion of germination by light has so far been found to be mediated only by the phytochromes. The loss of dormancy associated with certain situations found in the soil, or some incubation conditions in controlled environments, cause some seeds to display an extreme sensitivity to light (Casal and Sánchez, 1998). In those cases, millisecond exposures to sunlight are sufficient to induce germination; this is known as a very low fluence response (VLFR), which in *Arabidopsis* is mediated by phytochrome A (Botto et al., 1996; Shinomura et al., 1996). This response is saturated with very low levels of Pfr form of phytochrome (often less than 1 percent of total phytochrome [Pt] in the Pfr form) and does not display the classical red light (R)–far-red (FR) light reversibility (Casal, Sánchez, and Botto, 1998; Mandoli and Briggs, 1981). This mode of action of the phytochromes allows the detection of the brief exposure to light the seeds experience during soil disturbances such as those occurring during agricultural tillage operations (Scopel, Ballaré, and Sánchez, 1991).

In other physiological conditions, termination of dormancy requires light establishing Pfr/Pt > 0.05, and the photocontrol of germination displays the classical R-FR reversibility (Borthwick et al., 1952). This is called the low fluence response (LFR). When influenced by this mode of action, germination depends on the R:FR ratio of the light reaching the seeds, which is a good signal of the density of the canopy covering the soil (Insausti, Soriano, and Sánchez, 1995; Vazquez-Yañez et al., 1996; Vázquez-Yañez and Smith, 1982). The photoreceptors of this mode of action identified in *Arabidopsis* seeds are phytochromes B and E (Hennig et al., 2001). Almost all the research so far published on the physiological and molecular processes involved in the promotion of germination by light has been done with seeds displaying the LFR.

Inhibition of germination can be produced by light in the FR or blue (B) spectral regions (Bewley and Black, 1978). With few exceptions, inhibition by FR requires prolonged exposures to continuous light (λ_{max} 710 to 720 nm) or very frequent pulses and is irradiance dependent (Hartmann, 1966; Mancinelli, 1980). This effect is mediated by the high irradiance response (HIR) mode of action of phytochrome. In tomato seeds it has been shown that phyA is the photoreceptor (Shichijo et al., 2001). The HIR can both inhibit germination of dark-germinating seeds and antagonize the promotion of germination initiated by an LFR or a VLFR (Burgin et al., 2002; de Miguel et al., 1999). A continuous FR treatment can block germination even if given many hours after the R pulse starting the LFR (even very close to the time of radicle emergence). A subsequent pulse of R can relieve the inhibition imposed by the continuous FR treatment; therefore, the possibility of the antagonism LFR-HIR is there for much of the duration of the germination process (de Miguel et al., 2000). Blue light can also inhibit germination of many species, and R antagonizes its effect (Gwynn and Schiebe, 1972; Malacoste et al., 1972). The photoreceptor for the blue light action in seeds has not been identified so far.

It is clear that the control of germination is influenced by different photoreceptors interacting in a variety of ways. We are only beginning to understand the complex set of coordinated physiological and biochemical events involved in the photocontrol of germination and their regulation by cross talk between transduction chains initiated by endogenous and environmental signals.

EMBRYO GROWTH POTENTIAL

The promotion of germination is commonly associated with an increase in embryo growth potential. Although in seeds with coat-imposed dormancy the isolated embryos can grow without any particular stimulus, it has been observed that the embryos from seeds that are less dormant or have been exposed to a promotive treatment (i.e., light) have a greater growth rate than those from nonstimulated seeds (Figure 7.1a and b). The difference in growth potential is more evident when the incubation medium contains a factor opposing embryo expansion, such as an osmoticum. Growth of embryos isolated from lettuce (Carpita, Ross, and Nabors, 1979; Carpita et al., 1979) and *Datura ferox* seeds (de Miguel and Sánchez, 1992) is promoted by an R pulse, and the promotion by R is reversed if immediately followed by an FR pulse, displaying a typical LFR (Figure 7.2). Abscisic acid (ABA) antagonizes the LFR promotion of growth potential, whereas exogenous gibberellins (GAs) enhance growth potential in dark-incubated seeds (Karssen and Lacka, 1986). On the other hand, lowering the external water potential interacts with phytochrome in a complex fashion; small reductions in water potential enhance the phytochrome promotion of embryo growth potential, whereas large reductions block the LFR (de Miguel and Sánchez, 1992).

Although, as discussed earlier, an increase in the embryo growth potential accompanies the stimulus of germination, and the available evidence supports the contention that it may be necessary for radicle protrusion, the physiological and molecular bases of the enhancement in the embryo's expansion capacity are poorly understood. In the pioneering work of Carpita and colleagues (1979) with lettuce seeds it was shown that phytochrome-mediated K^+ transport could led to a decrease in the osmotic potential sufficient to explain a good part of the growth response. However, as the authors pointed out, it was likely that a change in the cell wall extensibility might also be involved. That modification of wall extensibility plays a predomi-

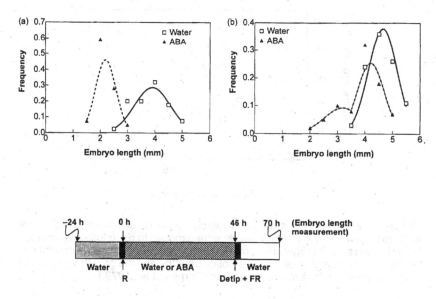

FIGURE 7.1 The effect of ABA on embryo length frequency distribution. Seeds were incubated in water or 100 μM ABA for 46 h after R. At the end of the incubation the seeds were detipped, irradiated with FR, and transferred to water. After further 24h incubation at 25°C in the dark, embryo length was measured.The seed pool in (a) has a higher degree of dormancy (12 months of dry storage) than seeds in (b) (24 months in dry storage). Data from ABA-treated seeds in (b) could not be fitted to a normal distribution. Assuming two subpopulations with different sensitivity to ABA allowed a good fit. (*Source:* de Miguel, L., Burgin, J., Casal, J. J., and Sánchez, R. A., unpublished results.)

nant role in ABA inhibition of embryo growth and germination is supported by studies with *Brassica napus* seeds (Schopfer and Plachy, 1985). The possibility that wall extensibility changes may participate in the expansion capacity of the embryo is also suggested by changes in expression of expansin genes. The temporal and spatial pattern of expression of two expansin genes in tomato seeds, *LeEXP8* and *LeEXP10*, are consistent with a role for expansins, and GA promotes the transcription of both genes in gibberellin-deficient *gib1* seeds (Chen, Dahal, and Bradford, 2001). On the other hand, low water potential blocks expression of *LeEXP8*, which is also inhibited by ABA (Chen, Dahal, and Bradford, 2001). Expansin transcript levels are also increased by R in the embryos of *D. ferox* seeds, an FR-reversible effect that is in good agreement with the photocontrol of germination (Mella et al., 2004). Although these results do not establish a direct causal connection be-

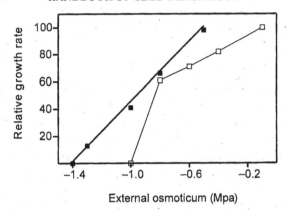

FIGURE 7.2. Growth in osmotica of radicles from R- and FR-treated lettuce embryos. 'Grand Rapids' lettuce seeds were exposed to red (closed squares) or far-red (open squares) light. The embryos were removed and placed in solutions of polyethylene glycol of different osmotic strengths and the growth rate over 24 h was determined. (*Source:* Adapted from Carpita et al., 1979.)

tween expansin gene expression and embryo growth potential, they certainly support an intervention of expansins in this process.

Taken together, the results obtained with seeds of different species suggest that the increase in embryo growth potential associated with germination may include a decrease in osmotic potential, consequently a rise in turgor pressure, and an increase in cell wall extensibility, probably with the participation of expansins.

ENDOSPERM WEAKENING

In seeds whose embryos are completely surrounded by a rigid endosperm, radicle emergence requires a significant reduction in the physical restriction that the endosperm opposes to embryo expansion. In particular, the micropylar endosperm (named the endosperm cap), which is the region directly opposed to the embryo radicle, must be weakened. Endosperm cap weakening has been shown to precede radicle protrusion (Groot and Karssen, 1987; Sánchez et al., 1990; Watkins and Cantliffe, 1983) (Figure 7.3) and is accompanied by extensive structural changes. In addition to changes in the cell walls, profound alterations are produced in other parts of the cells, protein and lipid reserves are degraded, and extensive vacuolation takes place; the structure changes from that typical of reserve cells to metabolically active ones (Figures 7.4 and 7.5) (Mella, Maldonado, and Sánchez,

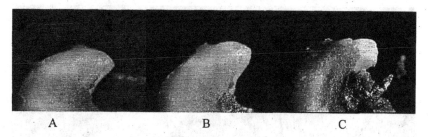

FIGURE 7.3. Photographs of the external aspect of the micropylar region of decoated *Datura ferox* L. seeds. (A) 46 h after a noninductive far-red pulse; (B) 46 h after an inductive R pulse (red light) showing signs of weakening; (C) germinated seed, with the protruded radicle and broken micropylar endosperm.

1995; Psaras, Georghiu, and Mitrakos, 1981; Sánchez et al., 1990). This preradicle protrusion syndrome is restricted to the micropylar region; the rest of the endosperm (frequently called the lateral endosperm) remains unchanged.

In relation to weakening, the attention has been focused on changes in the cell walls. The reduction in the mechanical resistance of the endosperm cap has been associated with the hydrolysis of cell wall polysaccharides, primarily of the main component: a mannose polymer, probably a β-(1,4)-mannan (Table 7.1) (Sánchez et al., 1990). A large increase in the activity of mannan-degrading enzymes precedes radicle protrusion when germination is promoted by light, through a LFR of the phytochromes, or GA in tomato (Groot and Karssen 1987; Nonogaki and Morohashi, 1996) and *D. ferox* (de Miguel and Sánchez 1992; Sánchez et al., 1990) seeds. In tomato seeds, a endo-β-mannanase gene *(LeMAN2)* is exclusively expressed in the endosperm cap prior to radicle emergence (Nonogaki, Gee, and Bradford, 2000). Consistently, red light strongly increases, in an FR-reversible fashion, the transcript level of *DfMAN1* (which shows high homology with *LeMAN2*) in *D. ferox* seeds only in the endosperm cap (Burgin et al., 2000). Although direct evidence is not yet available, the information from tomato and *Datura* strongly supports the possibility that cell wall mannan degradation is part of the mechanism of endosperm cap softening (Figure 7.6). This does not imply that mannan degradation or high mannanase activity is sufficient for endosperm softening or, even less, for germination. As described in the previous section, inhibiting embryo expansion may by itself block germination even in cases when the surrounding tissues offer little or no resistance to embryo growth (Schopfer and Plachy, 1993). Endosperm weakening may also involve changes in cell wall components other than mannose polymers. Several hydrolases (xyloglucan endotransglycosylase, arabinosidase, etc.)

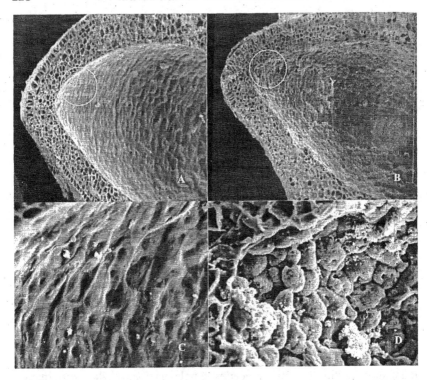

FIGURE 7.4. Scanning electron micrographs of median sections of micropylar and adjacent bulk endosperm of *Datura* seeds glutaraldehyde-osmium fixed and critical point dried. (A, C) FR-irradiated seeds; (B, D) FR+R-irradiated seeds; sampling at 38 hours after irradiation. Micrographs C and D are enlargements corresponding to the encircled areas of A and B. (*Source:* After Sánchez et al., 1990. Reprinted with permission of National Research Council of Canada.)

have been found to be expressed during tomato germination and could contribute to cell separation (Bradford et al., 2000). In tobacco seeds, a good correlation has been found between the increase in the activity of a class I β-1,3-glucanase and the promotion of germination. The gene coding for that glucanase is expressed specifically in the endosperm cap, and its overexpression alleviates the inhibitory action of ABA on tobacco seeds (Leubner-Metzger, Fründt, and Meins, 1996; Leubner-Metzger et al., 1995). However, it has been shown in tomato seeds that endosperm cap weakening begins before the expression of the β-1,3-glucanase gene can be detected and although ABA effectively inhibited β-1,3-glucanase gene expression, it did not affect endosperm softening (Wu et al., 2001). In the same work no

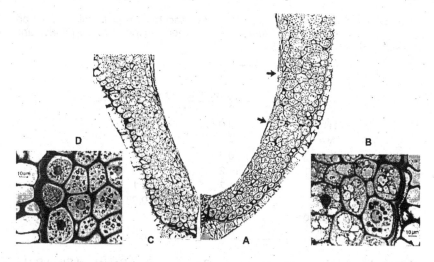

FIGURE 7.5. Median and longitudinal sections of the micropylar portions of *Datura ferox* L. seeds, 48 h after R light (A, B) or FR light (C, D) treatment. (A, C) low magnification view—the arrows in (A) indicate the zone with extensive degradation of protein bodies; (B) an enlargement of a portion between the arrows—note the vacuolation of protein bodies; (D) view of an area with similar localization of (B) but in a FR-treated seed—note the abundance of storage material and the absence of vacuoles. (*Source:* After Mella, Maldonado, and Sánchez, 1995, © American Society of Plant Biologists. Reprinted with permission.)

evidence was found of a substrate for the enzyme in endosperm cap cell walls; therefore, the involvement of the β-1,3-glucanase in the weakening process in tomato seeds was not supported and other functions for this enzyme were considered more likely. In addition, it has been suggested that expansins could play a role in endosperm weakening either by facilitating the access of hydrolases to their substrates or by loosening hemicellulosic bonds (Bradford et al., 2000; Chen, Dahal, and Bradford, 2001). This proposition is supported by the results of Chen, Dahal, and Bradford (2001) showing that a specific expansin gene, *LeEXP4*, is expressed exclusively in the tomato endosperm cap prior to radicle emergence and is up-regulated by GA.

Low water potential prevents endosperm weakening (Chen and Bradford, 2000; de Miguel and Sánchez, 1992; Sanchez et al., 2002). In *D. ferox* seeds it has been shown that germination is inhibited by water potentials that do not reduce phytochrome promotion of embryo growth potential but prevent endosperm weakening (de Miguel and Sánchez, 1992). Therefore,

TABLE 7.1. Sugar composition of cell wall polysaccharides in the micropylar portion of the endosperm of *Datura ferox* L. seeds induced to germinate and noninduced controls

	R	FR
	(mg/micropylar portion)	
Rha	0.5	0.35
Rib	0.025	0.04
Ara	4.4	3.2
Xyl	0.6	0.6
Man	11.2	3.3
Gal	1.7	1.1
Glc	0.9	0.65
Uronic acid	2.3	1.75
Cell	4.2	3.3

Source: Adapted from Sánchez et al., 1990, © American Society of Plant Biologists. Reprinted with permission. *Note:* Sampling was made 38 h after irradiation. The data are averages of two determinations.

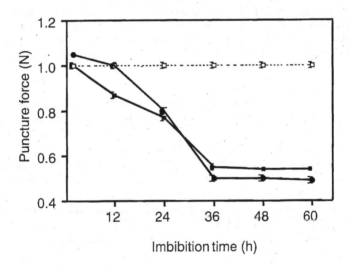

Imbibition time (h)

FIGURE 7.6. Puncture force analysis of wild-type ('Money Maker') seeds in water (closed circles) and *gib-1* mutant seeds in water (open squares) and 100 mM GA (closed squares). Error bars indicate standard errors ($n = 24$) when larger than the symbols. (*Source:* Chen and Bradford, 2000, © American Society of Plant Biologists. Reprinted with permission.)

low water potential, at least at certain values, can prevent the induction of germination mainly through its effect on endosperm softening. The mechanisms involved, however, are not clear yet. A relationship does exist between low water potential effects on the phytochrome-induced reduction in mannose content of the cell walls and endosperm cap weakening in *D. ferox* seeds (Sánchez et al., 2002); in addition, the inclusion of an osmoticum in the incubation medium reduced the release of mannose by tomato endosperm caps (Dahal, Nevins, and Bradford, 1997). These results suggest that interference with mannan degradation could be one of the ways through which water availability influences endosperm resistance to embryo penetration. On the other hand, low water potential does not prevent the increase in mannanase activity in phytochrome-promoted *D. ferox* (Sanchez et al., 2002) or in tomato seeds (although see Toorop, van Aelst, and Hilhorst, 1998). In *D. ferox* low water potential inhibits the increase in mannosidase associated with germination (and this could indirectly hamper mannan degradation), whereas in tomato it has been found that *LeEXP4* expression is reduced proportionally to the inhibition of endosperm cap weakening (Chen and Bradford, 2000). If *LeXP4* facilitates the access of the hydrolases to their substrates, inhibiting its production may be an obstacle to weakening, even if the activity of some of the hydrolases (i.e., mannanase) remains high. It seems likely, then, that a restriction in water supply may interfere with endosperm weakening, down-regulating the genes encoding some of the proteins contributing to the cell wall changes, but not necessarily affecting all of them.

Endosperm cap weakening is also blocked when the promotion of germination by an LFR of the phytochromes is antagonized by exposure to continuous FR through an HIR (de Miguel et al., 2000; Mella et al., 2002). Since the HIR does not affect embryo growth potential, it should influence germination interfering with endosperm cap softening. The continuous FR treatment can block germination of a part of the seed population even if it is applied when endosperm weakening has already advanced perceptibly. Interestingly, in a similar part of the population it provokes a sharp reduction in mannanase activity (Figure 7.7) (down to the values typical of noninduced seeds) and the interruption of the weakening process. The HIR can disengage the weakening process even at relatively advanced stage, but, apparently, the antagonism does not involve every process promoted by the LFR, and even some of the ones affected may be influenced in different ways by each of the phytochromes' modes of action. While the LFR promotes mannanase activity and increases the transcripts level of *DfMAN1* and *DfEXP2* in *D. ferox* seeds, the HIR inhibits mannanase activity but does not modify the transcript levels of neither of them (Burgin et al., 2000; Mella et al., 2004).

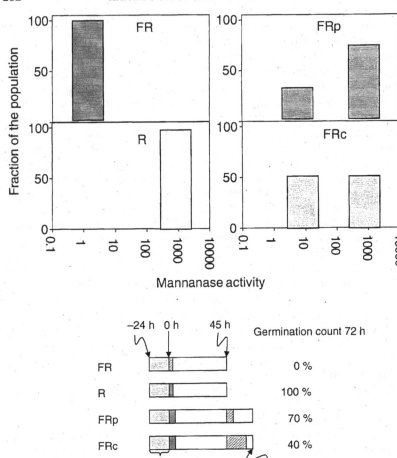

FIGURE 7.7. Endo-β-mannanase activity measured seed by seed in the population of *Datura ferox* induced by LFR or by LFR-HIR. R light promotes germination and the activity of endo-β-mannanase. In the noninduced seeds treated with FR immediately after R (FR) the values are substantially lower than in those seeds induced by R (R). The activity was evaluated 45 h after light treatments in darkness (20 to 30°C) (FRp). When an FR pulse is given 45 h after R (FR pulse 3 min·h⁻¹, 300 mmol·m⁻²·s⁻¹), and the activity is measured 6 h later, mannanase activity decreases in part of the population. The endo-β-mannanse activity was evaluated 51 h after initial light treatment (FRc). If instead of a pulse, FR is given continuously during 6 h (15 mmo·l⁻²·s⁻¹), mannanase activity decreased in a larger fraction of seeds. The endo-β-mannanase activity was evaluated 51 h after initial light treatment. (*Source:* Modified from Mella et al., 2002).

Whether the inhibition of germination by ABA includes an effect on endosperm softening is not completely clear. Although some experiments show that part of endosperm cap weakening is blocked by ABA in tomato seeds (Toorop, van Aelst, and Hilhorst, 2000), in other studies, with the same genotype and similar methods, no effect of ABA was found (Chen and Bradford, 2000). So far none of the physiological steps that are thought to be related to endosperm softening, such as mannanase and *LeEXP4*, are affected by exogenous applications of ABA; if there is an ABA-sensitive phase in the weakening process it depends on a process that has so far eluded us (Chen and Bradford, 2000; Nonogaki, Gee, and Bradford, 2000).

Although we do not yet have definite knowledge on the process of endosperm softening, it is apparent that it may depend on a variety of mechanisms and not every regulatory factor affects all of them.

Other tissues that restrict the expansion of the embryos are the testa, as shown in *Arabidopsis* (Debeaujon and Koornneef, 2000), and the perisperm of muskmelon (Welbaum et al., 1995).

TERMINATION OF DORMANCY: ITS RELATIONSHIP WITH THE SYNTHESIS AND SIGNALING OF GIBBERELLINS AND ABA

The paramount importance of GA for germination has been recognized for a long time (Bewley and Black, 1994) and is most clearly demonstrated in studies with mutants of *Arabidopsis* and tomato that are severely GA deficient (Groot and Karssen, 1987; Koornneef and van der Veen, 1980). The induction of germination by light requires GA synthesis (Derkx and Karssen, 1993; Yang et al., 1995), and in lettuce seeds an R pulse causes a significant increase in GA content; the effect of R is FR-reversible as in the classical LFR (Toyomasu et al., 1993). In addition to increasing the GA content, the reversible R-FR response enhances the sensitivity to GA in seeds (Yang et al., 1995). In lettuce and *Arabidopsis*, R promotes the expression of genes encoding GA 3-β-hydroxylase, a key enzyme in the active GA biosynthetic pathway (Toyomasu et al., 1998; Yamaguchi et al., 1998). In *Arabidopsis*, two GA 3-β-hydroxylase genes, *GA4* and *GA4H*, are under the control of the phytochromes. The Pfr of phyB promotes *GA4H*, whereas *GA4* is controlled by some other stable phytochrome (Yamaguchi et al., 1998). Whether promotion of germination by the VLFR, which in *Arabidopsis* is mediated by phyA, is also related to the up-regulation of these genes is still unknown.

Although the increase in GA levels affects processes in both the embryo and the endosperm, the available evidence suggests that the synthesis of GA takes place only in the embryo. Light promotes changes only when the embryo is in contact with the endosperm cap (Sánchez and de Miguel, 1997), and the presence of the embryo can be replaced by supplying the endosperm caps with exogenous GA (Groot et al., 1988; Sánchez and de Miguel, 1997). Moreover, the transcripts of *DfHydrox*, a GA 3-β-hydroxylase gene, are only found in the embryo and in significantly higher amounts after an R pulse than when R is immediately followed by FR (Burgin et al., 2000). The GA synthesized in the embryo would migrate to the endosperm cap where it induces weakening. In the micropylar endosperm of tomato, GA promotes the expression of a number of cell wall hydrolases and related proteins: endo-β-mannanase, cellulase, arabinosidase, xyloglucan endotransglyco-sylase, expansin, etc. (Bradford et al., 2000). Because the path of GA from the embryo to the micropylar endosperm includes an apoplastic segment, it seems likely that the GA may reach other endosperm cells in addition to those in the micropylar region. However, only the cells of the endosperm cap respond to GA (before radicle protrusion), suggesting a greater sensitivity of these cells to GA.

Recent work has shown the participation in the control of germination of two GA-response regulators: RGL1 and RGL2 (Peng and Harberd, 2002). Both are repressors of GA responses; loss-of-function mutations in *RGL2* completely restored germination to *ga1-3*, eliminating the requirement for exogenous GA (Lee et al., 2002). In wt *Arabidopsis*, *RGL2* transcript levels show a transitory increase after the initiation of imbibition followed by a decrease with the advance of germination; interestingly, in the GA-deficient *ga1-3* the *RGL2* transcripts remained at high levels throughout the incubation period unless exogenous GA was supplied (Lee et al., 2002). It has also been shown that SPY (O-GlcNAc transferase) influences seed germination. In *Arabidopsis*, *SPY* alleles confer resistance of germination to the GA-synthesis inhibitor paclobutrazol and restore the germination capacity to *ga1-2* (Jacobsen and Olszewski, 1993). On the basis of the available evidence, Peng and Harberd (2002) have proposed the following scheme. In dormant seeds upon imbibition proteins are expressed that repress germination (Peng and Harberd, 2002). In the proper physiological scenario phytochrome activation by light induces GA synthesis. The increased GA level down-regulates the expression of repressors such as RGL2 and SPY that might increase the germination potential of the embryo. At the same time the increase in GA reaches the endosperm cap where signaling factors (GCR1, SLY, and CTS are candidates) would induce the expression of the proteins (e.g., mannanase, expansins, XET, etc.) related to weakening. Although testing of several of these propositions is still pending, it is consistent with most of the information at hand and is useful to guide the design of future experiments.

ABA has an essential role in establishing dormancy during seed development (Hilhorst and Karssen, 1992; Hilhorst, 1995), which influences the responses of mature seeds to various environmental factors. In addition, synthesis of ABA during imbibition has been shown to be important for germination. ABA levels increase upon imbibition different in dormant and nondormant seeds (Grappin et al., 2000; Le Page-Degivry and Garello, 1992), and inhibitors of ABA biosynthesis promote germination (Grappin et al., 2000). In the same line are observations of a decline in ABA content after a R pulse promoting germination (Tillberg and Björkman, 1993) and the prevention of the decline of the ABA content normally occurring in germinating lettuce seeds by high-temperature treatments inhibitory of germination (Yoshioka, Endo, and Satoh, 1998). Mutations in several genes change the sensitivity of germination to ABA. The *abi* and *era* mutants have reduced and enhanced responses respectively (Koornneef, Bentsink, and Hilhorst, 2002), the *ethylene insensitive 2 (ein2)* and *ethylene response (etr)* mutants of *Arabidopsis* are hypersensitive, while the *ctr* genes have less sensitivity to ABA. Moreover, ABA and ethylene signaling interact with sugar signaling (Finkelstein, Gampala, and Rock, 2002) and mutants deficient *(det2-1)* or insensitive *(bri1-1)* to brassinosteroids are also more sensitive to ABA, suggesting that the BR signal participates in the control of germination opposing ABA inhibition (Steber and McCourt, 2001). Although the description of the components of the web of signaling networks modulated by different regulators is incomplete, it shows the variety and complexity of the system that permits the integration of environmental and internal signals which modulate germination.

CONCLUDING REMARKS

Germination is a crucial process for the adjustment of many plant populations to their environment. Taking into account the variety of environmental scenarios and plant genotypes, it is not surprising that the diversity of external factors and their combinations can control seed responses. Germination itself is a complex process involving a number of finely coordinated changes in the embryo and the surrounding tissues with the participation of a large number of genes. Congruent regulation of the several molecular, biochemical, and physiological events leading to germination requires cross talk between endogenous and environmental signals. The elements of this dialogue (several phytochromes, GA, ABA, ethylene, BR, sugars, etc.) are part of an intricate network with versatile switches and multiple pathways. In this context it is to be expected that a particular factor may not always have the same relevance and that correlations could vary according to the physiological and environmental scenario.

In the interplay between the embryo and the enveloping tissues, the role of the embryo seems to be central. It is in the embryos where the light signals are perceived (and the same is probably the case with other environmental signals) and there where GA and likely other regulators are produced that profoundly influence the activity in the surrounding tissues. Particularly when the endosperm is the main barrier to embryo growth, the GA synthesized in the embryo provokes changes in the expression of genes coding for cell wall hydrolases and other proteins involved in weakening. When the testa is limiting embryo expansion, as in *Arabidopsis,* it is also thought that testa weakening may depend on GA action (Debeaujon and Koornneef, 2000). Figure 7.8 shows a tentative scheme in which data from several species (we do not have the whole picture in just one species) and some components of the processes with different degrees of experimental support are included depicting some of the connections between the embryo and the endosperm and the points of action of some internal and external factors. The data available seem to indicate that not all of these factors influence the same processes. No master switch appears to be in control of the whole system. Once germination is promoted (e.g., by light) a number of activities are set in motion in the embryo and the endosperm. The action of some of the factors which can block that response may affect only part of them. For instance, ABA can sharply decrease embryo growth potential without affecting most of the endosperm softening; in contrast, the HIR of the phytochromes interferes with endosperm softening but does not seem to affect embryo growth potential. However, we have still a long way to go before being able to put the central pieces of the puzzle in their proper places. It would be helpful to have integrated information about more than one system. This is currently limited by the problem that knowledge on certain parts of the system has advanced more in some species than in others and, in fact, all the processes and their interactions may not be identical in the different species from which most of the information has been gathered so far.

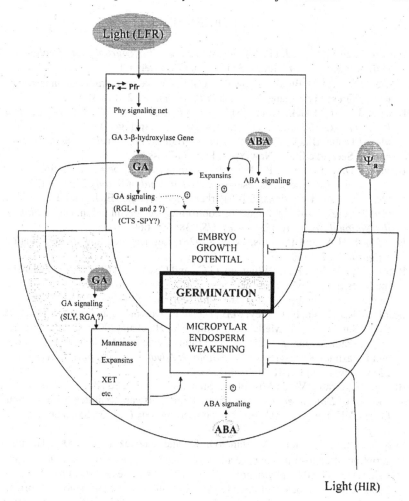

FIGURE 7.8. Diagramatic representation of part of the molecular signaling and components acting on the termination of coat-imposed dormancy and the induction of germination. Broken lines and question marks represent probable but unconfirmed interactions.

REFERENCES

Baskin, C. and Baskin, J. (1998). *Seeds: Ecology, Biogeography and Evolution of Dormancy and Germination.* San Diego, CA: Academic Press.

Bewley, J.D. (1997). Breaking down the walls—A role for endo β mannanase in release from seed dormancy? *Trends in Plant Science* 2: 464-469.

Bewley, J.D. and Black, M. (1978). *The Physiology and Biochemistry of Seeds in Relation to Germination,* Volume 1. New York: Springer-Verlag.

Bewley, J.D. and Black, M. (1994). *Seeds: Physiology of Development and Germination,* Second edition. New York: Plenum Press.

Bewley, J.D. and Halmer, P. (1980/1981). Embryo-endosperm interactions in the hydrolysis of lettuce seeds reserves. *Israel Journal of Botany* 29: 118-132.

Borthwick, H.A., Hendricks, S.B., Parker, M.W., Toole, E.H., and Toole, V.K. (1952). A reversible photoreaction controlling seed germination. *Proceedings of the National Academy of Sciences, USA* 38: 662-666.

Borthwick, H.A. and Robbins, W.W. (1928). Lettuce seed and its germination. *Hilgardia* 3: 275-305.

Botto, J.F., Sánchez, R.A., Whitelam, G.C., and Casal, J.J. (1996). Phytochrome A mediates the promotion of seed germination by very low fluences of light and canopy shade light in *Arabidopsis. Plant Physiology* 110: 439-444.

Bradford, K.J., Chen, F., Cooley, M.B., Dahal, P., Downie, B, Fukunaga, K.K., Gee, O.H., Gurusinghe, S., Mella, R.A., Nonogaki, H., et al. (2000). Gene expression prior to radicle emergence in imbibed tomato seed. In M. Black, K.J. Bradford, and J. Vazquez-Ramos (Eds.), *Seed Biology: Advances and Applications* (pp. 231-251). New York: CABI Publishing.

Burgin, M., Arana, V., de Miguel, L., Staneloni, R., and Sanchez, R. (2002). Responses of *Datura ferox* seeds to far red light pulse are inhibited by continuous far red. VII International Workshop on Seed Biology (p. 113), Salamanca, Spain, May 12-16.

Burgin, M.J.P.F.L., Mella, A., Staneloni, R., and Sánchez, R.A. (2000). The transcription of endo-β-mannanase and GA 3β-hydroxylase genes of *Datura ferox* seeds is regulated by phytochrome. Report No. 146. Presented at Plant Biology, 2000, American Society of Plant Physiology, July 15-19, San Diego, California.

Carpita, N.C., Nabors, M.W., Ross, C.W., and Peteric, N.L. (1979). The growth physics and water relations of red light induced germination in lettuce seeds: III. Changes in the osmotic and pressure potential in the embryonic axes of red and far-red treated seeds. *Planta* 144: 217-224.

Carpita, N.C., Ross, C., and Nabors, M. (1979). The influence of plant growth regulators on the growth of the embryonic axes of red- and far-red-treated lettuce seeds. *Planta* 145: 511-516.

Casal, J.J. and Sánchez, R.A. (1998). Phytochromes and seed germination. *Seed Science Research* 8: 317-329

Casal, J.J., Sánchez, R.A., and Botto, F.J. (1998). Modes of action of phytochromes. *Journal of Experimental Botany* 49: 127-138.

Chen, F. and Bradford, K.J. (2000). Expression of an expansin is associated with endosperm weakening during tomato seed germination. *Plant Physiology* 124: 1265-1274.

Chen, F., Dahal, P., and Bradford, K.J. (2001). Two tomato expansins genes show divergent expression and localization in embryos during seed development and germination. *Plant Physiology* 127: 928-936.

Dahal, P., Nevins, D.J., and Bradford, K.J. (1996). Endosperm cell wall sugar composition and sugars released during tomato seed germination. *Plant Physiology* 111: 160.

Dahal, P., Nevins, D.J., and Bradford, K.J. (1997). Relationship of endo-β-mannananse activity and cell wall hydrolysis in tomato endosperm to germination rates. *Plant Physiology* 113: 1243-1252.

de Miguel, L., Burgin, J., Casal, J.J., and Sánchez, R.A. (2000). Antagonistic action of low-fluence and high-irradiance modes of response of phytochrome on germination and β-mannanase activity in *Datura ferox* seeds. *Journal of Experimental Botany* 51: 1127-1133.

de Miguel, L.C. and Sánchez, R.A. (1992). Phytochrome-induced germination, endosperm softening and embryo growth potential in *Datura ferox* seeds: Sensitivity to low water potential and time to escape to FR reversal. *Journal of Experimental Botany* 43: 969-974.

Debeaujon, I. and Koornneef, M. (2000). Gibberellin requirement for *Arabidospsis thaliana* seed germination is determined both by testa characteristics and embryonic ABA. *Plant Physiology* 122: 415-424.

Derkx, M.P.M. and Karssen, C.M. (1993). Effects of light and temperature on seed dormancy and gibberellin-stimulated germination in *Arabidopsis thaliana*—Studies with gibberellin-deficient and gibberellin-insensitive mutants. *Physiologia Plantarum* 89: 360-368.

Finkelstein, R.R., Gampala, S.S.L., and Rock, C.D. (2002). Abscisic acid signaling in seeds and seedlings. *Plant Cell* 14: S15-S45.

Grappin, P., Bouinot, D., Sotta, B., Miginiac, E, and Jullien, M. (2000). Control of seed dormancy in *Nicotiana plumbaginifolia:* Post imbibition abscisic acid synthesis imposes dormancy maintenance. *Planta* 210: 279-285.

Groot, S.P.C. and Karssen, C.M. (1987). Gibberellins regulate seed germination in tomato by endosperm weakening: A study with gibberellin-deficient mutants. *Planta* 172: 525-531.

Groot, S.P.C. and Karssen, C.M. (1992). Dormancy and germination of abscisic acid-deficient tomato seeds: Studies with the sitiens mutant. *Plant Physiology* 99: 952-958.

Groot, S.P.C., Kieliszewska-Rockika, B., Vermeer, E., and Karssen, C.M. (1988). Gibberellin-induced hydrolysis of endosperm cell walls in gibberellin-deficient tomato seeds prior to radicle protrusion. *Planta* 174: 500-504.

Gwynn, D. and Schiebe, J. (1972). An action spectra for inhibition of lettuce seed. *Planta* 144: 121-124.

Hartmann, K.M. (1966). A general hypothesis to interpret "high energy phenomena" of photomorphogenesis on the basis of phytochrome. *Photochemistry and Photobiology* 5: 349-366.

Hennig, L., Stoddart, W.M., Dieterle, M., Whitelam, G.C., and Schäfer, E. (2001). Phytochrome E control light-induced germination of *Arabidopsis*. *Plant Physiology* 128: 194-200.

Hilhorst, H.W.M. (1995). A critical update on seed dormancy: I. Primary dormancy. *Seed Science Research* 5: 61-74.

Hilhorst, H.W.M. and Karssen, C. (1992). Seed dormancy and germination: The role of abscisic acid and gibberellins and the importance of hormone mutants. *Plant Growth Regulation* 11: 225-238.

Insausti, P., Soriano, A., and Sánchez, R.A. (1995). Effects of flood-related factors on seed germination of *Ambrosia tenuifolia*. *Oecologia* 103: 127-132.

Jacobsen, S. and Olszewski, N. (1993). Mutation at the SPINDLY locus alter gibberellin signal transduction. *Plant Cell* 5: 87-896.

Karssen, C. and Lacka, E. (1986). A revision of the hormone balance theory of seed dormancy: Studies on gibberellin and/or abscisic acid-deficient mutants of *Arabidopsis thaliana*. In M. Bopp (Ed.), *Plant Growth Substances 1985* (pp. 315-323). Heidelberg, Germany: Springer-Verlag.

Koornneef, M., Bentsink, L., and Hilhorst, H. (2002). Seed dormancy and germination. *Current Opinion in Plant Biology* 5: 33-36.

Koornneef, M. and van der Veen, J. (1980). Induction and analysis of gibberellin sensitive mutants in *Arabidopsis thaliana* (L.) Heynh. *Theoretical Applied Genetics* 58: 257-263.

Le Page-Degivry, M. and Garello, G. (1992). In situ abscisic acid synthesis: A requirement for induction of embryo dormancy in *Helianthus annuus*. *Plant Physiology* 98: 1386-1390.

Lee, S., Cheng, H., King, K., Wang, W., He, Y., Hussain, A., Lo, J., Harberd, N., and Peng, J. (2002). Gibberellin regulates *Arabidopsis* seed germination via RGL2, a GAI/RGA-like gene whose expression is up regulated following imbibition. *Genes and Development* 16: 646-658.

Leubner-Metzger, G., Fründt C., and Meins, F., Jr. (1996). Effects of gibberellins, darkness and osmotica on endosperm rupture and class I β-1,3-glucanase induction in tobacco seed germination. *Planta* 199: 282-288.

Leubner-Metzger, G., Fründt, C., Voegeli-Lange, R., and Meins, F., Jr. (1995). Class I β-1,3-glucanases in the endosperm of tobacco during germination. *Plant Physiology* 109: 751-759.

Malacoste, R., Tzanni, H., Jaques, R., and Rollin, P. (1972). The influence of blue light on dark red germinating seeds of *Nemophyla insignis*. *Planta* 103: 24-34.

Mancinelli, A.L. (1980). The photoreceptors of the high irradiance responses of plant photomorphogenesis. *Photochemistry and Photobiology* 32: 853-857.

Mandoli, D.F. and Briggs, W.R. (1981). Phytochrome control of two low-irradiance responses in etiolated oat seedlings. *Plant Physiology* 67: 733-739.

Mella, R.A., Burgin, M.J., and Sánchez, R.A. (2004). Expansin gene expression in *Datura ferox* L. seeds is regulated by the low-fluence response, but not by the high-irradiance response, of phytochromes. *Seed Science Research* 14: 61-72.

Mella, R., de Miguel, L., Garzarón, I., and Sánchez, R. (2002). Single seed studies of germination inhibition by continuous far-red light in *Datura ferox* L. seeds. VII International Workshop on Seed Biology (pp. 113), Salamanca, Spain, May 12-16.

Mella, R.A., Maldonado, S., and Sánchez, R.A. (1995). Phytochrome-induced structural changes and protein degradation prior to radicle protrusion in *Datura ferox* seeds. *Canadian Journal of Botany* 73: 1371-1378.

Ni, B.R. and Bradford, K.J. (1993). Germination and dormancy of abscisic acid- and gibberellin-deficient mutant tomato *(Lycopersicon esculentum)* seeds. *Plant Physiology*. 101: 607-617.

Nonogaki, H., Gee, O., and Bradford, K. (2000). A germination specific endo-β-mannanase gene is expressed in the micropylar endosperm cap of tomato seeds. *Plant Physiology* 123: 1235-1245.

Nonogaki, H. and Morohashi, Y. (1996). An endo-β-mannanase develops exclusively in the micropylar endosperm of tomato seeds prior to radicle emergence. *Plant Physiology* 110: 555-559.

Nonogaki, H. and Morohashi, Y. (1999). Temporal and spatial patterns of endo-β-mannanase expression in lettuce seeds. IV International Workshop on Seeds Biology, Merida, Mexico, January.

Nonogaki, H., Nomaguchi, M., and Morohashi, Y. (1995). Endo-β-mannanases in the endosperm of germinated tomato seeds. *Physiologia Plantarum* 94: 328-334.

Nonogaki, H., Nomaguchi, M., and Morohashi, Y. (1998). Temporal and spatial pattern of the biochemical activation of the endosperm during and following imbibition of tomato seeds. *Physiologia Plantarum* 102: 236.

Peng, J. and Harberd, N.P. (2002). The role of GA-mediated signaling in the control of seed germination. *Current Opinion in Plant Biology* 5: 376-371.

Psaras, G. and Georghiu, K. (1983). Gibberellic acid-induced structural alterations in the endosperm of germinating *Latuca sativa* L. achenes. *Zeitschrift für Pfanzenphysiologie* 112: 15-19.

Psaras, G., Georghiu, K., and Mitrakos, K. (1981). Red-light induced endosperm preparation for radicle protrusion of lettuce embryos. *Botanical Gazette* 142: 13-18.

Sánchez, R.A. and de Miguel, L. (1985). The effect of red light, ABA and K+ on the growth rate of *Datura ferox* embryos and its relations with the photocontrol of germination. *Botanical Gazette* 146: 472-476.

Sánchez, R.A. and de Miguel, L. (1997). Phytochrome promotion of mannan-degrading enzyme activities in the micropylar endosperm of *Datura ferox* seeds requires the presence of the embryo and gibberellin synthesis. *Seed Science Research* 7: 27-33.

Sánchez, R., de Miguel, L., Lima, C., and Lederkremer, R. (2002). Effect of low water potential on phytochrome-induced germination, endosperm softening and cell wall mannan degradation. *Seed Science Research* 12: 155-163.

Sánchez, R.A., Sunell, L., Labavitch, J., and Bonner, B.A. (1990). Changes in endosperm cell walls of two *Datura* species before radicle protrusion. *Plant Physiology* 93: 89-97.

Schopfer, P. and Plachy, C. (1985). Control of seed germination by abscisic acid: II. Effect on embryo growth potential (minimum turgor pressure) and growth coefficient (cell wall extensibility) in *Brassica napus* L. *Plant Physiology* 77: 676-686.

Schopfer, P. and Plachy, C. (1993). Photoinhibition of radish (*Raphanus sativus* L.) seed germination: Control of growth potential by cell-wall yielding in the embryo. *Plant, Cell and Environment* 16: 223-229.

Scopel, A.L., Ballaré, C.L., and Sánchez, R.A. (1991). Induction of extreme light sensitivity in buried weed seeds and its role in the perception of soil cultivations. *Plant, Cell and Environment* 14: 501-508.

Shichijo, C., Katada, K., Tanaka, O., and Hashimoto, T. (2001). Phytochrome A-mediated inhibition of seed germination in tomato. *Planta* 213: 764-769.

Shinomura, T., Nagatani, A., Hanzawa, H., Kubota, M., Watanabe, M., and Furuya, M. (1996). Action spectra for phytochrome A- and phytochrome B-specific photoinduction of seed germination in *Arabidopsis thaliana*. *Proceedings of the National Academy of Sciences, USA* 93: 8129-8133.

Steber, C. and McCourt, P. (2001). A role for brassinosteroids in germination in *Arabidopsis*. *Plant Physiology* 125: 763-769.

Tillberg, E. and Björkman, P.-O. (1993). Effect of red and far-red irradiation on ABA and IAA content in *Pinus sylvestris* L. seeds during the escape time period from photocontrol. *Plant Growth Regulation* 13: 1-6.

Toorop, P.E., van Aelst, A.C., and Hilhorst, H.W.M. (1998) Endosperm cap weakening and endo-β-mannanse activity during priming of tomato (*Lycopersicon esculentum* Mill cv. Money Maker) are initated upon crossing a threshold water potential. *Seed Science Research* 8: 483-491.

Toorop, P.E., van Aelst, A.C., and Hilhorst, H.W.M. (2000). The second step of the biphasic endosperm cap weakening that mediates tomato (*Lycopersicon esculentum*) seed germination is under control of ABA. *Journal of Experimental Botany* 51: 1371-1379.

Toyomasu, T., Kawaide, H., Mitsihashi, W., Inoue, Y., and Kamiya, Y. (1998). Phytochrome regulates gibberillin biosynthesis during germination of photoblastic lettuce seeds. *Plant Physiology* 118: 1517-1523.

Toyomasu, T., Tsuji, H., Yamane, H., Nakayama, M., Yamaguchi, I., Murofushi, N., Takahashi, N., and Inoue, Y. (1993). Light effects on endogenous levels of gibberellins in photoblastic lettuce seeds. *Journal of Plant Growth Regulation*. 12: 85-90.

Vázquez-Yañez, C., Rojas-Aréchiga, M., Sánchez-Coronado, M.E., and Orozco-Segovia, A. (1996). Comparison of light-regulated seed germination in *Ficus* spp. and *Cecropia obtusifolia:* Ecological implications. *Tree Physiology* 16: 871-875.

Vázquez-Yañez, C. and Smith, H. (1982). Phytochrome control of seed germination in the tropical rain forest pioneer trees *Cecropia obtusifolia* and *Piper auritum* and its ecological significance. *New Phytologist* 92: 477-485.

Watkins, J.T. and Cantliffe, D.J. (1983). Mechanical resistance of the seed coat and endosperm during germination of *Capsicum annuum* at low temperature. *Plant Physiology* 72: 146-150.

Watkins, J.T., Cantliffe, D.J., Huber, D.J., and Nell, T. (1985). Gibberellic acid stimulated degradation of endosperm in pepper. *Journal of the American Society of Horticultural Science* 110: 61-65.

Welbaum, G.E., Muthui, W.J., Wilson, J.H., Grayson, R.L., and Fell, R.D. (1995). Weakening of muskmelon perisperm envelope tissue during germination. *Journal of Experimental Botany* 46: 391-400.

Wu, C., Leubner-Metzger, G., Meins, J.G., and Bradford. K.J. (2001). Class I β-1,3-glucanase and chitinase are expressed in the micropylar endosperm of tomato seeds prior to radicle emergence. *Plant Physiology* 126: 1299-1313.

Yamaguchi, S., Smith, M.W., Brown, R.G.S., Kamiya, Y., and Sun, T. (1998). Phytochrome regulation and differential expression of gibberillin 3β-hydroxylasegenes in germinating *Arabidopsis* seeds. *Plant Cell* 10: 2115-2126.

Yang, Y.Y., Yamaguchi, I., Takenowada, K., Suzuki, Y., and Murofushi, N. (1995). Metabolism and translocation of gibberellins in seedlings of *Pharbitis nil:* 1. Effect of photoperiod on stem elongation and endogenous gibberellins in cotyledons and their phloem exudates. *Plant Cell Physiology* 36: 221-227.

Yoshioka, T., Endo, T., and Satoh, S. (1998). Restoration of seed germination at supraoptimal temperatures by fluridone, an inhibitor of abscisic acid biosynthesis. *Plant Cell Physiology* 210: 307-312.

Chapter 8

Modeling Changes in Dormancy in Weed Soil Seed Banks: Implications for the Prediction of Weed Emergence

Diego Batlla
Betina Claudia Kruk
Roberto L. Benech-Arnold

INTRODUCTION

The seedling stage is a common target of many weed mechanical control and herbicide methods because of its high vulnerability (Fenner, 1987). The success of control methods targeted at weed seedlings depends upon reaching the highest number of individuals at this developmental stage. However, it is practically impossible to determine which proportion of a certain weed population is being reached by a control method. Indeed, the number of emerged seedlings can be counted, but we do not really know which fraction of the population they represent. Construction of weed seed germination models that predict which proportion of the seed bank germinates at a certain time would be useful tools for determining the most suitable time for seedling control and, consequently, should result in a higher efficacy of controls methods (Benech-Arnold and Sánchez, 1995). Although many models that successfully predict seed germination have been developed, one of the most important limitations for the formulation of such models in many common weed species is the existence of dormancy. The lack of detailed research intending to understand and quantify how environmental factors regulate dormancy status in field situations probably prevented the elaboration of an adequate theoretical framework for the construction of predictive models addressing dormancy changes in weed seed bank populations.

In the first section of this chapter we discuss the different environmental factors affecting dormancy in weed seed banks and present a general framework for classifying and understanding the effects of these factors on weed seed dormancy changes under field situations. The aim of this classification is merely to facilitate the conceptualization of the whole system. The remainder of the chaper will discuss some attempts to model weed seed dormancy in relation to the effect of those environmental factors.

DORMANCY: DEFINITIONS AND CLASSIFICATION

Although many studies have been published concerning weed seed dormancy, the definition of dormancy is still a controversial subject. A new general definition of dormancy was recently proposed by Benech-Arnold and colleagues (2000, p. 106): "Dormancy is an internal condition of the seed that impedes its germination under otherwise adequate hydric, thermal and gaseous conditions." This implies that once the impedance has been removed seed germination would proceed under a wide range of environmental conditions (Benech-Arnold et al., 2000).

Karssen (1982) suggested that dormancy could be classified into primary and secondary dormancy. Primary dormancy refers to the innate dormancy possessed by seeds when they are dispersed from the mother plant. Secondary dormancy refers to a dormant state that is induced in non-dormant seeds by unfavorable conditions for germination, or reinduced in once-dormant seeds after a sufficiently low dormancy has been attained. Thus, it is by no means a classification referring to mechanisms or location, but one of timing of occurrence.

The release from primary dormancy followed by subsequent entrance into secondary dormancy (whenever conditions are given for this entrance) may lead to dormancy cycling. Evidence for dormancy cycling has been obtained for seeds of many weed species, but it is not the only possibility. For the case of species that present dormancy cycling under natural field conditions, the transition into and out of dormancy may continue to cycle for several years before the seeds germinate, decay, or are otherwise lost from the soil seed bank (Karssen, 1980/1981; Baskin and Baskin, 1985). In general, seeds are released from dormancy during the season preceding the period with favorable conditions for seedling development, whereas dormancy is induced in the season preceding the period with harmful conditions for plant survival. For example, several summer annual species present a high dormancy level in autumn; during winter they undergo dormancy relief but dormancy increases again during summer. Winter annual species show the reverse dormancy pattern. Therefore, the patterns of dormancy are of high survival value to weed species determining germination under environmental conditions that will ensure species growth and perpetuation (Karssen, 1982).

HOW IS DORMANCY LEVEL EXPRESSED?

Dormancy is not an all-or-nothing seed property; seed dormancy status can vary over a continuous dormancy degree scale between some point at which environmental conditions permissive for seed germination are narrowest and some point where environmental conditions permissive for seed germination are widest. Vegis (1964) was the first to introduce this concept of degrees of relative dormancy from the observation that as dormancy is released, the temperature range permissive for germination widens until it is maximal; in contrast, as dormancy is induced, the range of temperatures over which germination can proceed narrows, until germination is no longer possible at any temperature and full dormancy is reached. Clearly, Vegis's view relates the degree of dormancy of a seed population to the width of the thermal range permissive for germination. Karssen (1982) agreed with that view and emphasized that seasonal periodicity in the field emergence of annual weeds is the combined result of seasonal periodicity in field temperatures and the width of the temperature range permissive for germination. Germination in the field is therefore restricted to the period when the field temperature and the temperature range over which germination can proceed overlap.

The concept of base water potential for seed germination (Ψ_b) is fully developed in Chapter 1. Many experimental data show that dormancy alleviation could also be correlated with a decrease of the Ψ_b of the seed population (i.e., more negative values), whereas dormancy induction could be associated with an increase of the Ψ_b of the seed population (i.e., less negative or even positive values) (Dahal, Bradford, and Haigh, 1993; Ni and Bradford, 1992, 1993; Bradford and Somasco, 1994). Based on these findings, Bradford (1995) proposed that changes in the dormancy level of seed populations could be associated with changes in the seed population Ψ_b required for germination. Thus, analogous to the thermal range permissive for germination, the dormancy status of a seed population could also be evaluated by monitoring changes in the range of water potentials permissive for seed germination.

In many weed species, dormancy must be terminated by the effect of light, nitrate, or fluctuating temperatures to allow the germination process to proceed. In those cases, changes in degree of dormancy comprise changes not only in temperature requirements for germination (and eventually in Ψ_b), but also in sensitivity to the effect of dormancy terminating factors (Benech-Arnold et al., 2000).

Seedling emergence of a particular weed in the field occurs when environmental conditions are within the range of conditions permissive for seed germination (temperature range, water potential range, and requirement of dormancy terminating factors stimuli). Since permissive conditions for seed germination change together with changes in dormancy level of seed populations, in order to construct dynamic models for predicting seedling emergence we should understand the way in which the environment determines changes in seed environmental requirements for germination as dormancy is released or enforced. To achieve this aim, it is necessary (1) to identify the environmental factors involved in the modulation of dormancy level of seed populations and those necessary for dormancy termination, and (2) to establish functional relationships between these factors and the rates of these processes.

ENVIRONMENTAL FACTORS AFFECTING DORMANCY LEVEL OF SEED POPULATIONS

Dormancy cycles observed in some species are known to be regulated mainly by temperature in temperate environments in which water is not seasonally restricted (Baskin and Baskin, 1977, 1984; Kruk and Benech-Arnold, 1998). For example, in some summer annual species dormancy relief is produced by low temperatures experienced during winter, and their dormancy level is enhanced by high temperatures experienced during summer (Bouwmeester and Karssen, 1992, 1993a). Several winter annual species show the reverse dormancy pattern. Hence, high temperatures during summer result in dormancy relief, and low temperatures during winter can induce secondary dormancy (Baskin and Baskin, 1976; Karssen, 1982; Probert, 1992). As mentioned previously, these changes in dormancy level can be expressed through narrowing and widening of the temperature range permissive for germination. Thus, when soil temperature enters that permissive range, germination in the field occurs (Figure 8.1). This range is characterized by two thermal parameters: (1) the mean lower limit temperature (Tl_{50}) and (2) the mean higher limit temperature (Th_{50}). It should be noted that these parameters are conceptually different from base temperature (Tb) and maximum temperature (Tm) (as described in Chapter 1). Indeed, while Tb and Tm are theoretical temperatures over and under which thermal time (θ) is accumulated toward germination, Tl_{50} and Th_{50} represent the temperatures under and over which dormancy is expressed for 50 percent of the population. A main difference between these two kinds of parameters is that Tb and Tm are regarded as unique for the whole population, whereas Tl and Th are normally distributed within the population (Washitani, 1987; Kruk and Benech-Arnold, 2000). The fact that each individual has a different value of Tl and Th is consistent with the idea that dormancy

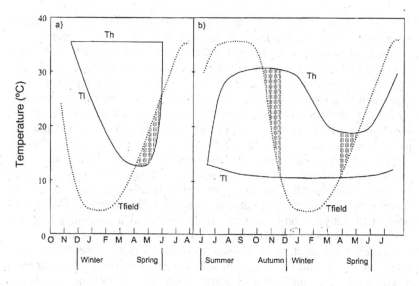

FIGURE 8.1. Seasonal changes in the permissive germination thermal range and its relation with soil temperature dynamics. Solid lines indicate lower *(Tl)* and higher *(Th)* limit temperatures allowing germination. Broken lines show mean daily maximum temperature in the field. The shaded area represents the period when field germination takes place due to overlapping of required and actual temperature. (a) Strict summer annual; (b) facultative winter species. (*Source:* Adapted from Benech-Arnold et al., 2000.)

level is different for each individual seed within the population. This concept will be further discussed in other sections of this chapter. Changes in the dormancy level (or variations in the thermal range in which germination can occur) in summer species are characterized by an increase or decrease of *Tl* (Baskin and Baskin, 1980). In contrast, in winter species changes in dormancy level are characterized by fluctuations in *Th* (Figure 8.1).

Although a good deal of experimental data support a main role of soil temperature as regulator of dormancy level in seed bank populations, some evidence indicates that the effect of temperature on dormancy release and induction may be modulated by soil moisture (Adámoli, Goldberg, and Soriano, 1973; de Miguel and Soriano, 1974; Reisman-Berman, Kigel, and Rubin, 1991; Christensen, Meyer, and Allen, 1996; Bauer, Meyer, and Allen, 1998). Some interactions with soil moisture were detected in *Polygonum aviculare* L. seeds; dormancy release occurred most rapidly when seeds were moist chilled at 4°C, but relief of dormancy was also possible with seeds dry stored at 4°C though at a much slower rate (Kruk and

Benech-Arnold, 1998). Batlla and Benech-Arnold (2000) also found that *P. aviculare* seeds buried in the field under contrasting soil water content conditions showed different annual pattern of changes in sensitivity to light, sensitivity to alternating temperatures, and the range of temperatures permissive for germination. Other interactions have been also reported for the light-requiring species *Sisymbrium officinale* L., for which a high sensitivity to light stimuli, usually occurring in buried seeds at the end of the winter, is not acquired if the seeds have been permanently water imbibed and subjected to low winter temperatures (Hilhorst, Derkx, and Karssen, 1996). On the other hand, seeds composing the seed bank would be normally subjected to dehydration-hydration cycles, particularly in the upper layers of the soil. As rehydration of seeds previously imbibed and then dried was found to break dormancy in many weed species, Bouwmeester (1990) proposed that dehydration-hydration cycles can act as a dormancy-breaking environmental factor affecting buried seeds under field conditions.

In the field, induction of secondary dormancy can proceed at temperatures that are within the range suitable for germination. In those cases it might result from inhibition of germination (i.e., germination-inhibitory water potentials or inhibition of germination under leaf canopies), or from a situation in which factors that terminate dormancy are not met (i.e., loss of sensitivity to light in light-requiring seeds held in darkness, loss of sensitivity to fluctuating temperatures in seeds held at low thermal amplitudes) (Benech-Arnold et al., 2000). In any case, the process itself should involve the narrowing of the range of suitable conditions for germination, ultimately leading to a state of relative or total dormancy, to be regarded as induction of secondary dormancy (Karssen, 1982).

FACTORS THAT TERMINATE DORMANCY

Once reaching a low dormancy level, several species require exposure to certain environmental stimuli for the termination of dormancy. Fluctuating temperatures and light are predominantly naturally occurring environmental factors that can complete exit from dormancy in many weed seeds (Benech-Arnold et al., 1990b; Scopel, Ballaré, and Sánchez, 1991; Ghersa, Benech-Arnold, and Martinez Ghersa, 1992), although other factors (i.e., CO_2, NO_3^-, O_2, and ethylene) can be involved in the termination of dormancy of buried seeds under field conditions (Benech-Arnold et al., 2000). An ecological interpretation of the requirement of light or fluctuating temperatures to complete exit from dormancy in certain weed species has been related to the possibility of detecting canopy gaps as well as depth of burial (Holmes and Smith, 1977; Frankland, 1981; Thompson and Grime, 1983;

Benech-Arnold et al., 1988; Deregibus et al., 1994; Batlla, Kruk, and Benech-Arnold, 2000). The requirement of fluctuating temperatures to terminate dormancy in some species has also been regarded as an effective mechanism for distributing germination over a longer period of time (Benech-Arnold et al., 1990a,b).

Several characteristics of diurnal temperature cycles could be responsible for its stimulatory activity (Roberts and Totterdell, 1981). Thermal amplitude is of paramount importance; in *Chenopodium album* L., the dormancy breakage response can increase from an amplitude of as little as 2.4°C up to about 15°C (Murdoch, Roberts, and Goedert, 1988). However, the response to a given amplitude is greater the higher the mean temperature of the cycle (i.e., average of lower and upper temperature) up to an optimum of about 25°C (Murdoch, Roberts, and Goedert, 1988). In some cases, diurnal temperature cycles with stimulatory characteristics tend to be additive in their effect. For example, in *Sorghum halepense* L. seeds, ten cycles with stimulatory characteristics release from dormancy twice the proportion of the population released with only five cycles (Benech-Arnold et al., 1990a).

In many weed species, dormancy is terminated when the hydrated seed is exposed to light, which is perceived through photoreceptors, particularly those from the phytochrome family. Phytochromes have two mutually photoconvertible forms: Pfr (considered the active form) with maximum absorption at 730 nm and Pr with maximum absorption at 660 nm. Phytochromes are synthesized as Pr, and the proportion of the pigment population (P) in the active form (Pfr/P) in a particular tissue depends on the light environment seeds are exposed. Exposure of seeds to light with a high red (R) to far-red (FR) ratio (R:FR) leads to larger Pfr/P determining, depending on seed dormancy level, breaking dormancy in many weed species. Phytochrome-mediated responses can be classified physiologically into three "action modes" (Kronenberg and Kendrick, 1986). Two of these action modes, the very low fluence response (VLFR) and the low fluence response (LFR), are characterized by the correlation between the intensity of the effect and level of Pfr predicted to be established by the light environment. They differ in that extremely low levels of Pfr saturate VLFRs, while higher Pfr levels are necessary to elicit LFRs (Casal and Sánchez, 1998). High irradiance responses (HIRs) are the third action mode; they show no simple relationship between Pfr levels and may involve additional components of the phytochrome system (Heim and Schäfer, 1982, 1984). Other particular characteristics of an HIR are that it has maximum activity at 710 to 720 nm (Hartman, 1966; Hendricks, Toole, and Borthwick, 1968; Mohr, 1972), and the inhibitory effect of continuous FR can be observed, even in R-promoted seed, after the escape time is over (Frankland and Taylorson, 1983).

Commonly crop or pasture leaf canopies reduce R:FR ratio perceived by weed seeds placed on the soil surface (Smith 1982; Pons 1992). This change in light quality would establish low Pfr/P, preventing weed seed germination. The LFR and/or HIR mode of action can mediate this type of inhibition by canopy presence (Deregibus et al., 1994; Batlla, Kruk, and Benech-Arnold, 2000). On the contrary, reductions in canopy density, for example by grazing, mainly lead to increases in R:FR ratios which consequently raises Pfr/P seed level promoting germination by an LFR of many weed seeds disposed on the soil surface (Deregibus et al., 1994; Insausti, Soriano, and Sánchez, 1995). An example of a VLFR is typically observed when soil is disturbed by agricultural practices. Seeds of several species may acquire a very great sensitivity to light as dormancy is released through a period of burial in the soil (Scopel, Ballaré, and Sánchez, 1991). These seeds can respond to exposures of the order of sub-milliseconds of sunlight determining the germination of a large number of seeds from the seed bank when soil is being disturbed by tillage practices.

CONCEPTUALIZING THE SYSTEM WITH MODELING PURPOSES

Although real scenarios under field conditions are far more complicated, showing interactions of many kinds between relevant environmental factors affecting seed dormancy, the classification carried out so far is useful for understanding how the environment controls dormancy in weed soil seed banks and, eventually, for developing simulation and predictive models. Thus, on the basis of the classification into two different kinds of environmental factors that affect dormancy as described in previous paragraphs, namely, (1) those that govern changes in the degree of dormancy of a seed population (i.e., temperature and its interactions with soil hydric conditions) and (2) those that remove the ultimate constraints for seed germination (i.e., mainly light and fluctuating temperatures), Benech-Arnold and colleagues (2000) proposed the diagram shown in Figure 8.2. It illustrates the conceptual framework derived from the definitions of the different factors that affect dormancy in weed seed populations. It should be noted that passage along the whole flowchart is by no means the only possibility for a seed population. On the contrary, the chart aims to illustrate the different pathways that a seed population could undergo. For example, a population might be dispersed with a low level of dormancy and might or might not require limited stimuli for dormancy termination. In that case, the population

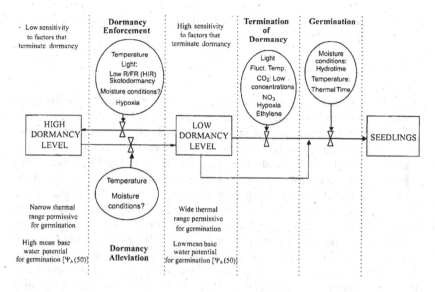

FIGURE 8.2. Flowchart representing changes in dormancy level and termination of dormancy in seed populations and the factors that most likely affect each process (*Source:* Reprinted from *Field Crops Research* 67(2), R. L. Benech-Arnold et al., Environmental control of dormancy in weed seed banks in soil, pp. 105-122, copyright 2000, with permission from Elsevier Science.)

would not experience the left side of the flowchart (unless induction of secondary dormancy takes place) and may or may not bypass the zone of dormancy termination.

MODELING DORMANCY CHANGES IN WEED SEED BANKS AS AFFECTED BY THE ENVIRONMENT

Dormancy is probably the most important of a series of components and processes that affect seedling emergence in weed species (Forcella et al., 2000). Thus, in order to predict time and proportion of weed seed bank emergence, we should consider changes in dormancy as affected by environmental factors in the construction of our germination models. Abundant information has been published concerning dormancy in weed species as described in the first part of this chapter, but few attempts have been made to model seed dormancy changes as affected by the environment. Although a vast review of existing models regarding seed dormancy is not intended in this section, some relevant attempts to model dormancy changes in weed species will be discussed.

The Role of Temperature

As mentioned earlier, soil temperature is one of the most important environmental factors controlling annual dormancy cycles of buried weed seeds in the field. Therefore, it is not surprising that almost all attempts to model dormancy changes in weed seeds use temperature as the key factor driving changes in seed population dormancy status. Totterdell and Roberts (1979) were probably the first investigators to establish functional relationships between temperature and dormancy changes in weed seed populations. These authors hypothesized that temperature-dependent changes in dormancy of *Rumex crispus* L. and *Rumex obtusifolius* L. seeds result from two simultaneous independent processes: (1) relief of primary dormancy and (2) induction of secondary dormancy. They suggest that relief of primary dormancy would occur only as a result of exposure of seeds to temperatures under a critical ceiling temperature for dormancy loss to occur. According to these authors, such a process proceeds at a constant rate, independently from the actual temperature, as long as it is below that ceiling temperature. For the case of *Rumex* species the authors estimate a ceiling temperature for dormancy loss of 15°C. On the other hand, they suggest that induction of secondary dormancy would occur at all temperatures at a rate that would rise concomitantly with temperature.

Using Permissive Germination Thermal Range
for Dormancy Modeling

Based on the hypothesis proposed by Totterdell and Roberts (1979) and the concept of dormancy introduced by Vegis (1964), Bouwmeester and Karssen (1992) developed a descriptive simulation model that successfully predicted changes in dormancy of buried seeds of the summer annual *Polygonum persicaria* L. as a function of soil thermal conditions. The model considers seed population dormancy status as a function of cold (C) and heat (H) unit sums. C is related to dormancy release, and H is related to secondary dormancy induction. For simulation of seed population dormancy status, the value of C is raised by an arbitrary value of one unit for each period of ten days during which the mean soil temperature has been below the ceiling temperature for dormancy loss to occur, which for *P. persicaria* the authors determined to be 15°C. On the other hand, H is calculated by summing the mean soil temperature of each successive ten-day period. Thus, seed dormancy status depends on the balance between two processes: dormancy alleviation, quantified by C, and dormancy induction, quantified by H. The authors observed that for seeds exhumed from the field at regular intervals, the germination percentage (G_t) obtained at three incubation temperatures could be described by a quadratic function of the germination test temperature (T_g):

$$G_t = a \bullet T_g^2 + b \bullet T_g + c \qquad (8.1)$$

where a, b, and c are functions of C, H, the presence or absence of nitrates in the germination medium, and the mean soil temperature during 30 days prior to seed exhumation. Using Equation 8.1 annual changes in seed germination behavior in relation to soil temperature can be predicted. The model also allows the estimation of the width of the thermal range permissive for germination for seeds buried for different periods of time and exposed to a variable thermal environment. Narrowing or widening of the germination permissive thermal range was the result of changes in the minimum temperature for germination of 50 percent of the seed population (presumably analogous to the previously introduced Tl_{50}). Therefore, germination in the field is restricted to the period during which field temperature and the germination permissive thermal range simulated by the model overlap (Figure 8.3). Model performance showed good agreement between simulated emergence timing for 50 percent of the seed population and observed timing of germination of seeds disposed in petri dishes outdoors. The present model structure was also used by the same authors to predict timing of field emergence of *Sisymbrium officinale* L., *Chenopodium album* L., and *Spergula arvensis* L. (Bouwmeester, 1990; Bouwmeester and Karssen, 1993a,b,c). However, results obtained with simulations showed that for these weeds the model cannot give a description of dormancy relief as accurate as observed for *P. persicaria*, suggesting that temperature effects on dormancy changes would not be as simple as initially described by Totterdell and Roberts (1979) for *Rumex* species.

Results obtained with other weed species also contrast with Totterdell and Roberts's hypothesis (Pritchard, Tompsett, and Manger, 1996; Kebreab and Murdoch, 1999). For example, Batlla and Benech-Arnold (2003) observed no induction of secondary dormancy in *P. aviculare* seeds stratified for 110 days under constant temperatures of 1.6, 7, and 12°C. Moreover, exhumed seeds showed a progressive decrease in their dormancy level during the stratification period, verified by a widening of the thermal range permissive for germination. The rate of decrease in seed dormancy status was shown to be negatively related to the stratification storage temperature. On

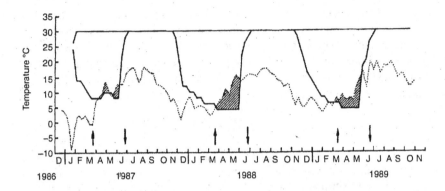

FIGURE 8.3. Simulated seasonal changes in the permissive thermal range for germination of 50 percent of *Polygonum persicaria* seed population (maximum and minimum temperatures indicated by solid lines). Air temperature at 1.5 m is represented by the dashed line. Hatched areas represent the period of overlap between simulated permissive germination thermal range and actual temperature. The arrows indicate the lapses at which germination of at least 50 percent of the population actually occurred in petri dishes placed outdoors. (*Source: Oceologia*, The dual role of temperature in the seasonal changes in dormancy and germination of seeds of *Polygonum persicaria* L., H. J. Bouwmeester and C. M. Karssen, Vol. 90, pp. 88-94, 1992, © Springer-Verlag. Reproduced with permission.)

the other hand, when seeds were stored for 12 days under a constant temperature of 22°C, secondary dormancy was rapidly induced. Exposed results contrast with Totterdell and Roberts's hypothesis, suggesting that for *P. aviculare* a threshold temperature for dormancy induction may exist and that the rate of dormancy release, under the ceiling temperature for this process to occur, would be dependent on the temperature at which seed after-ripening takes place.

Using Base Water Potential for Dormancy Modeling

Although many attempts to quantify dormancy changes in weed seed populations were done by assessing changes in the permissive thermal range for germination, Bradford (1996, 1997) proposed that dormancy changes can also be described and eventually predicted through changes in the seed population base water potential (Ψ_b) used, in this case, as an index of the seed population dormancy status. Working with the winter-annual weed *Bromus tectorum* L., Christensen, Meyer, and Allen, (1996) found that mean seed population base water potential [Ψ_b (50)] became more negative as seeds after-ripened under dry conditions. Based on these findings Bauer, Meyer, and Allen, (1998) derived a simulation model to predict *B. tectorum* dormancy loss by dry after-ripening in the field for four seed populations corresponding to contrasting habitats. The model allowed Ψ_b (50) to vary in relation to the accumulation of thermal time units (temperature above a threshold temperature for dormancy loss to occur), while other parameters of the hydrothermal time equation [the hydrothermal time constant (θ_{HT}), the standard deviation of Ψ_b ($\sigma_{\psi b}$), and the base temperature *(Tb)*] were held constant during after-ripening. Results showed that dormancy release was accompanied by a progressive decrease in Ψ_b (50) and that changes in Ψ_b (50) can be described by a linear negative relationship to thermal after-ripening time accumulation (Figure 8.4):

$$\Psi_b (50) = m \ [(T_s - T_l) \ (t_{ar})] + b \qquad (8.2)$$

where *b* is the estimated initial value of Ψ_b (50), *m* is the decrement in Ψ_b (50) per unit of thermal time, T_s is the after-ripening storage temperature, T_l is the threshold temperature at or below which after-ripening does not occur, and t_{ar} is the time required for full after-ripening.

Therefore, changes in Ψ_b (50) account for changes in seed germination time-course curves due to variations in dormancy status or incubation temperature. Model performance was evaluated against Ψ_b (50) values derived from incubation of previously buried seeds retrieved from the field at regular intervals. Hourly recorded seed-zone soil temperatures and estimated seed-zone water potential were inputs for the model. Thus, Ψ_b (50) decreases as a function of after-ripening thermal-time accumulation, if estimated seed-zone water potential during that hour is considered low enough for after-ripening to occur (below approximately –4 MPa). The process continues until the Ψ_b (50) value corresponding to fully after-ripened seeds is reached. Predictions of changes in Ψ_b (50) were generally close to that estimated for seeds retrieved from the field, suggesting that Ψ_b (50) can be used as a reliable index to quantify changes in seed dormancy status. Derived model parameters used in Equation 8.2 were different for each population, showing that model parameters should be adjusted for predicting dormancy changes in seed populations belonging to different habitats. Another interesting feature of the model is that the effect of variable times and after-ripening temperatures on seed dormancy status can be quantified as a thermal-time phenomenon. A similar thermal-time approach was successfully used by Pritchard, Tompsett, and Manger (1996) to quantify changes

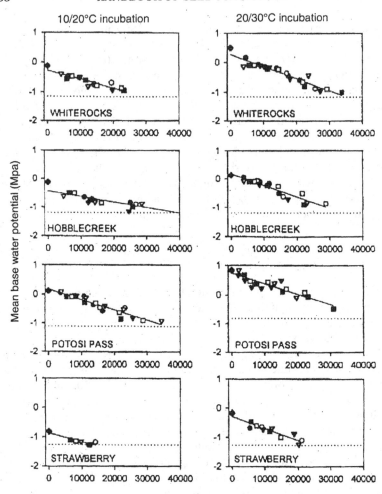

FIGURE 8.4. Dynamics of *Bromus tectorum* seed population Ψ_b (50) determined at two temperature regimes, plotted against thermal time accumulated during seeds dry after-ripening at different storage temperatures. Storage regimes are (♦) initial values, (•) 10°C, (o) 15°C, (▼) 20°C, (▽) 30°C, (■) 40°C, and (□) 50°C. Dotted horizontal lines correspond to Ψ_b (50) for fully after-ripened seeds. (*Source:* M. C. Bauer, S. E. Meyer, and P. S. Allen, A simulation model to predict seed dormancy loss in the field for *Bromus tectorum* L., *Journal of Experimental Botany*, 1998, 49(324): 1235-1244, by permission of Oxford University Press.)

in the germination percentage of *Aesculus hippocastanum* L. seeds in relation to after-ripening temperature. A more detailed review of the applications of the hydrothermal time concept to quantify and model seed dormancy was recently done by Bradford (2002).

Modeling Dormancy As a Population-Based Phenomenon

Although existing models for simulation of dormancy changes under field conditions can almost precisely predict the occurrence of the time window for weed emergence, attempts to predict the proportion of seeds that will germinate at that time window with reasonable accuracy have, so far, been unsuccessful (Vleeshouwers, 1997; Murdoch, 1998). For achieving this objective, we should be able to not only predict changes in mean population parameters that characterize seed dormancy status [namely, Tl_{50} and Th_{50} or Ψ_b (50)], but also account for changes in their distribution within the seed population. Several reports on germination behavior of different species indicate that limit temperatures and water potentials demarcating the germination range vary among individuals belonging to the same seed population (Washitani, 1987, Bradford, 1996). Thus, if these parameters characterize the dormancy level of an individual seed, its distribution is a measure of the spread of dormancy levels within the population. Quantification and inclusion of this variation in dormancy models is essential for allowing the estimation of germination percentage. The importance of accounting for this variation to predict germination percentage of a seed population can be exemplified by considering two populations with the same Tl_{50} (i.e., 15°C), but different standard deviation of Tl (σ_{Tl}) (i.e., batch (A) $\sigma_{Tl} = 5$; batch (B) $\sigma_{Tl} = 1$]: incubating seeds at 13.5°C will result in 35 percent of germination in batch (A), while only 5 percent of the seed population will germinate in batch (B) (Figure 8.5).

Recently, Batlla and Benech-Arnold (2003) developed a population-based threshold model for simulating *P. aviculare* seed dormancy loss in relation to stratification temperature. The model employs the mean lower limit temperature for germination (Tl_{50}) as an index of seed population dormancy status, also accounting for changes in the distribution of Tl within the population (σ_{Tl}) in relation to seed population dormancy changes. The authors suggest that distribution of Tl is associated with variations in the dormancy status among seeds belonging to the same population. The idea that dormancy is continuously distributed within individuals of a population, as referred to before, has been proposed frequently in the literature (Ransom, 1935; Probert, 1992; Bradford, 1996). Dynamic changes in these

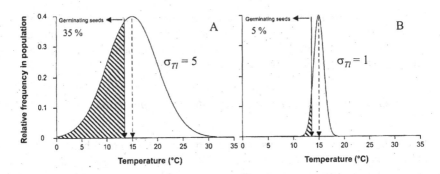

FIGURE 8.5. Hypothesized normal distributions of Tl values in two seed populations with the same Tl_{50} = 15°C (indicated by hatched arrows) but contrasting σ_{Tl} (σ_{Tl} A = 5; σ_{Tl} B = 1). Striped areas represent the fraction of the seed population that would germinate if the two seed populations were incubated at 13.5°C (incubation temperature is indicated by solid arrows).

two parameters (Tl and σ_{Tl}) as a function of a variable stratification thermal environment were predicted using thermal-time equations similar as those used by Bauer, Meyer, and Allen (1998) and Pritchard, Tompsett, and Mager (1996). Accumulation of stratification thermal time (Stt) began when temperature was below a threshold level for dormancy loss to occur (17°C). Interestingly, although changes in Tl_{50} were linearly and negatively related to accumulated Stt, the pattern of change in σ_{Tl} was different depending on the temperature at which accumulation of Stt takes place. This implies that under field situations changes in σ_{Tl} would depend on the prevailing winter temperature at which dormancy release occurs. Model performance was evaluated against data of two unrelated experiments, showing acceptable prediction of timing and percentage of germination of seeds exhumed from field and controlled stratification conditions. A population-based threshold model structured to assess the combined effects of temperature, water stress, and release from dormancy by cool-moist stratification on the germination of seeds of *Eucalyptus delegatensis* R. T. Baker was also developed by Battaglia (1997). The model satisfactorily describes germination of this species under a wide range of conditions, showing the ability of population-based threshold models to describe changes in dormancy and germination behavior in response to environmental factors. The author proposes that the model structure can be easily modified in order to include additional factors and factor interactions controlling seed dormancy and germination.

Modeling Changes in Sensitivity to Dormancy
Terminating Factors

As pointed out previously, in many weed seed populations, additional exposure to light, nitrate, or fluctuating temperatures may be required to allow the germination process to proceed. For example, *Sorghum halepense* L. seeds, a common summer weed in Argentinean pampas, present an absolute requirement of fluctuating temperatures for dormancy termination (Benech-Arnold et al., 1990b). Benech-Arnold and colleagues (1990a) developed a dynamic model for prediction of *S. halepense* seed germination in relation to soil temperature. The model basically states that two different fractions can be identified within a *S. halepense* seed population in relation to their dormancy level: those seeds that must be stimulated by fluctuating temperatures to terminate dormancy and those that will not be released from dormancy after exposure to fluctuating temperatures. Based on this classification, the model calculates the proportion of the seed population whose dormancy is terminated after experiencing a certain number of cycles of fluctuating temperatures with stimulatory characteristics. To be stimulatory, a cycle must have a defined composition in terms of thermal amplitude and upper temperature. For example, for freshly dispersed seeds, effective cycles must have at least 15°C of thermal amplitude and 30°C or more of upper temperature. Cycles with stimulatory characteristics have additive effects, each cycle releasing an additional proportion of the population from dormancy. The model assumes that cycle requirements for dormancy breakage can be satisfied even if those cycles are not met in a continuous sequence. However, as stated previously, changes in degree of dormancy comprise changes not only in temperature or water potential requirements for germination, but also in sensitivity to the effects of dormancy-terminating factors (Forcella et al., 2000). Thus, changes in the degree of dormancy in seeds that require fluctuating temperatures to terminate dormancy are likely to comprise changes in sensitivity to such fluctuations. To account for these changes, the model "looses" the thermal amplitude and upper temperature requirements for a cycle to produce a stimulatory effect on seeds that had after-ripened in the soil for a winter. Model performance was successfully validated against independent field data, using either freshly harvested or after-ripened seeds. Figure 8.6 shows the importance of considering the dormancy terminating effect of fluctuating temperatures when modeling weed seedling emergence. Two models were run to simulate seedling emergence under bare and shaded soil thermal conditions: (1) a simple thermal time model that ignores the seed requirements in terms of fluctuating temperatures, and (2) the previously-described model that

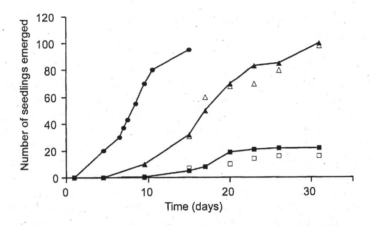

FIGURE 8.6. Cumulative number of *Sorghum halepense* L. seedlings recorded in field plots under two different soil regimes, bare soil (open triangles) and shaded soil surface (open squares). The observed data are compared with data corresponding to simulations done with a simple thermal-time model that ignores seed requirements of fluctuating temperature for dormancy breaking (full circles), and with a model that considers the effect of fluctuating temperature on dormancy breaking (full triangles and full squares).

considers the effect of fluctuating temperatures on seed dormancy. The thermal-time model not only predicted anticipated seedling emergence in relation to observed data, but also failed to distinguish between bare soil and shaded soil; clearly, seedling recruitment was much higher under bare soil because of the higher number of stimulatory cycles experienced by the seeds under this situation.

Although the model recognizes changes in seed sensitivity to the effects of fluctuating temperature as seeds are released from dormancy, it does it only discretely (i.e., freshly harvested seeds or seeds buried in the soil for a winter), and does not account for continuous dynamic changes in the response to alternating temperatures as seeds are released from or forced into dormancy. Moreover, changes in sensitivity to fluctuating temperatures should be different between years or locations, depending on soil temperatures experienced by the seeds during winter. Recently, Batlla, Verges, and Benech-Arnold (2003) showed that, for the summer annual *P. aviculare,* dormancy loss is indeed accompanied by changes in the sensitivity to fluctuating temperatures. These changes were characterized by a decrease in the number of cycles required to achieve maximal germination response and a progressive loss in the requirement of temperature fluctuations for dormancy breaking in further fractions of the seed population, as dormancy relief progresses. The authors relate seed response to the effect of cycle doses of fluctuating temperatures to accumulated Stt and observed that seeds stratified at different temperatures which had accumulated equal values of Stt showed similar response curves. Interestingly, as previously shown for the prediction of changes in Tl_{50} by the same authors (Batlla and Benech-Arnold, 2003), these preliminary results suggest that changes in seed sensitivity to cycle doses of fluctuating temperatures as a function of stratification temperature could be also predicted using Stt.

Derkx and Karssen (1994) showed that changes in dormancy of *Arabidopsis thaliana* L. seeds buried in the field were associated with changes in the sensitivity of the population to light stimuli. Sensitivity to light was shown to increase when dormancy was alleviated and to decrease when dormancy was enforced. Modeling changes in light sensitivity as a result of changes in seed dormancy level would be useful for predicting the time at which seed maximum sensitivity to light is attained in the field. Accurate prediction of maximum sensitivity to light in relation to soil temperature would permit to plan soil disturbance by tillage practices in order to produce the germination of an important fraction of the seed bank and, consequently, increase the efficacy of herbicide applications or cultural weed control methods. Recently, Vleeshouwers and Bouwmeester (2001) developed a mechanistic model for the simulation of dormancy changes in buried light-requiring weed seeds that is based on a physiological model concerning the action of phytochrome in the seed. Dormancy changes are driven by seasonal changes in soil temperature; this part of the model is based on relationships between temperature and changes in dormancy previously derived by Bouwmeester (1990) for *C. album, S. arvensis,* and *P. persicaria* seeds. In the model, the rate of dormancy release has a species-specific optimum temperature, ranging from 0 to 15°C, and decreases linearly at both sides of this optimum, limited by a minimum and a maximum dormancy release temperature. On the other hand, the rate of dormancy induction in *C. album* and *P. persicaria* increases linearly as soil temperature increases over a minimum temperature for the process to occur. For *S. arvensis* the rate of dormancy induction was characterized through an optimum, minimum, and maximum temperature. In contrast to the hypothesis of Totterdell and Roberts (1979), according to which dormancy induction and release are simultaneous processes, in this model an internal switch determines whether the prevailing temperature has a dormancy-relieving or -inducing effect. Thus, in this model periods of dormancy release and induction are strictly separated depending on the prevailing temperature. The dormancy model is coupled to a germination model that calculates germination percentages of seed samples irradiated with red light and tested for germination at different temperatures. In general, annual dormancy changes are fairly well predicted by the model for the three tested species. However, comparison with experimental data shows that, as yet, the model is not accurate enough to be used in the prediction of field emergence patterns (Vleeshouwers and Kropff, 2000).

Including Changes in Soil Water Content As a Factor That Can Modulate Changes in Dormancy Status

Although models presented so far are basically based on the effect of temperature on seed dormancy changes, changes in soil water content can affect dormancy status in many weed seeds under field situations (Egley, 1995). An example of the inclusion of soil water content as an environmental factor affecting dormancy status of buried weed seeds in predictive models was the previously described model for dry after-ripening of *B. tectorum* developed by Bauer, Meyer, and Allen (1998). In *B. tectorum* seeds, dry after-ripening occurs only when soil water content is below –4 MPa (Christensen, Meyer, and Allen, 1996). The authors include this restriction for the dry after-ripening process to occur in the model; therefore, stratification thermal time is accumulated only during hourly periods when the soil water content is below the threshold for dry after-ripening to proceed. Although weed seeds composing the seed bank are commonly subjected to soil water content fluctuations, models addressing seed dormancy changes rarely consider the effects of this factor. Future research to understand and quantify the effect of soil water status on dormancy of weed seeds will be essential for accurately modeling weed emergence under real field situations.

CONCLUDING REMARKS

Although several attempts were made to model dormancy changes of weed seed banks, further research is needed in order to successfully achieve this goal. Based on results obtained so far, important considerations for further research directions are as follows:

1. Although temperature has been shown to be the key factor regulating dormancy changes of weed seed banks in the field, effects of other environmental factors, such as changes in soil water status, nitrate content, oxygen concentration, etc., should be considered under certain situations. Experiments conducted to understand and quantify the ef-

fects of environmental factors, other than temperature, on seed dormancy status changes would permit their inclusion in future simulation models addressing dormancy and germination.

2. Many environmental factors naturally occurring under field situations, such as soil moisture content fluctuations, fluctuating temperatures, light, nitrates, etc., can affect seed dormancy status and consequently weed emergence. This generates a difficult scenario for modeling dormancy changes due to multifactorial effects and interactions that should be considered in order to accurately predict dormancy and germination in real field situations. Including as many environmental factors effects as possible (at least those considered relevant for the process) in our simulation models will lead to a better prediction of weed emergence. For this purpose, strong quantitative relationships between changes in the sensitivity to environmental factors and changes in seed population dormancy status have to be derived.

3. Seed dormancy is a population-based phenomenon (Bradford, 2002). Thus, quantifying seed-to-seed variation in the response to environmental factors is a key feature for understanding and modeling seed dormancy (Murdoch, 1998). Population-based threshold models, such as those proposed by Bradford (1996), Battaglia (1997), and Batlla and Benech-Arnold (2003), are encouraging attempts for modeling dormancy as a population phenomenon. In many cases, seed population responses to environmental factors can be characterized by a mean population threshold response, which corresponds to 50 percent of the population, its standard deviation (dispersion around mean response), and an associated development-time constant (Battaglia, 1997). Quantifying changes in these parameters for the response to naturally occurring environmental factors, such as light, temperature, etc., in relation to seed dormancy changes would permit the integration of seed population responses to the effect of many environmental factors in population-based threshold models.

4. The effect of temperature on seed dormancy status can be easily quantified using thermal-time equations, as done by Bauer, Meyer, and Allen (1998), Pritchard, Tompsett, and Manger (1996), and Batlla and Benech-Arnold (2003). This would permit referral of changes in the response to different environmental factors to a common thermal-time scale.

REFERENCES

Adámoli, J.M., Goldberg, A.D., and Soriano, A. (1973). El desbloqueo de las semillas de chamico (*Datura ferox* L.) enterradas en el suelo: Análisis de los factores causales. *Revista de Investigaciones Agropecuarias,* Serie 2, 10: 209-222.

Baskin, J.M. and Baskin, C.C. (1976). High temperature requirement for after-ripening in seeds of winter annuals. *New Phytologist* 77: 619-624.

Baskin, J.M. and Baskin, C.C. (1977). Role of temperature in the germination ecology of three summer annual weeds. *Oecologia* 30: 377-382.

Baskin, J.M. and Baskin, C.C. (1980). Ecophysiology of secondary dormancy in seeds of *Ambrosia artemisiifolia. Ecology* 61: 475-480.

Baskin, J.M. and Baskin, C.C. (1984). Role of temperature in regulating timing of germination in soil seed reserves of *Lamium purpureum* (L.). *Weed Research* 30: 341-349.

Baskin, J.M. and Baskin, C.C. (1985). The annual dormancy cycle in buried weed seeds: A continuum. *BioScience* 35: 392-398.

Batlla, D. and Benech-Arnold, R.L. (2000). Effects of soil water status and depth of burial on dormancy changes of *Polygonum aviculare* L. seeds. In *Abstracts of the Third International Weed Science Congress* (p. 22). Corvallis, OR: International Weed Science Society.

Batlla, D. and Benech-Arnold, R.L. (2003). A quantitative analysis of dormancy loss dynamics in *Polygonum aviculare* L. seeds: Development of a thermal time model based on changes in seed population thermal parameters. *Seed Science Research* 13: 55-68.

Batlla, D., Kruk, B., and Benech-Arnold, R.L. (2000). Very early detection of canopy presence by seeds through perception of subtle modifications in R:FR signals. *Functional Ecology* 14: 195-202.

Batlla, D., Verges, V., and Benech-Arnold, R.L. (2003). A quantitative analysis of seed responses to cycle-doses of fluctuating temperatures in relation to dormancy: Development of a thermal time model for *Polygonum aviculare* L. seeds. *Seed Science Research* 13: 197-207.

Battaglia, M. (1997). Seed germination model for *Eucalyptus delegatensis* provenances germinating under conditions of variable temperature and water potential. *Australian Journal of Plant Physiology* 24: 69-79.

Bauer, M.C., Meyer, S.E., and Allen P.S. (1998). A simulation model to predict seed dormancy loss in the field for *Bromus tectorum* L. *Journal of Experimental Botany* 49: 1235-1244.

Benech-Arnold, R.L., Ghersa, C.M., Sánchez, R.A., and García Fernandez, A. (1988). The role of fluctuating temperatures in the germination and establishment of *Sorghum halepense* (L.) Pers. regulation of germination under leaf canopies. *Functional Ecology* 2: 311-318.

Benech-Arnold, R.L, Ghersa, C.M., Sánchez, R.A., and Insausti, P. (1990a). A mathematical model to predict *Sorghum halepense* germination in relation to soil temperature. *Weed Research* 30: 81-89.

Benech-Arnold, R.L., Ghersa, C.M., Sánchez, R.A., and Insausti, P. (1990b). Temperature effects on dormancy release and germination rate in *Sorghum halepense* (L.) Pers. seeds: A quantitative analysis. *Weed Research* 30: 91-99.

Benech-Arnold, R.L. and Sánchez, R.A. (1995). Modeling weed seed germination. In Kigel, J. and Galili, G. (Eds.), *Seed Development and Germination* (pp. 545-566). New York: Marcel Dekker.

Benech-Arnold, R.L., Sánchez, R.A., Forcella, F., Kruk, B.C., and Ghersa, C.M. (2000). Environmental control of dormancy in weed seed banks in soil. *Field Crops Research* 67: 105-122.

Bouwmeester, H.J. (1990). The effect of environmental conditions on the seasonal dormancy pattern and germination of weed seeds. PhD thesis. Wageningen Agricultural University, The Netherlands.

Bouwmeester, H.J. and Karssen, C.M. (1992). The dual role of temperature in the regulation of the seasonal changes in dormancy and germination of seeds of *Polygonum persicaria* L. *Oecologia* 90: 88-94.

Bouwmeester, H.J. and Karssen, C.M. (1993a). Annual changes in dormancy and germination in seeds of *Sisymbrium officinale* (L.) Scop. *New Phytologist* 124: 179-191.

Bouwmeester, H.J. and Karssen, C.M. (1993b). The effect of environmental conditions on the dormancy pattern of *Spergula arvensis*. *Canadian Journal of Botany* 71: 64-73.

Bouwmeester, H.J. and Karssen, C.M. (1993c). Seasonal periodicity in germination of seeds of *Chenopodium album* L. *Annals of Botany* 72: 463-473.

Bradford, K.J. (1995). Water relations in seed germination. In Kigel, J. and Galili, G. (Eds.), *Seed Development and Germination* (pp. 351-395). New York: Marcel Dekker.

Bradford, K.J. (1996). Population-based models describing seed dormancy behaviour: Implications for experimental design and interpretation. In Lang, G.A. (Ed.), *Plant Dormancy: Physiology, Biochemistry, and Molecular Biology* (pp. 313-339). Wallingford, UK: CAB International.

Bradford, K.J. (1997). The hydrotime concept in seed germination and dormancy. In Ellis, R.H., Black, M., Murdoch, A.J., and Hong, T.D. (Eds.), *Basic and Applied Aspects of Seed Biology* (pp. 349-360). Dordrecht, the Netherlands: Kluwer Academic Publishers.

Bradford, K.J. (2002). Applications of hydrothermal time to quantifying and modeling seed germination and dormancy. *Weed Science* 50: 248-260.

Bradford, K.J. and Somasco, O.A. (1994). Water relations of lettuce seed thermoinhibition: I. Priming and endosperm effects on base water potential. *Seed Science Research* 4: 1-10.

Casal, J.J. and Sánchez, R.A. (1998). Phytochromes and seed germination. *Seed Science Research* 8: 317-329.

Christensen, M., Meyer, S.E., and Allen, P.S. (1996). A hydrothermal time model of seed after-ripening in *Bromus tectorum* L. *Seed Science Research* 6: 147-153.

Dahal P., Bradford, K.J., and Haigh, A.M. (1993). The concept of hydrothermal time in seed germination and priming. In Côme, D. and Corinbeau, F. (Eds.),

Fourth International Workshop on Seeds: Basic and Applied Aspects of Seed Biology, Volume 3 (pp. 1009-1014). Paris: ASFIS.

de Miguel, L.C. and Soriano, A. (1974). The breakage of dormancy in Datura ferox seeds as an effect of water absorption. Weed Research 14: 265-270.

Deregibus, V.A., Casal, J.J., Jacobo, E.J., Gibson, D., Kauffman, M., and Rodriguez, A.M. (1994). Evidence that heavy grazing may promote the germination of Lolium multiflorum seeds via phytochrome-mediated perception of high red/far-red ratios. Functional Ecology 8: 536-542.

Derkx, M.P.M. and Karssen, C.M. (1994). Are seasonal dormancy patterns in Arabidopsis thaliana regulated by changes in seed sensitivity to light, nitrate and gibberellin? Annals of Botany 73: 129-136.

Egley G. (1995). Seed germination in soil. In Kigel, J. and Galili, A. (Eds.), Seed Development and Germination (pp. 529-543). New York: Marcel Dekker.

Fenner, M. (1987). Seedlings. New Phytologist 106: 35-47.

Forcella, F., Benech-Arnold, R.L., Sánchez, R.A., and Ghersa, C.M. (2000). Modeling seedling emergence. Field Crops Research 67: 123-139.

Frankland, B. (1981). Germination in shade. In Smith, H. (Ed.), Plants and the Daylight Spectrum (pp. 187-204). London: Academic Press.

Frankland, B. and Taylorson, R.B. (1983). Light control of seed germination. In Shropshire, W. and Mohr, H. (Eds.), Encyclopedia of Plant Physiology, Volume 16A (pp. 428-456). New York: Springer Verlag.

Ghersa, C.M., Benech-Arnold, R.L., and Martinez Ghersa, M.A. (1992). The role of fluctuating temperatures in germination and establishment of Sorghum halepense (L.) Pers: II. Regulation of germination at increasing depths. Functional Ecology 6: 460-468.

Hartmann, K.M. (1966). A general hypothesis to interpret "high-energy phenomena" of photomorphogenesis on the basis of phytochrome. Photochemistry and Photobiology 5: 349-366.

Heim, B. and Schäfer, E. (1982). Light-controlled inhibition of hypocotyl growth in Sinapis alba L. seedlings. Planta 154: 150-155.

Heim, B. and Schäfer, E. (1984). The effect of red and far-red light in the high irradiance reaction of phytochrome (hypocotyl growth in dark-grown Sinapis alba L.). Plant, Cell and Environment 7: 39-44.

Hendricks, S.B., Toole, V.K., and Borthwick, H.A. (1968). Opposing action of light in seed germination of Poa pratensis and Amaranthus arenicola. Plant Physiology 43: 2023-2028.

Hilhorst, H.W.M., Derkx, M.P.M., and Karssen, C.M. (1996). An integrating model for seed dormancy cycling: Characterization of reversible sensitivity. In Lang, G.A. (Ed.), Plant Dormancy: Physiology, Biochemistry, and Molecular Biology. Wallingford, UK: CAB International.

Holmes, M.G. and Smith, H. (1977). The function of phytochrome in the natural environment: II. The influence of vegetation canopies on the spectral energy distribution of natural daylight. Photochemistry and Photobiology 25: 539-545.

Insausti, P., Soriano, A., and Sánchez, R.A. (1995). Effects of flood-influenced factors on seed germination of Ambrosia tenuifolia. Oecologia 103: 127-132.

Karssen C.M. (1980/1981). Patterns of change in dormancy during burial of seeds in soil. *Israel Journal of Botany* 29: 65-73.

Karssen, C.M. (1982). Seasonal patterns of dormancy in weed seeds. In Khan, A.A. (Ed.), *The Physiology and Biochemistry of Seed Development, Dormancy and Germination* (pp. 243-270). Amsterdam: Elsevier.

Kebreab, E. and Murdoch, A.J. (1999). A quantitative model for loss of primary dormancy and induction of secondary dormancy in imbibed seeds of *Orobanche* spp. *Journal of Experimental Botany* 50: 211-219.

Kronenberg, G.H.M and Kendrik, R.E. (1986). The physiology of action. In Kendrik, R.E. and Kronenberg, G.H.M. (Eds.), *Photomorphogenesis in Plants* (pp. 99-114). Dordrecht, the Netherlands: Marthinus Nijhoff/Dr. W. Junk Publishers.

Kruk, B.C. and Benech-Arnold, R. (1998). Functional and quantitative analysis of seed thermal responses in prostate knotweed *(Polygonum aviculare)* and common purslane *(Portulaca oleracea)*. *Weed Science* 46: 83-90.

Kruk, B.C. and Benech-Arnold, R.L. (2000). Evaluation of dormancy and germination responses to temperature in *Carduus acanthoides* and *Anagallis arvensis* using a screening system, and relationship with field-observed emergence patterns. *Seed Science Research* 10: 77-88.

Mohr, H. (1972). *Lectures on Photomorphogenesis*. Berlin: Springer Verlag.

Murdoch, A.J. (1998). Dormancy cycles of weed seeds in soil. *Aspects of Applied Biology* 51: 119-126.

Murdoch, A.J., Roberts, E.H., and Goedert, C.O. (1988). A model for germination responses to alternating temperatures. *Annals of Botany* 63: 97-111.

Ni, B.R. and Bradford, K.J. (1992). Quantitative models characterizing seed germination responses to abscisic acid and osmoticum. *Plant Physiology* 98: 1057-1068.

Ni, B.R. and Bradford, K.J. (1993). Germination and dormancy of abscisic acid- and gibberellin-deficient mutant tomato *(Lycopersicon esculentum)* seeds. *Plant Physiology* 101: 607-617.

Pons, T.L. (1992). Seed response to light. In Fenner, M. (Ed.), *The Ecology of Regeneration in Plant Communities* (pp. 259-284). Melksham, UK: CAB International.

Pritchard, H.W., Tompsett, P.B., and Manger, K.R. (1996). Development of a thermal time model for the quantification of dormancy loss in *Aesculus hippocastanum* seeds. *Seed Science Research* 6: 127-135.

Probert, R.J. (1992). The role of temperature in germination ecophysiology. In Fenner, M. (Ed.), *The Ecology of Regeneration in Plant Communities* (pp. 285-325). Melksham, UK: CAB International.

Ransom, E.R. (1935). The inter-relations of catalase, respiration, after-ripening, and germination in some dormant seeds of the Polygonaceae. *American Journal of Botany* 22: 815-825.

Reisman-Berman, O., Kigel, J., and Rubin, B. (1991). Dormancy patterns in buried seed of *Datura ferox* L. *Canadian Journal of Botany* 69: 173-179.

Roberts, E.H. and Totterdell, S. (1981). Seed dormancy in *Rumex* species in response to environmental factors. *Plant, Cell and Environment* 4: 97-106.

Scopel, A.L., Ballaré, C.L. and Sánchez, R.A. (1991). Induction of extreme light sensitivity in buried weed seeds and its role in the perception of soil cultivations. *Plant, Cell and Environment* 14: 501-508.

Smith, H. (1982). Light quality, photoperception, and plant strategy. *Annual Review of Plant Physiology* 33: 481-518.

Thompson, K. and Grime, J. (1983). A comparative study of germination responses to diurnally fluctuating temperatures. *Journal of Applied Ecology* 20: 141-156.

Totterdell, S. and Roberts, E.H. (1979). Effects of low temperatures on the loss of innate dormancy and the development of induced dormancy in seeds of *Rumex obtusifolius* L. and *Rumex crispus* L. *Plant, Cell and Environment* 2: 131-137.

Vegis, A. (1964). Dormancy in higher plants. *Annual Review of Plant Physiology* 15: 185-224.

Vleeshouwers, L.M. (1997). Modeling weed emergence patterns. PhD thesis, Agricultural University, Wageningen, The Netherlands.

Vleeshouwers, L.M. and Bouwmeester, H.J. (2001). A simulation model for seasonal changes in dormancy and germination of seeds. *Seed Science Research* 11: 77-92.

Vleeshouwers, L.M. and Kropff, M.J. (2000). Modeling field emergence patterns in arable weeds. *New Phytologist* 148: 445-457.

Washitani, I. (1987). A convenient screening test system and a model for thermal germination responses of wild plant seeds: Behaviour of model and real seed in the system. *Plant, Cell and Environment* 10: 587-598.

SECTION III:
SEED LONGEVITY AND STORAGE

Chapter 9

Orthodox Seed Deterioration and Its Repair

Miller B. McDonald

INTRODUCTION

Seed deterioration can be defined as deteriorative changes occurring with time that increase the seed's vulnerability to external challenges and decrease the ability of the seed to survive. Three general observations can be made about seed deterioration. First, seed deterioration is an undesirable attribute of agriculture. Annual losses of revenue from seed/grain products due to deterioration can be as much as 25 percent of the harvested crop. This value would be in the billions of U.S. dollars (McDonald and Nelson, 1986). An understanding of seed deterioration, therefore, provides a template for improved crop production as well as increased agricultural profits. Second, the physiology of seed deterioration is a separate event from seed development and/or germination. Thus, the knowledge gained from understanding these events likely does not apply to what occurs during deterioration. Third, seed deterioration is cumulative. As seed aging increases, seed performance is increasingly compromised. With these tenets in mind, what causes seeds to die? An understanding of this process might begin with an understanding of seed evolution—a topic seldom discussed.

THE FIRST SEED

What was the first seed like? Since no humans were present at the time the first seed was formed, it is difficult to answer this question with certainty. However, trying to understand that first seed might help in a quest to increase our knowledge of seed aging. In general, seeds are divided into two categories based on their storage characteristics: recalcitrant and orthodox (although gradations may exist—see Pammenter and Berjak, 2000). Recal-

Salaries and research support were provided by state and federal funds appropriated to the Ohio Agricultural Research and Development Center, Ohio State University.

citrant seeds are desiccation intolerant (cannot be dried below approximately 40 percent seed moisture content without damage) and are typically characterized as large seeds with small embryos from tropical trees and shrubs (Chin and Roberts, 1980). Orthodox seeds, in contrast, are desiccation tolerant (can be dried to 5 percent seed moisture content without damage), often manifest some type of dormancy, and are characteristic of most agriculturally important crops found worldwide. Orthodox seeds represent most of the seeds found in the world and are among the most agriculturally important species. As a result of their common occurrence, it is tempting to speculate that the first seed possessed orthodox behavior.

To determine whether the first seed was recalcitrant or orthodox, it is important to remember the agricultural definition of a seed which emphasizes that the propagule must be a reproductive unit. Thus, the first seed was one produced on the parent plant and later dispersed followed by germination, successfully regenerating the species. It is generally agreed that plant life first originated in the tropics, an area where seasons are consistent and temperatures conducive for maximum plant growth. If plants first appeared there, seeds could be shed at any time from the plant and, presuming adequate moisture, immediately resume growth without the need for dormancy or a quiescent period (Garwood, 1989; Vazquez-Yanes and Orozco-Segovia, 1993). Recalcitrant seeds, in contrast, do not possess dormancy but, instead, must continue their development and progress toward germination (Berjak et al., 1990). Recalcitrant seeds are not desiccation tolerant and, if dried, will die. Although this trait led to successful plant reproduction, it made agricultural planning difficult because seed life span was short. However, as successful plants expanded their range beyond the tropics, they encountered differing seasons that required seed adaptations for survival. These included the ability to dry down so that respiration and physiology associated with growth were reduced during unfavorable seasonal climes as well as the imposition of dormancy. Therefore, desiccation tolerance and dormancy may be acquired traits. The ability of seeds to reduce their metabolism following maturation, however, was a significant advantage for successful agriculture. For the first time, humans were able to harvest and store seeds for long durations without loss in seed quality. This permitted shipment of seeds to other locations as well as planting of seeds in subsequent seasons following long-term storage. The imposition of dormancy mechanisms that "sensed" the environment continues to be an agricultural challenge because the depth of dormancy likely varies from seed to seed in a seed lot. As a result, both desiccation tolerance and dormancy have been adaptive advantages that contributed to the prevalence of orthodox seeds around the world.

Therefore, it appears that the first seed was a recalcitrant seed, a case advocated by Pammenter and Berjak (2000). It was characterized by high moisture, high respiration, short life span, and an inability to dry down without seed damage. Orthodox seeds likely followed recalcitrant seeds, with the significant advantage that they developed desiccation-tolerant mechanisms which reduced respiration and ceased active embryo growth. Simultaneously, orthodox seeds increased their life span well beyond recalcitrant seeds. Why is that so, and what are the physiological mechanisms that govern and regulate orthodox seed deterioration? The purpose of this chapter is to describe the processes that cause orthodox seed deterioration and the physiological mechanisms that permit these seeds to survive long-term storage. From such knowledge, we gain insights into approaches for enhancing seed storage potential and improving seed quality.

SEED DETERIORATION

Seed deterioration is inexorable, and the best that can be done is to control its rate. Many factors contribute to seed deterioration. These include genetics, mechanical damage, relative humidity and temperature of the storage environment, seed moisture content, presence of microflora, seed maturity, etc. Of these, relative humidity and temperature are the two most important. Relative humidity is important because it directly influences the moisture content of seeds in storage as they come to equilibrium with the amount of gaseous water surrounding them. Temperature is important because it (1) determines the amount of moisture the air can hold (higher temperatures holding more water than lower temperatures) and (2) enhances the rate of deteriorative reactions occurring in seeds as temperature increases. These relationships are so important that Harrington (1972) identified the following two rules of thumb describing seed deterioration:

> *Rule 1:* Each 1 percent reduction in seed moisture content doubles the life of the seed.
> *Rule 2:* Each 5°C reduction in seed temperature doubles the life of the seed.

Harrington (1972) recognized that some qualifications to these rules were needed for them to be applied successfully. First, rule one does not apply above 14 or below 5 percent seed moisture content. Seeds stored at moisture contents above 14 percent begin to exhibit increased respiration, heating, and fungal invasion which destroy seed viability more rapidly than indicated by the moisture content rule. Below 5 percent seed moisture, a breakdown of membrane structure hastens seed deterioration (probably a consequence of reorientation of hydrophyllic membranes due to the loss of the water molecules necessary to retain their structural configuration).

For the second rule, for temperatures below 0°C the rule may not apply because many biochemical reactions associated with seed deterioration do not occur and further reductions in temperature have only a moderate effect in extending seed longevity. Finally, it should not be forgotten that these two factors, seed moisture content and temperature, interact. This was captured in another equation suggested by Harrington (1972) in which the sum of the temperature in degrees Farenheit and the percentage relative humidity should not exceed 100. From this equation, one can see that as the temperature of the storage environment increases, the relative humidity must decrease. The influence of seed moisture content, temperature, and orthodox seed deterioration were demonstrated in hypothetical deterioration curves presented in Figure 9.1 (Ellis, Osei-Bonsu, and Roberts, 1982).

This complex milieu of interacting factors makes the study of seed deterioration and its underlying physiology difficult. It is beyond the purview of this chapter to consider each of these factors, and the reader is encouraged to examine a book (Priestley, 1986) as well as a comprehensive chapter (Copeland and McDonald, 2001) and reviews (Halmer and Bewley, 1984; McDonald, 1985; McDonald and Nelson, 1986; Smith and Berjak, 1995; McDonald, 1999) on the subject. This chapter will consider orthodox seed deterioration from a physiological perspective. Starting with a high-quality seed under optimum storage conditions, what happens to the seed as its quality is reduced?

Seed Deterioration Is Not Uniform

A general assumption is that seed deterioration occurs uniformly throughout a seed, but a seed is a composite of tissues that differ in their chemistry and proximity to the external environment. Thus, it should not be assumed that seed deterioration occurs uniformly throughout the seed. Perhaps the best example that this does not occur comes from the use of the tetrazolium chloride (TZ) test which causes living tissues in a seed to turn red (Association of Official Seed Analysts [AOSA], 2000; Society of Commercial Seed Technologists [SCST], 2001). The challenge to the seed researcher/analyst is to decipher how important the living (or dead) tissues are to successful seedling establishment. When studies have been conducted on seeds using controlled natural and artificial aging conditions, differences in the deterioration of seed tissues have been observed. For example, in wheat seeds, deterioration begins with the root tip and progressively moves upward through the radicle, scutellum, and ultimately the leaves and coleoptile (Das and Sen-Mandi, 1988, 1992). Similar findings have been reported in maize in which root tip cells are the first to be damaged (Berjak, Dini, and Gevers, 1986) which causes the rate of radicle extension to be lower than coleoptile extension following aging (Bingham, Harris, and MacDonald,

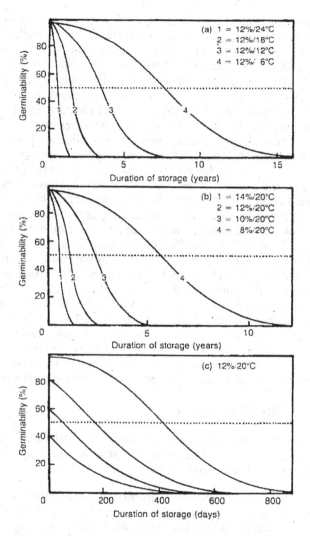

FIGURE 9.1. Theoretical distribution curves for soybean. Curves were computed for various constant storage conditions: (1) different temperatures at 12 percent moisture content with initial germinability 98 percent, (b) different moisture contents at 20°C with initial germinability 98 percent, and (c) different initial germinabilities (98, 80, 60, and 40 percent) at 12 percent moisture content and 20°C. (*Source:* R. H. Ellis, K. Osei-Bonsu, and E. H. Roberts, The influence of genotype, temperature, and moisture on seed longevity in chickpea, cowpea, and soya bean, *Annals of Botany,* 1982, 50: 69-82, by permission of Oxford University Press.)

1994). Similarly, in dicot seeds such as soybean, root growth is more sensitive to accelerated aging than shoot growth (Hahalis and Smith, 1997) and the embryonic axis more sensitive to deterioration than the cotyledons (Chauhan, 1985; Seneratna, Gusse, and McKersie, 1988; Tarquis and Bradford, 1992). These findings demonstrate that the embryonic axis is more prone to aging in monocot and dicot orthodox seeds, and, of the axis structures, the radicle axis is more sensitive to deterioration than the shoot axis.

Why this is so is not known, but at least two studies on the sequence of water uptake in soybean and maize seeds may provide an indication. McDonald, Vertucci, and Roos (1988) showed that the soybean axis hydrated more rapidly than the cotyledons. This was attributed to the presence of a radicle pocket in the seed coat that possessed large hourglass cells with a low matric potential for water (Figure 9.2). This external radicle pocket is in intimate contact with the radicle and ensures that water absorbed during imbibition is attracted preferentially to the radicle axis. Similarly, McDonald, Sullivan, and Lauer (1994) described the uptake of water in a maize seed beginning with hydration in the radicle followed by the scutellum and then the shoot axis and coleoptile (Figure 9.3). They attributed this route to the open pores present in the remnants of the funiculus or collapsed black layer which afforded rapid water penetration into the seed. Although these results were for free-flowing water, it is likely that water present in the gaseous phase of air would also be attracted by the same matric forces present in the seed coat. This would result in higher water content in the embryo (and radicle) than in storage reserves (and other embryonic structures) which could selectively facilitate the events that cause seed deterioration in certain seed parts.

Physiology of Seed Deterioration

Our understanding of the events that cause seed deterioration remains incomplete. McDonald (1999) identified at least six reasons why it is difficult to critically evaluate seed deterioration studies:

1. The physiological processes governing seed deterioration vary. For example, short-term deterioration in the field is likely a different physiological event than long-term deterioration in storage.
2. Seed researchers use different methods to study seed deterioration. They can precisely control short-term seed deterioration under high temperature, high relative humidity accelerated aging conditions, but is this process physiologically equivalent to the conditions occurring in natural, long-term storage conditions?
3. The rate of seed deterioration is influenced by confounding environmental and biological factors such as growth of storage fungi that create their own biological niche.

4. Seed treatments influence seed deterioration, and, when applied, their impact on seed quality must be recognized.

5. Most seed deterioration studies examine whole seeds. As emphasized, seed deterioration is not uniform within a seed and any study of seed deterioration should begin with an understanding of where seed deterioration occurs first.

6. Most seed deterioration studies report effects on a seed lot, but seed deterioration is an individual event occurring in a population of seeds composing the seed lot. Studies using bulk seeds are inappropriate.

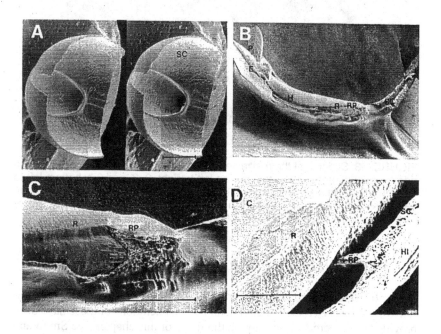

FIGURE 9.2. Scanning electron micrograph showing the radicle pocket. (A) Sterographic micrograph of the concave seed coat surface with embryo removed demonstrating the radicle pocket; (B) Cross section of the seed illustrating the radicle pocket enveloping the radicle of the embryonic axis; (C) magnified view of the radicle and radicle pocket; (D) cross section of the radicle pocket and radicle. Abbreviations: SC = seed coat, RP = radicle pocket, P = plumule, E = epicotyl, H = hypocotyl, R = radicle, HI = hilum, and C = cotyledon. Bars represent 1.0 mm. (*Source:* M. B. McDonald, C. W. Vertucci, and E. E. Roos, 1988, Soybean seed imbibition: Water absorption of seed parts, *Crop Science,* 28: 993-998. Reproduced with permission from the Crop Science Society of America.)

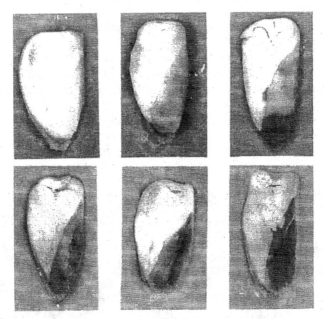

FIGURE 9.3. Staining of the maize seed embryo with nitroblue tetrazolium chloride at various intervals of soaking. Top left to right, 0, 3, and 6 h; bottom left to right, 15, 24, and 48 h. (*Source:* M. B. McDonald, J. Sullivan, and M. J. Lauer, 1994, The pathway of water uptake in maize (*Zea mays* L.) seeds, *Seed Science and Technology* 22:79-90. Reproduced with permission.)

MECHANISMS OF ORTHODOX SEED DETERIORATION

Our quest to better understand orthodox seed deterioration has led to a variety of proposals. (Excellent and detailed considerations of these have been provided elsewhere and will not be the focus of this chapter; see Smith and Berjak, 1995; McDonald, 1999). These include changes in the following:

- *Enzyme activities:* Most of these studies search for markers of germination such as increases in amylase activity or changes in free radical scavenging enzymes such as superoxide dismutase, catalase, peroxidase, and others.
- *Protein or amino acid content:* The consensus is that overall protein content declines while amino acid content increases with seed aging.
- *Nucleic acids:* A trend of decreased DNA synthesis and increased DNA degradation has been reported. It is widely believed that degra-

dation of DNA would lead to faulty translation and transcription of enzymes necessary for germination.

* *Membrane permeability:* Increased membrane permeability associated with increasing seed deterioration has been consistently observed and is the foundation for the success of the conductivity test as a measure of seed quality.

FREE RADICAL PRODUCTION

Each of these general findings represent the result, not the cause, of seed deterioration. As evidence mounts, the leading candidate causing seed deterioration increasingly appears to be free radical production. Free radical production, primarily initiated by oxygen, has been related to the peroxidation of lipids and other essential compounds found in cells. This causes a host of undesirable events including decreased lipid content, reduced respiratory competence, and increased evolution of volatile compounds such as aldehydes (Wilson and McDonald, 1986b).

Free Radicals—What Are They and Why Are They Important?

All atoms that make up molecules contain orbitals that occupy zero, one, or two electrons. An unpaired electron in an orbital carries more energy than each electron of a pair in an orbital. A molecule that possesses any unpaired electrons is called a free radical. Some free radicals are composed of only two atoms ($O_2^{\bullet-}$) while others can be as large as protein or DNA molecules. Why is the free radical important in biological systems? The energetic "lonely electron" (1) can detach from its host atom or molecule and move to another atom or molecule or (2) can pull another electron (which may not have been lonely) from another atom or molecule. The most common free radical reaction is when one free radical and one non-free radical transfer one electron between them, leaving the free radical as a non-free radical, while the non-free radical is now a free radical. This initiates a chain of similar reactions which cause substantial damage in the interval that the reactions are occurring. Thus, free radicals can react with one another and with non-free radicals to change the structure and function of other atoms and molecules. If these are proteins (enzymes), lipids (membranes), or nucleic acids (DNA), normal biological function is compromised and deterioration increased. The positive association of free radicals with animal aging has recently been reviewed (Beckman and Ames, 1998). What still remains uncertain is their role in orthodox seed aging.

FREE RADICALS AND THEIR EFFECTS ON LIPIDS

Lipid peroxidation begins with the generation of a free radical (an atom or a molecule with an unpaired electron) either by autoxidation or enzymatically by oxidative enzymes such as lipoxygenase present in many seeds. Various forms of free radicals have been observed or detected in living tissue, each with a differing capability for cell damage (Gille and Joenje, 1991; Larson, 1997).

Superoxide anion ($O_2^{\bullet-}$). Superoxide anion is produced by autoxidation of hydroquinones, leukoflavins, and thiols as well as enzymatically by flavoprotein dehydrogenases such as mitochondrial NADH dehydrogenase.

Hydrogen peroxide (H_2O_2). Hydrogen peroxide is produced by the spontaneous or enzyme-catalyzed dismutation of $O_2^{\bullet-}$ or by two-electron reduction of O_2. Flavoenzymes such as monoamine oxidase present on the outer mitochondrial membrane of virtually all cells are probably the most important contributors to intracellular generation of H_2O_2. These enzymes which normally use O_2 as a substrate catalyze two-electron transfer reactions that produce H_2O_2.

Hydroxyl radicals ($\bullet OH$). Hydroxyl radicals are formed from $O_2^{\bullet-}$ and H_2O_2 in the presence of iron which catalyzes the reaction. In cells, iron can be bound to compounds such as adenosine triphosphate (ATP), guanidin triphosphate (GTP), and citrate, thereby forming a more soluble iron-chelate complex. $\bullet OH$ is by far the most reactive oxygen radical, and it reacts almost immediately with any molecule at the site where it is generated. The main mechanism of toxicity of H_2O_2 and $O_2^{\bullet-}$ may be their ability to combine to form $\bullet OH$.

Singlet oxygen ($^1O_2$). During lipid peroxidation, $^1O_2$ can be generated in the termination step:

$$LOO\bullet + LOO\bullet \rightarrow LO + LOH + {}^1O_2$$

and in the reaction with triplet carboxyls formed during lipid peroxidation:

$$RO^* + O_2 \rightarrow LO + {}^1O_2$$

Singlet oxygen combines with DNA bases causing genetic damage.

HOW DO FREE RADICALS CAUSE LIPID PEROXIDATION?

The mechanism of lipid peroxidation is often initiated by oxygen around unsaturated or polyunsaturated fatty acids such as oleic and linoleic acids found commonly in seed membranes and storage oils. The result is the release of a free radical, often hydrogen (H•), from a methylene group of the fatty acid adjacent to a double bond. In other cases, the free radical hydrogen may combine with other free radicals from carboxyl groups (ROOH) leaving a peroxy-free radical (ROO•). Once these free radicals are initiated, they continue to propagate other free radicals that ultimately combine, terminating the destructive reactions. In their wake, they create profound damage to membranes and changes in oil quality. As a result, long-chain fatty acids are broken into smaller and smaller compounds, some of these being released as volatile hydrocarbons (Wilson and McDonald, 1986a; Esashi, Kamataki, and Zhang, 1997). The final consequence is loss of membrane structure, leakiness, and an inability to complete normal metabolism.

WHAT IS THE INFLUENCE OF SEED MOISTURE CONTENT ON FREE RADICAL ASSAULT?

Lipid peroxidation occurs in all cells, but in fully imbibed cells, water acts as a buffer between the autoxidatively generated free radicals and the target macromolecules, thereby reducing damage. Thus, as seed moisture content is lowered, autoxidation is more common and is accelerated by high temperatures and increased oxygen concentrations. Lipid autoxidation may be the primary cause of seed deterioration at moisture contents below 6 percent. Above 14 percent moisture content, lipid peroxidation may again be stimulated by the activity of hydrolytic oxidative enzymes such as lipoxygenase, becoming more active with increasing water content. Between 6 and 14 percent moisture content, lipid peroxidation is likely at a minimum because sufficient water is available to serve as a buffer against autoxidatively generated free radical attack, but not enough water is present to activate lipoxygenase-mediated free radical production.

Lipoxygenases may contribute to cell degradation by modifying cell membrane composition. In higher plants, two major pathways involving lipoxygenase activity have been described for the metabolism of fatty acid hydroperoxides (Figure 9.4, Loiseau et al., 2001). One pathway produces traumatic acid, a compound that may be involved in plant cell wound response (Zimmerman and Coudron, 1979) and volatile C_6-aldehydes and C_6-alcohols shown to be correlated with seed deterioration (Wilson and McDonald, 1986a). The other pathway produces jasmonic acid, a molecule that may play a regulatory role in plant cells (Staswick, 1992; Sembdner and Parthier, 1993). Lipoxygenases have been identified and associated with almost every subcellular body in plants (Losieau et al., 2001), so it is likely that they have important regulatory roles in development. This may

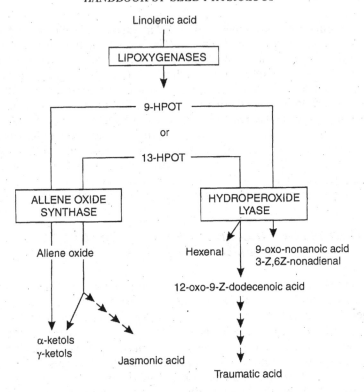

FIGURE 9.4. Overview of the lipoxgenase pathway. 9-HPOT, 9*(S)*-hydroperoxy-*trans*-10,*cis*-12,*cis*-15-octadecatrienoic acid; 1-HPOT, 1*(S)*-hydroperoxy-*cis*-9, *trans*-11,*cis*-15-octadecatrienoic acid. (*Source:* From Loiseau et al., 2001. Reprinted by permission of CABI Publishing, Wallingford, Oxon, UK.)

include the deterioration of hydrated seeds through free radical production. For example, Zacheo and colleagues (1998) found increased lipoxygenase activity at high relative humidity (80 percent) and temperature (20°C) during natural aging of almond seeds. Other correlative studies implicating lipoxygenases have been identified (McDonald, 1999). A direct study of the importance of lipoxygenases during orthodox seed deterioration employing mutants was reported by Suzuki and colleagues (1996, 1999). They found that a rice mutant deficient in lipoxygenase-3 had fewer peroxidative products and fewer volatile compounds during seed aging compared to the wild type.

Thus, the mechanism of lipid peroxidation may be different under long-term aging (autoxidation) compared to accelerated aging (e.g., lipoxygenase) conditions. This is consistent with the proposals by Wilson and Mc-Donald (1986b) and Smith and Berjak (1995) that seeds are exposed to separate lipid peroxidative events during storage and during imbibition. It should be noted that oxygen is deleterious to seed storage based on this proposal, which is consistent with the success of hermetic seed storage and lipid peroxidation as a cause of membrane integrity loss.

DO FREE RADICALS ATTACK ONLY LIPIDS?

Free radicals attack compounds other than lipids. Changes in protein structure of seeds have been observed and attributed to free radicals (McDonald, 1999). Soluble proteins may be attacked by different classes of oxidants (and be protected by different classes of antioxidants) than membrane proteins. The most reactive amino acids susceptible to oxidative damage appear to be cysteine, histidine, tryptophan, methionine, and phenylalanine, usually in that order (Larson, 1997).

Free radicals are also suspected of assault on chromosomal DNA. Potential targets for oxidative damage in the DNA chain include the purine and pyrimidine bases as well as the deoxyribose sugar moieties (Larson, 1997). Specific damage to the bases may leave the strand intact, but modification of sugar residues can also lead to strand breakage. This may explain the increased propensity for genetic mutations as seeds age. Many of these mutations are first detected as chromosomal aberrations that delay the onset of mitosis necessary for cell division and germination.

WHY SUSPECT FREE RADICAL ATTACK ON MITOCHONDRIA?

Three reasons exist to believe that free radical attack on mitochondria may be a prime cause of seed deterioration. First, mitochondria are the site of aerobic respiration. Thus, they are the prime "sink" for oxygen, some of which can leak from the membranes during respiration to create free radicals. Second, mitochondria are indispensable to normal cell function. They use oxygen and substrates to generate energy. Third, an important manifestation of seed deterioration is reduced seedling growth, perhaps a consequence of less efficient mitochondrial function.

Mitochondria contain an inner membrane encased in another outer membrane, and both membranes differ in many important ways. The inner membrane is intricately folded (structures called cristae) and has a much greater surface area than the outer membrane. The cristae are also the site of electron transport where lonely electrons can leak and cause damage to the extensive membrane surface, thereby compromising essential energy production necessary for germination. The space enclosed by the inner membrane is called the mitochondrial matrix. This matrix is high in protein concentration, containing many enzymes as well as their cofactors critical for oxidative phosphorylation. The matrix also contains a small amount of DNA (mtDNA) and ribosomes for decoding the DNA. The outer membrane is not folded and has large holes in it that permit the passage of many large proteins.

Of these compounds and structures, mtDNA is the most critical for maintaining normal cell function, and a review of its structure and function in plants has been provided (Hanson and Otto, 1992). To better understand this important role, it should be noted that mtDNA differs from nuclear DNA in two important ways. First, when a cell divides, both nuclear and mtDNA are separately replicated. Mitochondria can also divide in an active cell, requiring the creation of a new copy of mtDNA; mtDNA is important for the production of new mitochondria in rapidly dividing and physiologically active cells such as those that occur during germination. Second, the enzymes encoded by mtDNA are absolutely essential for oxidative phosphorylation. Thus, maintenance of mtDNA is vital for actively respiring cells, the cells responsible for seedling growth. As a result, any challenges to mtDNA would surely disrupt normal cellular growth and division.

Since it is now clear that mtDNA and mitochondria are essential for maintenance of cells during dry storage and growth of cells during germination, an essential question is whether mtDNA or nuclear DNA are more prone to free radical attack. Studies have now documented that mtDNA suffers more spontaneous changes in its DNA sequence compared to nuclear DNA in animal cells which results in the production of incorrect or truncated proteins (DeGrey, 1999). This greater susceptibility is attributed to the following:

- *mtDNA being more exposed to free radical attack than nuclear DNA:* Mitochondria are the principal site of oxygen utilization which results in a greater level of free radical production.
- *mtDNA being "naked":* Nuclear DNA is protected by special proteins called histones that must be degraded by free radicals prior to nuclear DNA exposure. mtDNA is not surrounded by these protective structures.
- *The repair of nuclear DNA is more successful than mtDNA:* Fewer repair enzymes exist around mtDNA.
- *mtDNA being circular and containing a high level of short sequences which appear twice, some distance apart:* If the circular mtDNA

wraps around itself to form a figure 8, the two identical sequences may end up next to each other and a "crossover" may occur at which the strands come apart and join each other. This causes the circular DNA to possess two circles, each with only a subset of the necessary genetic material. In addition, one of these circles will be without the necessary D-loop, a small portion of the DNA that contains no genes but is essential for initiating replication of the molecule. Thus, the circle without the D-loop will never be replicated and will eventually be lost during division, with the final result being the deletion of critical mtDNA information.

Oxidative damage is an important contributor to mutations in mtDNA. Various types of deletion mutations have been reported to increase with aging (Kang et al., 1998). Accumulation of specific mutations in somatic cells with aging may be due to either mutations occurring at particular sites (e.g., hot spots) or randomly throughout the genome. In humans, these mutations have been associated with specific mitochondrial diseases (Kang et al., 1998). In plants, early studies with plant mtDNA rearrangements have shown clear associations with abnormal growth mutants, cytoplasmic male sterility (Newton and Gabay-Laughman, 1998), as well as abnormal pollen development (Conley and Hanson, 1995) and protein formation (Lu et al., 1996). Whether these same changes occur in seeds has not been determined. However, with the advent of the polymerase chain reaction (PCR), it has become possible to extract low levels of mtDNA and amplify it for the determination of scissions possibly caused by free radical attack. In addition, a technique called long PCR now allows the identification of different mtDNA deletions present in the whole mtDNA (Barnes, 1994) in contrast to the original technique in which only a small proportion of sequence variants were identified (Cheng, Higuchi, and Stoneking, 1994; Reynier and Mathiery, 1995). Using these PCR refinements, it has been demonstrated that point mutations can occur in mtDNA in animal tissues and accumulate with age (Munscher, Muller-Hocker, and Kadenbach, 1993; Kadenbach, Munscher, and Frank, 1995).

How Are Free Radicals Produced in Mitochondria?

Mitochondria are the major source of reactive oxygen species (ROS) (Kang et al., 1998). Under normal physiological conditions, more than 1 percent of the oxygen consumed by cells is converted to ROS. In animal systems, this amounts to about 10^7 ROS molecules/mitochondrion/day. Mitochondrial respiration accounts for 90 percent of the cellular oxygen consumed, and the respiratory chain in mitochondria is principally responsible for the production of ROS (Kang et al., 1998). Free radicals are easily formed during oxidative phosphorylation where the consumed oxygen is turned into water by the addition of four protons and four electrons. The protons and electrons reach their targets by different routes. The electrons are carried one by one along a chain of molecules (cytochromes, etc.). Each time this is done, the electron carrier is turned into a free radical. If they can pass their free radical onto the next carrier, no harm is done. However, this is not a perfect system and sometimes the electrons escape at some stage in the chain. These loose electrons can form potentially toxic free radicals. The majority of free radicals during oxidative phosphorylation are accepted by one omnipresent, molecular oxygen (O_2). The resulting molecule is called superoxide with a chemical formula of $O_2^{\bullet-}$. The negative charge indicates that it has one more electron than proton, making it an anion, and the "$\bullet$" indicates it is a free radical.

Although superoxide is a free radical, it is one that does not contribute to significant cellular damage. Instead, superoxide is converted (usually by superoxide dismutase) into hydrogen peroxide (H_2O_2) that can accept an electron from $Fe^{2+\bullet}$ (or $Cu^{+\bullet}$) and, in so doing, splits in two to form $HO\bullet$ and water. $HO\bullet$ is vastly more reactive than superoxide and will readily initiate lipid peroxidation in the mitochondrial cristae.

In animals, the mitochondrial free radical theory of aging has become the leading candidate to explain cell aging (DeGrey, 1999). In particular, because of free radical attack, the integrity of mtDNA becomes increasingly damaged with age. For example, the tissues that exhibited the greatest levels of mtDNA damage were those that utilized the most energy per unit volume and/or generated the most reactive molecules. In other words, high metabolic rate shortens life span. Thus, because orthodox seeds are low in moisture content and their metabolic rates are consequently low, they are able to survive for longer durations.

Peroxidation proceeds as a chain reaction so the products of peroxidation in an intact mitochondrion will be concentrated around the occasional point at which a reaction was initiated. These locally high levels of membrane damage constitute pinpricks in the membrane through which protons flow rapidly.

HOW ARE SEEDS PROTECTED
AGAINST FREE RADICAL ATTACK?

Seeds contain a complex system of antioxidant defenses to protect against the harmful consequences of activated oxygen species. At least three defenses in seeds protect against free radical attack.

The first is an array of enzymes to neutralize activated oxygen species. Although these are unlikely to operate in dry seeds, their activity would be vital during imbibition. Specific enzymes exist that detoxify $O_2^{\bullet-}$, H_2O_2, and organic peroxides. No enzymes have yet been found that detoxify $\bullet OH$ or $^1O_2$. Examples (Gille and Joenje, 1991) include the following.

Superoxide dismutase (SOD). Superoxide dismutase enzymes catalyze the dismutation reaction:

$$2\ O_2^{\bullet-} + 2H^+ \rightarrow H_2O_2 + O_2$$

SOD enzymes have been found in the cellular cytoplasm and matrix space of mitochondria.

Catalase (CAT). Catalase catalyzes the decomposition of hydrogen peroxide to oxygen and water:

$$2\ H_2O_2 \rightarrow 2\ H_2O + O_2$$

Catalase subunits are formed in the cytoplasm, and synthesis of the enzyme is completed in the peroxisome. Catalase is absent in the mitochondrial matrix of most cells.

Glutathione peroxidase (GP). Glutathione peroxidase catalyzes the removal of H_2O_2 and lipid peroxides:

$$LOOH + 2\ GSH \rightarrow LOH + H_2O + GSSG$$

Reduced glutathione (GSH) is then regenerated from the oxidized state (GSSG) by glutathione reductase, a reduction that consumes NADPH.

The second protective approach includes nonenyzmatic compounds that react with activated oxygen species and thereby block the propagation of free radical chain reactions. These include the following.

Glutathione (GSH). Glutathione is a water-soluble antioxidant found in the cytoplasm which reacts with $O_2^{\bullet-}$, $\bullet OH$, and $^1O_2$:

$$O_2^{\bullet-} + H^+ + GSH \rightarrow GS\bullet + H_2O_2$$
$$\bullet OH + GSH \rightarrow GS\bullet + H_2O$$
$$^1O_2 + GSH \rightarrow GS_{ox}\ (\text{cysteic acid})$$

The glutathione radical (GS$\bullet$) is considered relatively stable and causes little damage.

Vitamin E (tocopherol). Vitamin E or tocopherol nonenzymatically reduces polyunsaturated lipid peroxide free radicals:

$$\text{Vit E} + \text{LOO}\bullet \rightarrow \text{Vit E}\bullet + \text{LOOH}$$

Vitamin E also readily reacts with $O_2^{\bullet-}$ and $^1O_2$ and is a water-insoluble compound found in the lipid domains of membranes.

Vitamin C (ascorbic acid). Vitamin C is a water-soluble compound capable of reacting with free radicals ($R\bullet$) and $O_2^{\bullet-}$ and $\bullet OH$. Vitamin C is also thought to regenerate vitamin E by accepting the activated electron from Vit E$\bullet$.

The third type of defense is enzymes that specifically fix damage created by free radicals. These include DNA repair enzymes that involve a combination of base excision, nucleotide excision, or DNA mismatch repair activity. These function in the following way (Rasmussen and Singh, 1998).

Base excision repair pathway. In the nucleus, the first step in the base excision repair process involves the removal of damaged bases by a damage-specific DNA glycolase. Removal of the damaged base by glycolase creates an apurinic or apyrimidinic site in the DNA. In the next step, AP-endonulcease cleaves the phosphodiester bond, forming a nucleotide gap. The gap is then filled and sealed by DNA polymerase and DNA ligase, respectively.

Nulecotide excision repair pathway. Oxidative lesions are removed by hydrolyzing phosphodiester bonds on both sides of the lesion. Two excision mechanisms accomplish this removal: the endonuclease-exonuclease and the excision nuclease mechanisms. A concerted action of at least 16 polypeptides is involved in the repair.

DNA mismatch repair pathway. Oxidative damage to mtDNA can lead to misincorporation of nucleotides during replication of mitochondria. In the nucleus, DNA mismatch repair corrects these types of mutations in duplex DNA during DNA replication.

Based on these findings, various approaches to protect orthodox seeds against antioxidant free radical scavengers exist. For example, one tocopherol molecule may afford antioxidant protection to several thousand fatty acid molecules (Bewley, 1986). Soybean seeds have a lower tocopherol content following aging, suggesting that tocopherol is consumed and protects the seed against free radical damage (Seneratna, Gusse, and McKersie, 1988). Superoxide dismutase increases with accelerated aging in pigeonpea seeds (Kalpana and Madhava Rao, 1994). Other enzymes, such as glutathione reductase, are antioxidants on the one hand but sources of free radicals on the other, so it is difficult to determine their protective qualities

(DeVos, Kraak, and Bino, 1994). However, glutathione is an efficient antioxidant and has such a role in aged sunflower (DePaula et al., 1996) and watermelon (Hsu and Sung, 1997) seeds. Thus, the addition of antioxidants might afford seeds protection against free radical attack. For example, the addition of 0.04 M ferrous sulfate reduced lipid peroxidation and increased the quenching of free radicals in soybean axes during the first hour of imbibition (Hailstones and Smith, 1991). Pretreatment of seeds with compounds such as dikegulac-sodium, ascorbic acid, cinnamic acid, and α-tocopherol prior to accelerated and natural aging improved seed vigor and seed storage of rice (Bhattacharjee and Bhattacharyya, 1989), maize and mustard (Dey and Mukherjee, 1988), sunflower (Bhattacharjee and Gupta, 1985), French bean, pea, lentil, and millet (Chhetri, Rai, and Bhattacharjee, 1993), and jute (Bhattacharjee, Chowdhury, and Choudhuri, 1986; Chowdhury and Choudhuri, 1994).

RAFFINOSE OLIGOSACCHARIDES AND THEIR PROTECTIVE ROLE

Raffinose oligosaccharides (RFOs) have been implicated as important components of cell membranes that maintain membrane integrity during drying and storage of orthodox seeds (Crowe, Crowe, and Hoekstra, 1989; Hoekstra, Crowe, and Crowe, 1992; Oliver, Crowe, and Crowe, 1998; Peterbauer and Richter, 2001). Sugars are also involved in maintaining the three-dimensional structure of proteins that prevent their unfolding and denaturation due to loss of associated water during seed dry down (Crowe, Hoekstra, and Crowe, 1992; Wolkers et al., 1998). RFOs also appear to be involved with sucrose in the formation of glasses in dry seeds (Burke, 1986), which are highly viscous solids that retard molecular diffusion and slow deteriorative reactions (Buitink et al., 2000). Glass formation in orthodox seeds as they dry down during seed maturation may protect both lipids and proteins against free radical attack. All of these events enhance membrane and protein stability leading to increased seed longevity (Brenac et al., 1997; Obendorf, 1997; Obendorf et al., 1998). In fact, it has been proposed that seeds having RFOs smaller than 1.0 tend to possess shorter seed storage lives while those with RFOs greater than 1.0 have longer seed storage lives (Horbowicz and Obendorf, 1994).

Despite these studies, increasing evidence suggests that RFOs may not be involved in increasing orthodox seed storage longevity. For example, no unique relationship was found between tomato seed longevity and sucrose or oligosaccharide content (Gurusinghe and Bradford, 2001). Studies of *Arabidopsis* seed longevity using recombinant inbred lines differing in

RFO content also failed to find a relationship between RFO content and desiccation tolerance to controlled deterioration (Groot et al., 2000). Intracellular glass stability of impatiens and pepper seeds using electron spin probes showed no change before and after priming despite a reduction in RFO content (Buitink, Hemminga, and Hoekstra, 2000). Since priming reduces storage life (Argerich and Bradford, 1989), these data suggest that the stability of intracellular glasses may not be involved in increasing orthodox seed longevity.

REPAIR OF SEED DAMAGE

Considerable evidence indicates that repair of DNA (Rao, Roberts, and Ellis, 1987; Sivritepe and Dourado, 1994; Dell'Aquila and Tritto, 1990), RNA (Kalpana and Madhava Rao, 1997), protein (Dell'Aquila and Tritto, 1991; Petruzzeli, 1986), membranes (Petruzzeli, 1986; Tilden and West, 1985; Powell and Harman, 1985), and enzymes (Jeng and Sung, 1994) occurs during imbibition. Increasing seed moisture content hastens the repair process (Ward and Powell, 1983). Oxygen also increases the repair of high-moisture (27 to 44 percent) lettuce (Ibrahim, Roberts, and Murdoch, 1983) and high-moisture (24 to 31 percent) wheat (Petruzzeli, 1986) seeds, suggesting that respiratory activity is an essential component of repair. This knowledge that repair occurs during imbibition has been practically adapted by the seed industry for many crops through seed priming. As a result, studies examining the physiological advantages/disadvantages in extending seed performance are appropriate. In general, it is accepted that repair of seeds deteriorated by lipid peroxidation occurs during hydration (priming). The repaired seed is then dried for normal handling and the benefits of repair retained as the primed seed completes germination. It should be noted, however, that the physiological improvements gained by priming are not solely attributable to repair since newly harvested muskmelon seeds show a dramatic improvement in seed performance following osmopriming (Welbaum and Bradford, 1991).

Most studies conclude that "repair" has occurred, but when (during priming or after), where (what seed part is repaired, if any), and how (what is the mechanism) repair occurs is still not known.

When Does Repair Occur?

The time when the beneficial effects of priming are achieved is unknown. It is generally thought that the hydration phase causes activation of essential metabolism associated with germination and the production of repair enzymes. These remain potentially active following subsequent drying and are quickly reactivated on imbibition, culminating in more rapid and uniform completion of germination. Other studies, however, suggest that the maximum beneficial effects of priming are achieved during the drying phase when enzymes are afforded sufficient time to effect repair and physiologically stabilize the seed. For example, the optimum effects of wheat seed osmopriming are observed two weeks after drying (Dell'Aquila and Tritto, 1990). Dell'Aquila and Bewley (1989) showed that protein synthesis is reduced in the axes of pea seeds imbibed in polyetheleneglycol (PEG), dried, and then increased on their return to imbibition. Further research is necessary to clarify whether the benefits of priming are achieved during the hydration or drying phases, or both.

Where Is the Location of Repair?

The location of the beneficial priming response still needs clarification. Reversal of seed deterioration by priming generally occurs in the meristematic axis or the radicle tip, e.g., peanut (Fu et al., 1988). Sivritepe and Dourado (1994) found that controlled humidification of aged pea seeds to 16.3 to 18.1 percent just prior to sowing decreases chromosomal aberrations, reduces imbibitional injury, and improves seed viability. Rao, Roberts, and Ellis (1987) reported a reversal of chromosomal damage (induced during seed aging) with partial hydration of lettuce seeds by osmopriming to 33 to 44 percent. This treatment also increases the rate of root growth and decreases the frequency of abnormal seedlings. In tomato, artificial aging increases the percentage of aberrant anaphases in seedling root tips (Van Pijlen et al., 1995). However, although osmopriming partially counteracts the detrimental effects of artificial aging on germination rate, uniformity, and normal seedlings, it does not influence the frequency of aberrant anaphases in seedling root tips.

Priming also appears to increase germination metabolism in aged axes more than those that are not aged. For example, Dell'Aquila and Taranto (1986) demonstrated that primed embryos of aged wheat seeds have a faster resumption of cell division and DNA synthesis on subsequent imbibition. Clarke and James (1991) showed that accelerated aging has an adverse effect on endosperm cells of leek seeds which results in their degradation and an overall loss in seed viability during osmopriming. During germination, however, those seeds that were aged and then osmoprimed showed an increase in RNA species in the whole seeds and their embryos.

What Is the Mechanism of Repair?

Priming appears to reverse the detrimental effects of seed deterioration. In sweet corn, osmo- and matripriming results in decreased conductivity, free sugars, and DNA content, while RNA content increased (Sung and Chang, 1993). Natural aging of French bean seeds stored for up to four years induced membrane disruption and leakage of UV-absorbing substances, which was ameliorated by hydropriming (Pandey, 1988, 1989). Lower electrical conductivity readings following hydropriming indicated reduced membrane leakage for eggplant and radish (Rudrapal and Nakamura, 1988a) and onion (Choudhuri and Basu, 1988) seeds. These beneficial effects may be due to the flushing of solutes from the seed during the priming procedure and prior to determination of leaked substances. As a practical result, primed seeds often perform better in disease-infested soils because of decreased electrolyte leakage and faster germination which reduce the window of opportunity for fungal attack (Osburn and Schroth, 1988). Osmopriming increased respiration in tabasco and jalapeno seeds (Sundstrom and Edwards, 1989; Halpin-Ingham and Sundstrom, 1992), although respiratory rates in −1.35 MPa NaCl osmoprimed pepper seeds were the same as in raw seeds (Smith and Cobb, 1992).

Priming is also thought to increase enzyme activity and counteract the effects of lipid peroxidation. Saha, Mandal, and Basu (1990) showed that matripriming caused increased amylase and dehydrogenase activity in aged soybean seeds compared to raw seeds. In wheat, osmopriming increased protein and DNA synthesis (Dell'Aquila and Tritto, 1990). L-isoaspartyl methyltransferase enzymes were reported to initiate the conversion of detrimental L-isoaspartyl residues to normal L-isoaspartyl residues that accumulate in naturally aged wheat seeds (Mudgett and Clarke, 1993; Mudgett, Lowenson, and Clarke, 1997). This enzyme is present in seeds of 45 species from 23 families representing most of the divisions of the plant kingdom (Mudgett, Lowenson, and Clarke, 1997). Osmoprimed tomato seeds subjected to accelerated aging showed restored activity of L-isoaspartyl methyltransferase to levels similar to nonaged controls, leading Kester, Geneve, and Houtz (1997) to suggest that this enzyme is involved in early repair of deteriorated seeds. Osmopriming reverses the loss of lipid-peroxidation-detoxifying enzymes, such as superoxide dismutase, catalase, and glutathione reductase, in aged sunflower seeds, and these enzymes are present at the same activities as in unaged seeds (Bailly et al., 1997).

Priming also reduces lipid peroxidation during subsequent seed storage. In onion seed, Choudhuri and Basu (1988) demonstrated that hydropriming treatments effectively slowed physiological deterioration under natural (15 months) and accelerated aging conditions, with the effect being dependent on seed vigor. This improved storability was associated with greater dehydrogenase activity and appreciably lower peroxide formation in cells. Similar findings were reported for hydroprimed eggplant and radish seeds with the conclusion that hydropriming reduces free radical damage to cellular components (Rudrapal and Nakamura, 1988). Jeng and Sung (1994) found that free radical scavenging enzymes such as superoxide dismutase, catalase, and peroxidase and glyoxysome enzymes such as isocitrate lyase and malate synthase were increased by increasing hydration of artificially aged peanut seeds. Chang and Sung (1998) also showed that martripriming with vermiculite of sweet corn seeds enhanced the activities of several lipid peroxide scavenging enzymes. Chiu, Wang, and Sung (1995) found that increasing hydration enhanced membrane repair in watermelon seeds and attributed this to the stimulation of peroxide scavenging enzymes that produced reduced glutathione which may control aging by counteracting lipid peroxidation. Another possible antioxidant is glutathione whose content has been shown to decrease with watermelon seed aging as seeds are hydrated (Hsu and Sung, 1997).

MODEL OF SEED DETERIORATION·AND REPAIR DURING PRIMING/HYDRATION

As seeds deteriorate, a cascade of disorganization ensues, ultimately leading to complete loss of cell function. The current model of seed deterioration accepts lipid peroxidation as a central cause of cellular degeneration through free radical assault on important cellular molecules and structures. Figure 9.5 demonstrates some proposed events associated with seed deterioration during storage and their repair, or lack of repair, during hydration that can occur during imbibition or priming and seeds contain a variety of antioxidants including vitamins, polyphenols, and flavonoids (Larson, 1997).

Storage

Low seed moisture content during storage favors free radical production by autoxidation. Through lipid peroxidation, these free radicals either directly or indirectly cause four types of cellular damage: mitochondrial dysfunction, enzyme inactivation, membrane perturbations, and genetic damage. Thus, the amount of antioxidants in seeds might reduce the incidence of cellular damage due to free radical assault during seed storage.

Imbibition and Priming

As time of seed storage increases, so does cellular damage. Imbibition and priming of the seed allows two events to occur. As imbibition proceeds,

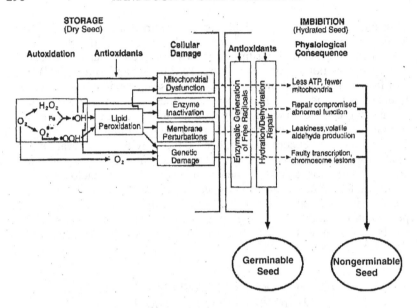

FIGURE 9.5. A model of seed deterioration and its physiological consequences during seed storage and imbibition (*Source:* M. B. McDonald, 1999, Seed deterioration: Physiology, repair and assessment, *Seed Science and Technology* 27: 177-237. Reproduced with permission.)

the cascade of cellular damage caused by autoxidation is furthered by free radical damage, induced less by autoxidation and more by free-radical-generating hydrolytic enzymes such as lipoxygenase. The presence of antioxidants may ameliorate this damage. In addition, upon hydration, anabolic enzymes associated with repair of cellular constituents counter these degenerative events. Their success determines whether a seed is capable of germinating and performing optimally. If unsuccessful, the cellular damage established during storage leads to unalterable detrimental physiological consequences resulting in a nongerminable seed.

CONCLUSIONS

In conclusion, this chapter has emphasized that many factors (external and internal) contribute to orthodox seed deterioration. Of these, seed moisture content and temperature have important roles that directly influence the biochemistry of deterioration. It is also apparent that seed deterioration is uniform neither among seeds nor among seed parts (membranes being more prone to deteriorative events). At the cellular level, the mitochondria may be a central organelle susceptible to deteriorative event, and their further study is warranted. As oxygen "sinks" that contain extensive membrane structure for respiratory events, they are particularly prone to free radical assault and lipid peroxidation. If these events occur, seed germination as measured by speed and uniformity of emergence would certainly be compromised. Fortunately, evidence exists that free radical attack can be reduced by free radical scavenger and antioxidant compounds found in seeds. In addition, specific repair enzymes have been identified that potentially function during hydration, perhaps providing a mechanism for the success of priming as a seed-enhancement technology. Clearly, all of this demonstrates that further studies are necessary to better understand the mechanism(s) of orthodox seed deterioration and its repair. Hopefully, this chapter has provided the foundation to initiate this quest.

REFERENCES

Argerich, C.A. and Bradford, K.J. (1989). The effects of priming and aging on seed vigor in tomato. *Journal of Experimental Botany* 40: 599-607.

Association of Official Seed Analysts (AOSA) (2000). *Tetrazolium Testing Handbook.* Contribution No. 29. Lincoln, NE: Author.

Bailly, C., Benamar, A., Corbineau, F., and Côme, D. (1997). Changes in superoxide dismutase, catalase and glutathione reductase activities in sunflower seeds during accelerated aging and subsequent priming. In Ellis, R.H., Black, M., Murdoch, A.J., and Hong, T.D. (Eds.), *Basic and Applied Aspects of Seed Biology* (pp. 665-672). Boston: Kluwer Academic Publishers.

Barnes, W.M. (1994). PCR amplification of up to 35-kb DNA with high fidelity and high yield from lambda bacteriophage templates. *Proceedings of the National Academy of Sciences, USA* 91: 2216-2220.

Beckman, K.B. and Ames, B.N. (1998). The free radical theory of aging matures. *Physiological Reviews* 78: 547-581.

Berjak, P., Dini, M., and Gevers, H.O. (1986). Deteriorative changes in embryos of long-stored, uninfected maize caryopses. *South African Journal of Botany* 52: 109-116.

Berjak, P., Farrant, J.M., Mycock, D.J., and Pammenter, N.W. (1990). Recalcitrant (homoiohydrous) seeds: The enigma of their desiccation sensitivity. *Seed Science and Technology* 18: 297-310.

Bewley, J.D. (1986). Membrane changes in seeds as related to germination and the perturbations resulting from deterioration in storage. In McDonald, M.B. and Nelson, C.J. (Eds.), *Physiology of Seed Deterioration* (pp. 27-47). Madison, WI: Crop Science Society of America.

Bhattacharjee, A. and Bhattacharyya, R.N. (1989). Prolongation of seed viability of *Oryza sativa* L. cultivar Ratna by dikegulac-sodium. *Seed Science and Technology* 17: 309-316.

Bhattacharjee, A., Chowdhury, R.S., and Choudhuri, M.A. (1986). Effects of CCC and Na-dikegulac on longevity and viability of seeds of two jute cultivars. *Seed Science and Technology* 14: 127-139.

Bhattacharjee, A. and Gupta, K. (1985). Effect of dikegulac-sodium, and growth retardant, on the viability of sunflower seeds. *Seed Science and Technology* 13: 165-174.

Bingham, I.J., Harris, A., and MacDonald, L. (1994). A comparative study of radicle and coleoptile extension in maize seedlings from aged and unaged seeds. *Seed Science and Technology* 22: 127-139.

Brenac, P., Horbowics, M., Downer, S.M., Dickerman, A.M., Smith, M.E., and Obendorf, R.L. (1997). Raffinose accumulation related to desiccation tolerance during maize (*Zea mays* L.) seed development and maturation. *Journal of Plant Physiology* 15: 481-488.

Buitink, J., Hemminga, M.A., and Hoekstra, F.A. (2000). Is there a role for oligosaccharides in seed longevity? An assessment of intracellular glass stability. *Plant Physiology* 122: 1217-1224.

Buitink, J., Leprince, O., Hemminga, M.A., and Hoekstra, F.A. (2000). Molecular mobility in the cytoplasm: An approach to describe and predict lifespan of dry germplasm. *Proceedings of the National Academy of Sciences, USA* 97: 2385-2390.

Burke, M.J. (1986). The glassy state and survival of anhydrous biological systems. In Leopold, A.C. (Ed.), *Membranes, Metabolism and Dry Organisms* (pp. 358-363). Ithaca, NY: Cornell University Press.

Chang, S.M. and Sung, J.M. (1998). Deteriorative changes in primed sweet corn seeds during storage. *Seed Science and Technology* 26: 613-626.

Chauhan, K.P.S. (1985). The incidence of deterioration and its localization in aged seeds of soybean and barley. *Seed Science and Technology* 13: 769-773.

Cheng, S., Higuchi, R., and Stoneking, M. (1994). Complete mitochondrial genome amplification. *Nature Genetics* 7: 350-351.

Chhetri, D.R., Rai, A.S., and Bhattacharjee, A. (1993). Chemical manipulation of seed longevity of four crop species in an unfavorable storage environment. *Seed Science and Technology* 21: 31-44.

Chin, H.F. and Roberts, E.H. (1980). *Recalcitrant Crop Seeds*. Kuala Lumpur, Malaysia: Tropical Press.

Chiu, K.Y., Wang, C.S., and Sung, J.M. (1995). Lipid peroxidation and peroxide-scavenging enzymes associated with accelerated aging and hydration of watermelon seeds differing in ploidy. *Physiologia Plantarum* 94: 441-446.

Choudhuri, N. and Basu, R.N. (1988). Maintenance of seed vigour and viability of onion (*Allium cepa* L.). *Seed Science and Technology* 16: 51-61.

Chowdhury, S.R. and Choudhuri, M.A. (1994). Effects of seed pretreatment with CCC, cinnamic acid, and Na-dikegulac on germination and early seedling growth performance from ageing jute seeds under water deficit stress. *Seed Science and Technology* 22: 203-208.

Clarke, N.A. and James, P.E. (1991). The effects of priming and acclerated ageing upon the nucleic acid content of leek seeds and their embryos. *Journal of Experimental Botany* 42: 261-268.

Conley, C.A. and Hanson, M.R. (1995). How do alterations in plant mitochondrial genomes disrupt pollen development? *Journal of Bioenergetics and Biomembranes* 27: 447-457.

Copeland, L.O. and McDonald, M.B. (2001). *Principles of Seed Science and Technology.* New York: Kluwer Academic Press.

Crowe, J.H., Crowe, L.M., and Hoekstra, F.A. (1989). Phase transitions and permeability changes in dry membranes during rehydration. *Journal of Bioenergetics and Biomembranes* 21: 77-91.

Crowe, J.H., Hoekstra, F.A., and Crowe, L.M. (1992). Anhydrobiosis. *Annual Reviews of Physiology* 54: 579-599.

Das, G. and Sen-Mandi, S. (1988). Root formation in deteriorated (aged) wheat embryos. *Plant Physiology* 88: 983-986.

Das, G. and Sen-Mandi, S. (1992). Triphenyl tetrazolium chloride staining pattern of differentially aged wheat embryos. *Seed Science and Technology* 20: 367-373.

DeGrey, A.D.W.J. (1999). *The Mitochondrial Free Radical Theory of Aging.* Austin, TX: R.G. Landes Company.

Dell'Aquila, A. and Bewley, J.D. (1989). Protein synthesis in the axes of polyethylene glycol treated pea seed and during subsequent germination. *Journal of Experimental Biology* 40: 1001-1007.

Dell'Aquila, A. and Taranto, G. (1986). Cell division and DNA synthesis during osmoconditioning treatment and following germination in aged wheat embryos. *Seed Science and Technology* 14: 333-341.

Dell'Aquila, A. and Tritto, V. (1990). Ageing and osmotic priming in wheat seeds: Effects upon certain components of seed quality. *Annals of Botany* 65: 21-26.

Dell'Aquila, A. and Tritto, V. (1991). Germination and biochemical activities in wheat seeds following delayed harvesting, ageing and osmotic priming. *Seed Science and Technology* 19: 73-82.

DePaula, M., Perez-Otaola, M., Darder, M., Torres, M., Frutos, G., and Martinez-Honduvilla, C.J. (1996). Function of the ascorbate-glutathione cycle in aged sunflower seeds. *Physiologia Plantarum* 96: 543-550.

DeVos, C.H.R., Kraak, H.L., and Bino, R.J. (1994). Ageing of tomato seeds involves glutathione oxidations. *Physiologia Plantarum* 92: 131-139.

Dey, P.G. and Mukherjee, R.K. (1988). Invigoration of dry seeds with physiologically active chemicals in organic solvents. *Seed Science and Technology* 16: 145-153.

Ellis, R.H., Osei-Bonsu, K., and Roberts, E.H. (1982). The influence of genotype, temperature, and moisture on seed longevity in chickpea, cowpea, and soya bean. *Annals of Botany* 50: 69-82.

Esashi, Y., Kamataki, A., and Zhang, M. (1997). The molecular mechanism of seed deterioration in relation to the accumulation of protein-acetaldehyde adducts. In Ellis, R.H., Black, M., Murdoch, A.J., and Hong, T.D. (Eds.), *Basic and Applied Aspects of Seed Biology* (pp. 489-498). Boston: Kluwer Academic Publishers.

Fu, J.R., Lu, X.H., Chen, R.Z., Zhang, B.Z., Liu, Z.S., Ki, Z.S., and Cai, C.Y. (1988). Osmoconditioning of peanut (*Arachis hypogaea* L.) seeds with PEG to improve vigour and some biochemical activities. *Seed Science and Technology* 16: 197-212.

Garwood, N.C. (1989). Tropical soil seed banks: A review. In Leck, M.A., Parker, V.T., and Simpson, R.L. (Eds.), *Ecology of Soil Seed Banks* (pp. 149-209). San Diego, CA: Academic Press.

Gille, J.J.P. and Joenje, H. (1991). Biological significance of oxygen toxicity: An introduction. In Vigo-Pelfrey, C. (Ed.), *Membrane Lipid Oxidation* (pp. 1-32). Boca Raton, FL: CRC Press.

Groot, S.P.C., van der Geest, A.H.M., Tesnier, K., Alonso-Blanco, C., Bentsink, L., Donkers, H., Koornneef, M., Vreugdenhil, D., and Bino, R.J. (2000). Molecular genetic analysis of *Arabidopsis* seed quality. In Black, M., Bradford, K.J., and Vazquez-Ramos, J. (Eds.), *Seed Biology: Advances and Implications* (pp. 123-132). Wallingford, UK: CABI Publishing.

Gurusinghe, S. and Bradford, K.J. (2001). Galactosyl-sucrose oligosaccharides and potential longevity of primed seeds. *Seed Science Research* 11: 121-133.

Hahalis, D.A. and Smith, M.L. (1997). Comparison of the storage potential of soyabean *(Glycine max)* cultivars with different rates of water uptake. In Ellis, R.H., Black, M., Murdoch, A.J., and Hong, T.D. (Eds.), *Basic and Applied Aspects of Seed Biology* (pp. 507-514). Boston: Kluwer Academic Publishers.

Hailstones, M.D. and Smith, M.T. (1991). Soybean seed invogoration by ferrous sulfate: Changes in lipid peroxidation, conductivity, tetrazolium reduction, DNA and protein synthesis. *Journal of Plant Physiology* 137:.307-311.

Halmer, P. and Bewley, J.D. (1984). A physiological perspective on seed vigour testing. *Seed Science and Technology* 12: 561-575.

Halpin-Ingham, B. and Sundstrom, F.J. (1992). Pepper seed water content, germination response and respiration following priming treatments. *Seed Science and Technology* 20: 589-596.

Hanson, M.R. and Otto, F. (1992). Structure and function of the higher plant mitochondrial genome. *International Review of Cytology* 141: 129-172.

Harrington, J.F. (1972). Seed storage and longevity. In Kozlowski, T.T. (Ed.), *Seed Biology,* Volume 3 (pp. 145-240). New York: Academic Press.

Hoekstra, F.A., Crowe, J.H., and Crowe, L.M. (1992). Germination and ion leakage are linked with phase transitions of membrane lipids during imbibition of *Typha latifolia* pollen. *Physiologia Plantarum* 84: 29-34.

Horbowicz, M. and Obendorf, R.L. (1994). Seed desiccation tolerance and storability: Dependence on flatulence-producing oligosaccharides and cyclitols—Review and survey. *Seed Science Research* 4: 385-405.

Hsu, J.L. and Sung, J.M. (1997). Antioxidant role of glutathione associated with accelerated aging and hydration of triploid watermelon seeds. *Physiologia Plantarum* 100: 967-974.

Ibrahim, A.E., Roberts, E.H., and Murdoch, A.J. (1983). Viability of lettuce seeds: II. Survival and oxygen uptake in somatically controlled storage. *Journal of Experimental Botany* 34: 631-640.

Jeng, T.L. and Sung, J.M. (1994). Hydration effect on lipid peroxidation and peroxide-scavenging enzymes activity of artificially-aged peanut seed. *Seed Science and Technology* 22: 531-539.

Kadenbach, B., Munscher, C., and Frank, V. (1995). Human aging is associated with stochastic somatic mutations of mitochondrial DNA. *Mutation Research* 338: 161-172.

Kalpana, R. and Madhava Rao, K.V. (1994). Absence of the role of lipid peroxidation during accelerated aging of seeds of pigeonpea [*Cajanus cajan* (L.) Millsp.] cultivars. *Seed Science and Technology* 22: 253-260.

Kalpana, R. and Madhava Rao, K.V. (1997). Nucleic acid metabolism of seeds of pegeonpea (*Cajanus cajan* L. Millsp.) cultivars during accelerated ageing. *Seed Science and Technology* 25: 293-301.

Kang, D., Takeshige, K., Sekiguchi, M., and Singh, K.K. (1998). Introduction. In Singh, K.K. (Ed.), *Mitochondrial DNA Mutations in Aging, Disease and Cancer* (pp. 1-15). New York: Springer.

Kester, S.T., Geneve, R.L., and Houtz, R.L. (1997). Priming and accelerated aging affect L-isoaspartyl methyltransferase activity in tomato (*Lycopersicon esculentum* Mill) seed. *Journal of Experimental Botany* 48: 943-949.

Larson, R.A. (1997). *Naturally Occurring Antioxidants*. Boca Raton, FL: Lewis Publishers.

Loiseau, J., Benoit, L.V., Macherel, M.H., and Deunff, Y.L. (2001). Seed lipoxygenases: Occurrence and functions. *Seed Science Research* 11: 199-211.

Lu, B., Wilson, R.K., Phreaner, C.G., Mulligan, M.R., and Hanson, M.R. (1996). Protein polymorphism generated by differential RNA editing of a plant mitochondrial rps12 gene. *Molecular and Cellular Biology* 16: 1543-1549.

McDonald, M.B. (1985). Physical seed quality of soybean. *Seed Science and Technology* 13: 601-628.

McDonald, M.B. (1999). Seed deterioration: Physiology, repair and assessment. *Seed Science and Technology* 27: 177-237.

McDonald, M.B. and Nelson, C.J. (1986). *Physiology of Seed Deterioration*. CSSA Special Publication No. 11. Madison, WI: Crop Science Society of America.

McDonald, M.B., Sullivan, J., and Lauer, M.J. (1994). The pathway of water uptake in maize (*Zea mays* L.) seeds. *Seed Science and Technology* 22: 79-90.

McDonald, M.B., Vertucci, C.W., and Roos, E.E. (1988). Soybean seed imbibition: Water absorption of seed parts. *Crop Science* 28: 993-998.

Mudgett, M.B. and Clarke, S. (1993). Characterization of plant L-isoaspartyl methyltransferases that may be involved in seed survival: Purification, cloning, and sequence analysis of the wheat germ enzyme. *Biochemistry* 32: 111000-111111.

Mudgett, M.B., Lowenson, J.D., and Clarke, S. (1997). Protein repair L-isoaspartyl methyltransferase in plants: Phylogenetic distribution and the accumulation of substrate proteins in aged barley seeds. *Plant Physiology* 114: 1481-1489.

Munscher, C., Muller-Hocker, J., and Kadenbach, B. (1993). Human aging is associated with various point mutations in tRNA genes of mitochondrial DNA. *Biological Chemistry Hoppe Seyler* 374: 1099-1104.

Newton, K.J. and Gabay-Laughman, S.J. (1998). Abnormal growth and male sterility associated with mitochondrial DNA arrangements in plants. In Singh, K.K. (Ed.), *Mitochondrial DNA Mutations in Aging, Disease, and Cancer* (pp. 365-381). New York: Springer.

Obendorf, R.L. (1997). Oligosaccharides and galactosyl cyclitols in seed desiccation tolerance. *Seed Science Research* 7: 63-74.

Obendorf, R.L., Dickerman, A.M., Pflum, T.M., Kacalanos, M.A., and Smith, M.E. (1998). Drying rate alters soluble carbohydrates, desiccation tolerance, and subsequent seedling growth of soybean (*Glycine max* L. Merrill) zygotic embryos in vitro maturation. *Plant Science* 132: 1-12.

Oliver, A.E., Crowe, L.M., and Crowe, J.H. (1998). Methods for dehydration-tolerance: Depression of the phase transition temperature in dry membranes and carbohydrate vitrification. *Seed Science Research* 8: 211-221.

Osburn, R.M. and Schroth, M.N. (1988). Effect of osmopriming sugar beet seed on exudation and subsequent damping-off caused by *Pythium ultimum*. *Phytopathologie* 78: 1246-1250.

Pammenter, N.W. and Berjak, P. (2000). Evolutionary and ecological aspects of recalcitrant seed biology. *Seed Science Research* 10: 301-306.

Pandey, K.K. (1988). Priming induced repair in French bean seeds. *Seed Science and Technology* 16: 527-532.

Pandey, K.K. (1989). Priming induced alleviation of the effects of natural ageing derived selective leakage of constituents in French bean seeds. *Seed Science and Technology* 17: 391-397.

Peterbauer, T. and Richter, A. (2001). Biochemistry and physiology of raffinose family oligosaccharides and galactosyl cyclitols in seeds. *Seed Science Research* 11: 185-197.

Petruzzeli, L. (1986). Wheat viability at high moisture content under hermetic and aerobic storage conditions. *Annals of Botany* 58: 259-265.

Powell, A.A. and Harman, G.E. (1985). Absence of a consistent association of changes in membranal lipids with the ageing of pea seeds. *Seed Science and Technology* 13: 659-667.

Priestley, D.A. (1986). *Seed Ageing: Implications for Seed Storage and Persistence in the Soil*. Ithaca, NY: Cornell University Press.

Rao, N.K., Roberts, E.H., and Ellis, R.H. (1987). Loss of viability in lettuce seeds and the accumulation of chromosome damage under different storage conditions. *Annals of Botany* 60: 85-96.

Rasmussen, L.J. and Singh, K.K. (1998). Genetic integrity of the mitochondrial genome. In Sing, K.K. (Ed.), *Mitochondrial DNA Mutations in Aging, Disease, and Cancer* (pp. 115-127). New York: Springer.

Reynier, P. and Mathiery, Y. (1995). Accumulation of deletions in mtDNA during tissue aging: Analysis by long PCR. *Biochemical and Biophysical Research Communications* 217: 59-67.

Rudrapal, D. and Nakamura, S. (1988a). The effect of hydration-dehydration pretreatments on eggplant and radish seed viability and vigour. *Seed Science and Technology* 16: 123-130.

Rudrapal, D. and Nakamura, S. (1988b). Use of halogens in controlling eggplant and radish seed deterioration. *Seed Science and Technology* 16: 115-122.

Saha, R., Mandal, A.K., and Basu, R.N. (1990). Physiology of seed invigoration treatments in soybean (*Glycine max* L.). *Seed Science and Technology* 18: 269-276.

Sembdner, G. and Parthier, B. (1993). The biochemistry and the physiological and molecular actions of jasmonates. *Annual Review of Plant Physiology and Plant Molecular Biology* 44: 569-589.

Seneratna, T., Gusse, J.F., and McKersie, B.D. (1988). Age-induced changes in cellular membranes of imbibed soybean axes. *Physiologia Plantarum* 73: 85-91.

Sivritepe, H.O. and Dourado, A.M. (1994). The effects of humidification treatments on viability and the accumulation of chromosomal aberrations in pea seeds. *Seed Science and Technology* 22: 337-348.

Smith, M.T. and Berjak, P. (1995). Deteriorative changes associated with the loss of viability of stored desiccation-tolerant and desiccation-sensitive seeds. In Kigel, J. and Galili, G. (Eds.), *Seed Development and Germination* (pp. 701-746). New York: Marcel Dekker.

Smith, P.T. and Cobb, B.G. (1992). Physiological/enzymatic characteristics of primed, redried, and germinated pepper seeds (*Capsicum annuum* L.). *Seed Science and Technology* 20: 503-513.

Society of Commercial Seed Technologists (SCST) (2001). *Seed Technologists Training Manual*. Lincoln, NE: Author.

Staswick, P.E. (1992). Jasmonate, genes and fragrant signals. *Plant Physiology* 99: 804-807.

Sundstrom, F.J. and Edwards, R.L. (1989). Pepper seed respiration, germination, and seedling development following seed priming. *HortScience* 24: 343-345.

Sung, F.J.M. and Chang, Y.H. (1993). Biochemical activities associated with priming of sweet corn seeds to improve vigor. *Seed Science and Technology* 21: 97-105.

Suzuki, Y., Ise, K., Li, C.Y., Honda, I., Iwai, Y., and Matsukura, U. (1999). Volatile components in stored rice [*Oryza sativa* (L.)] of volatiles with and without lipoxygenase-3 in seeds. *Journal of Agricultural and Food Chemistry* 47: 1119-1124.

Suzuki, Y., Yasui, T., Matsukura, U., and Terao, J. (1996). Oxidative stability of bran lipids from rice variety [*Oryza sativa* (L.)] lacking lipoxygenase-3 in seeds. *Journal of Agricultural and Food Chemistry* 44: 3479-3483.

Tarquis, A.M. and Bradford, K.J. (1992). Prehydration and priming treatments that advance germination also increase the rate of deterioration of lettuce seeds. *Journal of Experimental Botany* 43: 307-317.

Tilden, R.L. and West, S.H. (1985). Reversal of the effects of ageing in soybean seeds. *Plant Physiology* 77: 584-586.

Van Pijlen, J.G., Kraak, H.L., Bino, R.J., and De Vos, C.H.R. (1995). Effects of ageing and osmoconditioning on germination characteristics and chromosome aberrations of tomato (*Lycopersicon esculentum* Mill.) seeds. *Seed Science and Technology* 23: 823-830.

Vazquez-Yanes, C. and Orozco-Segovia, A. (1993). Patterns of seed longevity and germination in the tropical rainforest. *Annual Review of Ecology and Systematics* 24: 69-87.

Ward, F.H. and Powell, A.A. (1983). Evidence for repair processes in onion seeds during storage at high seed moisture contents. *Journal of Experimental Botany* 34: 277-282.

Welbaum, G.E. and Bradford, K.J. (1991). Water relations of seed development and germination in muskmelon (*Cucumis melo* L.): VI. Influence of priming on germination responses to temperature and water potential during seed development. *Journal of Experimental Botany* 42: 393-399.

Wilson, D.O. and McDonald, M.B. (1986a). A convenient volatile aldehyde assay for measuring seed vigour. *Seed Science and Technology* 14: 259-268.

Wilson, D.O. and McDonald, M.B. (1986b). The lipid peroxidation model of seed deterioration. *Seed Science and Technology* 14: 269-300.

Wolkers, W.F., Bochicchio, A., Selvaggi, G., and Hoekstra, F.A. (1998). Fourier transform infrared microspectroscopy detects changes in protein secondary structure associated with desiccation tolerance in developing maize embryos. *Plant Physiology* 116: 1169-1177.

Zacheo, G., Cappello, A.R., Perrone, L.M., and Gnoni, G.V. (1998). Analysis of factors influencing lipid oxidation of almond seeds during accelerated aging. *Lebensmittel-Wissenschaft und Technologie* 31: 6-9.

Zimmerman, D.C. and Coudron, C.A (1979). Identification of traumatin, a wound hormone, as 12-oxo-*trans*-10-dodecenoic acid. *Plant Physiology* 63: 536-541.

Chapter 10

Recalcitrant Seeds

Patricia Berjak
Norman W. Pammenter

SEED CHARACTERISTICS—THE BROAD PICTURE

Aside from the provision of food and feedstock from one season to the next, seeds are stored as base and active collections, as a means of long-term conservation of valuable genetic resources representing species biodiversity, and to provide planting stock for subsequent seasons. However, their conservation in seed banks or gene banks, or in commercial storage, makes the assumption that seeds are storable in the first place, which in turn is based on the premise that they show orthodox postharvest behavior (Roberts, 1973). By this it is meant that the period for which the seeds may be stored without loss of quality is predictable under defined conditions of storage temperature and seed water (moisture) content, the longevity, within limits, increasing logarithmically with decreasing water content (Ellis and Roberts, 1980). Orthodox seeds are, or can be, dehydrated to low water contents, which is a consequence of their having acquired the property of desiccation tolerance relatively early during their preshedding development (e.g., Bewley and Black, 1994; Vertucci and Farrant, 1995). The property of desiccation tolerance and its maintenance in dry orthodox seeds is based on the presence and interplay of a suite of mechanisms and processes expressed during development (Pammenter and Berjak, 1999).

However, not all seeds are orthodox—i.e., some seeds do not fully acquire the property of desiccation tolerance. Indeed, the responses to dehydration of mature seeds of some species (e.g., *Avicennia marina*, Farrant, Berjak, and Pammenter, 1993; Farrant, Pammenter, and Berjak, 1993) indicate that few, if any, of the mechanisms and processes allowing tolerance of the loss of more than the slightest proportion of tissue water are operational. Such seeds are highly *recalcitrant*—this being the term introduced by Roberts (1973) for seeds that cannot be stored at low water contents. Since the publication by Chin and Roberts (1980), the list of species recorded as producing recalcitrant seeds—or, at any rate, seeds that are nonorthodox (orthodox seeds being those that are desiccation tolerant)—has steadily lengthened. Some examples of species producing recalcitrant seeds that have been recorded over the past decade alone include *Machilus thunbergii* and *M. kusanoi* (Lin and Chen, 1995 and Chien and Lin, 1997, respectively); *Aporusa lindleyana* (Kumar, Thomas, and Pushpangadan, 1996); *Garcinia gummi-gutta* (Chacko and Pillai, 1997); *Litsea acuminata* (Chien and Yang, 1997); *Euterpe edulis* (de Andrade and Pereira, 1997); possibly some species of dryland palms from Africa and Madagascar (Davies and Pritchard, 1998); *Carapa guianensis* and *C. procera* (Connor et al., 1998); *Guarea guidonia* (Connor and Bonner, 1998); *Inga uruguensis* (Bilia, Marcos Filho, and Novembre, 1999); *Bhesa indica* (Kumar and Chacko, 1999); and *Boscia senegalensis* (an arid-zone species), *Butyrospermum parkii, Cordyla pinnata,* and *Saba senegalensis* (Danthu et al., 2000).

Several points arise from this list: first, the genera and species concerned will be largely unfamiliar; second, although not stated, their provenance is tropical or subtropical; and third—also not immediately obvious—is that, as far as can be ascertained, all are tree species. These points, in turn, underlie three generalizations: (1) almost all the knowledge amassed to date about seed biology and physiology has been derived from work on cultivated crops plus a few woody species, representing less than 0.1 percent of the higher plants, which is hardly a representative sample of the more than 250,000 documented species of spermatophytes; (2) probably many tropical and subtropical species produce nonorthodox seeds; and (3) such seeds appear to be produced predominantly by trees (Berjak and Pammenter, 2001), although nonorthodox seeds produced by herbaceous species have been recorded, especially among the Amaryllidaceae.

SEED BEHAVIOR

Evolutionary and Taxonomic Considerations

There seems to be little correlation between the occurrence of seed recalcitrance—considered simply as desiccation sensitivity—and taxonomic status, as the phenomenon is widespread across families. Although there are dicotyledonous families in which apparently no species produces recalcitrant seeds, in others the phenomenon is common (e.g., the Dipterocarpaceae, Tompsett, 1992). In an in-depth review, von Teichman and van Wyk (1994) associated recalcitrance across 45 dicotyledonous families with large seeds developing from bitegmic, crassinucellate ovules showing nuclear endosperm development. Those authors also drew attention to the woody habitat and tropical habitat being associated features which, together with the ovule/seed characteristics, are generally considered as ancestral states. However, recalcitrant seeds are also produced by species in relatively advanced dicotyledonous families, as well as by some that are monocotyledonous. Despite the opinion that characteristics of contemporary seeds probably reflect an evolutionary history incorporating parallelism, convergence, and reversion, when the evidence is weighed up recalcitrance is considered to be the ancestral seed condition in the angiosperms (von Teichman and van Wyk, 1994; Pammenter and Berjak, 2000).

It seems probable that production of recalcitrant seeds (whether as an ancestral or relictual trait or by reversion) has been favored in environments where there would be little selective advantage to the acquisition of desiccation tolerance, e.g., in the humid tropics or other regions where no seasonal constraints prevent immediate seedling establishment. Nevertheless, recalcitrant seeds are produced by some species in dry environments, e.g., *Boscia senegalensis* from the Sahelian zone (Danthu et al., 2000), *Vitellaria paradoxum (Butyrospermum paradoxum)* from Burkina Faso (Gamene, 1997), and possibly some dryland palms (Davies and Pritchard, 1998), although surprisingly, of 87 aquatic species only 6.9 percent were unequivocally established as producing seeds showing recalcitrant storage behavior (Hay et al., 2000). Furthermore, a few temperate tree species also produce recalcitrant seeds which, in some cases, may overwinter in a dormant condition (e.g., *Aesculus hippocastanum*, Pritchard, Tompsett, and Manger, 1996; Pritchard et al., 1999). These observations emphasize that a far deeper appreciation of seed behavior than indicated only by desiccation sensitivity is required and that the seed biology of many more species across the range of families needs to be characterized in fine detail.

Although perusal of the literature reveals that differences are apparent in the degree of dehydration tolerated among recalcitrant seeds of different species, it is difficult to make direct comparisons, as the conditions under which drying was carried out lack consistency. However, major differences are also apparent in responses of seeds of different species to dehydration under similar conditions. A documented case involves a comparison of those of *Araucaria angustifolia* (a gymnosperm), *Scadoxus membranaceus* (monocotyledonous), and a dicotyledonous vine, *Landolphia kirkii* (Farrant, Pammenter, and Berjak, 1989), all of which graphically illustrated lethal intracellular responses to dehydration under identical conditions occurring at considerably different water contents among the species. That report compared seeds of completely unrelated taxa, but other record differences in response among species of the same genus. For example, in a comparison of intact oak seed drying responses, Connor and Bonner (1996) found that acorns of *Quercus alba* were considerably more desiccation sensitive than were those of *Q. nigra*. Similarly, Normah, Ramiya, and Gintangga (1997) reported noteworthy differences in the lowest water content to which seeds of two species of *Baccaurea* would survive. There are also examples of more extreme divergence in seed behavior among species of a single genus. In the case of *Acer,* seeds of *A. pseudoplatanus* (sycamore) are sufficiently desiccation sensitive to be unequivocally categorized as recalcitrant, whereas those of *A. platanoides* (Norway maple) are orthodox (Hong and Ellis, 1990). We have found a similarly wide divergence from orthodox to recalcitrant seed characteristics among southern African species of the gymnosperm *Podocarpus* (our unpublished observations). For species of *Coffea,* Hong and Ellis (1995) describe seeds of *C. liberica* as showing recalcitrant postharvest behavior, with those of *C. canephora* (robusta coffee) fitting the categorization of intermediate, originally described for *C. arabica* (i.e., being somewhat less desiccation sensitive than orthodox seeds, but showing chilling sensitivity when dehydrated, Ellis and Hong, 1990). Similarly, Eira and colleagues (1999) described seeds of *C. liberica* as being the least tolerant to dehydration with another species studied, *C. racemosa,* being relatively the most tolerant. In a study on *Coffea* seeds of nine species from various central African provenances, Dussert and colleagues (2000) suggested that the varying degrees of desiccation sensitivity occurring among them could be an adaptive feature, related to the mean number of dry months typical of each habitat.

Variability

It is very difficult to define succinctly the exact nature of recalcitrance—or, more generally, of nonorthodoxy—because of the marked differences in behavior of seeds among species, although all are shed at relatively to very high water contents and all are metabolically active when shed. Inherent variation among seeds of different species—including size, structure, the nature of the testa/pericarp, and chemical makeup—contributes to the differences in their responses to dehydration. For seeds of any one angiosperm species, differences may also exist between the axis and cotyledons. Axes have been shown to be more sensitive than cotyledons in *Quercus robur* (Finch-Savage et al., 1992) and *Theobroma cacao* (Li and Sun, 1999), while the reverse was found for *Castanea sativa* (Leprince, Buitink, and Hoekstra, 1999). Our unpublished data show that for recalcitrant seeds from a spectrum of species, the axes are generally at a higher water content than are the storage tissues, and Pritchard and colleagues (1995) reported uneven distribution of water within the component tissues of embryos of the gymnosperm *Araucaria hunsteinii*. In a biophysical consideration of *Coffea* spp., Eira, Walters, and Caldas (1999) found that although the heats of sorption calculated for whole seeds were similar to those of orthodox seeds, at the same relative humidity (RH) heats of sorption for excised embryonic axes were intermediate between values for orthodox and recalcitrant axes. However, the situation is further complicated because, for seeds of any one species, there is also intra- and interseasonal variation!

Intraseasonal variation will be considered later, but in terms of interseason variation we have found that for ostensibly mature seeds of *Camellia sinensis*, embryonic axis water content varied from 2.0 ± 0.3 to 4.4 ± 2.4 g per g dry mass (g·g⁻¹) for harvests made in different years (Berjak et al., 1996). Many recalcitrant seeds will show visible signs of germination in storage at the water content at which they are shed. However, seeds of *Q. robur* collected in one particular year (from the same parent tree used previously and since) had water contents lower than normal and did not germinate in storage (Finch-Savage et al., 1993; Finch-Savage, 1996). Interseasonal differences in germination capacity following dormancy-breaking chilling have been recorded for *Aesculus hippocastanum* seeds, this effect being ascribed to differences in mean temperature during seed filling (Pritchard et al., 1999). Interseasonal variability is also a common feature of recalcitrant seeds from species of tropical provenance. In two recently reported examples, del Carmen Rodriguez and colleagues (2000) related effects of dehydration on germination with seasonality for neotropical rain forest species in Mexico, and differences in a variety of traits among seed lots of *Euterpe edulis* from one season to the next have also been described (Martins, Nakagawa, and Bovi, 2000).

The State of Metabolic Activity May Vary, but It Is Continuous

Familiarity with the history of any batch of recalcitrant seeds from harvest is essential in assessing their germination performance, as well as when attempting to explain the results of any manipulations carried out on those seeds (Berjak, Farrant, and Pammenter, 1989). This is because such seeds are not only hydrated, but also metabolically active. Their status is inexorably changing, and the state of development of recalcitrant seeds—both before and after harvest—influences their desiccation sensitivity. For most species of recalcitrant seeds, the least desiccation-sensitive stage occurs when the metabolic rate is at its lowest (which generally coincides with natural shedding), but they are *always* metabolically active (reviewed by Pammenter and Berjak, 1999). If germination commences rapidly after harvest, the seeds will show heightened sensitivity to water loss in a very short time. This is because as germinative metabolism progresses to the stage when mitosis and cellular expansion by vacuolation occur, the seeds must take up water from an exogenous supply—and so the minimum water level commensurate with viability retention increases (Farrant, Pammenter, and Berjak, 1986; Berjak, Farrant, and Pammenter, 1989). Enhanced desiccation sensitivity as germination progresses has been demonstrated for a variety of nonorthodox seed species, including *Coffea arabica* (Ellis and Hong, 1991), *Landolphia kirkii* (Berjak, Pammenter, and Vertucci, 1992), *Camellia sinensis* (Berjak, Vertucci, and Pammenter, 1993), *Quercus robur* (Finch-Savage, Blake, and Clay, 1996), and *Aesculus hippocastanum* (Tompsett and Pritchard, 1998).

In general, the more heightened the state of metabolism, the greater will be the desiccation sensitivity; this underlies not only the decreased tolerance to water loss as germination in storage progresses, but equally, in the earlier stages of seed ontogeny desiccation damage occurs readily. Sensitivity to water loss decreases with development in orthodox seeds; nonorthodox seeds never become desiccation tolerant in the strict sense. Decreasing desiccation sensitivity with development has been demonstrated for robusta coffee *(Coffea canephora)* (Hong and Ellis, 1995), for *Aesculus hippocastanum* (Tompsett and Pritchard, 1993), and for *Clausena lansium* and *Litchi chinensis* (Fu et al., 1994). One of the problems posed in work on recalcitrant seeds is that there are no outwards signs by which absolute seed maturity can be gauged; generally, development grades virtually imperceptibly into germination. Familiarity with individual species facilitates recognition of seed maturity, generally in terms of fruit developmental changes, but this lacks the precision conferred by shedding of orthodox seeds only after the termination of maturation drying.

Intraseasonal Differences

Certain intraseasonal effects—which are at present largely inexplicable—impose degrees of variability upon seeds of individual species. We have consistently found that the water content of ostensibly mature seeds of individual species varies depending on the stage in a season at which they are harvested (our unpublished data). Furthermore, for most species, seed-to-seed variability in the axis water content within any single harvest is significant (Berjak and Pammenter, 1997). Our other unpublished observations on recalcitrant species show that fruits produced late in a season either will abort or will not abscise, instead withering and dying while remaining attached to the parent. Also, late-season seeds are of very inferior quality, often showing extremely high fungal infection levels. Observations on *Garcinia gummi-gutta* have shown that cumulative germination values fell slightly as the seeds matured from cotyledon colors described as "light-cream" to "new-marigold," and that on further darkening (to "cherry") all germination potential was lost (Chacko and Pillai, 1997). For *Machilus kusanoi,* Chien and Lin (1997) reported that the later seeds were harvested, the greater was the rate of deterioration on dehydration.

Consideration of the range of variability presented by nonorthodox seeds underscores the difficulties in working with such seeds. No a priori assumptions can be made about either the inherent properties or the reactions that might occur in response to the imposition of any particular experimental parameter or set of parameters.

Seed Categorization—Discrete Behavioral Groups or a Continuum?

The inherent variability among seeds of the broadest range of species and their responses to dehydration and other manipulations questions whether seeds should be classed according to the discrete categories—orthodox, intermediate, and recalcitrant—or whether a more fluid basis of categorization would not be more appropriate. Although it is unquestionably convenient to be able to slot the seeds of individual species into a discrete category, this constrains investigators to use such categorizations although the seeds being described may not conform in all respects to the definitions. It is clear that there are various degrees of recalcitrant behavior—loosely described as maximal and minimal, and separated by a series of gradations (e.g., Farrant, Pammenter, and Berjak, 1988). Furthermore, it has been shown that the rate at which recalcitrant seeds lose water determines the degree of dehydration they will tolerate (Pammenter et al., 1998; Pammenter and Berjak, 1999). Similarly, seeds of all orthodox species are not equally desiccation tolerant (Walters, 1998). Intermediate storage behavior is taken to mean those seeds that are shed at relatively high water content and will withstand substantial dehydration, but not to the degree tolerated by orthodox seeds (Hong and Ellis, 1996).

In view of the extreme variability in seed postharvest behavior which has gradually emerged, it seems that many factors must be considered in any categorization scheme (Berjak and Pammenter, 1994). It has been suggested that the postharvest behavior of seeds should be considered as constituting a continuum, subtended by extreme orthodoxy at the one end and the highest degree of recalcitrance at the other, with subtle gradations between the two extremes (Berjak and Pammenter, 1997, 2001).

THE SUITE OF INTERACTING PROCESSES
AND MECHANISMS INVOLVED
IN DESICCATION TOLERANCE

Many factors have been implicated in the acquisition and maintenance of seed desiccation tolerance, and assuredly, the list is not yet complete. While individual processes and mechanisms have enjoyed and fallen from favor as *the* factor facilitating desiccation tolerance, it has become apparent that orthodox behavior is the outcome of the complete expression of a suite of interacting mechanisms and processes. As such, acquisition and maintenance of the desiccated state in seeds must be the outcome of coordinated multigenic control.

In orthodox seeds profound changes accompany the acquisition of desiccation tolerance during development and the ability to survive in the desiccated state when mature. In the case of nonorthodox seeds the expression and interaction of the factors concerned is incomplete (Pammenter and Berjak, 1999). Recalcitrant behavior is thus inevitably a product of seed development.

Intracellular Physical Characteristics

Desiccation-tolerant plant cells must withstand the physical strains that accompany the volume reduction associated with the loss of considerable portions of cellular water. The water in fluid-filled spaces is generally replaced by space-occupying insoluble material, generally protein in the vacuoles and starch and/or lipids external to the vacuole (reviewed by Vertucci and Farrant, 1995). The cytoskeleton must have the ability to dissociate in an orderly manner, and, although there is no direct evidence, it is also possible that the cell walls are plastic and can readily fold. In addition, although direct data are lacking, the chromatin must assume a conformation that will protect the integrity of the genome in the desiccated condition (Osborne and Broubiak, 1994). The nucleoskeleton, which determines nuclear architecture, must be modified in a strictly controlled manner.

These characteristics develop prior to or concomitant with maturation drying in orthodox seeds but are lacking or only partially manifested in recalcitrant seeds. Even from the relatively few studies that have been done, there appears to be a correlation between the degree of recalcitrance and the manifestation of some of these features. For example, in *Avicennia marina, Ekebergia capensis,* and *Aesculus hippocastanum* there is a decreasing degree of vacuolation in embryo cells and an decreasing sensitivity to dehydration under similar conditions (Farrant et al., 1997; Pammenter et al.,

1998). Nevertheless, *A. hippocastanum* seeds are desiccation sensitive, indicating that although protection against physical strains is necessary, it is not adequate in itself in conferring desiccation tolerance.

In parentheses, it should be noted that some of the ultrastructural "damage" observed on drying recalcitrant seeds, such as the withdrawal of the plasmalemma from the cell wall, may in fact be an artifact of fixing partially dry tissue in an aqueous medium (Wesley-Smith, 2001).

Intracellular Dedifferentiation and Metabolic "Switch Off"

Intracellular dedifferentiation accompanies the onset of maturation drying in orthodox seeds. Mitochondria and plastids lose internal structure, and endomembranes such as the rough endoplasmic reticulum (ER) become substantially reduced and the cisternae of Golgi bodies disassociated (Bain and Mercer, 1966; Klein and Pollock, 1968; Hallam, 1972). These changes imply minimization of metabolic activity, including respiration and membrane synthesis and processing. The reduction in membrane surface area also reduces the sites that would undergo substantial, often deleterious, changes upon desiccation. A further indication of metabolic "switch off" is the cessation of DNA replication and the arrest of most embryo cells in the G_1 phase (prereplication; DNA in the undoubled 2C form) with the onset of maturation drying (Brunori, 1967). Dehydrated orthodox seeds are effectively ametabolic, not simply because no water is available, but because of a controlled shutdown of activity and dismantling of structures preceding or accompanying maturation drying.

Marked dedifferentiation does not occur in any of the recalcitrant embryos of a variety of species examined to date, and respiration rates remain high. Interestingly, although the mitochondria of the recalcitrant embryos of both *Avicennia marina* and *Aesculus hippocastanum* remain highly differentiated, the mitochondria occupy a greater proportion of cell area in the more desiccation-sensitive species, *A. marina* (Farrant et al., 1997). We lack sufficient data on cell cycling in recalcitrant embryos to comment unequivocally. However, in *A. marina* seeds only the most transient cessation of DNA replication occurs, with resumption of DNA synthesis (the S phase) resulting in its entering the more vulnerable doubled state (4C) soon after shedding and the DNA of newly shed seeds is severely damaged by only a slight degree of dehydration (Boubriak et al., 2000). In the temperate recalcitrant species *Acer pseudoplatanus,* on the other hand, more than 60 percent of the embryo cells have been reported to be arrested in the 2C state, although this might be associated with the dormancy of these seeds (Finch-Savage et al., 1998).

Free Radicals, Reactive Oxygen Species, and Antioxidant Systems

Free radicals, which include reactive oxygen species (ROS), are strong oxidants and can cause, inter alia, peroxidation of membrane lipids leading to impairment of membrane structure and function. Reactive oxygen species are naturally produced during normal metabolism, but tissues contain a range of both enzymatic and nonenzymatic antioxidants which function to prevent injurious consequences of "escaped" ROS. However, if metabolism is disturbed (by, for example, dehydration) there is the potential for unregulated ROS production with deleterious consequences such as oxidation of macromolecules and membrane deterioration. To prevent damage during the early stages of maturation drying, the presence and optimal operation of antioxidants is essential.

In developing recalcitrant seeds metabolism is not programmed to be "switched off"; it continues, if not unabated, certainly at a relatively high level. It appears that although recalcitrant seeds/embryos do possess antioxidants (and there are variations among species), these may become impaired or otherwise unable to cope with the level of ROS generation accompanying slow dehydration (Hendry et al., 1992; Finch-Savage et al., 1993; Côme and Corbineau, 1996; Tommasi, Paciolla, and Arrigoni, 1999).

The Presence and Operation of Putatively Protective Molecules

Sucrose with Certain Oligosaccharides or Sugar Alcohols

Maturing orthodox seeds accumulate considerable amounts of sucrose and oligosaccharides (usually raffinose and/or stachyose) (Koster and Leopold, 1988), or sucrose coaccumulates with galactosyl cyclitols (Obendorf, 1997), depending on the species. As dehydration proceeds, these mixtures form a highly viscous supersaturated solution known as a glass. This glassy (vitrified) state can become so viscous as to curtail molecular diffusion (Williams and Leopold, 1989; Koster 1991; Leopold, Sun, and Bernal-Lugo, 1994; Bryant, Koster, and Wolfe, 2001), thus minimizing unregulated metabolism and its deleterious consequences. It is possible that the life span of mature orthodox seeds under defined storage conditions is influenced by the stability of the glassy state (Leopold, Sun, and Bernal-Lugo, 1994). Because they are metastable, glasses tend to break down, and this phenomenon may underlie the inevitable deterioration of orthodox seeds during storage.

Interestingly, the recalcitrant seeds of several species do accumulate sucrose and oligosaccharides (Farrant, Pammenter, and Berjak, 1993; Finch-Savage and Blake, 1994; Lin and Huang, 1994; Steadman, Pritchard, and Dey, 1996), which may be present in mass ratios conducive to glass formation (Horbowicz and Obendorf, 1994). However, vitrification would occur only at water contents lower than those at which recalcitrant seeds lose viability on slow drying (whether naturally after being shed or experimentally after harvest).

Late Embryogenic Accumulating/Abundant Proteins (LEAs)

Synthesis of the set of robust, hydrophilic proteins termed LEAs (dehydrin-like proteins) precedes maturation drying during orthodox seed development (Galau, Hughes, and Dure, 1986; Kermode, 1990). Convincing evidence indicates that LEAs are somehow involved in the acquisition and maintenance of desiccation tolerance in orthodox seeds, perhaps because their amphipathic nature facilitates interaction with a wide range of macromolecules and ions, thus preventing denaturation of the macromolecules under dehydrating conditions (Blackman, Obendorf, and Leopold, 1995; Close, 1997).

The situation with respect to the occurrence and possible role of LEAs in recalcitrant seeds is equivocal. Such proteins occur in recalcitrant seeds of a variety of species, from grasses to trees, from a range of habitats (Finch-Savage, Pramanik, and Bewley, 1994; Gee, Probert, and Coomber, 1994; Farrant et al., 1996). On the other hand, LEAs were conspicuously absent from recalcitrant seeds of ten tropical wetland species tested (Farrant et al., 1996). Similarly, although seed small heat-shock proteins have been implicated in desiccation tolerance, a homologous protein isolated from the cotyledons of recalcitrant *Castanea sativa* seeds obviously does not confer tolerance (Collada et al., 1997).

The evidence of the occurrence of both appropriate sugar/oligosaccharide combinations and LEAs in recalcitrant seeds underscores the contention that no one factor can be considered to be *the* key factor in either the acquisition or the maintenance of desiccation tolerance. The phenomenon must be the result of the interplay of a variety of mechanisms and processes which will surely emerge as being under multigenic control.

Amphipathic Substances

It has been suggested that endogenous amphipathic substances may partition into membrane lipid bilayers during dehydration, preventing the formation of the gel phase in the desiccated state (Hoekstra et al., 1997; Golovina, Hoekstra, and Hemminga, 1998). On rehydration, the amphipaths have been shown to partition back into the cytoplasm. Although this may be a further mechanism involved in desiccation tolerance in orthodox seeds, the status of amphipathic molecules in recalcitrant seeds has not yet been ascertained.

The Ability for Damage Repair on Rehydration

Storage of orthodox seeds at high temperatures and water contents causes damage that decreases vigor and brings about viability loss. However, before viability is reduced, the decreased vigor is manifested as an increasing time lag between seed imbibition and radicle extension. During this period, intracellular repair mechanisms become operational and repair must be effected before germination can occur (e.g., Osborne, 1983). Repair during this lag phase in orthodox seeds occurs at the level of protein macromolecules (Mudgett, Lowensen, and Clarke, 1997), membranes (Berjak and Villiers, 1972), and nucleic acids (Elder et al., 1987). In fact, the efficacy of osmopriming of low-vigor orthodox seeds is because repair processes occur while the seeds are held at water potentials that allow this metabolism but preclude germination (Bray, 1995).

There are very few studies of repair by damaged recalcitrant seeds. However, following rehydration of the highly recalcitrant seeds of *Avicennia marina*, no DNA repair is possible once 22 percent of the originally present water has been lost, suggesting a very inadequate DNA repair system compared with orthodox seeds (Boubriak et al., 2000). In terms of free-radical scavenging processes, evidence suggests that antioxidant systems fail during dehydration of desiccation-sensitive seeds and seedlings and are assumed to remain ineffective on rehydration. It appears that the repair mechanisms of recalcitrant seeds are as sensitive to water loss as all other processes.

The mechanisms and processes outlined constitute some of a suite of protective mechanisms, probably all of which must be present, that act together to confer tolerance to dehydration and the ability to survive for extended periods in the dry state. However, the list is probably by no means complete, with essential developmental phenomena remaining to be identified. In seeds that are not orthodox the features are represented to differing extents, and some may not be present at all. This may be the underlying cause of the differing degrees of recalcitrance observed among species.

DRYING RATE AND CAUSES OF DAMAGE
IN RECALCITRANT SEEDS

It is now well established that the response of recalcitrant seeds, or axes excised from seeds, to drying depends on the rate at which water is lost (Normah, Chin, and Hor, 1986; Pammenter, Vertucci, and Berjak, 1991; Pammenter et al., 1998; Pritchard, 1991; Kundu and Kachari, 2000; Potts and Lumpkin, 2000). Although in a few exceptions drying rates intermediate between slow and rapid appear to favor survival to relatively low water contents (e.g., cacao, Liang and Sun, 2000; *Warburgia salutaris,* Kioko et al., 1999), seeds or axes that are dried very rapidly can survive to the lowest water contents, most probably because insufficient time is allowed for the accumulation of damage that occurs when the material is dried slowly. However, no matter how fast the water loss, recalcitrant material cannot be dried to as low a water content as orthodox seeds; there is an absolute lower limit below which recalcitrant seeds will not survive. These data have been interpreted as suggesting at least two types of damage can occur on drying desiccation-sensitive seeds. At higher water contents (above the lower limit) aqueous-based degradative oxidative processes, initiated because of disturbance of ongoing metabolism, lead to the accumulation of damage. This kills the seeds if drying is slow, but if dehydration is sufficiently rapid this damage does not accumulate to lethal levels. However, if material is subjected to rapid drying, at lower water contents (below the lower limit) direct damage consequent upon removing water from membrane and macromolecular surfaces occurs virtually instantaneously and rapidly kills the tissue. These types of damage have been referred to as "metabolism-derived damage" and "desiccation damage *sensu stricto,*" respectively (Pammenter et al., 1998; Walters et al., 2001).

Whatever the details of the primary events initiating metabolism-linked damage as a consequence of water stress, opinion generally favors oxidative processes to be a major cause of lethal damage. In particular, evidence that uncontrolled generation of reactive oxygen species occurs, leading ultimately to peroxidation, has been linked to observable or measurable membrane damage and cell death (Hendry et al., 1992; Finch-Savage et al., 1993; Finch-Savage, Blake, and Clay, 1996; Chaitanya and Naithani, 1994; Li and Sun, 1999; Leprince et al., 2000). During dehydration of the highly recalcitrant seeds of *Avicennia marina,* although data indicate responses typical of oxidative stress, these seem to be eclipsed by catastrophic physical damage (Greggains et al., 2001).

There are marked differences in the desiccation sensitivity of recalcitrant seeds of different species, when dehydrated under identical conditions, which are due to inherent properties of the seeds themselves (e.g., Farrant, Pammenter, and Berjak, 1989). However, the drying rate (Pammenter, Vertucci, and Berjak, 1991; Pammenter et al., 1998; Berjak, Vertucci, and Pammenter, 1993), and probably also the temperature under which dehydration occurs (Kovach and Bradford, 1992; Vertucci et al., 1995; Ntuli et al., 1997) and the maturation status of the seeds (Berjak, Pammenter, and Vertucci, 1992; Berjak, Vertucci, and Pammenter, 1993; Vertucci et al., 1994, 1995) are related to the water content at which damage occurs or is lethal. It is thus obvious that one cannot define *unqualified* "critical water contents" at which viability will be lost without specifying parameters relating both to the seeds/axes and the experimental conditions (Pammenter et al., 1998; Pammenter and Berjak, 1999).

The Practicalities of Handling Recalcitrant and Other Nonorthodox Seeds

The natural tendency among those who harvest seeds has long been to spread these out to dry before placing them into whatever storage facility is locally used. When sophisticated drying rooms are not available, seed drying is often done by equilibration with the ambient RH, generally in the shade. This has been the common practice, as is described, for example, by Kioko, Albrecht, and Uncovsky (1993) for seeds from pulpy fruits in Kenya. Inevitably seeds not (immediately) recognized as being recalcitrant sustain lethal damage during this very slow method of drying, resulting in complete non-availability of planting stocks. It is highly probable that as a result of such practices worldwide, species producing recalcitrant seeds would have been propagated vegetatively whenever possible. For example, propagation by layering in the humid tropics might have arisen as an almost natural consequence of plant interactions with the environment for trees such as *Mangifera indica* (mango), *Litchi chinensis* (lychee), *Durio zibethinus* (durian), *Nephelium lappaceum* (rambutan), *Euphoria longan* (longan), and *Artocarpus altilis* (breadfruit) (Hartmann et al., 1997). The case of breadfruit is historically interesting for another reason, but also to do with the short-lived, recalcitrant seeds: one of the contributing factors to the mutiny on the Bounty was that established young plants of this species, which were being carried from Tahiti for cultivation in the British West Indies, deprived the sailors of a significant proportion of the ship's water supply!

Almost without exception, the crop species used in agriculture produce orthodox seeds. It is tempting to speculate that the common crop species cultivated worldwide were domesticated because they were useful, and the fact that the seeds could be stored greatly facilitated their domestication.

This would have been important not only for maintaining planting stocks from one season to the next but also because of the maintenance of quality of the seeds as food and feed. Furthermore, most of the domesticated crops cultivated globally have temperate origins, where species producing nonorthodox seeds are uncommon. One exception is *Zizania palustris*, North American wild rice, which is a commercially grown, aquatic species that produces seeds recognized to be recalcitrant (Probert and Brierley, 1989; Probert and Longley, 1989; Vertucci et al., 1995). Under natural conditions, however, the deeply dormant caryopses are shed into water, where they overwinter and cannot dry out. Nevertheless, long-term storage of the submerged seeds at low temperatures, which essentially constitutes dormancy-breaking stratification, is not an option, and thus developing methods for conservation of the seeds or excised embryonic axes of *Zizania* spp. has been the subject of considerable research. Those investigations considered various aspects of the responses to water loss, which were found to differ physiologically, ultrastructurally, and biochemically depending on the temperature during dehydration (Kovach and Bradford, 1992; Berjak et al., 1994; Ntuli et al., 1997), with the deleterious effects being increasingly severe with declining temperature. The most telling studies on *Zizania* spp., however, were carried out by Vertucci and colleagues (1995) who reported that although critical water contents for desiccation damage under defined dehydration conditions varied with developmental status and temperature, all equated to a common water *activity* value (a_w) of 0.90. Those authors, in presenting a model interrelating this water activity value with water content and temperature, proposed that optimum storage conditions could be predicted for caryopses of *Zizania* spp. from different populations, and that long-term conservation should be possible at −20°C. Since then, Touchell and Walters (2000) have employed the optimal drying conditions elucidated for *Zizania palustris* embryos and have achieved success in their cryopreservation.

Caryopses of *Zizania* spp. probably proved amenable to cultivation in the first place because of their dormancy and aquatic habitat, and there are other dormant, recalcitrant seeds (e.g., *Wasabia japonica,* Nakamura and Sathiyamoorthy, 1990; *Aesculus hippocastanum,* Pritchard et al., 1999), although away from the temperate regions, dormancy seldom seems to coexist with recalcitrance. Some informally grown "vegetable" species in the tropics produce recalcitrant seeds. One such example is *Telfairia occidentalis,* an annual cucurbit native to tropical Africa, which provides a popular leaf and stem vegetable (Akoroda, 1986). However, as the seeds are edible and a source of oil, as well as being extremely short-lived when dehydrated, planting stock is scarcely available. Successful cultivation of *T. occidentalis* and species producing similar seeds will probably rely heavily on methods of vegetative propagation and/or germplasm cryopreservation, if these can be developed.

Whether one is considering the tropics or the temperate zones, the phenomenon of seed recalcitrance (or indeed, of any degree of nonorthodox behavior) is predominantly a characteristic of tree species. Hence the handling of such seeds is more properly in the province of silviculture or horticulture than of agriculture—although it needs to be borne in mind that thousands of thousands of species of unknown seed behavior might prove immensely valuable as cultivated crops in the future.

Harvest and Transport Criteria

Because recalcitrant seeds are hydrated and metabolically active, it is vital that they are harvested in the best possible condition that will optimize not only any storage period that might be necessitated, but also germination performance if planted immediately. Unless such seeds are enclosed in fleshy or impermeable fruits, they are liable to start losing water as soon as they are shed. This necessitates daily harvests throughout the season. Even when immediate dehydration does not occur, the physiological status of the seeds changes, more or less rapidly depending on the species, as a consequence of their ongoing development or progress into germination, which in turn influences their desiccation sensitivity (e.g., Farrant, Pammenter, and Berjak, 1986; Berjak, Farrant, and Pammenter, 1989; Berjak, Pammenter, and Vertucci, 1992; Berjak, Vertucci, and Pammenter, 1993; Finch-Savage, Blake, and Clay, 1996; Tompsett and Pritchard, 1998), and this is also the case for seeds categorized as intermediate (e.g., Ellis and Hong, 1991). An additional, very significant problem in collecting fruits or seeds from the ground is that they will have been exposed to a spectrum of microorganisms in addition to any they already harbored before being shed. For example, in the case of acorns, their infection by the aggressive pathogen *Sclerotinia pseudotuberosa* Rehm. *(Ciboria batschiana)* occurs only as a consequence of their contact with the ground (Kehr and Schroeder, 1996). For all these reasons, it is preferable to harvest fruits or seeds directly from the parent plant. However, this practice is not without other problems. Because recalcitrant and other nonorthodox, seeds do not undergo maturation drying, there are few clear indications of the state of maturity of the seeds. Outward criteria such as fruit color or, e.g., in the case of *Avicennia marina,* how readily the structure will detach from the peduncle are used, but these are approximate rather than definite indicators of seed maturity. Intra- and interseasonal variability in seed properties add to the difficulties of predicting their postharvest responses. Nevertheless, harvesting directly from the parent plant remains the procedure of choice, as this practice avoids the more serious problems of seed dehydration, increasing desiccation sensitivity, and further infection.

For these same reasons, it is essential that the seeds are transported under optimal conditions to the central facility, whether this be a repository or a laboratory. The means and duration of transport will determine what precautions need to be taken for nonorthodox seeds. If it is merely a matter of transporting the seeds rapidly over a short distance, then it is sufficient to enclose them in plastic bags or other containers that will not permit water loss. However, when long intervals intervene between collection and delivery of seeds, then further precautions are necessary. When possible, transport of seeds within the fruits is best, despite the fact that this necessitates greater volumes and weights of the consignments. If this is impossible, then the seeds themselves need to be treated with fungicide before packaging, to minimize fungal proliferation in the high humidity conditions of the containers necessary to prevent seed water loss. However, as surface applications of fungicide are not effective in curtailing the activity of mycelium below the pericarp/testa, recourse to seed treatments with systemic fungicides (our unpublished data) may be necessary. Obviously, however, the efficacy of such fungicides in curtailing proliferation of the specific fungi involved, as well as prior establishment that they cause no seed damage, is necessary. The effects of the seed-associated fungi need to be eliminated, or at least minimized, for two major reasons, the first being the obvious deterioration of the seeds by fungal degradation and toxin production. The second point is that fungal respiration produces metabolic water; thus, even if the fungi themselves are relatively benign, the seeds are provided with an additional source of water and consequently are likely to become more metabolically active. This, in turn, could result in earlier radicle emergence than would otherwise occur in hydrated storage, rendering the resultant seedlings useless for storage and of dubious value as planting stock.

The fact that recalcitrant and other nonorthodox seeds types are actively metabolic and require transportation in closed containers to minimize water loss imposes the problem of the storage atmosphere becoming anoxic (Smith, 1995). To counteract this, periodically containers need to be opened briefly and the seeds mixed around. As an alternative to closed containers, transport of recalcitrant seeds may be in moist medium, as described by Kioko, Albrecht, and Uncovsky (1993). Those authors advocate the use of sawdust, peat, vermiculite, or sand as providing a suitable medium, which also has been the practice in Kenya for moist storage of such seeds. Use of moist-medium packaging for transport does, however, lend further bulk and weight to the consignment. Nevertheless, convenience may have to be forfeited, as the objective is to transport the seeds under the best possible conditions for vigor and viability retention, as well as to minimize any (additional) infection. Although more difficult to achieve, temperature during transport of recalcitrant seeds should be as low as can be managed, but not low enough to damage those that are chilling sensitive.

If transport of recalcitrant seeds over long distances—whether enclosed within the fruits or not—is to be most efficient, then the time factor needs to be minimized. Thus, at least for experimental purposes, consignment by air freight is generally used. Under these conditions additional precautions need to be taken. First, one needs to be sure that the fruits or seeds are stowed in the temperature- and pressure-controlled hold of the aircraft, often loosely described as the "live animal" hold. This is because at high altitudes hold temperature can be so low that freezing damage to the seeds might occur, especially if the flight duration is several hours. Second, it is imperative that the packages are not labelled as "perishable material," as this inevitably will result in the consignment being held under refrigerated conditions, especially prior to the flight and on receipt at the destination. Third is the matter of the international rules governing plant and seed importation and quarantine if the material is to be sent from one country to another. It is essential that the appropriate documentation accompanies the consignment, which means that the receiver must have arranged for an importation permit that must be valid for the date of receipt of the fruits or seeds and that the exporter has provided the necessary phytosanitary certificate for the species concerned. Although it may seem simple in theory to fulfill these conditions, in practice there are frequently difficulties, with the loss of precious material as a result of a inappropriate handling or substantial delays. We have found the cost of employing the services of a reliable international firm of couriers to be well worthwhile in minimizing the trouble and losses that can otherwise be incurred.

In some cases—particularly when seeds are very short-lived and/or collecting missions are protracted and transport difficulties acute—in vitro collecting techniques can be employed (Pence, 1996; Engelmann, 1997). Englemann (1997) describes in vitro field collection of coconut and cacao to involve extraction of embryonic axes (in a plug of endosperm for these species), surface sterilization of plugs with calcium hypochlorite or commercial bleach, dissection out of the embryos under a makeshift "hood," and their inoculation onto semisolid medium. As an alternative, Engelmann cites transport of the endosperm plugs to the laboratory before embryo excision and notes the efficiency of such in vitro field collecting if suitable precautions are taken. In these cases, however, whether the material is to be stored or not, germination followed by initial onward growth needs to be under in vitro conditions, as will be discussed as follows. It should be noted that similar field collection technology applied to vegetative propagatory material can be very successful.

Short- to Medium-Term Storage

To be successful, any storage regime must ensure that seeds retain unimpaired (or only minimally impaired) vigor and viability for a practically useful time period, which means from harvest until they are required for planting. In these terms, storage of recalcitrant seeds in the short to medium term is fraught with difficulties. Examination of the data collected by King and Roberts (1980) shows that some two decades ago moist-stored recalcitrant seeds were recorded as retaining viability for periods from days to weeks for tropical types and months to, at most, two and one-half years, for temperate species, with lowered temperature being an important parameter particularly for the latter. With few exceptions, for those species for which viability was recorded it was not high—and certainly would not satisfy the stringent requirements of the international seed trade. In addition, little was recorded regarding the vigor of the surviving seeds. Seeds of *Q. robur* were recorded from data collated by King and Roberts (1980) as being among the longest surviving of temperate recalcitrant types. However, it has been our experience (unpublished data) that *Q. robur* seeds obtained from an impeccable source after they had been stored for a few months, although still viable, were extremely debilitated and almost all were fungally infected.

In general, some progress has recently been made in extending the storage longevity of recalcitrant seeds, and it has emerged that not only their initial quality but also the prestorage manipulation will have a marked influence on the success of their short- to medium-term storage. An a priori requirement for viability retention of stored recalcitrant seeds is the maintenance of high tissue water content, as was discussed previously. This is generally achieved in a variety of ways. The Kenya Forestry Research Institute (KEFRI) advocates the use of moisture-retaining packing media (as for seed transport), with sand noted to be the least effective because of its porosity (Kioko, Albrecht, and Uncovsky, 1993). Twice the volume of peat, sawdust, vermiculite, or sand to seeds is used, moistened with distilled or deionized water, thus lessening the chances of microorganisms being introduced, and storage temperature is kept as low as will not cause chilling damage. There are, however, problems even with this relatively straightforward procedure, including the necessity of keeping the seeds aerated by turning the mixture regularly, which could cause mechanical damage and must increase the chances of microorganisms being introduced. In addition, the periodic remoistening of the packing medium may be necessary, with the amount of water used being carefully controlled, otherwise the seed water content will increase with the concomitant speeding up of the germination processes. In this regard, for storage of *Camellia sinensis* seeds Bhattacharya, Rahman, and Basu (1994) have suggested use of moist sand incubation with polyethylene glycol (PEG) to maintain the water potential at –0.5 MPa, together with the use of 0.05 percent mercuric chloride.

The storage containers used in Kenya vary from glass to plastic jars which may be rubber lined, with tightly closing lids, or heavy-gauge polythene bags (Kioko, Albrecht, and Uncovsky, 1993). Those authors record nonorthodox seeds of a spectrum of chilling-*tolerant* species to survive for up to a year at 1 to 4°C when stored in moist medium. Other investigators have reported using similar means for short- to medium-term storage of recalcitrant seeds. For example, high-quality seeds of *Inga uruguensis* harvested directly from the tree maintained 80 percent viability for up to 80 d when cold stored in moist vermiculite, but germinated in storage within 20 to 30 d when maintained at ambient temperature (Barbedo and Cicero, 2000). Those authors reported that inclusion of abscisic acid (ABA) in the moistening water improved the storage capacity, especially of immature seeds. The success of ABA application in extending storage longevity of recalcitrant seeds is not likely to be universal, as this will depend not only on maturity status but also on whether seeds of particular species are responsive to this growth regulator. In the case of the highly recalcitrant, chilling-sensitive (our unpublished data) seeds of *Avicennia marina,* inclusion of ABA in encapsulating gel designed to take the place of the pericarp had no beneficial effects in extending storage longevity in addition to those afforded by the gel alone (Pammenter, Motete, and Berjak, 1997).

Other authors, too, have reported marked success in extending storage longevity of recalcitrant or otherwise nonorthodox seeds, even those that are chilling sensitive. For *Aporusa lindleyana,* Kumar, Thomas, and Pushpangadan (1996) report that seeds that lose viability within a few days in slowly dehydrating conditions retained better than 90 percent viability after one year of storage in airtight polycarbonate bottles maintained constantly at 30°C or at a range between 20/30°C, although time taken to germinate increased by an order of magnitude, indicating a substantial loss of vigor. However, the data of Kumar, Thomas, and Pushpangadan (1996) indicate that the *A. lindleyana* seeds selected for storage must have been of high quality at the outset.

Achievement of a useful storage period for hydrated, recalcitrant seeds (wet or hydrated storage) depends critically on the quality of the seeds at harvest, which includes their infection status. In terms of inherent seed quality, *Trichilia dregeana* seeds of poor quality (as a result of a presumed heat stress after shedding) when wet stored declined in viability from 100 percent to about 20 percent over three weeks at 16°C, and to 0 percent within two weeks at 25°C (Drew, Pammenter, and Berjak, 2000). In contrast, high-quality, uninfected seeds of the same species have retained viability for eight months and longer (our unpublished observations).

In our laboratory, where seeds are stored for experimental purposes, we generally use plastic buckets with sealing lids. The clean containers are sterilized with sodium hypochlorite before use, after which distilled water to a depth of about 10 to 20 mm is introduced. A wick of sterile paper toweling is used to line the lower section of the bucket wall, in order to achieve and retain a saturated atmosphere. The size of the bucket is chosen to complement the number and dimensions of the seeds, which are placed (usually in a monolayer) on a grid suspended over the water in the base of the bucket. The seeds themselves will have been surface sterilized before storage, blotted dry with sterile paper towel, and usually dusted with a benomyl-based fungicide. Our recent (as yet unpublished) studies have indicated that an additional prestorage step of great value to storage of the seeds of several species is treatment with a systemic fungicide "cocktail." As an alternative to use of the buckets, similarly pretreated seeds of some species have been found to survive well in polythene bags. This has the advantage of a far more efficient use of space in the seed store, as bulky buckets take up a lot of space. The storage temperatures used in our laboratory are either $16 \pm 2°C$, for seeds suspected to be chilling sensitive, or $6 \pm 2°C$. Interestingly, although Smith (1995) cautions against allowing anoxic conditions to build up, we have seldom encountered problems with hydrated seeds stored in polythene bags, despite the limited airspace.

Whatever the precautions taken for hydrated storage of recalcitrant seeds, retention of vigor and viability for the longest period possible (which varies greatly depending on the species) depends critically not only on initial seed quality, but also the elimination—or at least the minimization—of the associated mycoflora. In early work on *Avicennia marina* seeds, it was noted that as the intact propagules became increasingly dehydrated, they became more and more resistant to water uptake when reimbibed (Berjak, Dini, and Pammenter, 1984). We later became aware that it was not that the large embryo would not take up water, but that a thick, interwoven mat of fungal mycelium—which after desiccation appeared to have become hydrophobic—had insidiously replaced the pericarp. In an attempt to eliminate the source of peripheral inoculum, we removed the pericarp from fresh seeds and, to retain tissue water, encapsulated them in a crude, slightly acidic, potassium alginate gel. This study, aimed at investigating aspects of metabolism and ultrastructure during hydrated storage of *A. marina* seeds, produced surprising results in the highly significant extension of their storage longevity (Pammenter, Motete, and Berjak, 1997; Motete et al., 1997).

Although gel encapsulation did not appear to offer any metabolic advantages to the seeds, compared with the decoated condition (pericarp removed, but not encapsulated), the efficacy of the gel as a fungistatic treatment suggested itself. The beneficial effects of similar gel encapsulation on storage longevity of recalcitrant seeds of several species have, however, been equivocal; possibly this is because, as the preparation is a crude extract, its properties are not consistent. Nevertheless, our current studies have shown the freshly prepared alginate gel to be strongly fungicidal and, at least in the case of seeds of *A. marina, Artocarpus integer,* and *Trichilia dregeana,* to extend hydrated storage life span.

Work on *A. marina* seeds involving an aerosol application of fungicide to seeds after pericarp removal and surface sterilization and periodically during storage has shown clearly that when the effects of the mycoflora are curtailed in this way, the longevity of the hydrated seeds is significantly extended (Calistru et al., 2000). Coupled with this were the observations that subcellular integrity was maintained while deteriorative ultrastructural changes within the tissues were associated with the presence of fungal structures on the surfaces of cotyledons and axis in seeds that had been inoculated after surface sterilization. As a result of these observations, and on the basis that all recalcitrant seeds are metabolically active and those of many species can be considered to be developing seedlings, work is now proceeding on the evaluation of systemic fungicidal cocktails to be used as pretreatments before hydrated storage of recalcitrant seeds. The unpublished results show for several species that storage longevity can be significantly extended. However, it would be entirely incorrect to assume that elimination of the effects of seed-associated fungi would confer indefinite storability on recalcitrant and other nonorthodox propagules. Antifungal treatments appear to make possible considerably more effective periods of short- to medium-term storage, but ultimately it is the demands of ongoing germinative metabolism in the absence of additional water that will curtail longevity (Pammenter et al., 1994). There is evidence that this is the result of water-stress-related generation of highly destructive reactive oxygen species which will rapidly bring about deterioration and death of the seeds in hydrated storage (Hendry et al., 1992; Finch-Savage et al., 1993; Finch-Savage, Blake, and Clay, 1996; Chaitanya and Naithani, 1994; Li and Sun, 1999; Leprince et al., 2000).

Clearly, if metabolic rate can be kept to a minimum by lowered storage temperatures, the timing of the onset of the germination phase that demands additional water can also be postponed. This is the basis of storage at the lowest temperature that recalcitrant seeds of individual species will tolerate. For example, Suszka and Tylkowski (1980) reported considerably increased storage longevity of *Quercus robur* seeds at around 0°C, and Pritchard and colleagues (1995) have shown significant extension of viability of *Araucaria hunsteinii* seeds in cold storage where temperatures were sufficiently low to preclude radicle emergence. However, use of lowered temperatures that would be effective is precluded, especially in the case of some tropical species, e.g., *Theobroma cacao, Nephelium lappaceum, Hopea* spp., and others (King and Roberts, 1980). This is vividly illustrated in the case of *Hevea brasiliensis* seeds, for which Normah and Chin (1991) found that deterioration was slowest at 27°C and most rapid at 10°C. Another approach that has been periodically suggested for prolonging storage life span of recalcitrant seeds is to lower the water content to a point that might curtail microorganism proliferation and would limit metabolism to the point of precluding the progress of germination (e.g., Hong and Ellis, 1996); this is termed "subimbibed" storage. The water content of axes and storage tissues of recalcitrant seeds at shedding is variable both inter- and intraseasonally, as discussed previously, but is presumably correlated with the physiological status of the seeds at that time. If improved storage longevity is observed for seeds of particular species at water contents that are somewhat lower than might be expected, this could well be a function of the combination of physiological status and water content at harvest not yet having reached its lowest level.

Nevertheless, King and Roberts (1980) recorded that although subimbibed storage had been used for seeds of a variety of truly recalcitrant and other nonorthodox species, this did not seem to be effective even in the relatively short term. Since then, data have further indicated that partial drying has adverse consequences for seeds of tropical species (King and Roberts, 1982; Corbineau and Côme, 1986a,b, 1988; Xia, Chen, and Fu, 1992) as well as those of temperate origin (Tompsett and Pritchard, 1993; Pritchard et al., 1995). Drew, Pammenter, and Berjak (2000), working with *T. dregeana* showed that for seeds of this species, lowering the water content by less than 15 percent was associated with gross ultrastructural deterioration that occurred far more rapidly than in seeds stored in parallel at the original water content. Furthermore, fungal proliferation posed a considerably greater problem on and in the subimbibed *T. dregeana* seeds. The deleterious effects of subimbibed storage are interpreted in terms of our current view that the time for which recalcitrant seeds are held at unfavorable water contents, even if these are relatively high, greatly exacerbates metabolism-linked damage (Pammenter et al., 1998; Walters et al., 2001).

There is no doubt that if only good-quality seeds are selected for storage, and if these can be suitably treated to curtail—or even eliminate—the internal as well as the surface-associated microflora, then hydrated storage in scrupulously clean containers held at the lowest temperature commensurate with vigor and viability retention will achieve the longest possible useful life span. It is, however, essential that the stored seeds be monitored periodically, so that any which show signs of deterioration, fungal growth, or visible germination (i.e., radicle protrusion) can be removed from the container. At any signs of fungal proliferation, it is advisable to retreat the remaining seeds. However, it must be appreciated that hydrated storage offers only an interim measure to conserve planting resources of recalcitrant seeds.

Seedling Slow Growth As a Means of Storage

Although traditionally slow growth storage is used for material cultured in vitro, including embryogenic callus, plantlets, shoot apices, etc. (Engelmann, 1997), there is also the possibility of maintaining seedlings under growth-restricting conditions, instead of contending with the difficulties inherent in short- to medium-term storage of recalcitrant or other nonorthodox seeds (Chin, 1996). Chin records survival for nine months of seedlings of *Shorea* sp., maintained at 16°C and 70 percent RH, with an 8 h low-light photoperiod, while those of *Calamus* sp. (rattan palm) could be maintained for two years. As suggested by Chin (1996), seedling slow growth in shaded conditions under natural canopies is an inexpensive alternative, based on conditions in forests where light limitation restricts seedling growth until a gap occurs.

Cryostorage

On the basis that hydrated storage of intact seeds is essentially only a short-term option, other methodology must be employed to ensure the conservation of the genetic resources of species producing recalcitrant—or other nonorthodox—seeds. From our current viewpoint, it appears that cryostorage is the only option. This involves storage of the material at temperatures between –80 and –196°C, using ultra-deep-freeze facilities or liquid nitrogen, respectively. However, to cryostore hydrated, metabolically active material in such a way that lethal damage is obviated requires considerable manipulation both before and after the storage phase.

It is obvious that the intact seed would be the best form of the germplasm to cryostore, as theoretically it should be merely a matter of thawing the propagules after storage and planting them out. However, it is impossible to freeze seeds at the high water contents characteristic of the newly shed or harvested condition, as cooling would be slow and accompanied by lethal ice formation. In addition, the great majority of recalcitrant seeds cannot be dried sufficiently rapidly to maintain viability at the low water contents that are required to prevent ice crystal formation on freezing. Even if this were possible, they are generally just too large to cool (freeze) successfully. However, cryostorage of intact seeds has been achieved for isolated species where the propagules are small enough and dry sufficiently rapidly to employ this technological approach. This is the case for *Azadirachta indica* (neem), a species for which seed postharvest behavior appears to vary from orthodox through intermediate to recalcitrant, perhaps as a function of provenance (Chaudhury and Chandel, 1991; Berjak and Dumet, 1996).

Neem seeds, previously shown to be chilling sensitive and to exhibit nonorthodox (tending to recalcitrant) behavior (Berjak et al., 1995) were able to be dried to very low water contents within 1 to 2 d when maintained on a layer of activated silica gel (Berjak and Dumet, 1996). The exo- and mesocarp had been removed prior to desiccation, as retention of these coverings was associated with lethally slow dehydration. After cooling at an intermediate rate in cryotubes plunged into liquid nitrogen, 70 to 75 percent of the endocarp-enclosed seeds had retained viability when sampled over the four months that they were stored in this cryogen (Berjak and Dumet, 1996). Whole seeds of *Wasabia japonica* have also survived cryopreservation in liquid nitrogen after rapid dehydration to low water content (Potts and Lumpkin, 2000), and, similarly, those authors found that slow dehydration was deleterious. Hu, Guo, and Shi (1993) cryopreserved seeds of tea *(Camellia sinensis)* after dehydration to low water content and documented that equilibration on a low-moisture-content sand bed after retrieval from liquid nitrogen favored germination. In the studies described the seeds survived desiccation to 0.16 g·g⁻¹ *(Azadirachta indica)*, 0.17 g·g⁻¹ *(Wasabia japonica)*, and 0.16 g·g⁻¹ *(Camellia sinensis)*. As has been shown by Wesley-Smith et al. (1992), it is only at such low water contents that the specimens will survive the relatively slow cooling in liquid nitrogen achieved for the seeds in the studies outlined. This presupposes that the tissues of nonorthodox seeds of individual species will withstand such drastic dehydration and retain viability at ambient temperature (i.e., before introduction into liquid nitrogen) even for very short periods. It is also significant that two of these species, *W. japonica* and *C. sinensis,* are not of tropical provenance, as it seems that generally temperate provenance is correlated with the ability of seeds/seed organs to withstand—even temporarily—such extreme dehydration.

However, this generalization is not without exception. Kioko and colleagues (1999) dehydrated seeds of the tropical species *Warburgia salutaris* relatively rapidly to about 0.1 g·g⁻¹. In that study, 30 percent of the seeds survived cryostorage after relatively slow cooling in liquid nitrogen, and in subsequent experiments (unpublished), healthy saplings have been produced from 65 percent of cryostored seeds of the related species *W. ugandensis.* Slow dehydration is invariably lethal for seeds of *Warburgia* spp.,

. thus providing an additional example of the practical value of rapid drying in retaining viability of nonorthodox seeds to low water contents. Viability retention at low water contents of seeds of *Warburgia* spp. is, however, only transient, and these seeds will die within two weeks if maintained at ambient temperature or in cold storage (unpublished data). Their survival of rapid dehydration, however, provides viable material for cryostorage.

In the studies cited previously the seeds shared two characteristics: (1) they tolerated desiccation to water contents that would be lethal for most recalcitrant seeds, however rapidly they could be dried, and (2) they were relatively small. Large seeds—the norm for most recalcitrant species—cannot be rapidly dehydrated, simply because of their size. The solution to the problem of providing a suitable form of the germplasm for cryopreservation for most large-seeded species is the use of excised embryonic axes (as originally demonstrated by Normah, Chin, and Hor, 1986) which can be dehydrated extremely rapidly by the process of flash drying (Berjak et al., 1990). Zygotic embryonic axes offer the same advantages to conservation of genetic diversity as do intact seeds, but their use involves several procedures, all of which can cause problems in achieving the ultimate objective, the production of normal, vigorous plants (e.g., Berjak et al., 1996; Engelmann, 1997; Dumet, Berjak, and Engelmann, 1997). However, these can be overcome in most cases by careful experimentation—presently on a species-specific basis.

Excision of embryonic axes removes them from the stored nutrients vital for their ongoing development, imposing the necessity of germination in vitro on a medium supplying a carbon source and other factors. It is imperative that appropriate in vitro conditions be established as the first step in developing a cryopreservation protocol for axes of any species. Because of the often short fruiting season and the fact that the seeds cannot be stored effectively, frequently a season is lost before any further trials can be carried out. A major problem with germination of excised axes in vitro is that associated microorganisms (usually fungi) must be successfully eliminated, as these will proliferate swiftly and vigorously while the explant is still recovering from excision and other manipulations essential for cryostorage.

Fungi pose a major problem in tissue culture irrespective of the origin of the explants, thus their removal—and exploration of the use of antibiotics to eliminate any bacterial contaminants if necessary—are a priori requirements for work with zygotic axes. The same is true for the production of somatic embryos or culture of shoot apices, which offer alternate genetic resources for conservation by cryopreservation. When zygotic axes are used as explants, it is imperative that they are surface sterilized after excision from the seeds, which have usually been similarly treated themselves. As surface sterilants kill microorganisms, they are potentially injurious to the excised axes as well; therefore, the least injurious but effective treatment must be identified on a species-specific basis. In general, however, the axes of most species can withstand immersion for 10 to 15 min in a 1 percent sodium hypochlorite solution containing a few drops of a wetting agent such as Tween 20 (Dumet, Berjak, and Engelmann, 1997). Those authors have also reviewed cases in which, because sodium (or calcium) hypochlorite is too harsh, ethanol or a 0.1 percent solution of mercuric chloride has been employed. In some isolated instances (e.g., *Warburgia* spp., unpublished observations) all surface sterilants have proved so injurious, that use of the axes as explants is precluded. In such cases, except in isolated instances in which whole seeds can be cryopreserved (e.g., *Warburgia* spp.), use of explants alternative to zygotic axes, particularly somatic embryos or shoot apices, generally is necessary. An additional problem posed by the seed-associated fungi is that the inoculum may be internal and, in such cases, surface sterilization alone will be ineffective. We are presently experimenting with the use of solutions of systemic fungicides, on the basis that recalcitrant seeds, being hydrated and metabolic, are in many cases more like seedlings and should assimilate these without deleterious consequences.

The water content of axes excised from fresh seeds is generally far too high for cryopreservation without lethal ice formation during the cooling step. This is at least partly because the wetter the tissues, the slower the cooling rate and thus the longer the time for passage of the axis through the temperature range that facilitates ice crystallization (Wesley-Smith, Walters, et al., 2001). Thus the axes need to be dehydrated as rapidly as possible to water contents that will not cause dehydration damage but will facilitate noninjurious freezing. This was first achieved by placing the excised axes in a laminar flow cabinet (*Hevea brasiliensis,* Normah, Chin, and Hor, 1986; *Araucaria hunsteinii,* Pritchard and Prendergast, 1986) and has been employed since (e.g., *Camellia sinensis,* Chaudhury, Radhamani, and Chandel, 1991, *Cocos nucifera,* Assy-Bah and Engelmann, 1992; *Quercus faginea,* Gonzales-Benito and Perez-Ruiz, 1992; *Q. robur,* Poulsen, 1992). However, generally more rapid and thus effective dehydration is obtained by flash drying (Berjak et al., 1990) using a relatively simple apparatus in which a flow of air passes through a grid from below, with the axes positioned on the grid. The air is vented through holes in the lid of the small-volume container. Since then, the flash drying apparatus has been improved, with the use of a small fan and the introduction of activated silica gel (Wesley-Smith, Pammenter, et al., 2001). Flash drying achieves, in a matter of minutes to an hour or two, water contents that will facilitate rapid cooling (freezing).

Wesley-Smith and colleagues (Wesley-Smith et al., 1992; Wesley-Smith, Pammenter, et al., 2001) have presented data for *Camellia sinensis* and

Aesculus hippocastanum showing that the more rapidly axes are cooled, the higher is the water content at which they can be successfully frozen, and similar results have been obtained for *Quercus robur* axes (Berjak et al., 1999). The success of rapid freezing is explained as the minimization of the time that the axes spend in the temperature range in which ice crystallization occurs during cooling to the temperature of the cryogen. The general means of achieving this is by rapid introduction of unenclosed (naked) axes into subcooled liquid nitrogen (–210°C), although cooling enclosed axes at somewhat lower rates has been reported as successful for *Hevea brasiliensis* (Normah, Chin, and Hor, 1986), *Cocos nucifera* (Assy-Bah and Engelmann, 1992), and *Coffea* spp. (Abdelnour-Esquivel, Villalobos, and Engelmann, 1992). However, cooling very rapidly appears to be the optimal procedure for the axes of most, but perhaps not all, species (J. Wesley-Smith, personal communication). For *Euphoria longan,* successful cryopreservation has been reported for axes subjected to precooling to –18°C before being introduced into the cryogen (Fu, Xia, and Tang, 1993). In general, however, slow cooling will be successful only when intracellular viscosity has been increased by prior dehydration to very low water contents. These low water contents will either cause desiccation damage *sensu stricto* (Pammenter et al., 1998; Walters et al., 2001) or poise the axes perilously close to the point where this will occur. Considering the various manipulations to which excised zygotic axes are subjected during the cryopreservation protocol, application of excessive stress—even if it is non-lethal in itself—is tantamount to predisposing the tissues to further injury during subsequent steps of the procedure (Berjak et al., 1999). Thus a balance needs to be sought—presently on a species-specific basis—between the least extent of dehydration commensurate with the rapid cooling rate required to achieve successful freezing of excised axes. So far, no mention has been made of the use of cryoprotectants—which are generally osmotica that may or may not penetrate the tissues but have in common the effect of decreasing water contents. It has been our experience that their use, although effective with somatic embryos, appears highly injurious to excised zygotic axes (unpublished observations).

Cryopreservation in liquid nitrogen at –196°C has the potential to conserve germplasm indefinitely, although, as occurs under any storage conditions, extraneous factors causing free-radical generation cannot be obviated. However, at the temperature of liquid nitrogen, metabolism is suspended, as would be any associated deleterious reactions, and, within the reinforced cryocontainers, there should be minimal effects of any extraneous factors. It is when cryostored germplasm is retrieved from the cryogen, however, that the potential for damage again becomes a reality.

It has long been known that to avert damage rewarming of frozen tissues needs to be rapid (e.g., Sakai, Otsuka, and Yoshida, 1968). When partially hydrated axes are retrieved from cryostorage, passage through the temperature range that promotes crystallization events must be as rapid as possible, as is the case during cooling. Thus thawing is best achieved by immediate immersion at about 40°C. Naked axes have usually been plunged into distilled water around that temperature, while cryovials containing axes are immediately introduced into a water bath. Although there is no doubt about the requirement for rapid warming (thawing), we have considerable disquiet about plunging naked, partially dehydrated axes into distilled water, as it seems impossible that imbibitional damage and/or considerable leakage of solutes from the partially dehydrated tissues can be avoided. However, trials using liquid medium (i.e., the growth medium to be used for in vitro germination, minus the gelling agent) have not been encouraging (unpublished data).

Work with successfully cryopreserved zygotic axes of *Quercus robur* that were warmed by plunging into distilled water showed that while roots developed strongly they showed no gravitropic response, and shoot ultrastructure showed a progression through derangement to necrosis (Berjak et al., 2000). Both abnormalities were obviated by warming the axes in a solution containing calcium and magnesium ions calculated to favor intracellular skeleton reconstitution. Analyses showed that normal shoot apical meristem structure and function were promoted and that statoliths developed strongly in the root cap columella, where these geosensors were notably absent after water thawing (Berjak et al., 2000). Following in vitro recovery of axes, germination, and hardening off, vigorous saplings resulted. By using an interim recovery period in the dark for retrieved *Zizania palustris* axes, Touchell and Walters (2000) showed that light is another important parameter that might generally impede axis recovery.

It should be realized, however, that the science of cryopreservation of zygotic axes is still in its formative stages, but, by painstaking species-by-species experimentation, there is a gradual emergence of general principles. However, many confounding factors still prevent success with individual species, factors which are presently largely unknown. Probably these are not technological but reside in the seeds/axes themselves—in terms of the marked diversity of species producing recalcitrant and other nonorthodox seed types—and in the significant inter- and intraseasonal variability within any one species.

There are, however, examples of such seeds in which the zygotic axis is entirely unsuitable for cryopreservation, generally because it is far too large (Berjak et al., 1996; Dumet, Berjak, and Engelmann, 1997). In such cases alternative explants need to be developed or identified to enable conservation of the germplasm. There are two common routes for this, namely, development of somatic embryos and use of shoot apices, both adding considerably to the time-consuming phases of in vitro experimentation or practice. Elaboration on the detailed methodology of either is beyond the scope of this chapter, for which the interested reader is directed to a recent review on somatic embryogenesis by Ibaraki and Kurata (2001) or, for propagation from shoot apices (meristem tip cultures), to George (1993/1996).

Whether the genetic resources of species producing essentially unstorable seeds are cryopreserved as shoot apices, zygotic axes, somatic embryos—or embryogenic callus from which these are usually produced—the practical matter of their distribution and onward propagation has commonality. The most straightforward way of achieving this would be despatch of the cryostored explants in liquid nitrogen, within specially constructed dewar containers. However, this would require the availability at the receiving end of all the necessary in vitro facilities and the expertise necessary to retrieve the explants from cryostorage, thaw them without damage, and propagate plantlets which would then need to be hardened off. These requirements clearly will seldom be able to be met. At the other end of the limited spectrum of possibilities is the transport of small, hardened-off plants or, less conveniently, of in vitro plantlets, each in a sterile polythene bag. The most practical approach, if it can be achieved, would be the production of so-called artificial or synthetic seeds.

Synthetic seeds are usually produced by encapsulating propagatory material in alginate beads, in which a variety of additives may be included. This technology is most frequently employed for propagation of somatic embryos (e.g., Bajaj, 1995a,b). Shoot apices, too, may be successfully encapsulated in alginate beads and subsequently stored, including those of some tropical forest trees (Maruyama et al., 1997). Elaborations of this artificial seed technology are easily achieved; e.g., Patel and colleagues (2000) encapsulated plant material in a solution of carboxymethylcellulose and calcium chloride, which maintained a liquid core within the alginate beads.

It is conceivable that a similar technology may be used for the distribution of germplasm of species producing recalcitrant seeds. Although cryostorage of already encapsulated zygotic axes or somatic embryos is possible, this would exacerbate the problems of efficient cooling, as the alginate bead would necessarily be of significantly greater volume than the propagatory unit. Hence, encapsulation after safe retrieval from cryostorage is envisaged as the better approach. It would be necessary that the alginate bead imposed conditions slowing ongoing germinative metabolism. Although use of crude potassium alginate has achieved this for whole seeds of *Avicennia marina* (Motete et al., 1997), treatment with mannitol or ABA has been reported to retard ongoing development of recalcitrant somatic nucellar embryos of *Mangifera indica* (mango), although not in the context of cryopreservation (Pliego Alfaro et al., 1996).

Although considerable research will be necessary, it seems possible that artificial seeds could be produced containing zygotic axes, somatic embryos, or apical meristems of species producing recalcitrant seeds. Such units would probably have to include appropriate fungicides and antibiotics, as well as a nutrient source in the case of axes or somatic embryos to sustain germination if the propagatory unit is to be planted in soil, rather than being developed further in vitro.

Although this concluding thought is conceptual rather than reflecting a current reality, if the production of such artificial seeds encapsulating propagatory material retrieved from cryostorage can be realized, then practical methods of conservation, distribution, and exchange of recalcitrant— and other nonorthodox—germplasm will have been achieved.

REFERENCES

Abdelnour-Esquivel, A., Villalobos, V., and Engelmann, F. (1992). Cryopreservation of zygotic embryos of *Coffea* spp. *CryoLetters* 13: 297-302.

Akaroda, M.O. (1986). Seed desiccation and recalcitrance in *Telfairia occidentalis*. *Seed Science and Technology* 14: 327-332.

Assy-Bah, B. and Engelmann, F. (1992). Cryopreservation of mature embryos of coconut (*Cocos nucifera* L.) and subsequent regeneration of plantlets. *Cryo-Letters* 13: 117-126.

Bain, J. and Mercer, F.V. (1966). Subcellular organisation of the developing cotyledons of *Pisum sativum* L. *Australian Journal of Biological Sciences* 19: 49-67.

Bajaj, Y.P.S. (1995a). *Somatic Embryogenesis and Synthetic Seed I. Biotechnology in Agriculture and Forestry*, Volume 30. Berlin, Heidelberg, New York: Springer.

Bajaj, Y.P.S. (1995b). *Somatic Embryogenesis and Synthetic Seed II. Biotechnology in Agriculture and Forestry*, Volume 31. Berlin, Heidelberg, New York: Springer.

Barbedo, C.J. and Cicero, S.M. (2000). Effects of initial quality, low temperature and ABA on the storage of seeds of *Inga uruguensis*, a tropical species with recalcitrant seeds. *Seed Science and Technology* 28: 793-808.

Berjak, P., Bradford, K.J., Kovach, D.A., and Pammenter, N.W. (1994). Differential effects of temperature on ultrastructural responses to dehydration in seeds of *Zizania palustris*. *Seed Science Research* 4: 111-121.

Berjak, P., Campbell, G.K., Farrant, J.M., Omondi-Oloo, W., and Pammenter, N.W. (1995). Responses of seeds of *Azadirachta indica* (neem) to short-term storage under ambient or chilled conditions. *Seed Science and Technology* 23: 779-792.

Berjak, P., Dini, M., and Pammenter, N.W. (1984). Possible mechanisms underlying the differing dehydration responses in recalcitrant and orthodox seeds: Desiccation-associated subcellular changes in propagules of *Avicennia marina*. *Seed Science and Technology* 12: 365-384.

Berjak, P. and Dumet, D. (1996). Cryopreservation of seeds and isolated embryonic axes of neem *(Azadirachta indica)*. *CryoLetters* 17: 99-104.

Berjak, P., Farrant, J.M., Mycock, D.J., and Pammenter, N.W. (1990). Recalcitrant (homoiohydrous) seeds: The enigma of their desiccation-sensitivity. *Seed Science and Technology* 18: 297-310.

Berjak, P., Farrant, J.M., and Pammenter, N.W. (1989). The basis of recalcitrant seed behavior. In Taylorson R.B. (Ed.), *Recent Advances in the Development and Germination of Seeds* (pp. 89-108). New York: Plenum Press.

Berjak, P., Mycock, D.J., Walker, M., Kioko, J.I., Pammenter, N.W., and Wesley-Smith, J. (2000). Conservation of genetic resources naturally occurring as recalcitrant seeds. In Black, M., Bradford, K.J., and Vázquez-Ramos, J. (Eds.), *Seed Biology: Advances and Applications* (pp. 223-228). Wallingford, UK: CABI Publishing.

Berjak, P., Mycock, D.J., Wesley-Smith, J., Dumet, D., and Watt, M.P. (1996). Strategies for in vitro conservation of hydrated germplasm. In Normah, M.N., Narimah, M.K., and Clyde, M.M. (Eds.), *In Vitro Conservation of Plant Genetic Resources* (pp. 19-52). Kuala Lumpur, Malaysia: Percetakan Watan Sdn.Bhd.

Berjak, P. and Pammenter, N.W. (1994). Recalcitrance is not an all-or-nothing situation. *Seed Science Research* 4: 263-264.

Berjak, P. and Pammenter, N.W. (1997). Progress in the understanding and manipulation of desiccation-sensitive (recalcitrant) seeds. In Ellis, R.H., Black, M., Murdoch, A.J., and Hong, T.D. (Eds.), *Basic and Applied Aspects of Seed Biology* (pp. 689-703). Dordrecht, the Netherlands: Kluwer Academic Publishers.

Berjak, P. and Pammenter, N.W. (2001). Recalcitrance—Current perspectives. *South African Journal of Botany* 67: 79-89.

Berjak, P., Pammenter, N.W., and Vertucci, C.W. (1992). Homoiohydrous (recalcitrant) seeds: Developmental status, desiccation sensitivity and the state of water in axes of *Landolphia kirkii* Dyer. *Planta* 186: 249-261.

Berjak, P., Vertucci, C.W., and Pammenter, N.W. (1993). Effects of developmental status and dehydration rate on characteristics of water and desiccation-sensitivity in recalcitrant seeds of *Camellia sinensis*. *Seed Science Research* 3: 155-166.

Berjak, P. and Villiers, T.A. (1972). Ageing in plant embryos: II. Age-induced damage and its repair during early germination. *New Phytologist* 71: 135-144.

Berjak, P., Walker, M., Watt, M.P., and Mycock, D.J. (1999). Experimental parameters underlying failure or success in plant germplasm conservation: A case study on zygotic axes of *Quercus robur* L. *CryoLetters* 20: 251-262.

Bewley, J.D. and Black, M. (1994). *Seeds: Physiology of Development and Germination*, Second Edition. New York: Plenum Press.

Bhattacharya, A.K., Rahman, F., and Basu, R.N. (1994). Preservation of tea *(Camellia sinensis* L.) seed viability. *Seed Research* 22: 108-111.

Bilia, D.A.C., Marcos Filho, J., and Novembre, A.D.C.L. (1999). Desiccation tolerance and seed storability of *Inga uruguensis* (Hook. Et Arn.). *Seed Science and Technology* 27: 77-89.

Blackman, S.A., Obendorf, R.L., and Leopold, A.C. (1995). Desiccation tolerance in developing soybean seeds: The role of stress proteins. *Physiologia Plantarum* 93: 630-638.

Boubriak, I., Dini, M., Berjak, P., and Osborne, D.J. (2000). Desiccation and survival in the recalcitrant seeds of *Avicennia marina:* DNA replication, DNA repair and protein synthesis. *Seed Science Research* 10: 307-315.

Bray, C.M. (1995) Biochemical processes during the osmopriming of seeds. In Kigel, J. and Galili, G. (Eds.), *Seed Development and Germination* (pp. 767-789). New York: Marcel Dekker, Inc.

Brunori, A. (1967). A relationship between DNA synthesis and water content during ripening of *Vicia faba* seeds. *Caryologia* 20: 333-338.

Bryant, G., Koster, K.L., and Wolfe, J. (2001). Membrane behavior in seeds and other systems at low water content: The various effects of solutes. *Seed Science Research* 11: 17-25.

Calistru, C., McLean, M., Pammenter, N.W., and Berjak, P. (2000). The effects of mycofloral infection on the viability and ultrastructure of wet-stored recalcitrant seeds of *Avicennia marina* (Forssk.) Vierh. *Seed Science Research* 10: 341-353.

Chacko, K.C. and Pillai, P.K.C. (1997). Seed characteristics and germination of *Garcinia gummi-gutta* (L.) Robs. *Indian Forester* 123: 123-126.

Chaitanya, K.S.K. and Naithani, S.C. (1994). Role of superoxide, lipid peroxidation and superoxide dismutase in membrane perturbation during loss of viability in seed of *Shorea robusta* Gaertn. *New Phytologist* 126: 623-627.

Chaudhury, R. and Chandel, K.S. (1991). Cryopreservation of desiccated seeds of neem (*Azadirachta indica* A. Juss) for germplasm conservation. *Indian Journal of Plant Genetic Resources* 4: 67-72.

Chaudhury, R., Radhamani, J., and Chandel, K.P.S. (1991). Preliminary observations on the cryopreservation of desiccated embryonic axes of tea [*Camellia sinensis* (L.) O. Kuntze] seeds for genetic conservation. *CryoLetters* 12: 31-36.

Chien, C.T. and Lin, P. (1997). Effects of harvest date on the storability of desiccation-sensitive seeds of *Machilus kusanoi* Hay. *Seed Science and Technology* 25: 361-371.

Chien, C.T. and Yang, J.J. (1997). Effect of seed maturity on storability of *Litsea acuminata* seeds. *Taiwan Journal of Forest Science* 12: 369-372.

Chin, H.F. (1996). Strategies for conservation of recalcitrant species. In Normah, M.N., Narimah, M.K., and Clyde, M.M. (Eds.), *In Vitro Conservation of Plant Genetic Resources* (pp. 203-215). Kuala Lumpur, Malaysia: Percetakan Watan Sdn.Bhd.

Chin, H.F. and Roberts, E.H. (1980). *Recalcitrant Crop Seeds.* Kuala Lumpur, Malaysia: Tropical Press SDN.BDH.

Close, T.J. (1997). Dehydrins: A commonality in the response of plants to dehydration and low temperature. *Physiologia Plantarum* 100: 291-296.

Collada, C., Gomez, L., Cosada, R., and Aragoncillo, C. (1997). Purification and in vitro chaperone activity of a class I small heat-shock protein abundant in recalcitrant chestnut seeds. *Plant Physiology* 115: 71-77.

Côme, D. and Corbineau, F. (1996). Metabolic damage related to desiccation sensitivity. In Ouédraogo, A.-S., Poulsen, K., and Stubsgaard, F. (Eds.), *Intermediate/Recalcitrant Tropical Forest Tree Seeds* (pp. 83-97). Rome: International Plant Genetic Resource Institute (IPGRI).

Connor, K.F. and Bonner, F.T. (1996). Effects of desiccation on temperate recalcitrant seeds: Differential scanning calorimetry, gas chromatography, electron microscopy, and moisture studies on *Quercus nigra* and *Quercus alba. Canadian Journal of Forest Research* 26: 1813-1821.

Connor, K.F. and Bonner, F.T. (1998). Physiology and biochemistry of recalcitrant *Guarea guidonia* (L.) Sleumer seeds. *Seed Technology* 20: 32-42.

Connor, K.F., Kossmann Ferraz, I.D., Bonner, F.T., and Vozzo, J.A. (1998). Effects of desiccation on the recalcitrant seeds of *Carapa guianensis* Aubl. and *Carapa procera* DC. *Seed Technology* 20: 71-82.

Corbineau, F. and Côme, D. (1986a). Experiments on germination and storage of seeds of two dipterocarp: *Shorea roxburghii* and *Hopea odorata. The Malaysian Forester* 49: 371-381.

Corbineau, F. and Côme, D. (1986b). Experiments on the storage of seeds and seedlings of *Symphonia globulifera* L.f. (Guttiferae). *Seed Science and Technology* 14: 585-591.

Corbineau, F. and Côme, D. (1988). Storage of recalcitrant seeds of four tropical species. *Seed Science and Technology* 16: 97-103.

Danthu, P., Gueye, A., Boye, A., Bauwens, D., and Sarr, A. (2000). Seed storage behavior of four Sahelian and Sudanian tree species (*Boschia senegalensis, Butyrospermum parkii, Cordyla pinnata* and *Saba senegalensis*). *Seed Science Research* 10: 183-187.

Davies, R.I. and Pritchard, H.W. (1998). Seed conservation of dryland palms of Africa and Madagascar: Needs and prospects. *Forest Genetic Resources* 26: 37-44.

de Andrade, A.C.S. and Pereira, T.S. (1997). Storage behavior in palm heart (*Euterpe edulis* Mart.) seeds. *Pesquisa Agropecuaria Brasileira* 32: 987-991.

del Carmen Rodriguez, M., Orozco-Sergovia, A., Sanchez-Coronado, M.E., and Vazquez-Yanes, C. (2000). Seed germination of six mature neotropical rain forest species in response to dehydration. *Tree Physiology* 20: 693-699.

Drew, P.J., Pammenter, N.W., and Berjak, P. (2000). "Sub-imbibed" storage is not an option for extending longevity of recalcitrant seeds of the tropical species, *Trichilia dregeana. Seed Science Research* 10: 355-363.

Dumet, D., Berjak, P., and Engelmann, F. (1997). Cryopreservation of zygotic and somatic embryos of tropical species producing recalcitrant or intermediate seeds. In Razdan, M.K. and Cocking, E.C. (Eds.), *Conservation of Plant Genetic Resources in Vitro* (pp. 153-17). Enfield, NH: Science Publishers, Inc.

Dussert, S., Chabrillange, N., Engelmann, F., Anthony, F., Louarn, J., and Hamon, S. (2000). Relationship between seed desiccation sensitivity, seed water content at maturity and climatic characteristics of native environments on nine *Coffea* species. *Seed Science Research* 10: 293-300.

Eira, M.T.S., Walters, C., and Caldas, L.S. (1999). Water sorption properties in *Coffea* spp. seeds and embryos. *Seed Science Research* 9: 321-330.

Eira, M.T.S., Walters, C., Caldas, L.S., Fazuoli, L.C., Sampaio, J.B., and Dias, M.C.L.L. (1999). Tolerance of *Coffea* spp. seeds to desiccation and low temperature. *Revista Brasileira de Fisiologia Vegetal* 11: 97-105.

Elder, R.H., Dell'Aquila, A., Mezzina, M., Sarasin, A., and Osborne, D.J. (1987). DNA ligase in repair and replication in the embryos of rye, *Secale cereale*. *Mutation Research* 181: 61-71.

Ellis, R.H. and Hong, T.D. (1990). An intermediate category of seed storage behavior? I. Coffee. *Journal of Experimental Botany* 41: 1167-1174.

Ellis, R.H. and Hong, T.D. (1991). An intermediate category of seed storage behavior? II. Effect of provenance, immaturity, and imbibition on desiccation tolerance in coffee. *Journal of Experimental Botany* 42: 653-657.

Ellis, R.H. and Roberts, E.H. (1980). Improved equations for the prediction of seed longevity. *Annals of Botany* 45: 13-30.

Engelmann, F. (1997). In vitro conservation methods. In Callow, J.A., Ford-Lloyd, B.V., and Newbury, H.J. (Eds.), *Biotechnology and Plant Genetic Resources* (pp. 119-161). Wallingford, UK: CAB International.

Farrant, J.M., Berjak, P., and Pammenter, N.W. (1993). Studies on the development of the desiccation-sensitive (recalcitrant) seeds of *Avicennia marina* (Forsk.) Vierh.: The acquisition of germinability and response to storage and dehydration. *Annals of Botany* 71: 405-410.

Farrant, J.M., Pammenter, N.W., and Berjak, P. (1986). The increasing desiccation sensitivity of recalcitrant *Avicennia marina* seeds with storage time. *Physiologia Plantarum* 67: 291-298.

Farrant, J.M., Pammenter, N.W., and Berjak, P. (1988). Recalcitrance—A current assessment. *Seed Science and Technology* 16: 155-166.

Farrant, J.M., Pammenter, N.W., and Berjak, P. (1989). Germination-associated events and the desiccation sensitivity of recalcitrant seeds—A study on three unrelated species. *Planta* 178: 189-198.

Farrant, J.M., Pammenter, N.W., and Berjak, P. (1993). Seed development in relation to desiccation tolerance: A comparison between desiccation-sensitive (recalcitrant) seeds of *Avicennia marina* and desiccation-tolerant types. *Seed Science Research* 3: 1-13.

Farrant, J.M., Pammenter, N.W., Berjak, P., Farnsworth, E.J., and Vertucci, C.W. (1996). Presence of dehydrin-like proteins and levels of abscisic acid in recalcitrant (desiccation sensitive) seeds may be related to habitat. *Seed Science Research* 6: 175-182.

Farrant, J.M., Pammenter, N.W., Berjak, P., and Walters, C. (1997) Subcellular organization and metabolic activity during the development of seeds that attain different levels of desiccation tolerance. *Seed Science Research* 7: 135-144.

Finch-Savage, W.E. (1996). The role of developmental studies in research on recalcitrant and intermediate seeds. In Ouédraogo, A.-S., Poulsen, K., and Stubsgaard, F. (Eds.), *Intermediate/Recalcitrant Tropical Forest Tree Seeds* (pp. 83-97). Rome: IPGRI.

Finch-Savage, W.E., Bergervoet, J.H.W., Bino, R.J., Clay, H.A., and Groot, S.P.C. (1998). Nuclear replication activity during seed development, dormancy breakage and germination in three tree species: Norway maple (*Acer platanoides* L.), sycamore (*Acer pseudoplatanus* L.) and cherry (*Prunus avium* L.). *Annals of Botany* 81: 519-526.

Finch-Savage, W.E. and Blake, P.S. (1994). Indeterminate development in desiccation-sensitive seeds of *Quercus robur* L. *Seed Science Research* 4: 127-133.

Finch-Savage, W.E., Blake, P.S., and Clay, H.A. (1996). Desiccation stress in *Quercus robur* seeds results in lipid peroxidation and increased synthesis of jasmonates and abscisic acid. *Journal of Experimental Botany* 47: 661-667.

Finch-Savage, W.E., Clay, H.A., Blake, P.S., and Browning, G. (1992). Seed development in the recalcitrant species *Quercus robur* L.: Water status and endogenous abscisic acid levels. *Journal of Experimental Botany* 43: 671-679.

Finch-Savage, W.E., Grange, R.I., Hendry, G.A.F., and Atherton, N.M. (1993). Embryo water status and loss of viability during desiccation in the recalcitrant species *Quercus robur* L. In Côme, D. and Corbineau, F. (Eds.), *Basic and Applied Aspects of Seed Biology* (pp. 723-730). Paris: ASFIS.

Finch-Savage, W.E., Pramanik, S.K., and Bewley, J.D. (1994). The expression of dehydrin proteins in desiccation-sensitive (recalcitrant) seeds of temperate trees. *Planta* 193: 478-485.

Fu, J.-R., Jin, J.P., Peng, Y.F., and Xia, Q.H. (1994). Desiccation tolerance in two species with recalcitrant seeds: *Clausena lansium* (Lour.) and *Litchi chinensis* (Sonn.). *Seed Science Research* 4: 257-261.

Fu, J.-R., Xia, Q.H., and Tang, L.F. (1993). Effects of desiccation on excised embryonic axes of three recalcitrant seeds and studies on cryopreservation. *Seed Science and Technology* 21: 85-95.

Galau, G.A., Hughes, D.W., and Dure, L. (1986). Abscisic acid induction of cloned cotton late embryogenesis-abundant (*lea*) gene family during embryogenesis and germination. *Plant Molecular Biology* 7: 157-170.

Gamene, S. (1997). *Vitellaria paradoxum (Butyrospermum paradoxum). IPGRI/Danida Newsletter: The Project on Handling and Storage of Recalcitrant and Intermediate Tropical Forest Tree Seeds* 3: 9.

Gee, O.H., Probert, R.J., and Coomber, S.A. (1994). "Dehydrin-like" proteins and desiccation tolerance in seeds. *Seed Science Research* 4: 135-141.

George, E.F. (1993/1996). *Plant Propagation by Tissue Culture*, Part 2: *In Practice, Second Edition*. Edington, Wilts, England: Exegetics Limited.

Golovina, E.A., Hoekstra, F.A., and Hemminga, M.A. (1998). Drying increases intracellular partitioning of amphiphilic substances into the lipid phase: Impact on membrane permeability and significance for desiccation tolerance. *Plant Physiology* 118: 975-986.

Gonzales-Benito, M.E. and Perez-Ruiz, C. (1992). Cryopreservation of *Quercus faginea* embryonic axes. *Cryobiology* 29: 685-690.

Greggains, V., Finch-Savage, W.E., Atherton, N.M., and Berjak, P. (2001). Viability loss and free radical processes during desiccation of recalcitrant *Avicennia marina* seeds. *Seed Science Research* 11: 235-242.

Hallam, N.D. (1972). Embryogenesis and germination in rye *(Secale cereale):* I. Fine structure of the developing embryo. *Planta* 104: 157-166.

Hartmann, H.T., Kester, D.E., Davies, F.T., Jr, and Geneve, R.L. (1997). *Plant Propagation: Principles and Practices,* Sixth Edition. Englewood Cliffs, NJ: Prentice-Hall.

Hay, F.R., Probert, R., Marro, J., and Dawson, M. (2000). Toward the ex situ conservation of aquatic angiosperms: A review of seed storage behavior. In Black, M., Bradford, K.J., and Vázqez-Ramos, J. (Eds.), *Seed Biology—Advances and Applications* (pp. 161-177). Wallingford, UK: CABI Publishing.

Hendry, G.A.F., Finch-Savage, W.E., Thorpe, P.C., Atherton, N.M., Buckland, S.M., Nilsson, K.A., and Seel, W.E. (1992). Free radical processes and loss of seed viability during desiccation in the recalcitrant species *Quercus robur* L. *New Phytologist* 122: 273-279.

Hoekstra, F.A., Wolkers, W.F., Buitink, J., Golovina, E.A., Crowe, J.H., and Crowe, L.M. (1997). Membrane stabilization in the dry state. *Comparative Biochemistry and Physiology* 117A: 335-341.

Hong, T.D. and Ellis, R.H. (1990). A comparison of maturation drying, germination, and desiccation tolerance between developing seeds of *Acer pseudoplatanus* L. and *Acer platanoides* L. *New Phytologist* 116: 589-596.

Hong, T.D. and Ellis, R.H. (1995). Interspecific variation in seed storage behavior within two genera: *Coffea* and *Citrus*. *Seed Science and Technology* 23: 165-181.

Hong, T.D. and Ellis, R.H. (1996). A protocol to determine seed storage behavior. In Engels, J.M.M. and Toll, J. (Eds.), *IPGRI Technical Bulletin* No. 1. Rome: International Plant Genetic Resources Institute.

Horbowicz, M. and Obendorf, R.L. (1994). Seed desiccation tolerance and storability: Dependence on flatulence-producing oligosaccharides and cyclitols—Review and survey. *Seed Science Research* 4: 385-405.

Hu, J., Guo, G., and Shi, S.X. (1993). Partial drying and postthaw preconditioning improve survival and germination of cryopreserved seeds of tea *(Camellia sinensis)*. *Plant Genetic Resources Newsletter* No. 93: 1-4.

Ibaraki, Y. and Kurata, K. (2001). Automation of somatic embryo formation. *Plant Cell, Tissue and Organ Culture* 65: 179-199.

Kehr, R.D. and Schroeder, T. (1996). Long-term storage of oak seeds—New methods and mycological aspects. In *Proceedings of the ISTA Tree Seed Pathology Meeting* (pp. 50-61). Opocno, Czech Republic, October 1996. International Seed Testing Association.

Kermode, A.R. (1990). Regulatory mechanisms involved in the transition from seed development to germination. *Critical Reviews in Plant Science* 9: 155-195.

King, M.W. and Roberts, E.H. (1980). Maintenance of recalcitrant seeds in storage. In Chin, H.F., and Roberts, E.H. (Eds.), *Recalcitrant Crop Seeds* (pp. 53-89). Kuala Lumpur, Malaysia: Tropical Press SDN.BHD.

King, M.W. and Roberts, E.H. (1982). The imbibed storage of cocoa *(Theobroma cacao)* seeds. *Seed Science and Technology* 10: 535-540.

Kioko, J., Albrecht, J., and Uncovsky, S. (1993). Seed collection and handling. In Albrecht, J. (Ed.), *Tree Seed Handbook of Kenya* (pp. 34-54). Nairobi, Kenya: GTZ Forestry Seed Centre.

Kioko, J., Berjak, P., Pritchard, H., and Daws, M. (1999). Studies of postshedding behavior and cryopreservation of seeds of *Warburgia salutaris*, a highly endangered medicinal plant indigenous to tropical Africa. In Marzalina, M., Khoo, K.C., Jayanthi, N., Tsan, F.Y., and Krishnapillay, B. (Eds.), *Recalcitrant Seeds* (pp. 365-371). Kuala Lumpur, Malaysia: Forest Research Institute Malasya (FRIM).

Klein, S. and Pollock, B.M. (1968). Cell fine structure of developing lima bean seeds related to seed desiccation. *American Journal of Botany* 55: 658-672.

Koster, K.L. (1991). Glass formation and desiccation tolerance in seeds. *Plant Physiology* 96: 302-304.

Koster, K.L. and Leopold, A.C. (1988). Sugars and desiccation tolerance in seeds. *Plant Physiology* 88: 829-832.

Kovach, D.A. and Bradford, K.J. (1992). Temperature dependence of viability and dormancy of *Zizania palustris* var. *interior* seeds stored at high moisture contents. *Annals of Botany* 69: 297-301.

Kumar, C.A., Thomas, J., and Pushpangadan, P. (1996). Storage and germination of *Aporusa lindleyana* (Wight) Baillon, an economically important plant of Western Ghats (India). *Seed Science and Technology* 25: 1-6.

Kumar, K.K. and Chacko, K.C. (1999). Seed characteristics and germination of a 'shola' forest tree: *Bhesa indica* (Bedd.) Ding Hou. *Indian Forester* 125: 206-211.

Kundu, M. and Kachari, J. (2000). Desiccation sensitivity and recalcitrant behavior of seeds of *Aquilaria agallocha* Roxb. *Seed Science and Technology* 28: 755-760.

Leopold, A.C., Sun, W.Q., and Bernal-Lugo, I. (1994). The glassy state in seeds: Analysis and function. *Seed Science Research* 4: 267-274.

Leprince, O., Buitink, J., and Hoekstra, F.A. (1999). Axes and cotyledons of recalcitrant seeds of *Castanea sativa* Mill. exhibit contrasting responses of respiration to drying in relation to desiccation sensitivity. *Journal of Experimental Botany* 50: 1515-1524.

Leprince, O., Harren, F.J.M., Buitink, J., Alberda, M., and Hoekstra, F.A. (2000). Metabolic dysfunction and unabated respiration precede the loss of membrane integrity during dehydration of germinating radicles. *Plant Physiology* 122: 597-608.

Li, C.R. and Sun, W.Q. (1999). Desiccation sensitivity and activities of free radical-scavenging enzymes in recalcitrant *Theobroma cacao* seeds. *Seed Science Research* 9: 209-217.

Liang, Y.H. and Sun, W.Q. (2000). Desiccation tolerance of recalcitrant *Theobroma cacao* embryonic axes: The optimal drying rate and its physiological basis. *Journal of Experimental Botany* 51: 1911-1919.

Lin, T.-P. and Chen, M.-H. (1995). Biochemical characteristics associated with the development of the desiccation-sensitive seeds of *Machilus thunbergii* Sieb. and Zucc. *Annals of Botany* 76: 381-387.

Lin, T.-P. and Huang, N.-H. (1994). The relationship between carbohydrate composition of some tree seeds and their longevity. *Journal of Experimental Botany* 45: 1289-1294.

Martins, C.C., Nakagawa, J., and Bovi, M.L.A. (2000). Desiccation tolerance of four seedlots from *Euterpe edulis* Mart. *Seed Science and Technology* 28: 101-113.

Maruyama, E., Kinoshita, I., Ishii, K., Ohba, K., and Saito, A. (1997). Germplasm conservation of the tropical forest trees, *Cedrela odorata* L., *Guazuma crinita* Mart., and *Jacaranda mimosaefolia* D. Don., by shoot tip encapsulation in calcium-alginate and storage at 12-25°C. *Plant Cell Reports* 16: 393-396.

Motete, N., Pammenter, N.W., Berjak, P., and Frédéric, J.C. (1997). Response of the recalcitrant seeds of *Avicennia marina* to hydrated storage: Events occurring at the root primordia. *Seed Science Research* 7: 169-178.

Mudgett, M.B., Lowensen, J.D., and Clarke, S. (1997). Protein repair L-isoaspartyl methyltransferase in plants: Phylogenetic distribution and the accumulation of substrate proteins in aged barley seeds. *Plant Physiology* 115: 1481-1489.

Nakamura, S. and Sathiyamoorthy, P. (1990). Germination of *Wasabia japonica* Matsum. seeds. *Journal of the Japanese Society for Horticultural Science* 59: 573-577.

Normah, M.N. and Chin, H.F. (1991). Changes in germination, respiration rate and leachate conductivity during storage of *Hevea* seeds. *Pertanika* 14: 1-6.

Normah, M.N., Chin, H.F., and Hor, Y.L. (1986). Desiccation and cryostorage of embryonic axes of *Hevea brasiliensis* Muell.-Arg. *Pertanika* 9: 299-303.

Normah, M.N., Ramiya, S.D., and Gintangga, M. (1997). Desiccation sensitivity of recalcitrant seeds—A study on tropical fruit species. *Seed Science Research* 7: 179-183.

Ntuli, T.M., Berjak, P., Pammenter, N.W., and Smith, M.T. (1997). Effects of temperature on desiccation responses of seeds of *Zizania palustris*. *Seed Science Research* 7: 145-160.

Obendorf, R.L. (1997). Oligosaccharides and galactosyl cyclitols in seed desiccation tolerance. *Seed Science Research* 7: 63-74.

Osborne, D.J. (1983). Biochemical control of systems operating in the early hours of germination. *Canadian Journal of Botany* 61: 3568-3577.

Osborne, D.J. and Boubriak, I.I. (1994). DNA and desiccation tolerance. *Seed Science Research* 4: 175-185.

Pammenter, N.W. and Berjak, P. (1999). A review of recalcitrant seed physiology in relation to desiccation-tolerance mechanisms. *Seed Science Research* 9: 13-37.

Pammenter, N.W. and Berjak, P. (2000). Evolutionary and ecological aspects of recalcitrant seed biology. *Seed Science Research* 10: 301-306.

Pammenter, N.W., Berjak, P., Farrant, J.M., Smith, M.T., and Ross, G. (1994). Why do stored, hydrated recalcitrant seeds die? *Seed Science Research* 4: 187-191.

Pammenter, N.W., Greggains, V., Kioko, J.I., Wesley-Smith, J., Berjak, P., and Finch-Savage, W.E. (1998). Effects of differential drying rates on viability retention of recalcitrant seeds of *Ekebergia capensis*. *Seed Science Research* 8: 463-471.

Pammenter, N.W., Motete, N., and Berjak, P. (1997). The response of hydrated recalcitrant seeds to long-term storage. In Ellis, R.H., Black, M., Murdoch, A.J., and Hong, T.D. (Eds.), *Basic and Applied Aspects of Seed Biology* (pp. 671-687). Dordrecht, the Netherlands: Kluwer Academic Publishers.

Pammenter, N.W., Vertucci, C.W., and Berjak, P. (1991). Homeohydrous (recalcitrant) seeds: Dehydration, the state of water and viability characteristics in *Landolphia kirkii*. *Plant Physiology* 96: 1093-1098.

Patel, A.V., Pusch, I., Mix-Wagner, G., and Dunlop, K.D. (2000). A novel encapsulation technique for the production of artificial seeds. *Plant Cell Reports* 19: 868-874.

Pence, V.C. (1996). In vitro collection (IVC) method. In Normah, M.N., Narimah, M.K., and Clyde, M.M. (Eds.), *In Vitro Conservation of Plant Genetic Resources* (pp. 181-190). Kuala Lumpur, Malaysia: Percetakan Watan Sdn.Bhd.

Pliego Alfaro, F., Litz, R.E., Moon, P.A., and Gray, D.J. (1996). Effect of abscisic acid, osmolarity and temperature on in vitro development of recalcitrant mango nucellar embryos. *Plant Cell, Tissue and Organ Culture* 44: 53-61.

Potts, S.E. and Lumpkin, T.A. (2000). Cryopreservation of *Wasabia* spp. seeds. *CryoLetters* 18: 185-190.

Poulsen, K. (1992). Sensitivity to low temperatures (−196°C) of embryonic axes from acorns of the pedunculate oak (*Quercus robur* L.). *CryoLetters* 13: 75-82.

Pritchard, H.W. (1991). Water potential and embryonic axis viability in recalcitrant seeds of *Quercus rubra*. *Annals of Botany* 67: 43-49.

Pritchard, H.W. and Prendergast, F.G. (1986). Effects of desiccation and cryopreservation on the in vitro viability of embryos of the recalcitrant seed species *Araucaria hunsteinii*. *Journal of Experimental Botany* 37: 1388-1397.

Pritchard, H.W., Steadman, K.J., Nash, V.J., and Jones, C. (1999). Kinetics of dormancy release and the high-temperature germination response in *Aesculus hippocastanum* seeds. *Journal of Experimental Botany* 50: 1507-1514.

Pritchard, H.W., Tompsett, P.B., and Manger, K.R. (1996). Development of a thermal time model for the quantification of dormancy loss in *Aesculus hippocastanum* seeds. *Seed Science Research* 6: 127-135.

Pritchard, H.W., Tompsett, P.B., Manger, K., and Smidt, W.J. (1995). The effect of moisture content on the low temperature responses of *Araucaria hunsteinii* seed and embryos. *Annals of Botany* 76: 79-88.

Probert, R.J. and Brierley, E.R. (1989). Desiccation intolerance in seeds of *Zizania palustris* is not related to developmental age or the duration of postharvest storage. *Annals of Botany* 64: 669-674.

Probert, R.J. and Longley, P.L. (1989). Recalcitrant storage physiology in three aquatic grasses (*Zizania palustris, Spartina anglica* and *Porteresia coarctata*). *Annals of Botany* 63: 53-63.

Roberts, E.H. (1973). Predicting the storage life of seeds. *Seed Science and Technology* 1: 499-514.

Sakai, A., Otsuka, K., and Yoshida, S. (1968). Mechanism of survival in plant cells at super-low temperatures by rapid cooling and rewarming. *Cryobiology* 4: 165-173.

Smith, R.D. (1995). Collecting and handling seeds in the field. In Guarino L., Rao, V.R., and Reid, R. (Eds.), *Collecting Plant Genetic Diversity* (pp. 419-456). Wallingford, UK: CAB International.

Steadman, K.J., Pritchard, H.W., and Dey, P.M. (1996). Tissue-specific soluble sugars in seeds as indicators of storage category. *Annals of Botany* 77: 667-674.

Suszka, B. and Tylkowski, T. (1980). Storage of acorns of the English oak (*Quercus robur* L.) over 1-5 winters. *Arboretum Kórnickie* 25: 199-229.

Tommasi, E., Paciolla, C., and Arrigoni, O. (1999) The ascorbate system in recalcitrant and orthodox seeds. *Physiologia Plantarum* 105: 193-198.

Tompsett, P.B. (1992). A review of the literature on storage of dipterocarp seeds. *Seed Science and Technology* 20: 251-267.

Tompsett, P.B. and Pritchard, H.W. (1993). Water changes during development in relation to the germination and desiccation tolerance of *Aesculus hippocastanum* L. seeds. *Annals of Botany* 71: 107-116.

Tompsett, P.B. and Pritchard, H.W. (1998). The effect of chilling and moisture status on the germination, desiccation tolerance and longevity of *Aesculus hippocastanum* L. seed. *Annals of Botany* 82: 249-261.

Touchell, D. and Walters, C. (2000). Recovery of embryos of *Zizania palustris* following exposure to liquid nitrogen. *CryoLetters* 21: 261-270.

Vertucci, C.W., Crane, J., Porter, R.A., and Oelke, E.A. (1994). Physical properties of water in *Zizania* embryos in relation to maturity status, water content and temperature. *Seed Science Research* 4: 211-224.

Vertucci, C.W., Crane, J., Porter, R.A., and Oelke, E.A. (1995). Survival of *Zizania* embryos in relation to water content, temperature and maturity status. *Seed Science Research* 5: 31-40.

Vertucci, C.W. and Farrant, J.M. (1995). Acquisition and loss of desiccation tolerance. In Kigel, J. and Galili, G. (Eds.), *Seed Development and Germination* (pp. 237-271). New York: Marcel Dekker, Inc.

von Teichman, I. and van Wyk, A.E. (1994). Structural aspects and trends in the evolution of recalcitrant seeds in the dicotyledons. *Seed Science Research* 4: 225-239.

Walters, C.W. (1998). Ultra-dry seed storage. *Seed Science Research* 8 (Supplement no. 1).

Walters, C.W., Pammenter, N.W., Berjak, P., and Crane, J. (2001). Desiccation damage, accelerated ageing and respiration in desiccation tolerant and sensitive seeds. *Seed Science Research* 11: 135-148.

Wesley-Smith, J. (2001). Freeze-substitution of dehydrated plant tissues: Artefacts of aqueous fixation revisited. *Protoplasma* 218: 154-167.

Wesley-Smith, J., Pammenter, N.W., Berjak, P., and Walters, C. (2001). The effects of two drying rates on the desiccation tolerance of embryonic axes of recalcitrant jackfruit (*Artocarpus heterophyllus* Lamk.) seeds. *Annals of Botany* 88: 653-664.

Wesley-Smith, J., Vertucci, C.W., Berjak, P., Pammenter, N.W., and Crane, J. (1992). Cryopreservation of recalcitrant axes of *Camellia sinensis* in relation to dehydration, freezing rate and thermal properties of tissue water. *Journal of Plant Physiology* 140: 596-604.

Wesley-Smith, J., Walters, C., Pammenter, N.W., and Berjak, P. (2001). Interactions among water content, rapid (nonequilibrium) cooling to -196°C, and survival of embryonic axes of *Aesculus hippocastanum* L. seeds. *Cryobiology* 42: 196-206.

Williams, R.J. and Leopold, A.C. (1989). The glassy state in corn embryos. *Plant Physiology* 89: 977-981.

Xia, Q.H., Chen, R.Z., and Fu, J.R. (1992). Moist storage of lychee (*Litchi sinensis* Sonn.) and longan (*Euphoria longan* Steud.) seeds. *Seed Science and Technology* 20: 269-279.

SECTION IV:
INDUSTRIAL QUALITY OF SEEDS

Chapter 11

Processing Quality Requirements for Wheat and Other Cereal Grains

Colin W. Wrigley
Ferenc Bekes

INTRODUCTION

A seed is a plant's means of producing another plant, thereby perpetuating the species. The requirements for a seed to fulfill this role are for it to provide a good store of nutrients to supply the new plant in the early stages of its growth. Safe storage of these nutrients is essential throughout whatever conditions may occur until the right combination of moisture and temperature triggers the germination response. The ability of plants to provide seeds as stores of nutrients has also made them an attractive food source for humankind. Because grains have been recognized as an important food since prehistory, the first propagation of seed-bearing plants was an important phase in the development of humankind, marking the transition from hunter-gatherer to a settled agricultural existence. These developments led, in turn, to the building of permanent dwellings and to a wide range of cultural activities.

THE RANGE OF GRAIN SPECIES USED INDUSTRIALLY

Thousands of years of such agricultural practice have led to the selection of species that suit humankind's requirements. Further improvements have led to the development of cultivated varieties within those species with quality attributes that are even better suited to processing requirements. A short list of such species (Table 11.1) includes members of both monocotyledonous and dicotyledonous plants. The success of this approach to obtaining food is indicated by the worldwide cultivation of billions of seed-bearing plants annually. This activity yields well over 2 billion tonnes of

TABLE 11.1. Major grain species used for human food and for animal feed

Family	Common name	Genus and species
Monocotyledons		
Gramineae (Pooids)	Bread wheat	*Triticum aestivum*
	Durum wheat	*Triticum durum*
	Triticale	*Triticosecale* species
	Rye	*Secale cereale*
	Barley	*Hordeum vulgare*
	Tritordeum	Hybrid between *Hordeum chilense* and durum wheat
Bambusoids	Oats	*Avena sativa*
	Rice	*Oryza sativa*
	Wild rice	*Zizania aquatica*
Eu-panicoids	Pearl millet	*Pennisetum glaucum*
	Finger millet	*Eleusine coracana*
	Japanese millet	*Echinochloa* species
Andropogonoids	Sorghum	*Sorghum bicolor*
	Maize (corn)	*Zea mays*
Dicotyledons		
Amaranthaceae	Amaranth	*Amaranthus* species
Polygonaceae	Buckwheat	*Fagopyrum esculentum*
Malvaceae	Cottonseed	*Gossypium* species
Leguminosae	Lupin	*Lupinus* species
	Pea	*Pisum sativum*
	Peanut	*Arachis hypogaea*
	Soybean	*Glycine max*
Cruciferae	Rapeseed	*Brassica napus*
Linaceae	Linseed	*Linum usitatissimum*
Compositae	Sunflower	*Helianthus annuus*

Source: Adapted from Watson and Wrigley, 1984, and from Wrigley and Bekes, 2001.

grain of all species (counting only the production of those countries that register relevant statistics) (Wright, 2001). As a result, grains are the main source of protein and energy for humankind, either directly by processing the seeds into food or indirectly by the ingestion of animal foods (meat, milk, eggs) following the feeding of grains to animals.

Dicotyledonous Grains

The taxonomic diversity of grains is indicated in Table 11.1. The greatest diversity is found among the dicots, including various grain legumes (also called pulses) (including lupins, peas, peanuts), soybeans, and oilseeds such as rapeseed/canola, safflower, linseed/linola, sunflower, and cottonseed. Collectively, oilseed production totals about 300 million tonnes annually, with most of this (250 million tonnes) being crushed for oil production (Wright, 2001). The utilization of oilseeds is reviewed in Chapter 12. World production of pulses totals nearly 60 million tonnes, with about 25 percent of this being produced in India. The dicot group of grains also contains a range of less common grains, such as amaranth, for which valuable modes of utilization have been developed (Lehmann, 1996).

Monocotyledonous Grains—The Cereals

On the other hand, the range of monocot grains all belong to a single family of grassy plants (the Gramineae), being grouped under the general title "cereals" (Watson and Wrigley, 1984). Wheat is unique in its ability to form a viscoelastic dough after milling and mixing with water. Many of the range of wheat-based products in Figure 11.1 are generally familiar, especially the conventional leavened bread. Less familiar to Westerners are the various flat (Arabic) breads in Figure 11.1, the wide range of pasta and noodle types in Figure 11.2, and the Chinese steamed bread in Figure 11.3. However, all these wheaten products rely on the gluten-forming proteins of wheat. For this reason, wheat is the cereal grain for which markets are the most conscious of quality specifications. Wheat is thus the main topic of this chapter on quality requirements. World production of wheat totals about 600 million tonnes annually, coming from some 220 million hectares. World trade in wheat is about 100 million tonnes annually, the main trading nations being the United States, Canada, Australia, Argentina, and the countries of Europe. Wheat's close relative, rye, is minor by comparison; world rye production is about 22 million tonnes annually, the main production regions being in eastern parts of Europe, especially Poland, Germany, and Russia (Bushuk, 2001a,b).

Rice, too, has a unique place in the human diet, providing the major (sometimes, almost the sole) source of energy and protein for many cultures. World production rivals that of wheat, based on the volume of paddy rice (hulled). After milling to dehulled rice, world production is about 400 million tonnes. Maize fulfills a similar but distinct role for other cultures, in addition to being the major grain used in industrial processing. Its annual

FIGURE 11.1. Baked goods made from wheat flour include Arabic-style flat breads as well as the leavened square loaf of bread.

production (approaching 600 million tonnes) is included in the statisticians' term "coarse grains," which total 900 million tonnes annually, covering over 300 million hectares of agricultural production. World trade in coarse grain is about 100 million tonnes. This term also includes barley (about 150 million tonnes annually), sorghum (50 million tonnes), and oats (30 million tonnes) (Wright, 2001). Close behind wheat in having exacting quality requirements comes barley, with its unique ability to provide malt for brewing, as is reviewed in Chapter 13.

CEREAL GRAINS AND OUR DIET

Most forms of processing involve the application of moist heat to gelatinize starch (making it more readily digested), to denature proteins, and to inactivate antinutritional compounds (especially for some dicot grains). Heat processing may follow various forms of milling to remove

FIGURE 11.2. The diversity of noodle types made from wheat

outer husks or bran layers, thereby making the grain product more readily digestible. Grain-based foods therefore occur almost universally in the human diet, although in many different forms. However, for some individuals, cereal grains may cause dietary problems. One of the best characterized of these intolerances is celiac disease, a condition caused by the ingestion of wheat gluten protein and analogous grain proteins of rye, triticale, barley, and sometimes oats (Skerritt, Devery, and Hill, 1990; Kasarda, 2001). However, Table 11.1 shows that buckwheat is a distant relative of wheat (despite its misleading name), so buckwheat is not toxic to celiacs (Skerritt, 1986). Test kits have been developed for home use to detect the presence of gluten in foods to assist celiacs in diet control (Skerritt and Hill, 1991).

Whole Grain and Fiber

On the other hand, health benefits are provided by the inclusion of the outer layers of the cereal grains in whole-grain foods. This is because these parts of the grain (Figure 11.4) provide various vitamins and minerals, and because they contribute increased amounts of fiber to the diet (Malkki, 2001). The cereal grains offer a wide range of high-fiber ingredients in the

FIGURE 11.3. Chinese steamed bread, shown by Sidi Huang of BRI Australia Ltd. (Sydney). This type of bread is steam cooked, so that it does not develop the brown crust normally associated with baked products.

human diet, depending on the source of the fiber and the manner of its processing, as is reviewed by Nelson (2001). A further report (American Association of Cereal Chemists [AACC], 2001) reviews methods of defining and determining dietary fiber, as well as providing a list of commercially available sources of fiber from the various cereals.

The correspondence between milling fractions and grain morphology is illustrated in Figure 11.4 for the wheat grain. The positions of these tissues in a sectioned grain are shown diagrammatically in Figure 11.5. These diagrams illustrate that white flour consists primarily of the endosperm of the grain. Normal milling practice provides about 75 percent of the grain as white flour, but in some commercial operations the extraction rate may exceed 80 percent. For some special products, such as Japanese udon noodles, lower extraction rates are adopted (about 60 percent). Figure 11.6 shows how the range of nutrients changes with the extraction rate used in flour milling. Complex carbohydrates (and therefore energy) are little affected by extraction rate, and protein content is only slightly altered. However, the con-

volves the blending of a high proportion of wheat flour to improve loaf quality. Nevertheless, rye bread is a popular item in some cultures, especially in eastern Europe, the site of the greatest rye production. In addition to the use of rye for bread making, it is used to a limited extent for whiskey manufacture. Rye is well suited to cool, temperate regions and to high altitudes, even for poor soils. It has the agronomic advantage that the young plants can be grazed, with adequate grain production still being assured at maturity. The grain, however, is not well suited for animal feed, especially if there is contamination with ergot, which is a particular problem with rye.

The man-made hybrid triticale is an attempt to combine the agronomic advantages of rye with wheat. However, the production of triticale is not great compared to the wider range of cereals. Some varieties of triticale offer the advantage of better baking quality than rye alone, while providing some of the flavor characteristics of rye. The higher fiber content of triticale (compared to wheat) has been offered as a health advantage that should be considered for food use. In addition, claims have been made for the dietary use of triticale to reduce the risks of cancer and coronory heart disease (Slavin, Marquart, and Jacobs, 2000).

Barley

The premium use of barley is for malting and brewing, for which there may be specific segregation and marketing of individual varieties separately from others, because of the unique malting properties that are preferred by the maltster for one particular variety. This important use of barley is described in detail in Chapter 13 and also in dedicated monographs such as MacGregor and Bhatty (1993). The other major use of barley is as animal feed—both for ruminants and nonruminants. Smaller proportions of the barley crop are used for a range of human foods, mainly in the pearled form, resulting from abrasion to remove the outer lemma, palea, and bran layers. A novel cereal grain developed from barley is the man-made hybrid tritordeum, an amphiploid between *Hordeum chilense* and durum wheat (Martin et al., 1999). Tritordeum is morphologically and agronomically similar to wheat, and it shows promise of providing satisfactory baking quality.

Oats

Oats may suffer to some extent from the famous altercation between the Englishman Samuel Johnson and the Scot Boswell; the former describing oats as a grain fit only for horses in England, although eaten by men in Scotland. Boswell's retort was, "And pray, where do we see such horses as were

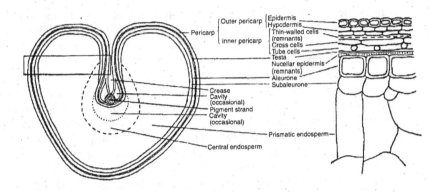

FIGURE 11.5. Diagrams of the transversally sectioned cereal grain (applicable to wheat, rye, or triticale), specifying the outer layers of bran and aleurone cells (*Source:* From W. K. Heneen and K. Brismar, 1987, Scanning electron microscopy of mature grains of rye, wheat, and triticale with emphasis on grain shriveling, *Hereditas* 107: 147-162. Reproduced with permission.)

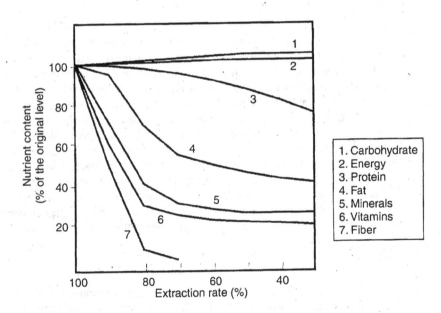

FIGURE 11.6. Progressive changes in the content of various nutrients in wheat flour as the milling extraction rate falls

USES OF CEREAL GRAINS

Wheat

The great diversity of food uses of wheat (Table 11.2) includes the many types of breads (leavened and unleavened), flat breads (Arabic/pocket types), pizza crust, tortillas, Chinese steamed breads, noodles of many types, breakfast cereals and porridge, cakes, biscuits/cookies, scones, muffins, chapatis, extruded snack foods, and the wide range of pasta (Faridi and Faubion, 1995). Some of these baked products are illustrated in Figures 11.1, 11.2, and 11.3. Flat breads are made in many types in Middle Eastern countries, in Turkey and surrounding countries, in southeastern Europe, and in the Indian subcontinent (Qarooni, Ponte, and Posner, 1992; Qarooni,

TABLE 11.2. Quality attributes preferred in wheats for specific products. In all cases, good milling is required, giving a high yield of white flour.

Product	Protein content (%)	Grain hardness	Dough strength
Breads			
Pan bread	>13	Hard	Strong
Flat bread	11-13	Hard	Medium
Chinese steamed bread			
Northern style	11-13	Hard	Medium/strong
Southern style	10-12	Medium/hard	Medium
Guangdong style[a]	9-11	Medium/hard	Medium
Guangdong style[b]	9	Soft/medium	Weak/medium
Noodles[c]			
Alkaline	11-13	Hard	Medium
White	10-12	Medium/soft	Medium
Instant	11-12	Medium	Medium
Biscuit/cake	8-10	Very soft	Weak
Pasta	>13	Very hard	Very strong
Starch/gluten manufacture[d]	>13	Hard (soft preferred)	Strong

Source: Adapted from Wrigley, 1994, and updated by S. Huang.
[a]Fat-containing formula (popular in Taiwan)
[b]Formula without fat
[c]Null-4A genotypes preferred for starch properties
[d]Genotypes preferred should have soft grain and high proportions of large (A type) starch granules

1996). These types include lavash, barbari, taftoon, sangak, baladi, pita, tanoor, and chapati, each having distinctive characteristics and a particular shape and texture to suit a variety of ethnic origins and preferences.

A wide range of noodle types are shown in Figure 11.2. Traditionally, noodles are made by cutting a sheet of dough into strips, the dough being made from a hexaploid wheat (*Triticum aestivum* in Table 11.1). Noodles may be white, cream, or yellow, as well as commonly being colored by additions such as buckwheat and spinach. Alkaline noodles have a yellow color from the incorporation of alkaline salts, e.g., sodium carbonate or bicarbonate (Moss, Miskelly, and Moss, 1986). In contrast to noodles, pastas are traditionally made by extruding a dry dough. Pasta products include macaroni, spaghetti, and many other geometric forms. Pasta dough is traditionally made from durum wheat—a tetraploid species that differs genetically from the more common hexaploid bread wheat (Table 11.1). Nevertheless, hexaploid wheat is sometimes blended with durum wheat for pasta production. In addition, there are the many pastry uses of wheat, including pies, fancy baked goods, grocery flour for many home-cooking uses, and, finally, animal feed and nonfood industrial uses.

A major industrial use of wheat flour is the separation of gluten and starch by the water washing of dough. The resulting starch goes into many food applications, especially those requiring thickening agents, and into a wide range of industrial uses, including adhesives. The resulting gluten has traditionally been dried and added to bread to increase dough strength and to facilitate the production of specialty products such as pizza crust and high-bran and fiber-increased breads. This is still the major use of vital dry gluten, but dry gluten is being increasingly used as a general food additive, finding its way into breakfast cereals, cheese, processed meats, snacks, and even chewing gum, as well as being formulated into fish and meat analogs. For nonhuman consumption, gluten is used in pet foods, animal feed pellets, and aquaculture feeds. Table 11.3 lists a wide range of innovative extensions of gluten utilization, being used "as is" and after various forms of modification (Bietz and Lookhart, 1996).

Most of these wheat-based foods are dependent on the unique dough-forming quality of wheat flour and the resulting ability of dough to retain the gas cells produced by yeast fermentation, leading to the familiar light texture of leavened bread. In other cases, the dough-forming quality is needed to permit machining or hand kneading followed by sheeting and cutting (for noodles) or stretching and covering (for many pastry goods). The strength and extensibility requirements differ for the various products, as listed for the range of products in Table 11.2. Dough-forming properties also depend on protein content, which is a prominent attribute determining market value. Grain hardness is also a critical attribute determining process-

volves the blending of a high proportion of wheat flour to improve loaf quality. Nevertheless, rye bread is a popular item in some cultures, especially in eastern Europe, the site of the greatest rye production. In addition to the use of rye for bread making, it is used to a limited extent for whiskey manufacture. Rye is well suited to cool, temperate regions and to high altitudes, even for poor soils. It has the agronomic advantage that the young plants can be grazed, with adequate grain production still being assured at maturity. The grain, however, is not well suited for animal feed, especially if there is contamination with ergot, which is a particular problem with rye.

The man-made hybrid triticale is an attempt to combine the agronomic advantages of rye with wheat. However, the production of triticale is not great compared to the wider range of cereals. Some varieties of triticale offer the advantage of better baking quality than rye alone, while providing some of the flavor characteristics of rye. The higher fiber content of triticale (compared to wheat) has been offered as a health advantage that should be considered for food use. In addition, claims have been made for the dietary use of triticale to reduce the risks of cancer and coronory heart disease (Slavin, Marquart, and Jacobs, 2000).

Barley

The premium use of barley is for malting and brewing, for which there may be specific segregation and marketing of individual varieties separately from others, because of the unique malting properties that are preferred by the maltster for one particular variety. This important use of barley is described in detail in Chapter 13 and also in dedicated monographs such as MacGregor and Bhatty (1993). The other major use of barley is as animal feed—both for ruminants and nonruminants. Smaller proportions of the barley crop are used for a range of human foods, mainly in the pearled form, resulting from abrasion to remove the outer lemma, palea, and bran layers. A novel cereal grain developed from barley is the man-made hybrid tritordeum, an amphiploid between *Hordeum chilense* and durum wheat (Martin et al., 1999). Tritordeum is morphologically and agronomically similar to wheat, and it shows promise of providing satisfactory baking quality.

Oats

Oats may suffer to some extent from the famous altercation between the Englishman Samuel Johnson and the Scot Boswell; the former describing oats as a grain fit only for horses in England, although eaten by men in Scotland. Boswell's retort was, "And pray, where do we see such horses as were

volves the blending of a high proportion of wheat flour to improve loaf quality. Nevertheless, rye bread is a popular item in some cultures, especially in eastern Europe, the site of the greatest rye production. In addition to the use of rye for bread making, it is used to a limited extent for whiskey manufacture. Rye is well suited to cool, temperate regions and to high altitudes, even for poor soils. It has the agronomic advantage that the young plants can be grazed, with adequate grain production still being assured at maturity. The grain, however, is not well suited for animal feed, especially if there is contamination with ergot, which is a particular problem with rye.

The man-made hybrid triticale is an attempt to combine the agronomic advantages of rye with wheat. However, the production of triticale is not great compared to the wider range of cereals. Some varieties of triticale offer the advantage of better baking quality than rye alone, while providing some of the flavor characteristics of rye. The higher fiber content of triticale (compared to wheat) has been offered as a health advantage that should be considered for food use. In addition, claims have been made for the dietary use of triticale to reduce the risks of cancer and coronory heart disease (Slavin, Marquart, and Jacobs, 2000).

Barley

The premium use of barley is for malting and brewing, for which there may be specific segregation and marketing of individual varieties separately from others, because of the unique malting properties that are preferred by the maltster for one particular variety. This important use of barley is described in detail in Chapter 13 and also in dedicated monographs such as MacGregor and Bhatty (1993). The other major use of barley is as animal feed—both for ruminants and nonruminants. Smaller proportions of the barley crop are used for a range of human foods, mainly in the pearled form, resulting from abrasion to remove the outer lemma, palea, and bran layers. A novel cereal grain developed from barley is the man-made hybrid tritordeum, an amphiploid between *Hordeum chilense* and durum wheat (Martin et al., 1999). Tritordeum is morphologically and agronomically similar to wheat, and it shows promise of providing satisfactory baking quality.

Oats

Oats may suffer to some extent from the famous altercation between the Englishman Samuel Johnson and the Scot Boswell; the former describing oats as a grain fit only for horses in England, although eaten by men in Scotland. Boswell's retort was, "And pray, where do we see such horses as were

produced in England and such men as were reared in Scotland." In most oat-growing countries, oats are mainly destined for animal feeding, often at the site of production, with about 20 percent of production going for human consumption. Webster's (1986) monograph is still a general source book on oats; it has been partly updated (Webster, 1996). These are complemented by a monograph on oat bran (Wood, 1993), produced to provide a reasoned source of information at a time when extreme nutritional claims were being made for oat bran.

A major use of oats is as a hot breakfast cereal in the form of porridge, produced from rolled oats. The rolling process is applied to the whole groat. Heat processing is essential to inactivate enzymes (lipase, lipoxygenase, and peroxidase), which would otherwise produce soapy bitter flavors. Further processing may involve sectioning the rolled groat into smaller pieces to permit quicker cooking. Oat flakes may also be incorporated into cold breakfast cereals, such as muesli, and into baked goods, such as muffins and cookies. Oatmeal and oat flour are major components of many infant foods.

Most genotypes of oats, as harvested, have the outer husks attached, so these must be removed during processing for human food. There are some hull-less oat genotypes. The dehulled grain is termed the "groat." Its aleurone layer, surrounding the starchy endosperm, is a good source of soluble dietary fiber in the form of beta-glucan. The cell walls of the endosperm are also rich in beta-glucans. Groats have relatively high levels of protein (range of 10 to 20 percent) and fat (5 to 10 percent), compared to other cereal grains. The balance of oleic and linoleic acids is a desirable blend for human nutrition. The effects of genotype and growth condition on fat composition have been studied using near-infrared spectroscopy (Krishnan et al., 2000).

The protein of oats has a high nutritional value, based on its content of essential amino acids, relative to other cereal grains, because oats have a lower content of the prolamin class of proteins and relatively more of the globulin class of proteins (Figure 11.7). The prolamins of oats, termed "avenins," are polymorphic—as is seen by SDS-gel electrophoresis or two-dimensional isoelectric-focusing electrophoresis. The alpha-avenins have higher isoelectric points than the gamma avenins. The amino-acid sequences are known for many of the avenins (Egorov et al., 1994; Shewry, 1999).

Oats (and also rice) are unusual among the cereal grains in that their major protein class is the globulins, rather than the prolamins (Figure 11.7). Globulins account for about 75 percent of the seed protein content (Shotwell, 1999). The oat-globulin protein is a hexamer with subunits of about 55,000 Daltons, which each comprise disulfide-linked polypeptides of 32,000 and 23,000 Daltons. Oat globulin is rich in glutamine and aspar-

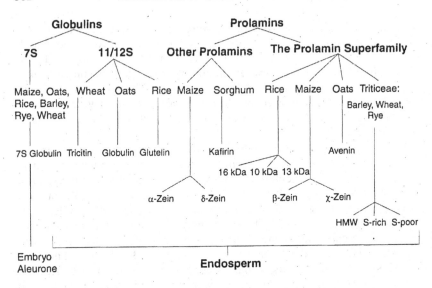

FIGURE 11.7. Classification of the cereals, according to the types of proteins in their grains (*Source:* Adapted from Shewry, 1996.)

agine, consistent with the role of storing nitrogen. However, oat-globulin protein is slightly deficient in the sulfur-containing amino acids cysteine and methionine, being similar to the globulins of the legumes in this respect. Oat globulin shares about 70 percent amino acid sequence similarity with the storage glutelin of rice, despite the differences in solubility of these two classes of protein (Shotwell, 1999).

Rice

The production and consumption of rice are mainly concentrated in the countries of Asia, the home of about 60 percent of the world's population. In Bangladesh, Cambodia, Indonesia, Laos, Myanmar, Thailand, and Vietnam, rice provides 55 to 80 percent of energy in the diet. Worldwide, rice provides about 20 percent of the total food calories consumed (Athwal, 1971). International rice trade accounts for only about 5 percent of world production, and much of this trade is in specialty grades. For example, the high-quality scented basmati rice of Pakistan and Northern India may command a fourfold price premium. The major exporters are Thailand, the United States, Vietnam, and Pakistan. Others include Australia, China, India, and Uruguay. Standard test procedures have been established to assess rice-cooking quality (Juliano, 1985a). The range of chemical and technological aspects of rice utilization is reviewed in Juliano (1985b).

As a protein source, milled rice contains the lowest amount of protein (ca. 5 percent) among the major cereals; moreover, this protein content is not easily digestible by humans or by monogastric animals. However, compared to other cereal proteins, the overall amino acid composition of rice protein is significantly better balanced, because of its relatively high level of lysine. The amino acid composition of rice is unusual among the cereal grains because rice is one of the few cultivated plants in which both the globulins and prolamins, the two classes of storage proteins of higher plants, are present in significant levels (Figure 11.7). Unlike most other cereals, which accumulate prolamins as their primary nitrogen reserve, the major storage proteins in rice are the glutelins, which are homologous at the primary-sequence level to the 11S globulin proteins, a class which is the dominant form of nitrogen deposition in legumes. Furthermore, the rice prolamin proteins have a number of characteristics that are different from the prolamins of most other cereals.

Based on the classic study of Betchel and Juliano (1980), the most important characteristics of the deposition of nitrogen in the rice kernel are three kinds of protein bodies in the rice endosperm, namely, large spherical protein bodies, small spherical ones, and crystalline protein bodies, each of them surrounded by a single continuous-unit membrane. The spherical protein bodies form within vacuoles, but the proteins are synthesized in the endoplasmatic reticulum and in the Golgi apparatus before being transported to the vacuoles by vesicles.

Removal of the husk from rough rice yields the kernel, which composes the pericarp, seed coat, the aleurone layer, the endosperm, and the germ; this form is known as *brown rice* (with a protein content of 9 to 10 percent). Brown rice has a significantly higher nutritional value than white polished (milled) rice, which is the most commonly utilized rice product (with a protein content of about 8 percent). The milling process for rice results in 40 to 55 percent white milled rice and three major by-products: husks (20 percent), bran (10 percent), and "brokens" (10 to 22 percent), with protein contents of 3, 17, and 8.5 percent, respectively. The aleurone layer, the tissue with the highest levels of protein and nutritionally important minor components, is removed during the process of rice milling.

The distribution of the protein fractions from the Osborne procedure of fractionation for brown or white rice (Table 11.4) reflects the observation that the levels of the albumin and globulin classes are significantly higher in the outer layers of the seed; they decrease toward the center of the grain while the proportion of glutelins has an inverse distribution (Bechtel and

TABLE 11.4. The distribution of protein classes for brown and white rice according to the Osborne fractionation method

	Albumin	Globulin	Prolamin	Glutelin
Brown rice	10.8	9.7	2.2	77.3
White rice	6.5	12.7	8.9	71.9

Pomeranz, 1980; Takaiwa, Ogawa, and Okita, 2000). The subaleurone region plays very significant role nutritionally; it is several cell layers thick and is rich in the globulin class of proteins. Its lysine content is much higher than the proteins located in the endosperm. It is therefore desirable to mill as lightly as possible to retain most of the subaleurone layer on white rice.

The albumin fraction isolated from rice is highly heterogeneous and contains many biologically important components. It can be separated into four subfractions, based on the molecular size of the protein components, which range from 10 to 200 kDa. More than 50 individual polypeptides have been observed in the albumin fraction, based on fractionation by isoelectric focusing. Detailed studies on many of these components have shown that rice albumins have mostly enzymic or enzyme-inhibitor activities. Compared to wheat, rye, and barley, rice contains significantly lower amounts of the high-pI α-amylases and much higher levels of the low pI α-amylases.

Maize (Corn)

Corn, indigenous to North America, was developed by Central American natives many centuries before Columbus arrived. Corn was the foundation of the extensive North and South American ancient civilizations and was important in the agriculture and nutrition of more recent American Indian populations in a unique form of treatment, lime cooking. This form of processing is still widely used today, with its nutritional advantages, in the making of corn-based products such as tacos and tortillas (Serna-Salvidar, Gomez, and Rooney, 1990). Columbus carried corn seed to Europe, where it became established as an important crop in southern latitudes. Nevertheless, U.S. corn production accounts for over half the total world production and 80 percent of the annual world corn exports (see monograph by Watson and Ramsrad, 1987).

In countries where corn is an important crop, it is the principal component of livestock feeds and most of it is fed to farm animals. In only a few countries is corn a major constituent of human diets. In developed countries, corn is consumed mainly as popcorn, sweet corn, corn snacks, and occasionally as corn bread. About one-third of processed corn is used to produce corn starch, sweeteners, corn oil, and various feed by-products. The remainder is utilized to prepare various food products and alcoholic beverages. Beyond alkali treatment, corn is prepared in several ways as human food: (1) parched to be eaten whole; (2) ground to make hominy, corn meal, or corn flour; and (3) converted to a variety of breakfast foods (Hoseney, 1986).

The corn kernel is the largest of all cereals. Kernels are usually flattened due to the pressure from adjacent kernels during growth. Dent maize is the most widely grown type of maize. The structure of the kernel is made up of four principal parts: (1) the epidermis and the seed coat (in practice called the "hull" or "bran"), (2) the endosperm (including the aleurone layer), (3) the germ, and (4) the tip cap. This last part is the point of attachment of the cob to the plant. It may or may not stay with the kernel during shelling. Table 11.5 shows typical values for the chemical composition of the various parts of the maize kernel.

The color of the kernels is quite variable; yellow to orange is the most typical. However, white or red-brown varieties are also known. The hull constitutes 5 to 6 percent of the kernel and consists mainly of cellulose and other insoluble polysaccharides. The proportion of germ is the highest among the cereal grains—about 10 percent of the kernel mass. Most of the lipids and minerals are present in the germ. The protein content of the embryo is also high. There are two major types of starchy endosperm for corn, either horny (hard, translucent) or floury (soft, opaque). The horny endosperm is tightly compact, with few or no air spaces. Its starch granules, polygonal in shape, are held together by a matrix protein. In the opaque endosperm, the starch granules are spherical and are covered with a protein matrix, and there are many air spaces between the starch granules. Flint corn varieties contain more horny than floury endosperm.

TABLE 11.5. Component parts of mature corn kernels and their chemical composition

Part of kernel	Dry weight of whole kernel (%)	Composition of kernel parts (% dry basis)				
		Starch	Fat	Protein	Ash	Sugar
Whole kernel	100.0	72.4	4.7	9.6	1.4	1.9
Germ	11.5	8.3	34.4	18.5	10.3	11.0
Endosperm	82.3	86.6	0.86	8.6	0.3	0.6
Tip cup	0.8	5.3	3.8	9.7	1.7	1.5
Pericarp	5.3	7.3	1.0	3.5	0.7	0.3

The amylose content in normal maize starch ranges from 25 to 30 percent but can vary among cultivars and especially in corns with mutant genes among the starch biosynthetic enzymes. The amylose content of starch from "high-amylose" (amylomaize) varieties can even reach up to 80 percent. In contrast, almost all of the starch derived from waxy corn, the mutant for the *wx* waxy gene, is amylopectin. A third mutant, called sugary corn, contains significantly more highly branched amylopectins than normal corn.

The protein content of the corn grain varies widely according to the variety, agronomical conditions, and other environmental factors. It ranges from 6 to 18 percent (Lasztity, 1999). The Osborne procedure has been widely used for the fractional extraction of the proteins, resulting in albumins (~4 percent), globulins (~8 percent), prolamins (called "zeins") (~50 percent), and the polymeric glutelin fraction (~40 percent). The amino acid composition of the total corn-protein fraction is characterized by low contents of lysine and tryptophan. Genotypes with high levels of lysine (opaque-2 maize) are also known (Mertz, Nelson, and Bate, 1964), but their acceptability has been hindered by their susceptibility to certain pests due particularly to their significantly higher moisture content. The nutritional advantages of some of the starch variants of maize endosperm have found new applications in various processed foods, such as are described by Branlard, Autran, and Monneveux (1989).

Genes coding the very polymorphic zein proteins are located on three different chromosomes. The genes of the Z19 polypeptides (containing 210 to 220 amino acid residues) are present in the region of about 30 crossover units on the short arms of chromosomes 7 and 9, while genes coding the Z22 polypeptides (with 240 to 245 amino acids) are scattered on both arms of chromosome 4. All of these polypeptides contain a 35 to 36 amino-acid-long N-terminal and a 10 amino-acid-long C-terminal region. The middle of the polypeptides consists of a repetitive region, built up of a 20-residue-long motif (Tatham, Shewry, and Belton, 1990).

Glutelin is a macromolecule of protein made up of diverse polypeptides (subunits) linked together via disulfide bonds. Based on solubility, the subunits are grouped into three classes: (1) The water- (and alcohol-) soluble subunits (ASG proteins) are proline-rich prolamin-like polypeptides, also called "gamma-zeins." (2) A second group of polypeptides (C- and D-zeins) are soluble in alcohols but insoluble in water. (3) Polypeptides of the third group are soluble in alkaline solutions. The most characterized polypeptide among the glutelin subunits is the ASG-1 protein. It consists of about 200 amino acids with a high proportion (8 percent) of cysteine residues. The 11 amino-acid-residue-long N-terminal region is followed by a repetitive domain consisting of a highly conserved hexapeptide motive.

The amylose content in normal maize starch ranges from 25 to 30 percent but can vary among cultivars and especially in corns with mutant genes among the starch biosynthetic enzymes. The amylose content of starch from "high-amylose" (amylomaize) varieties can even reach up to 80 percent. In contrast, almost all of the starch derived from waxy corn, the mutant for the *wx* waxy gene, is amylopectin. A third mutant, called sugary corn, contains significantly more highly branched amylopectins than normal corn.

The protein content of the corn grain varies widely according to the variety, agronomical conditions, and other environmental factors. It ranges from 6 to 18 percent (Lasztity, 1999). The Osborne procedure has been widely used for the fractional extraction of the proteins, resulting in albumins (~4 percent), globulins (~8 percent), prolamins (called "zeins") (~50 percent), and the polymeric glutelin fraction (~40 percent). The amino acid composition of the total corn-protein fraction is characterized by low contents of lysine and tryptophan. Genotypes with high levels of lysine (opaque-2 maize) are also known (Mertz, Nelson, and Bate, 1964), but their acceptability has been hindered by their susceptibility to certain pests due particularly to their significantly higher moisture content. The nutritional advantages of some of the starch variants of maize endosperm have found new applications in various processed foods, such as are described by Branlard, Autran, and Monneveux (1989).

Genes coding the very polymorphic zein proteins are located on three different chromosomes. The genes of the Z19 polypeptides (containing 210 to 220 amino acid residues) are present in the region of about 30 crossover units on the short arms of chromosomes 7 and 9, while genes coding the Z22 polypeptides (with 240 to 245 amino acids) are scattered on both arms of chromosome 4. All of these polypeptides contain a 35 to 36 amino-acid-long N-terminal and a 10 amino-acid-long C-terminal region. The middle of the polypeptides consists of a repetitive region, built up of a 20-residue-long motif (Tatham, Shewry, and Belton, 1990).

Glutelin is a macromolecule of protein made up of diverse polypeptides (subunits) linked together via disulfide bonds. Based on solubility, the subunits are grouped into three classes: (1) The water- (and alcohol-) soluble subunits (ASG proteins) are proline-rich prolamin-like polypeptides, also called "gamma-zeins." (2) A second group of polypeptides (C- and D-zeins) are soluble in alcohols but insoluble in water. (3) Polypeptides of the third group are soluble in alkaline solutions. The most characterized polypeptide among the glutelin subunits is the ASG-1 protein. It consists of about 200 amino acids with a high proportion (8 percent) of cysteine residues. The 11 amino-acid-residue-long N-terminal region is followed by a repetitive domain consisting of a highly conserved hexapeptide motive.

dia, and Uruguay. Standard test procedures have been established to assess rice-cooking quality (Juliano, 1985a). The range of chemical and technological aspects of rice utilization is reviewed in Juliano (1985b).

As a protein source, milled rice contains the lowest amount of protein (ca. 5 percent) among the major cereals; moreover, this protein content is not easily digestible by humans or by monogastric animals. However, compared to other cereal proteins, the overall amino acid composition of rice protein is significantly better balanced, because of its relatively high level of lysine. The amino acid composition of rice is unusual among the cereal grains because rice is one of the few cultivated plants in which both the globulins and prolamins, the two classes of storage proteins of higher plants, are present in significant levels (Figure 11.7). Unlike most other cereals, which accumulate prolamins as their primary nitrogen reserve, the major storage proteins in rice are the glutelins, which are homologous at the primary-sequence level to the 11S globulin proteins, a class which is the dominant form of nitrogen deposition in legumes. Furthermore, the rice prolamin proteins have a number of characteristics that are different from the prolamins of most other cereals.

Based on the classic study of Betchel and Juliano (1980), the most important characteristics of the deposition of nitrogen in the rice kernel are three kinds of protein bodies in the rice endosperm, namely, large spherical protein bodies, small spherical ones, and crystalline protein bodies, each of them surrounded by a single continuous-unit membrane. The spherical protein bodies form within vacuoles, but the proteins are synthesized in the endoplasmatic reticulum and in the Golgi apparatus before being transported to the vacuoles by vesicles.

Removal of the husk from rough rice yields the kernel, which composes the pericarp, seed coat, the aleurone layer, the endosperm, and the germ; this form is known as *brown rice* (with a protein content of 9 to 10 percent). Brown rice has a significantly higher nutritional value than white polished (milled) rice, which is the most commonly utilized rice product (with a protein content of about 8 percent). The milling process for rice results in 40 to 55 percent white milled rice and three major by-products: husks (20 percent), bran (10 percent), and "brokens" (10 to 22 percent), with protein contents of 3, 17, and 8.5 percent, respectively. The aleurone layer, the tissue with the highest levels of protein and nutritionally important minor components, is removed during the process of rice milling.

The distribution of the protein fractions from the Osborne procedure of fractionation for brown or white rice (Table 11.4) reflects the observation that the levels of the albumin and globulin classes are significantly higher in the outer layers of the seed; they decrease toward the center of the grain while the proportion of glutelins has an inverse distribution (Bechtel and

TABLE 11.4. The distribution of protein classes for brown and white rice according to the Osborne fractionation method

	Albumin	Globulin	Prolamin	Glutelin
Brown rice	10.8	9.7	2.2	77.3
White rice	6.5	12.7	8.9	71.9

Pomeranz, 1980; Takaiwa, Ogawa, and Okita, 2000). The subaleurone region plays very significant role nutritionally; it is several cell layers thick and is rich in the globulin class of proteins. Its lysine content is much higher than the proteins located in the endosperm. It is therefore desirable to mill as lightly as possible to retain most of the subaleurone layer on white rice.

The albumin fraction isolated from rice is highly heterogeneous and contains many biologically important components. It can be separated into four subfractions, based on the molecular size of the protein components, which range from 10 to 200 kDa. More than 50 individual polypeptides have been observed in the albumin fraction, based on fractionation by isoelectric focusing. Detailed studies on many of these components have shown that rice albumins have mostly enzymic or enzyme-inhibitor activities. Compared to wheat, rye, and barley, rice contains significantly lower amounts of the high-pI α-amylases and much higher levels of the low pI α-amylases.

Maize (Corn)

Corn, indigenous to North America, was developed by Central American natives many centuries before Columbus arrived. Corn was the foundation of the extensive North and South American ancient civilizations and was important in the agriculture and nutrition of more recent American Indian populations in a unique form of treatment, lime cooking. This form of processing is still widely used today, with its nutritional advantages, in the making of corn-based products such as tacos and tortillas (Serna-Salvidar, Gomez, and Rooney, 1990). Columbus carried corn seed to Europe, where it became established as an important crop in southern latitudes. Nevertheless, U.S. corn production accounts for over half the total world production and 80 percent of the annual world corn exports (see monograph by Watson and Ramsrad, 1987).

In countries where corn is an important crop, it is the principal component of livestock feeds and most of it is fed to farm animals. In only a few countries is corn a major constituent of human diets. In developed countries, corn is consumed mainly as popcorn, sweet corn, corn snacks, and oc-

sential first step in most studies. Whole-grain studies may be possible using a limited number of techniques, but it is difficult. For example, near-infrared spectroscopy is being used to estimate not only protein content but also other factors that may indicate further information about processing quality.

An ideal approach is the microscopic study of the grain after cutting sections or after breaking the grain in half for microscopic examination of morphological ultrastructure. This approach tells where the protein is located within the endosperm cells, its disproportionate distribution between the outer and inner layers of the endosperm cells, and possibly even information about where the different types of protein are laid down. Wheat-grain morphology has been reviewed by Simmonds and O'Brien (1981) and by Heneen and Brismar (1987). Figure 11.9 shows the ultrastructure of parts of the wheat grain. It should be compared to the diagram of grain structure in Figure 11.5. The outer layers of bran can be seen covering the layer of aleurone cells, with the endosperm cells on the inside, the endosperm being the material released by milling to produce white flour. The storage protein of the endosperm is the source of gluten in dough, produced by the wetting and mixing action of dough formation. However, it is difficult for microscopic examination of the intact grain to provide major information about what aspects of gluten structure account for its importance in the provision of dough properties.

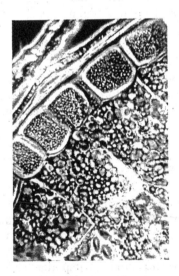

FIGURE 11.9. Light micrograph of mature wheat grain, showing the outer layers of bran and the aleurone cells surrounding the starchy endosperm (*Source:* Provided by W. P. Campbell, CSIRO, North Ryde, Australia.)

The Polypeptides of Gluten

When gluten is solubilized, even by the gentlest methods, disruption of many of the bonds that account for its cohesion occurs, particularly the noncovalent bonds such as hydrogen bonds, hydrophobic bonds, and van der Waals bonds. These many interactions have been reviewed in a book edited by Hamer and Hoseney (1997). Rupture of the disulfide bonds of gluten proteins releases the individual polypeptides, and their composition becomes accessible to study (by SDS gel electrophoresis or RP-HPLC) but we lose the information about which parts of the polypeptides are linked together by disulfide bonds. We also lose vital information about the sizes of the enormous polymers of the glutenin proteins. Figures 11.8 and 11.10 show that the resulting polypeptides may come from either gliadin or glutenin fractions. The disulfide bonds of the gliadin proteins are mainly intrachain, so that their rupture does not change their chain length. In contrast to the gliadin fraction, many of the disulfide bonds of the glutenin fraction are between the individual chains, holding them into very large polymer structures. The contrast between these two types of components is shown in the diagrammatic representation of the molecules that are present in dough (Figure 11.11).

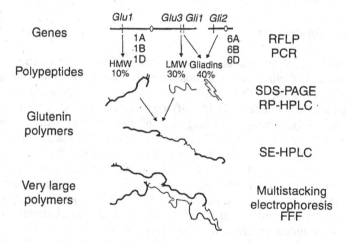

FIGURE 11.10. Diagram of the major components of the gluten complex, including locations of genes and methods of analysis (RFLP = restriction-length fragment polymorphism; PCR = polymerase chain reaction; SDS-PAGE = sodium dodecyl sulfate polyacrylamide gel electrophoresis; RP-HPLC = reversed phase high-performance liquid chromatography; SE-HPLC = size-exclusion high-performance liquid chromatography; FFF = field-flow fractionation)

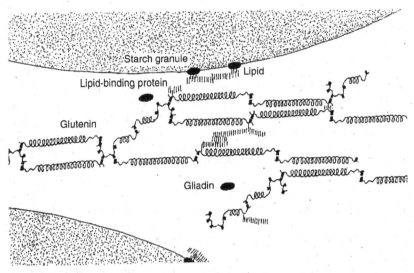

FIGURE 11.11. This diagrammatic representation of wheat-flour dough illustrates gluten as a combination of the smaller monomeric gliadin molecules con-
trasted with the very large polymeric structure of glutenin. (*Source:* Adapted from
Wrigley, 1996.)

The Gliadin Proteins

Nevertheless, determination of the composition and proportions of the
polypeptides of gluten provides considerable information about the quality
potential of the grain. The composition of the gliadin fraction has been used
for decades as a means of identifying varieties (Wrigley, Autran, and
Bushuk, 1982), thereby providing vital information about the type of processing quality that has been "built in" by the breeder. The electrophoretic
patterns for gliadin proteins from a range of wheat varieties from various
countries are compared in Figure 11.12. Gel electrophoresis of gliadin proteins has even been used to investigate the origins of archeological grain
specimens (Zeven, Doekes, and Kislev, 1975). Wheat grains likely to date
back to 500 B.C. were not suitable for analysis in this way, but useful results
could be obtained for grain of about 175 years old.

As Figure 11.10 indicates, reversed-phase high performance liquid chromatography (RP-HPLC) is also used to fractionate the polypeptides of the
gluten proteins and thus to provide information about genotype and variety.
The gliadin proteins are coded by the families of *Gli-1* genes on the short
arms of the group-1 chromosomes and by the *Gli-2* genes on the short arms
of the group-6 chromosomes. The chromosomal locations of the genes for

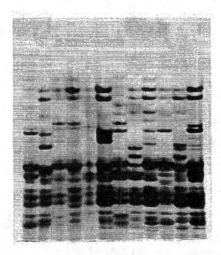

FIGURE 11.12. Gliadin-protein patterns for a series of varieties of hexaploid (bread) wheat, showing how cathodic gel electrophoresis at pH 3 can be used to distinguish between varieties. Gliadin proteins were extracted from crushed grains with 6 percent urea solution and the clarified extracts were applied to a polyacrylamide gel, whose concentration varied from 3 percent at the top to 13 percent acrylamide at the bottom (negative electrode). From left to right, the varieties are Scout 66, Inia 66R, Capitole, Diplomat, Marquis, Chinese Spring, Eagle (Australian version), Halberd, Millewa, Olympic, Jabiru, and Lance. (*Source:* From C. W. Wrigley, J.-C. Autran, and W. Bushuk, 1982, Identification of cereal varieties by gel electrophoresis of the grain proteins. *Advances in Cereal Science and Technology* 5: 211-259. Reproduced by permission.)

the respective gluten proteins have been used as a major basis for developing gluten-protein nomenclature (Wrigley, Bushuk, and Gupta, 1996). The subfractions of the gliadin class of monomeric gluten proteins have been distinguished according to their mobilities in acidic gel electrophoresis.

Sulfur Deficiency and the Omega-Gliadin Proteins

The slowest-moving proteins, the omega-gliadins, are distinguished from the alpha-, beta-, and gamma-gliadins by being almost completely lacking in the sulfur-containing amino acids cysteine and methionine (see Figure 11.8). This difference is especially evident when analyzing the gliadin proteins from grain grown under sulfur-deficient conditions (Randall and Wrigley, 1986), as is shown in Figure 11.13, using either single-dimension acidic gel electrophoresis or a two-dimensional combination of elec-

(a)   (b)

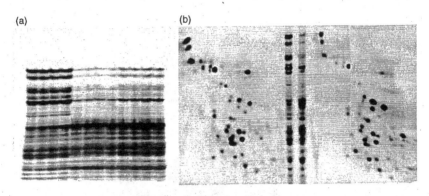

FIGURE 11.13. The gliadin proteins of hexaploid (bread) wheat, illustrating the effects of sulfur deficiency, indicated (a) by one-dimensional cathodic gel electrophoresis at pH 3, and (b) by two-dimensional isoelectric focusing followed by pH 3 gel electrophoresis. In part (a), the four lanes on the left correspond to grain grown with severe sulfur deficiency and adequate nitrogen fertilizer, the central group corresponds to grains fertilized with sulfur but not nitrogen, and the four lanes on the right are for grain with adequate nitrogen and sulfur nutrition. The pattern on the left of part (b) shows the effect of sulfur deficiency, compared to the control on the right.

trophoretic methods. Moss and colleagues (1981) showed that dough quality was severely impaired for grain from sulfur-deficient soil, especially with increased application of nitrogen fertilizer, due to changes in grain-protein composition resulting in the reduced synthesis of sulfur-rich proteins (Kettlewell et al., 1998; Wrigley et al., 1984). These differences relate not only to the gliadin fraction, but also to the subfractions of glutenin (MacRitchie and Gupta, 1993). For both types of analysis of the gliadins (parts a and b of Figure 11.13), sulfur deficiency has affected the proportions of the components of slowest mobility (the omega-gliadins). These are more prominent on the left of the two-dimensional patterns (for sulfur-deficient grain). The four patterns on the left of Figure 11.13a are from grain grown with severe sulfur deficiency with adequate nitrogen fertilizer; the omega-gliadin bands are much more prominent than for the other samples. These grain samples had low levels of sulfur (0.075, 0.081, 0.085, and 0.091 percent S, left to right). The central group of patterns in Figure 11.13a are for grain fertilized with sulfur but not nitrogen (0.107, 0.111, 0.116, and 0.122 percent S). The right-hand group of patterns are for "normal" grain with adequate nitrogen and sulfur nutrition (0.135, 0.146, 0.157, and 0.161 percent S). Studies of the sulfur-poor storage proteins of cereals have been extended to rye and barley (Tatham and Shewry, 1995).

Glutenin Subunits

Gel electrophoresis of flour proteins in the presence of sodium dodecyl sulfate (SDS) shows the glutenin proteins as an unresolved streak of protein staining (Wrigley, Gupta, and Bekes, 1993). This is understood to indicate that the disulfide-linked polymers of glutenin subunits cover a wide range of molecular dimensions and that there is no ordered grouping of polymer dimensions. The sizes of these glutenin polymers range up into the tens of millions of Daltons, as has been shown by flow field-flow fractionation (FFF in Figure 11.9) (Stevenson and Preston, 1996; Wrigley, 1996).

However, when a reducing agent is added in the process of extracting flour (generally in the presence of SDS), disulfide bonds are broken and SDS-gel electrophoresis demonstrates that there are many discrete polypeptide components, covering the molecular weight range from about 30,000 to 120,000 Daltons (Figure 11.14). The larger polypeptides (the

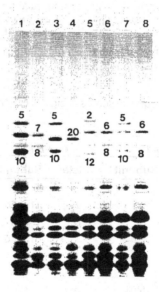

FIGURE 11.14. SDS-gel electrophoresis of reduced glutenin polypeptides from a series of durum-wheat genotypes and near-isogenic lines (NIL) in which pairs of HMW subunits (numbered) have been incorporated. (1) Svevo-derived NIL with 5+10; (2) Svevo; (3) Lira-derived NIL with 5+10; (4) Lira; (5) Zenit-derived NIL with 2+12; (6) Zenit; (7) Zenit-derived NIL with 5+10; (8) Zenit. (*Source:* From D. Lafiandra et al., 1999, The formation of glutenin polymer in practice, *Cereal Foods World* 44: 572-578. Reproduced with permission.)

high-molecular weight [HMW] subunits) stand out above all the other flour components; they have been allocated numbers according to their mobilities in the gel pattern, as is indicated in Figure 11.14. The list of numbered subunits in Table 11.6 also indicates the corresponding allele designation (as lowercase italic letters). The HMW subunits are synthesized under the control of *Glu-1* genes on the long arms of wheat chromosomes 1A, 1B, and 1D, as indicated in Figure 11.10. The HMW subunits often appear as pairs (e.g., 5 with 10, 2 with 12, and 17 with 18), because their synthesis is controlled by corresponding pairs of genes at the *Glu-1* locus. This is also indicated in Table 11.6, which lists the correspondence between the subunit numbers and the gene designations (alleles). For example, subunit 1 is synthesized under the control of allele *a* (also designated as *Glu-A1a*), and subunits 5 and 10, under the control of the allele *Glu-D1d*.

The subunits of glutenin have been reported to have a central domain with a highly stable spiral structure, stabilized by extensive inter-turn hydrogen bonding, involving the glutamine side chains of the repeating amino acid sequences which are rich in glutamine, proline, and glycine (Kasarda, King, and Kumosinski, 1994). This spiral structure is proposed as a means by which the glutenin subunits may confer elastic properties in dough; it is the reason that spiral structures have been used in Figure 11.11 to depict the main central region of the glutenin subunits.

TABLE 11.6. *Glu-1* dough-quality scores assigned to HMW-glutenin subunits and corresponding alleles

Glu-1 Score	Glu-A1 Allele	Glu-A1 Subunit	Glu-B1 Allele	Glu-B1 Subunit	Glu-D1 Allele	Glu-D1 Subunit
4					d	5+10
3	a	1	i	17+18		
3	b	2*	b	7+8		
3			f	13+16		
2					a	2+12
2					b	3+12
1	c	Null	a	7	c	4+12
1			d	6+8		
1			e	20		

Source: According to Payne, 1987.
Note: A high score (maximum of 10) indicates the prediction of strong dough properties.

The low-molecular weight (LMW) subunits, which outweigh the HMW subunits by three to one, appear farther down the SDS-gel electrophoretic pattern, but their presence is normally disguised by the overlapping bands of non-glutenin components, mainly gliadins. This difficulty has been overcome by a two-step method of electrophoresis or by selective extraction procedures (Gupta and Shepherd, 1990). The LMW subunits are coded by the families of *Glu-3* genes on the short arms of the group-1 chromosomes, tightly linked to the genes for some of the gliadin proteins (the *Gli-1* alleles, shown in Figure 11.10). Accordingly, a unified classification system has been proposed for the alleles of the gliadins *(Gli-1)* and LMW-subunits of glutenin *(Glu-3)* (Jackson et al., 1997).

Prediction of Dough Quality from Glutenin Allelic Constitution

The functional properties of the HMW subunits have been deduced by a range of methods. Initial indications came from quality analyses of progeny from crosses between wheats that differed in glutenin composition (Payne, Corfield, and Blackman, 1979). In addition, correlations for subunit composition to quality have been studied for many sets of wheat genotypes (e.g., Gupta, Bekes, and Wrigley, 1991; Gupta et al., 1994), and from durum lines into which were incorporated pairs of glutenin polypeptides (Lafiandra et al., 1999). A more direct approach has been the analysis of genotypes devised to have some or all of the HMW subunits absent (Lawrence, MacRitchie, and Wrigley, 1988), as is shown in Figure 11.15. In this case, the loss of any one of the subunits reduced dough quality, but of the three sets of glutenin subunits, the loss of the 5+10 pair (for lines corresponding to lanes b, c, and d in Figure 11.15) caused the most dramatic loss of dough properties. Loss of two or more subunits caused progressively further losses of quality. Dough-forming ability was lost completely with the loss of all HMW subunits, even though the full complement of LMW subunits was retained. Other approaches have involved the development of lines having specific interchanges of subunits and most recently by the isolation of individual subunits (or their genes) for direct testing by incorporation into the glutenin structure of a common parent dough (Bekes et al., 1994; Barro et al., 1997). In addition, confirmatory data about the contributions to dough quality of these proteins has been obtained by transformation studies, in which the genes for the polypeptides have been inserted into a common wheat background (e.g., one of the lines that lack the full complement of subunits in Figure 11.15) (Barro et al., 1997).

The contributions of the individual HMW subunits of glutenin have been ranked according to their additive effects by Payne (1987). These are set out

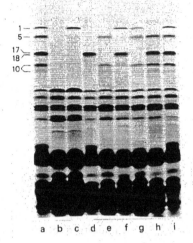

FIGURE 11.15. SDS-gel electrophoresis of reduced glutenin polypeptides from a series of wheat genotypes and near-isogenic lines (NIL) in which specific combinations of HMW subunits have been removed (*Source:* From G. J. Lawrence, F. MacRitchie, and C. W. Wrigley, 1988, Dough and baking quality of wheat lines deficient for glutenin subunits controlled by Glu-A1, Glu-B1, and Glu-D1 loci. *Journal of Cereal Science* 7: 109-112. Reproduced with permission from Academic Press.)

systematically in Table 11.6 as score numbers. The total score for a variety is based on the set of its three HMW subunits as the sum of the contribution from each subunit (one for each of the three genomes of wheat—A, B, and D). For example, a score of 10 (highest possible) would be obtained for alleles *a, i,* and *d* (in each of the A, B, and D genomes, corresponding to protein subunits 1, 17+18, and 5+10), derived from individual scores of 3, 3, and 4, respectively, according to Table 11.6. As these rankings indicate the importance of the subunits in relation to dough properties, they can be used in selecting breeding lines for processing quality. This strategy of predicting dough quality is valuable because the *Glu-1* constitution of many wheats is now known. Catalogs of these subunits have been reported for many national collections of wheats (e.g., for Canada by Lukow, Payne, and Tkachuk, 1989; for Australia by Lawrence, 1986), and the worldwide distribution of these alleles has been surveyed by Morgunov and colleagues (1993).

However, in several of these surveys of national wheats, the *Glu-1* quality score accounts for only part of the variation in dough or baking quality, e.g., for only about 60 percent of the variation in bread-making quality for 67 Canadian varieties (Lukow, Payne, and Tkachuk, 1989) and more or less for other national collections (MacRitchie, du Cros, and Wrigley, 1990). This is partly because the HMW-subunits of glutenin constitute only about one-quarter of the total glutenin protein; the remaining portion is made up of the LMW subunits. Nevertheless, the HMW subunits appear to have a disproportionately larger contribution to dough properties, probably because of their larger size. The LMW subunits of glutenin also confer a moderating effect on dough properties, but this contribution has been more difficult to quantify, due to their greater diversity and the relative difficulty in analyzing for LMW-subunit composition. Gupta and colleauges (1994) analyzed the HMW and LMW subunits of the glutenin proteins of 74 recombinant inbred lines that were homozygous for these alleles. They found that about 80 percent of the variability in dough strength (measured as Rmax in the Extensigraph) could be accounted for if both LMW and HMW subunit composition were considered, but that predictability was much worse if only one or other aspect of glutenin subunit composition were considered. Cornish, Panozzo, and Wrigley (1999) and Cornish, Griffin, and Wrigley (2000) have shown how breeding for specific quality traits can be manipulated by consideration of overall glutenin-subunit composition.

The pasta-cooking quality of the durum wheats might be expected to be reflected in the composition of their HMW glutenin subunits. However, most studies have indicated that the LMW subunits of durum glutenin have better predictive value than their HMW subunits (Branlard, Autran, and Monneveux, 1989; Feillet et al., 1989; Pogna et al., 1988).

The Size Distribution of Glutenin Polymers

The most recent studies aimed at relating gluten composition to functional properties in dough have concentrated on the size distribution of the native polymers of glutenin. Indeed, this concept may be extended to a consideration of size distribution for the full complement of gluten proteins, thereby including the monomeric gliadin proteins and the ratio of gliadin to glutenin proteins. Size distribution for the gluten proteins thus extends from the relatively small gliadin proteins up into the tens of millions of Daltons for the largest of the glutenin polymers (Southan and MacRitchie, 1999; Zhu and Khan, 2001). It appears to be these very large molecules that contribute the resistance to extension which is critical to dough strength, while the range of smaller proteins provides the balance of dough viscosity. Furthermore, some of the subunits of glutenin seem to be more effective than others in contributing to the functional properties of glutenin, possibly by providing extra size (length) to its polymeric structure. A reduction in size distribution of glutenin has been shown to be a critical factor in explaining the loss of dough strength (Lafiandra et al., 1999) that is often associated with heat stress during grain filling in the field (Blumenthal, Barlow, and Wrigley, 1993; Ciaffi et al., 1996; Corbellini et al., 1997, 1998).

GRAIN HARDNESS

The need for appropriate dough properties is complemented by that of grain hardness (Table 11.2). In the technological sequence of grain utilization, grain hardness has its first major role in grain milling. When soft wheats are milled, the endosperm falls apart readily, allowing the individual starch granules to separate from one another, with minimal damage to the surface of the granules. As a result, the remnants of the membranes surrounding the granules remain undamaged, providing a degree of resistance to the attack of starch-degrading enzymes. On the other hand, when hard wheats are milled, the endosperm tends to hold together, and fractures may occur right through the starch granules. As a result, hydrolytic enzymes have ready access to the starch polymers inside the granules during the later stages of processing, such as fermentation and baking. Furthermore, water absorption is likely to be higher due to the damage to the granules.

Grain hardness is largely determined by one major gene, *Ha*, located on the short arm of chromosome 5D (Turner et al., 1999). One of the proteins associated with the starch granules ("friabilin") has been implicated as being a potential marker of grain hardness (Morrison et al., 1992). This protein (of about 15,000 Daltons) is present on the surface of starch granules washed from soft wheats and is absent or present in smaller amounts on the surface of granules from hard wheats. It is a member of a family of hydrolase inhibitors and puroindolines (Rahman et al., 1994), particularly puroindolines *a* and *b* (Giroux and Morris, 1997, 1998). More recent results indicate that biochemical factors in addition to the puroindolines are involved in grain hardness, but the presence of puroindoline *b* has been found to correlate with bread-making quality (as loaf volume) (Igrejas et al., 2001).

STARCH PROPERTIES

Other specific proteins of the starch granule have been associated with variations in the functional properties of wheat starch, particularly the ratio of amylose to amylopectin (Rahman et al., 2000). The extent of synthesis of the smaller linear molecule, amylose, has been shown to be determined by three isoforms of the enzyme granule-bound starch synthase (GBSS) (Nakamura et al., 1993). These isoforms are coded by homologous genes on chromosomes 7A, 7D, and 4A, designated as the "waxy" genes *Wx-A1, Wx-D1,* and *Wx-B1,* respectively. A wheat genotype that lacks all three of these genes produces no amylose due to the absence of the GBSS enzymes; all its starch is of the amylopectin type—highly branched starch with large molecular weight. This is known as waxy wheat because of the translucent appearance of the endosperm (Zhao, Shariflou, et al., 1998). The production of the waxy genotype is new in wheat, whereas waxy types have long been known in rice, maize, barley, and sorghum. The starch from waxy wheat has distinct functional properties, namely, greater hot-paste viscosity when heated with water and better freeze-thaw stability for the resulting gel, which is optically clear.

The other extreme genotype, having all three isoforms of GBSS, is the most common in wheat; it produces 20 to 25 percent amylose, the remainder being amylopectin. All intermediate combinations of the three genes have been produced, lacking one or two of the genes; these contain intermediate levels of amylose (Yamamori and Quynh, 2000; Zhao, Shariflou, et al., 1998). The genotype lacking the 4A gene *(Wx-B1)* produces wheats having starch properties that are particularly suited to the production of various types of noodles (Zhao, Batey, et al., 1998). Null-4A *(Wx-B1)* genotypes have starch properties with higher viscosity (when heated in water) and greater swelling power. These properties are conducive to the production of many types of noodles, especially the Japanese udon type. The story of the development of the range of reduced-amylose wheats has been reviewed by Seib (2000).

A recent breakthrough in starch properties has been the ability to manipulate the proportions of the A-type (large) starch granules versus the small B granules (Stoddard, 1999). Earlier studies had been hampered by an apparent lack of genetic diversity with respect to this trait. In addition, it had been assumed to be of relatively little importance for baking quality, although granule-size distribution has been reported to contribute to dough properties and to water binding (Rasper and deMan, 1980). However, granule-size distribution is critical to the starch-gluten industry; a high proportion of A granules means a better yield of high-quality starch, and the consequent lower proportion of the small B granules means less starch appearing in effluent streams, thereby incurring reduced disposal problems (Rahman et al., 2000). It now appears likely that wheat cultivars can be developed with a wider range of size distribution in their starch granules (Stoddard, 1999).

CONCLUSION

Humankind is critically dependent on the cereal grains for nutrition—today as much as in the early days of civilization. Today, however, we have the great advantage of genotypes that are well adapted to the sites of cultivation, yielding grain suited to the specific requirements of processing and consumption. These advantages have been won over millennia of effort. In prehistory and in more recent times, this involved the careful selection of plants whose grain appeared to be better for grinding and baking. During the past century, the range of genetic diversity has been extended by cross-breeding, permitting desirable traits to be combined from a few parents into one genotype. These methods, now considered to be conventional, may next be extended by genetic-engineering technology, with even wider genetic diversity being possible from which to make selections for improvement. Nevertheless, even the imaginative use of the more conventional methods is still proving effective in producing improved genotypes, as indicated by the examples described for better dough and starch properties in wheat.

REFERENCES

American Association of Cereal Chemists (2001). The definition of dietary fiber. *Cereal Foods World* 46: 112-129.

Athwal, D.L. (1971). Semidwarf rice and wheat in global food needs. *Quarterly Reviews of Biology* 46: 1-34.

Bailey, H.C. (1941). A translation of Beccari's lecture "Concerning Grain" (1729). *Cereal Chemistry* 18: 555-561.

Barro, F., Rooke, L., Bekes, F., Gras, P., Tatham, A.S., Fido, R., Lazzeri, P.A., Shewrym, P.R., and Barcelo, P. (1997). Transformation of wheat with high molecular weight subunit genes results in improved functional properties. *Nature Biotechnology* 15: 1295-1299.

Bechtel, D.B. and Juliano, B.O. (1980). Protein body formation, the starchy endosperm of rice. *American Journal of Botany* 67: 966-973.

Bechtel, D.B. and Pomeranz, Y. (1980). The rice kernel. *Advances in Cereal Science and Technology* 3: 73-113.

Bekes, F., Gras, P.W., Gupta, R.B., Hickman, D.R., and Tatham, A.S. (1994). Effects of a high Mr glutenin subunit (1Bx20) on the dough mixing properties of wheat flour. *Journal of Cereal Science* 19: 3-7.

Bietz, J.A. and Lookhart G.L. (1996). Properties and non-food potential of gluten. *Cereal Foods World* 41: 376-382.

Blumenthal, C.S., Barlow, E.W.R., and Wrigley, C.W. (1993). Growth environment and wheat quality: The effect of heat stress on dough properties and gluten proteins. *Journal of Cereal Science* 18: 3-21.

Branlard, G., Autran, J.C., and Monneveux, P. (1989). High molecular weight glutenin subunit in durum wheat (*T. durum*). *Theoretical and Applied Genetics* 78: 353-358.

Bushuk, W. (2001a). Rye production and uses worldwide. *Cereal Foods World* 46(2): 70-73.

Bushuk, W. (2001b). *Rye: Production, Chemistry and Technology*, Second Edition. St. Paul, MN: American Association of Cereal Chemists.

Choct, M. and Annison, G. (1990). Anti-nutritive activity of wheat pentosans in broiler diets. *British Poultry Science* 31: 811-821.

Ciaffi, M., Tozzi, L., Borghi, B., Corbellini, M., and Lafiandra, D. (1996). Effect of heat shock during grain filling on the gluten protein composition of bread wheat. *Journal of Cereal Science* 24: 91-100.

Corbellini, M., Canevar, M.G., Mazza, L., Ciaffi, M., Lafiandra, D., and Borghi, B. (1997). Effect of the duration and intensity of heat shock during grain filling on dry matter and protein accumulation, technological quality and protein composition in bread and durum wheat. *Australian Journal of Plant Physiology* 24: 245-260.

Corbellini, M., Mazza, L., Ciaffi, M., Lafiandra, D., and Borghi, B. (1998). Effect of heat shock during grain filling on protein composition and technological quality of wheat. *Euphytica* 100: 147-154.

Cornish, G.B., Griffin, W.B., and Wrigley, C.W. (2000). Glutenin alleles and their influence on the quality of New Zealand wheats. In Panozzo, J., Radcliffe, M., Wootton, M., and Wrigley, C.W. (Eds.), *Cereals '99* (pp. 362-371). Melbourne: Royal Australian Chemical Institute.

Cornish, G.B., Panozzo, J.F., and Wrigley, C.W. (1999). Victorian wheat protein families. In O'Brien, L., Blakeney, A.B., Ross, A.S., and Wrigley, C.W. (Eds.), *Cereals '98* (pp. 183-188). Melbourne: Royal Australian Chemical Institute.

Egorov, T.A., Musolyamov, A.K., Anderson, J.S., and Roepstorff, P. (1994). The complete amino-acid sequence and disulphide arrangement of oat alcohol-soluble avenin-3. *European Journal of Biochemistry* 224: 631-638.

Faridi, H. and Faubion, J. (1995). *Wheat End Uses Around the World*. St. Paul, MN: American Association of Cereal Chemists.

Feillet, P., Ait-Mouh, O., Kobrehel, K., and Autran, J.-C. (1989). The role of low molecular weight glutenin proteins in the determination of cooking quality of pasta products: An overview. *Cereal Chemistry* 66: 26-30.

Giroux, M.J. and Morris, C.F. (1997). A glycine to serine change in puroindoline b is associated with hardness and low levels of starch-surface friabilin. *Theoretical and Applied Genetics* 95: 857-864.

Giroux, M.J. and Morris, C.F. (1998). Wheat grain hardness results from highly conserved mutations in the friabilin components puroindoline a and b. *Proceedings of the National Academy of Sciences, USA* 95: 6262-6266.

Gupta, R.B., Bekes, F., and Wrigley, C.W. (1991). Prediction of physical dough properties from glutenin subunit composition in bread wheats: Correlation studies. *Cereal Chemistry* 68: 328-333.

Gupta, R.B., Paul, J.G., Cornish, G.B., Palmer, G.A., Bekes, F., and Rathjen, A.J. (1994). Allelic variation at glutenin subunit and gliadin loci, Glu-1, Glu-3 and

Gli-1, of common wheats: 1. Its additive and interactive effects on dough properties. *Journal of Cereal Science* 19: 9-17.

Gupta, R.B. and Shepherd, K.W. (1990). Two-step one-dimensional SDS-PAGE analysis of LMW subiunits of glutelin: 1. Variation and genetic control of the subunits in hexaploid wheats. *Theoretical and Applied Genetics* 80: 65-74.

Hamer, R.J. and Hoseney, R.C. (1997). *Interactions: The Keys to Cereal Quality.* St. Paul, MN: American Association of Cereal Chemists.

Heneen, W.K. and Brismar, K. (1987). Scanning electron microscopy of mature grains of rye, wheat and triticale with emphasis on grain shriveling. *Hereditas* 107: 147-162.

Hoseney, R.C. (1986). *Principles of Cereal Science and Technology.* St. Paul, MN: American Association of Cereal Chemists.

Igrejas, G., Gaborit, T., Oury, F.-X., Ciron, H., Marion, D., and Marion, G. (2001). Genetic and environmental effects on puroindoline-a and puroindoline-b content and their relationship to technological properties in French bread wheats. *Journal of Cereal Science* 34: 37-47.

Jackson, E.A., Morel, M-H., Sontag-Strom, T., Branlard, G., Metakovsky, E.V., and Radaelli, R. (1997). Proposal for combining the classification systems of alleles of Gli-1and Glu-3 loci in bread wheat (*Triticum aestivum* L.). *Journal of Genetics and Breeding* 50: 321-336.

Juliano, B.O. (1985a). Cooperative tests on cooking properties of milled rice. *Cereal Foods World* 30: 651-656.

Juliano, B.O. (1985b). *Rice Chemistry and Technology,* Second Edition. St. Paul, MN: American Association of Cereal Chemists.

Kasarda, D.D. (2001). Grains in relation to celiac disease. *Cereal Foods World* 46: 209-210.

Kasarda, D.D., King, G., and Kumosinski, T.F. (1994). Comparison of spiral structures in wheat high molecular weight glutenin subunits and elastin by molecular modeling. In Kumosinski, T.F. and Lieberman, M.N. (Eds.), *ACS Symposium Series No. 576. Molecular Modelling: From Virtual Tools to Real Problems* (pp. 209-220).Washington, DC: American Chemical Society.

Kettlewell, P.S., Griffiths, M.W., Hocking, T.J., and Wallington, D.J. (1998). Dependence of wheat dough extensibility on flour sulphur and nitrogen concentrations and the influence of foliar-applied sulphur and nitrogen fertilisers. *Journal of Cereal Science* 28: 15-23.

Krishnan, P.G., Reeves, D.L., Kephart, K.D., Thiex, N., and Calimente, M. (2000). Robustness of near infrared reflectance spectroscopy measurement of fatty acids and oil concentrations in oats. *Cereal Foods World* 45: 513-519.

Lafiandra, D., Masci, S., Blumenthal, C., and Wrigley, C.W. (1999). The formation of glutenin polymer in practice. *Cereal Foods World* 44: 572-578.

Lasztity, R. (1999). *Cereal Chemistry.* Budapest, Hungary: Akadémiai Könyvkiadó.

Lawrence, G.J. (1986). The high molecular weight subunit composition of Australian wheat cultivars. *Australian Journal of Agricultural Research* 37: 125-133.

Lawrence, G.J., MacRitchie, F., and Wrigley, C.W. (1988). Dough and baking quality of wheat lines deficient for glutenin subunits controlled by Glu-A1, Glu-B1 and Glu-D1 loci. *Journal of Cereal Science* 7: 109-112.

Lehmann, J.W. (1996). Case history of grain amaranth as an alternative crop. *Cereal Foods World* 41: 399-411.

Leite, A., Neto, G.C., Vettore, A.L., Yunes, J.A., and Arruda, P. (1999). The prolamins of sorghum, Coix and millets. In Shewry, P.R. and Casey, R. (Eds.), *Seed Proteins* (pp. 141-157). Dordrecht, the Netherlands: Kluwer Academic Publishers.

Lukow, O.M., Payne, P.I., and Tkachuk, R. (1989). The HMW glutenin subunit composition of Canadian wheat cultivars and their association with bread-making quality. *Journal of the Science of Food and Agriculture* 47: 451-460.

MacGregor, A.W. and Bhatty, R.S. (1993). *Barley: Chemistry and Technology*. St. Paul, MN: American Association of Cereal Chemists.

MacRitchie, F., du Cros, D.L., and Wrigley, C.W. (1990). Flour polypeptides related to wheat quality. *Advances in Cereal Science and Technology* 10: 79-145.

MacRitchie, F. and Gupta, R.B. (1993). Functionality-composition relationships of wheat flour as a result of variation in sulfur availability. *Australian Journal of Agricultural Research* 44: 1767-1774.

Malkki, Y. (2001). Physical properties of dietary fiber as keys to physiological functions. *Cereal Foods World* 46: 196-199.

Martin, A., Alvarez, J.B., Martin, L.M., Barro, F., and Ballesteros, J. (1999). The development of tritordeum: A novel cereal for food processing. *Journal of Cereal Science* 30: 85-95.

Mertz, E., Nelson, O., and Bate, L.S. (1964). Mutant gene that changes composition and increases lysine content of maize endosperm. *Science* 154: 279-281.

Morgunov, A.I., Pena, R.J., Crossa, J., and Rajaram, S. (1993). Worldwide distribution of Glu-1 alleles in bread wheat. *Journal of Genetics and Breeding* 47: 53-60.

Morris, C.F. and Rose, S.P. (1996). Wheat. In Henry, R.J. and Kettlewell, P.S. (Eds.), *Cereal Grain Quality* (pp. 3-54). London: Chapman and Hall.

Morrison, W.R., Greenwell, P., Law, C.N., and Sulaiman, B.D. (1992). Occurrence of friabilin, a low molecular weight protein associated with grain softness, on starch granules isolated from some wheats and related species. *Journal of Cereal Science* 15: 143-149.

Moss, H.J., Miskelly, D.M., and Moss, R. (1986). The effect of alkaline conditions on the properties of wheat flour dough and Cantonese-style noodles. *Journal of Cereal Science* 4: 261-268.

Moss, H.J., Wrigley, C.W., MacRitchie, F., and Randall, P.J. (1981). Sulfur and nitrogen fertiliser effects: II. Influence on grain quality. *Australian Journal of Agricultural Research* 32: 213-226.

Nakamura, T., Yamamori, M., Hirano, H., and Hidaka, S. (1993). Decrease in waxy (Wx) in two common wheat cultivars with low amylose content. *Plant Breeding* 111: 99-105.

Nelson, A.L. (2001). Properties of high-fiber ingredients. *Cereal Foods World* 46: 93-97.

Osborne, T.B. and Vorhees, C.G. (1893). The proteids of the wheat kernel. *Journal of the American Chemical Society* 15: 392-471.

Payne, P.I. (1987). Genetics of wheat storage proteins and the effect of allelic variation on bread-making quality. *Annual Review of Plant Physiology* 38: 141-153.

Payne, P.I., Corfield, K.G., and Blackman, J.A. (1979). Identification of a high-molecular-weight subunit of glutenin whose presence correlates with bread-making quality in wheats of related pedigree. *Theoretical and Applied Genetics* 55: 153-159.

Pogna, N., Lafiandra, D., Feillet, P., and Autran, J.C. (1988). Evidence for a direct causal effect of low molecular weight subunits of glutenins on gluten viscoelasticity in durum wheats. *Journal of Cereal Science* 7: 211-214.

Qarooni, J. (1996). Wheat characteristics for flat breads. *Cereal Foods World* 41: 391-395.

Qarooni, J., Ponte, J.G., and Posner, E.S. (1992). Flat breads of the world. *Cereal Foods World* 37: 863-865.

Rahman, S., Jolly, C.J., Skerritt, J.H., and Walloscheck, A. (1994). Cloning of a wheat 15-kDa grain softness protein (GSP): GSP is a mixture of puroindoline-like polypeptides. *European Journal of Biochemistry* 223: 917-925.

Rahman, S., Li, Z., Batey, I., Cochrane, M.P., Appels, R., and Morell, M. (2000). Genetic alteration of starch functionality in wheat. *Journal of Cereal Science* 31: 91-110.

Randall, P.J. and Wrigley, C.W. (1986). Effects of sulfur deficiency on the yield, composition and quality of grain from cereals, oil seeds and legumes. *Advances in Cereal Science and Technology* 8: 171-206.

Rasper, V.F. and deMan, J.M. (1980). Effect of granule size of substituted starches on the rheological character of composite doughs. *Cereal Chemistry* 50: 331-340.

Ronalds, J.A. (1974). The determination of the protein content of wheat and barley by direct alkaline distillation. *Journal of the Science of Food and Agriculture* 25: 179-185.

Rooney, L. (1996). Sorghum and millets. In Henry, R.J. and Kettlewell, P.S. (Eds.), *Cereal Grain Quality* (pp. 153-177). London: Chapman and Hall.

Rooney, L., Kirleis, A.W., and Murty, D.S. (1987). Traditional foods for sorghum: Their production, evaluation, and nutritional value. *Advances in Cereal Science and Technology* 8: 317-353.

Seib, P.A. (2000). Reduced-amylose wheats and Asian noodles. *Cereal Foods World* 45: 504-512.

Serna-Salvidar, S.O., Gomez, M.H., and Rooney, L.W. (1990) Technology, chemistry and nutritional value of alkaline-cooked corn products. *Advances in Cereal Science and Technology* 10: 243-308.

Shewry, P.R. (1996). Cereal grain proteins. In Henry, R.J. and Kettlewell, P.S. (Eds.), *Cereal Grain Quality* (pp. 227-250). London: Chapman and Hall.

Shewry, P.R. (1999). Avenins: The prolamins of oats. In Shewry, P.R. and Casey, R. (Eds.), *Seed Proteins* (pp. 79-92). Dordrecht, the Netherlands: Kluwer Academic Publishers.

Shotwell, M.A. (1999). Oat globulins. In P.R. and Casey, R. (Eds.), *Seed Proteins* (pp. 389-400). Dordrecht, the Netherlands: Kluwer Academic Publishers.

Simmonds, D.H. and O'Brien, T.P. (1981). Morphological and biochemical development of the wheat endosperm. *Advances in Cereal Science and Technology* 4: 5-70.

Skerritt, J.H. (1986). Molecular comparison of alcohol-soluble wheat and buckwheat proteins. *Cereal Chemistry* 63: 365-369.

Skerritt, J.H., Devery, J.M., and Hill, A.S. (1990). Gluten intolerance: Chemistry, celiac toxicity, and detection of prolamins in foods. *Cereal Foods World* 35: 638-644.

Skerritt, J.H. and Hill, A.S. (1991). Self-management of dietary compliance in celiac disease by means of ELISA "home test" to detect gluten. *Lancet* 337: 379-382.

Slavin, J., Marquart, L., and Jacobs, D. (2000). Consumption of whole-grain foods and decreased risk of cancer: Proposed mechanisms. *Cereal Foods World* 45: 54-58.

Southan, M. and MacRitchie, F. (1999). Molecular weight distribution of wheat proteins. *Cereal Chemistry* 76: 827-836.

Stevenson, S.G. and Preston, K.R. (1996). Flow field-flow fractionation of wheat proteins. *Journal of Cereal Science* 23: 121-131.

Stoddard, F. (1999). Survey of starch particle-size distribution in wheat and related species. *Cereal Chemistry* 76: 145-149.

Takaiwa, F., Ogawa, M., and Okita, T.W. (2000). Rice glutelins. In Shewry, P.R. and Casey, R. (Eds.), *Seed Proteins* (pp. 93-108). Dordrecht, the Netherlands: Kluwer Academic Publishers.

Tatham, A.S. and Shewry, P.R. (1995). The S-poor prolamins of wheat, barley and rye. *Journal of Cereal Science* 22: 1-16.

Tatham, A.S., Shewry, P.R., and Belton, P.S. (1990). Structural studies of cereal prolamins, including wheat gluten. *Advances in Cereal Science and Technology* 10: 1-78.

Turner, M., Mukai, Y., Leroy, P., Charef, B., Appels, R., and Rahman, S. (1999). The Ha locus of wheat: Identification of a polymorphic region for tracing grain hardness in crosses. *Genome* 42: 1242-1250.

Watson, L. and Wrigley, C.W. (1984). Relationships between plants relevant to allergy. *Medical Journal of Australia* Supplement 141: S18.

Watson, S.A. and Ramsrad, P.E. (1987). *Corn Chemistry and Technology*. St. Paul, MN: American Association of Cereal Chemists.

Webster, F.H. (1986). *Oats: Chemistry and Technology*. St. Paul, MN: American Association of Cereal Chemists.

Webster, F.H. (1996). Oats. In Henry, R.J. and Kettlewell, P.S. (Eds.), *Cereal Grain Quality* (pp. 179-203). London: Chapman and Hall.

Wood, P.J. (1993). *Oat Bran*. St. Paul, MN: American Association of Cereal Chemists.

Wright, A. (2001). *Australian Commodities; Forecasts and Issues*. Canberra: Australian Bureau of Agricultural and Resource Economics (ABARE).

Wrigley, C.W. (1994). Developing better strategies to improve grain quality for wheat. *Australian Journal of Agricultural Research* 45: 1-17

Wrigley, C.W. (1996). Giant proteins with flour power. *Nature* 381: 738-739.

Wrigley, C.W., Autran, J.-C., and Bushuk, W. (1982). Identification of cereal varieties by gel electrophoresis of the grain proteins. *Advances in Cereal Science and Technology* 5: 211-259.

Wrigley, C.W. and Bekes, F. (2001). Cereal-grain proteins. In Sikorski, Z.E. (Ed.), *The Chemical and Functional Properties of Food Proteins*. Lancaster, PA: Technomic Publishing Co. Inc.

Wrigley, C.W., Bushuk, W., and Gupta, R. (1996). Nomenclature: Establishing a common gluten language. In Wrigley, C.W. (Ed.), *Gluten '96* (pp. 403-407). Melbourne: Royal Australian Chemical Institute.

Wrigley, C.W., du Cros, D.L., Fullington, J.G., and Kasarda, D.D. (1984). Changes in polypeptide composition and grain quality due to sulfur deficiency in wheat. *Journal of Cereal Science* 2: 15-24.

Wrigley, C.W., Gupta, R.B., and Bekes, F. (1993). Our obsession with high resolution in gel electrophoresis: Does it necessarily give the right answer? *Electrophoresis* 14: 1257-1258.

Yamamori, M. and Quynh, N.T. (2000). Differential effects of Wx-A1, -B1 and -D1 protein deficiencies on apparent amylose content and starch pasting properties in common wheat. *Theoretical and Applied Genetics* 100: 32-38.

Zeven, A.C., Dockes, G.J., and Kislev, M. (1975). Proteins in old grains of *Triticum* sp. *Journal of Archaeological Science* 2: 209-213.

Zhao, X.C., Batey, I.L., Sharp, P.J., Crosbie, G., Barclay, I., Wilson, R., Morell, M.K., and Appels, R. (1998). A single locus associated with starch granule properties and noodle quality in wheat. *Journal of Cereal Science* 27: 7-13.

Zhao, X.C., Shariflou, M.R., Good, G., and Sharp, P.J. (1998). Developing waxy wheat cultivars: Wx null alleles and molecular markers. In Slinkar, A.E. (Ed.), *Proceedings of the Ninth International Wheat Genetics Symposium*, Volume 1 (pp. 254-256). Saskatoon, Canada: University of Saskatchewan.

Zhu, J. and Khan, K. (2001). Effects of genotype and environment on glutenin polymers and breadmaking quality. *Cereal Chemistry* 78: 125-130.

Chapter 12

Grain Quality in Oil Crops

Leonardo Velasco
Begoña Pérez-Vich
José M. Fernández-Martínez

INTRODUCTION

Oil crops are domesticated plants whose seeds or fruits are valued mainly for the oils or fats that are extracted from them. The difference between oils and fats is merely the consistency at room temperature. We speak of an oil if it is liquid at the prevailing temperature of the region where it is produced and of a fat if it is normally solid (Hatje, 1989). Oil crops include both annual (usually called oilseeds) and perennial plants from a wide range of plant families. Table 12.1 lists the most important oil crops of the world as well as their most relevant properties and uses.

About 8 percent of the world production of oil crops is directly consumed as food (e.g., groundnuts), and about 6 percent is used for seed and animal feed. The remaining production is processed into oil (Food and Agriculture Organization of the United Nations [FAO], 2001). The oil/fat content of oil crops varies widely, from about 10 percent of the weight in coconuts to over 50 percent in palm kernels. An important aspect of oil crops is that they yield two products of economic value: the oil or fat and the oilmeal (also known as oilcake) that remains after oil extraction. Such oilmeals usually contain a high crude protein content (from 20 percent in palm kernel meal to almost 50 percent in soybean meal) and are mainly used as protein supplements for animal feed (Table 12.2). Oilseed meals are also used as fertilizers and soil improvers in many areas of the world (Bell, 1989). In most oilcrops the oil contributes a major percentage to the total value of the products. In soybean, however, the meal accounts for approximately 60 to 70 percent of the value of the seed (Smith and Huyser, 1987).

Vegetable oils and fats have two main uses: human consumption and technical or industrial applications. Vegetable oils account for about 70 percent of the world edible fat production, the rest coming from animal fats.

TABLE 12.1. Main oil crops of the world

Oil crop	Family	Plant habit	Plant part	Oil/fat	Annual production of seed/fruit (× 1000 t)	Annual production of oil/fat (× 1000 t)
Soybean [Glycine max (L.) Merr.]	Fabaceae	Annual	Seed	Oil	161,993	23,235
Palm (Elaeis guineensis Jacq.)	Palmae	Perennial	Fruit	Fat	116,315	21,951
Rapeseed (Brassica spp.)	Brassicaceae	Annual	Seed	Oil	40,193	12,362
Sunflower (Helianthus annuus L.)	Asteraceae	Annual	Seed	Oil	26,800	9,513
Groundnut[a] (Arachis hypogaea L.)	Fabaceae	Annual	Seed	Oil	34,507	4,557
Cottonseed (Gossypium spp.)	Malvaceae	Annual	Seed	Oil	54,143	3,849
Coconut (Cocos nucifera L.)	Palmae	Perennial	Seed	Oil[b]	46,482	3,319
Palm kernel[c] (Elaeis guineensis Jacq.)	Palmae	Perennial	Seed	Fat	6,318	2,695
Olive (Olea europaea L.)	Oleaceae	Perennial	Fruit	Oil	13,599	2,457

Source: FAO, 2001.
[a]In shell
[b]Liquid at the temperature of tropical areas, where it is produced, but solid in temperate regions
[c]Oil from palm fruits obtained both from the pulp or mesocarp (palm oil) and the kernel (palm kernel oil)

TABLE 12.2. Typical oil content of the seeds (percent dry seed weight), protein (N × 6.25), crude fiber (CF), and nitrogen-free extract (NFE) contents of the defatted meals (dry meal weight), and lysine and methionine + cysteine concentrations in the proteins (percent total protein) of the principal oilseeds

Oilseed	Oil	Protein*	CF*	NFE*	Lysine	Met+Cys
Canola	45	39	10	37	6.0	3.0
Cottonseed	16	42	13	35	4.1	3.3
Linseed	40	36	10	42	3.3	3.6
Peanut	50	47	14	28	3.8	2.7
Safflower	35	41	9	29	3.3	3.1
Soybean	18	49	6	34	6.2	2.0
Sunflower	45	47	12	27	3.0	2.8

Source: Data from Weiss, 1983; Bell, 1989; Vohra, 1989; Dorrell and Vick, 1997.
*Data for dehulled sunflower and safflower meals and partly dehulled peanut meal

Oils and fats are a vital component of the human diet because they are important sources of energy, act as carriers for fat-soluble vitamins, and provide the organism with essential fatty acids (Vles and Gottenbos, 1989). Human fat consumption has two main components: the so-called visible fat (butter, margarine, salad oil, cooking oil) and invisible fat (milk, meat, cheese, pastry, snacks, bread, nuts). Apart from food uses, large quantities of vegetable oils are directed to nonfood applications. They are used as motor fuels (biodiesel) and lubricants as well as for many applications in the oleochemical industry (detergents, soaps, surfactants, emulsifiers, cosmetics, etc.).

Breeding advances in the improvement of oil and meal properties of oil crops have had great market impacts. Rapeseed is probably the most remarkable example. It traditionally contained a toxic fatty acid in the oil (erucic acid) as well as antinutritive compounds in the meal (glucosinolates). In the 1970s, plant breeders were able to develop new types, later named canola, that essentially were free from both factors. Such improvement led to an enormous expansion of acreage and, consequently, to a considerable increase of rapeseed oil, meal, and products in the world market (Becker, Löptien, and Röbbelen, 1999). Similarly, the outstanding results of Russian breeders in raising the oil concentration of sunflower seeds was one of the keys for the expansion of sunflower as one of the most important oil crops in the world (Putt, 1997).

COMPONENTS OF GRAIN QUALITY IN OIL CROPS AND FACTORS INFLUENCING THEM

Oilseeds contain large amounts of food reserves which support the initial development of the seedling. Unlike cereals and legumes, in which the main food reserves are carbohydrates and proteins, respectively, most oilseeds contain oil as the main seed reserve (Table 12.2). The oil reserves are laid down in discrete subcellular organelles, the oil bodies, concentrated in the embryonic tissues. In the case of castor bean, however, oil bodies are mainly located in the endosperm (Bewley and Black, 1978).

Grain quality of oil crops has three main components: the oil content of the grain, the quality of the oil, and the quality of the oilmeal that remains after oil extraction. The quality of the oil is mainly determined by its composition in triacylglycerols and fatty acids as well as by the total content and characteristics of the antioxidant substances present in the oil. The quality of the meal is largely defined by the fiber content, protein content, and its nutritional value, as well as by the absence of toxic and antinutritional compounds. All these components will be described in detail in the corresponding sections of this chapter.

Grain quality of an oil plant is a combination of its genotypic constitution and the expression of the genotype in a given environment. The latter depends on not only environmental factors such as light and temperature, but also on intrinsic plant characteristics such as mode of reproduction (self-pollination versus outcrossing) and on the relative contribution of the parent genotypes to the trait (gametophytic versus sporophytic control). For example, oil and protein contents are mainly determined by the genotype of the grain-bearing plant (maternal or sporophytic control), whereas the fatty acid composition of seed oil is mainly determined by the genotype of the developing embryo (embryogenic or gametophytic control).

There is some variation in the degree to which the components of grain quality are affected by genotypic and environmental factors. As a general rule, grain quality traits can be divided into quantitative traits, if they are polygenic and their expression is largely affected by the environment in which the plants are grown, and qualitative traits, if their expression is relatively independent from the environment and determined by major genes. Examples of quantitative traits are the oil and protein contents and the total concentration of antioxidant or antinutritional compounds. Examples of qualitative traits are the oil fatty acid profile and the tocopherol profile.

In general, it can be stated that all factors affecting the general plant and grain development also influence grain quality. The influence of temperature (Canvin, 1965), light intensity (Dybing and Zimmerman, 1966), and environmental stress (Bouchereau et al., 1996; Velasco, Fernández-Martínez, and De Haro, 2001) on grain quality components is well documented. In addition, intrinsic grain characteristics such as hull percentage, grain size, and grain color can also affect grain quality. In sunflower, about two-thirds of the increase in achene oil content has resulted from a reduction in hull percentage, and about one-third from an increase in kernel oil content (Fick and Miller, 1997). In rapeseed, both yellow-coated grains and larger grains contain a greater proportion of meat to hull, which has been associated with greater oil and protein contents and a lower crude fiber content, resulting in a better digestibility of the oilmeal (Bell and Shires, 1982; Jensen, Liu, and Eggum, 1995).

OIL QUALITY

Basically, vegetable oils are made up of triacylglycerol molecules, which usually constitute more than 95 percent of the oil weight. The triacylglycerol contains one glycerol and three fatty acid molecules. There are several types of fatty acids, mainly differing in the number of carbon atoms and/or number and position of double bonds in the carbon chain. Both the fatty acid profile of the oil and the pattern of distribution of fatty acids within the triacylglycerol molecule constitute the principal factors determining the quality of vegetable oils, i.e., their physical, chemical, physiological, nutritional, and technological properties (Somerville, 1991; Padley, Gunstone, and Harwood, 1994).

Vegetable oils also contain a number of minor compounds, including lipids (polar lipids, mono- and diacylglycerols, free fatty acids, etc.) and lipid-soluble compounds. The most relevant of the latter are a series of derivatives of isoprene, comprising sterols, tocopherols, carotenoids, and chlorophylls, some of which are of paramount importance for oil quality because of their antioxidant properties.

Fatty Acid Composition of Oils

The most common classification of fatty acids is based on the number of double bonds present in the molecule. Thus, fatty acids are classified into saturated if they do not contain double bonds, monounsaturated if they contain one double bond, and polyunsaturated if they possess two or more double bonds in the molecule. Saturated fatty acids are major components of lipids that are solid at room temperature (fats), whereas unsaturated fatty acids are the major components of liquid lipids (oils). Fatty acids are represented by a system of abbreviated nomenclature that designates chain length and degree of unsaturation. For example 18:0 designates an 18-carbon saturated fatty acid (stearic acid) whereas 18:3 indicates three double bonds (Lobb, 1992). In addition, the abbreviated information for unsaturated fatty acids includes the (n-x) symbol, where x is the position of the first unsaturated carbon from the methyl end in the fatty acid molecule. This position is of utmost importance for the nutritional and pharmaceutical properties of fatty acids (Åppelqvist, 1989). With this nomenclature, the symbol 18:3 (n-3) refers to alpha-linolenic acid, whereas 18:3 (n-6) is used for gamma-linolenic acid. The main fatty acids in vegetable oils are listed in Table 12.3.

The formation of the major fatty acids in oilseeds starts by de novo synthesis of 16-carbon and 18-carbon fatty acids in the cell plastid (Somerville and Browse, 1991) through the combined activity of different enzymes using acetyl-CoA and malonyl-CoA as precursors (Figure 12.1). In addition, acyl carrier protein (ACP) is a required cofactor to which the intermediate metabolites in the plastid pathway are attached as thioesters. The first step in the pathway is the transfer of malonate from coenzyme A (CoA) to ACP.

TABLE 12.3. Main fatty acids present in vegetable oils

Common name	Systematic name	Symbol
Caprylic	Octanoic	8:0
Capric	Decanoic	10:0
Lauric	Dodecanoic	12:0
Myristic	Tetradecanoic	14:0
Palmitic	Hexadecanoic	16:0
Palmitoleic	cis-9-Hexadecenoic	16:1 (n-7)
Stearic	Octadecanoic	18:0
Oleic	cis-9-Octadecenoic	18:1 (n-9)
Ricinoleic	12D(R)-Hydroxy-9-Octadecenoic	18:1-OH
Linoleic	9,12-Octadecadienoic	18:2 (n-6)
Alpha-linolenic	9,12,15-Octadecatrienoic	18:3 (n-3)
Gamma-linolenic	6,9,12-Octadecatrienoic	18:3 (n-6)
Arachidic	Eicosanoic	20:0
Behenic	Docosanoic	22:0
Gadoleic	cis-9-Eicosenoic	20:1 (n-11)
Eicosenoic	cis-11-Eicosenoic	20:1 (n-9)
Erucic	cis-13-Docosenoic	22:1 (n-9)

Source: Data from Lobb, 1992 and Åppelqvist, 1989.

Three condensing enzymes then utilize malonyl-ACP as the 2-carbon donor for elongation of the growing acyl chain. The first condensation reaction is between malonyl-ACP and acetyl-CoA by the action of 3-ketoacyl-ACP synthase III (KAS III) (Jaworski, Clough, and Barnum, 1989). Subsequent condensations are between malonyl-ACP and acyl-ACP intermediates and are catalyzed by KAS I. The final 2-carbon elongation occurring in plastids is from 16:0 to 18:0 and requires KAS II. After each condensation, the 3-keto-acyl-ACP intermediates are reduced, dehydrated, and reduced again to yield the saturated acyl-ACP intermediates (Harwood, 1996). This fatty acid synthase (FAS) system is similar to the type II fatty acid synthase of *Escherichia coli*, as each of its component enzymes can be isolated separately. Finally, 18:0 is efficiently desaturated to 18:1 by the stearoyl-ACP desaturase (SAD).

The 16:0-ACP, 18:0-ACP, and 18:1-ACP formed in the plastid are hydrolized to free fatty acids by acyl-ACP thioesterases. The hydrolysis of the acyl-ACP thioester bond by the thioesterases implies the termination of the acyl chain elongation. Acyl-ACP thioesterases have been divided into FatA type (with oleoyl-ACP as preferred substrate) and FatB type (with saturated

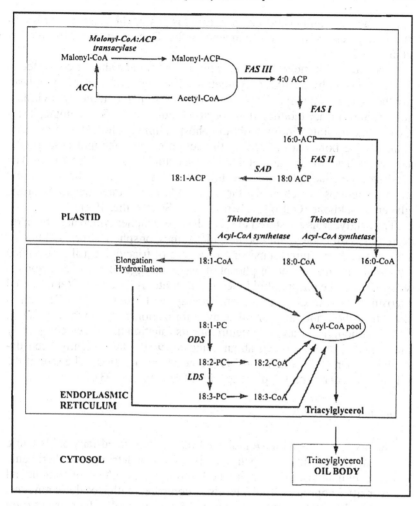

FIGURE 12.1. Schematic representation of triacylglycerol biosynthesis in developing seeds. ACC = acetyl-CoA caroxilase; FAS = fatty acid synthetase; SAD = stearoyl-ACP desaturase; ODS = oleoyl-phosphatidylcholine desaturase; LDS = linoleoyl-phosphatidylcholine desaturase; PC = phosphatidylcholine.

substrates preferred) (Jones, Davies, and Voelker, 1995). Thioesterases play an important role in determining the proportion of the different fatty acyl-CoAs that are produced, as different thioesterases show specificity for acyl-ACPs of different chain lengths and degree of saturation. For example, plant species accumulating short-chain fatty acids posses thioesterases with substrate preferences for short-chain acyl-ACPs (Dehesh et al., 1996; Voelker et al., 1997).

Free fatty acids move through the plastid membrane and are converted to CoA thioesters by acyl-CoA synthetase. The acyl-CoAs in the cytoplasm are then incorporated into lipids in the endoplasmic reticulum by acyltransferases and further modifications occur (Figure 12.1). For example, 18:1-CoA is incorporated into membrane phospholipids, where the second and third double bonds are added by the action of phospholipid desaturases. Desaturated acyl-CoAs are returned to the cytoplasmic acyl-CoA pool. The seed storage triacylglycerols are formed by the action of three different acyltransferases, which attach the acyl-CoAs to the three positions of the glycerol backbone (Ohlrogge, Browse, and Somerville, 1991).

Triacylglycerols are stored in specialized organelles which have been referred to as oil bodies, lipid bodies, oleosomes, or spherosomes. Oil bodies are spherical structures consisting of a core of triacylglycerol surrounded by a half-unit membrane of phospholipid (Ohlrogge, Browse, and Somerville, 1991). The phospholipid membrane contains specific proteins named oleosins and caleosins (Frandsen, Mundy, and Tzen, 2001). Oleosin is thought to be important for oil body stabilization in the cytosol (Huang, 1996), although neither its structure nor its function have been completely elucidated. Little is known about caleosin, which has recently been described (Chen, Tsai, and Tzen, 1999; Naested et al., 2000). The size of the oil bodies depends on the plant species (Tzen et al., 1993).

Saturated Fatty Acids

Dietary experiments have indicated that the saturated fatty acids lauric (12:0), myristic (14:0), and palmitic (16:0) have a detrimental atherogenic effect on human health by raising both serum total cholesterol content and low-density lipoprotein (LDL) levels as compared with isocaloric amounts of carbohydrates (Mensink, Temme, and Hornstra, 1994). Increased serum total and LDL cholesterol levels are a well-known risk factor for coronary heart disease. Conversely, neither saturated fatty acids with less than 12 carbon atoms nor stearic acid (18:0) have been found to be hypercholesterolemic.

The principal vegetable sources of saturated fatty acids in the world market are coconut and palm (fruit and kernel) oils, which mainly contain hypercholesterolemic saturated fatty acids. Coconut and palm kernel oil mainly contain lauric acid (12:0), whereas palm oil contains a high proportion of palmitic acid (16:0) (Table 12.4). Dietary guidelines recommend a

TABLE 12.4. Average composition of the principal vegetable oils and fats for major fatty acids

Fats and oils	8:0	10:0	12:0	14:0	16:0	16:1	18:0	18:1	18:1(OH)	18:2	18:3	20:0	20:1	22:0	22:1	24:0
Canola*					3.9	0.2	1.9	64.1		18.7	9.2	0.6	1.0	0.2		0.2
Castor		6.7	47.6	18.1	1.0		1.0	3.0	90.0	4.0	tr					
Coconut	7.8				8.8		2.6	6.2		1.6		0.1		0.1		
Cottonseed				0.8	24.0	0.8	2.6	19.0		52.5	tr	0.3				
Linseed					6.1	0.1	3.2	16.6		14.2	59.8					
Maize			0.1	0.2	13.0		2.5	30.5		52.0	1.0	0.5	0.2			
Olive					13.7	1.2	2.5	71.1		10.0	0.6	0.9				
Palm			0.3	1.1	45.1	0.1	4.7	38.8		9.4	0.3	0.2				
Palm kernel	2.5	4.0	49.0	16.0	9.0		2.0	14.0		2.0		1.0				
Peanut					12.5		2.5	37.0		41.0	0.3	1.2	0.7	2.5	1.0	1.3
Rapeseed					3.0		1.0	16.0		14.0	10.0	1.0	6.0	tr	49.0	
Safflower				0.1	6.5	0.1	2.9	13.8		75.3		0.4		0.2		
Soybean					11.0	0.5	4.0	22.0		53.0	7.5	1.0	1.0			
Sunflower				0.1	5.5	0.1	4.7	19.5		68.5	0.1	0.3	0.1	0.9		0.2

Source: Data from Padley, Gunstone, and Harwood (1994) and White (1992).
*Canola is the designation for rapeseed cultivars with no erucic acid in the seed oil and with low levels of glucosinolates in the oilmeal.

397

reduction in the consumption of saturated fats and oils and their replacement by unsaturated fatty acids, which are not considered to be hypercholesterolemic (U.S. Department of Agriculture [USDA], 1992).

Saturated fatty acids possess advantageous technological properties for some applications, for example, shortening and margarine manufacture. For these applications, liquid oils rich in unsaturated fatty acids must be converted to semisolid, plastic fats by means of the hardening process, which involves the conversion of part of the unsaturated fatty acids into saturated fatty acids. During this process, some double bonds change their position and/or stereochemical configuration producing *trans* and positional isomers, which are a major risk factor of heart disease (Willett and Ascherio, 1994). In consequence, semisolid fats with a high proportion of the saturated fatty acids with no detrimental health effects are required. Unfortunately, such fats are not available in the major vegetable sources (Table 12.4).

Two main breeding objectives must be outlined in relation to the previous discussion: the reduction in total saturated fatty acid content in edible oils and the increase of nondetrimental saturated fatty acids in liquid oils for using in margarine and shortening production. In the first case, lines producing oils with reduced levels of total saturated fatty acids have been developed in soybean (Erickson, Wilcox, and Cavins, 1988; Fehr et al., 1991; Takagi et al., 1995; Stojšin, Alblett, et al., 1998), safflower (Velasco and Fernández-Martínez, 2000), and sunflower (Miller and Vick, 1999). In the second case, lines with increased levels of stearic acid have been developed in soybean (Hammond and Fehr, 1983; Graef, Fehr, and Hammond, 1985; Bubeck, Fehr, and Hammond, 1989; Rahman et al., 1995) and sunflower (Osorio et al., 1995).

Unsaturated Fatty Acids

The degree of unsaturation is not only a useful criterion for fatty acid classification, but also one of the key aspects defining the properties of fatty acids. One of the most relevant aspects to take into account is that double bonds are the main centers of oil oxidation. The double bonds react with oxygen in the air in a process involving the production of free radicals, which are implicated in a number of diseases, in tissue injuries, and in the process of aging (Shahidi, 1997). Furthermore, the lipid oxidation products are the major source of off flavors in oils during storage (Tatum and Chow, 1992). Although intact polyunsaturated fatty acids are beneficial for human health (Horrobin, 1992), their high susceptibility to autoxidation make them undesirable at high levels in edible oils.

Oleic acid (18:1, n-9) is today the preferred fatty acid for edible purposes, as it combines a hypocholesterolemic effect (Mensink and Katan, 1989) with a much greater oxidative stability than polyunsaturated fatty acids (Yodice, 1990). High concentrations of oleic acid occur naturally in olive oil (Table 12.4). Oilseed breeding, however, has created additional oil sources with even higher oleic acid than olive oil. Cultivars with seed oil characterized by an exceptionally high oleic acid content (>75 percent) have been developed in canola (Auld et al., 1992; Rücker and Röbbelen, 1997), safflower (Knowles and Mutwakil, 1963; Fernández-Martínez, del Río, and de Haro, 1993), soybean (Kinney, 1997), and sunflower (Soldatov, 1976). In addition to the high nutritional value of high oleic acid oil, it also possesses important industrial applications (Friedt, 1988).

A series of monounsaturated fatty acids, some of them presenting functional groups (hidroxy, epoxy, etc.) in the carbon chain, possess important industrial applications but are not suitable for edible purposes because of toxic or antinutritional effects. Some of the most relevant are erucic acid (22:1, n-9), which is mainly present in seed oils from plants of the Brassicaceae family (e.g., rapeseed, mustards, and crambe) (Kumar and Tsunoda, 1980), petroselinic acid (18:1, n-12), present in the Apiaceae (e.g., coriander) (Knapp, 1990), vernolic acid (epoxy-18:1), found in some wild species (e.g., *Vernonia* spp., *Euphorbia* spp.) (Pascual-Villalobos et al., 1992; Thompson et al, 1994), or ricinoleic acid (hydroxy-18:1) characteristic of the castor bean (Canvin, 1963).

The most common polyunsaturated fatty acids in vegetable oils are linoleic acid (18:2, n-6) and alpha-linolenic acid (18:3, n-3). Both fatty acids are of great value from a nutritional point of view, as they are essential fatty acids. This means that they must be included in the diet because the human body is not able to manufacture them (Horrobin, 1992). Essential fatty acids have an important structural function in the cell membranes and also play a crucial role as precursors of metabolic regulators and other important metabolites such as prostaglandins (Vles and Gottenbos, 1989). In addition, polyunsaturated fatty acids have a hypocholesterolemic effect in humans (Chan, Bruce, and McDonald, 1991). Despite their high nutritional value, polyunsaturated fatty acids are undesired in edible oils because of their high susceptibility to oxidation during processes such as storage or heating. Alpha-linolenic acid, with three double bonds in the molecule, is much more susceptible to oxidation than linoleic acid, which possesses two double bonds.

Alpha-linolenic acid is present at high or relatively high proportions in some commercial vegetable oils, especially in linseed (60 percent of total fatty acids, see Table 12.4), rapeseed and canola (about 10 percent of total fatty acids), and soybean (about 8 percent of total fatty acids). Linseed oil is not suitable for edible purposes because of its high alpha-linolenic acid concentration (Frankel, 1991). However, it is precisely this characteristic that makes linseed oil unsurpassed as a drying oil for use in paints, varnishes, printing inks, etc. (McHughen, 1992). Successful breeding through mutagenesis led to the development of linseed mutants with seed oil containing less than 2 percent alpha-linolenic acid (Green, 1986; Rowland and Bhatty, 1991). Such an oil is of great value for edible purposes (Bickert, Lühs, and Friedt, 1994). The utilization of mutagenesis has also enabled important reductions of linolenic acid content in rapeseed/canola (Rakow, 1973; Röbbelen and Nitsch, 1975; Wong and Swanson, 1991; Auld et al., 1992; Hitz et al., 1995) and soybean (Hammond and Fehr, 1983; Wilcox, Cavins, and Nielsen, 1984; Takagi et al., 1990; Hitz et al., 1995; Stojšin, Luzzi, et al., 1998b).

Gamma-linolenic acid is a polyunsaturated fatty acid that attracts much interest because of its many health benefits (Fan and Chapkin, 1998). Current commercial sources of gamma-linolenic acid for the pharmaceutical industry are evening primrose (*Oenothera biennis* L.) and borage (*Borago officinalis* L.). Other interesting sources of this fatty acid are the fruits of currants and gooseberries (*Ribes* spp.), also rich in antioxidant compounds, which increases the biological value of gamma-linolenic acid (Goffman and Galletti, 2001).

Triacylglycerol Structure in Vegetable Oils

The functional and nutritional characteristics of an oil are affected not only by the fatty acid composition, but also by the triacylglycerol structure, i.e., the position of the fatty acids on the glycerol backbone (Reske, Siebrecht, and Hazebroek, 1997). The stereochemical positions of the three fatty acids in the glycerol molecule are designated *sn-1*, *sn-2*, and *sn-3* (Figure 12.2). The distribution of fatty acids within the triacylglycerol molecule is not random. Initial studies on seed oils concluded that saturated fatty acids were virtually excluded from the *sn-2* position and randomly distributed between *sn-1* and *sn-3* positions (van der Wal, 1960). Later studies, however, demonstrated that triacylglycerol stereospecificity was more complex than initially anticipated. In safflower seeds, the acylation of position *sn-1* has selectivity for saturated fatty acids, whereas position *sn-3* has no selectivity (Ichihara and Noda, 1982). Conversely, saturated fatty acids in sunflower showed preference for the *sn-3* over the *sn-1* position (Reske, Siebrecht, and Hazebroek, 1997).

Fatty acid stereospecificity within the triacylglycerol molecule plays an essential role in lipid nutritional value, as the absorption rates of fatty acids

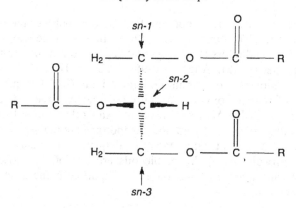

FIGURE 12.2. Structure of a triacylglycerol molecule. *sn-1, sn-2,* and *sn-3* refer to the carbon numbers of the glycerol; *sn-2* is a chiral center. R represents fatty acids.

depend on the location of fatty acids in the triacylglycerol (Small, 1991). In the case of atherogenic fatty acids (some of the saturated fatty acids, see section discussing fatty acids in this chapter), their absorption rate is higher when they are sterified at the central *sn-2* triacylglycerol position than when they are at the external *sn-1* and *sn-3* positions (Bracco, 1994). Thus, vegetable oils containing a high proportion of saturated fatty acids at the *sn-2* position are considered to be more atherogenic than those having similar total saturated fatty acid content but distributed in the external positions (Renaud, Ruf, and Petithory, 1995). Palm oil, widely used in food products, contains approximately 10 percent saturated fatty acids at the *sn-2* position (Padley, Gunstone, and Harwood, 1994). Recently, mutant lines of sunflower and soybean with increased levels of saturated fatty acids almost exclusively at the *sn-1, 3* positions have been developed (Álvarez-Ortega et al., 1997; Reske, Siebrecht, and Hazebroek, 1997). Besides the advantages derived from the positional distribution of saturated fatty acids, seed oils from these mutants possess adequate technological properties for margarine and other solid-fat substitute production without need of detrimental physical transformations such as hydrogenation or tranesterification (List et al., 1996; Kinney, 1999).

Natural Antioxidants in Vegetable Oils

Oxidative processes occur both in vitro and in vivo. Autoxidation of lipids during storage (in vitro) is one of the main factors diminishing food quality. This process affects not only fats and oils, but also feeds and foods containing them. The consequence of oxidation is the development of unpleasant tastes and odors (rancidity) and degradation of functional and nutritional properties (St. Angelo, 1996; Crapiste, Brevedan, and Carelli, 1999). Oxidation also occurs in the human body (in vivo), promoting the formation of reactive oxygen and nitrogen species (free radicals), which cause damage to DNA, lipids, proteins, and other biomolecules. Diet-derived antioxidants are of paramount importance in maintaining health, as endogenous antioxidant defenses are inadequate to prevent damage completely (Halliwell, 1996). Vegetable oils are one of the most important sources of natural antioxidants, the most important being described in the following sections.

Chromanols

Chromanols consist of a chroman head with two rings, one phenolic and one heterocyclic, the latter substituted with a phytyl tail (Kamal-Eldin and Åppelqvist, 1996). The most important chromanols are tocopherols, with saturated phytyl tails, tocotrienols, having unsaturated phytyl tails, and plastochromanol-8, with a saturated phytyl tail longer than that of tocopherols. Tocopherols and tocotrienols include mono (δ-), di (α- or γ-), and trimethyl (α-) tocol derivatives (Figure 12.3).

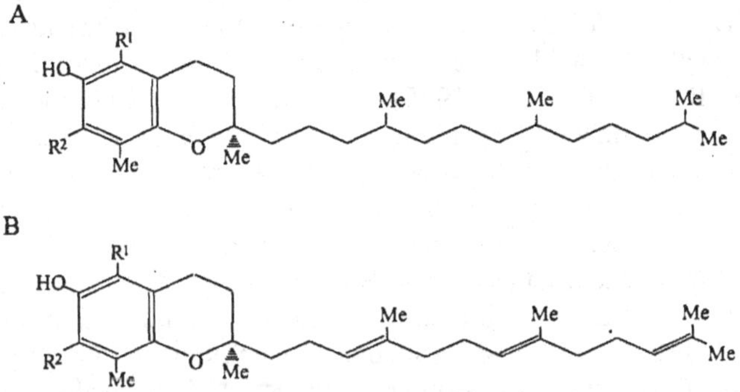

R^1=Me; R^2=Me: α-tocopherol/tocotrienol
R^1=Me; R^2=H: β-tocopherol/tocotrienol
R^1=H; R^2=Me: γ-tocopherol/tocotrienol
R^1=H; R^2=H: δ-tocopherol/tocotrienol

FIGURE 12.3. Chemical formulas of (A) tocopherols and (B) tocotrienols. Me = methyl groups.

The chromanols exhibit antioxidant activity both in vivo and in vitro. In vivo they exert vitamin E activity, protecting cellular membrane lipids against oxidative damage (Muggli, 1994). In vitro they inhibit lipid oxidation in oils and fats, as well as in foods and feeds containing them (Kamal-Eldin and Åppelqvist, 1996). The biologically most active chromanol form is α-tocopherol (Traber and Sies, 1996). According to Padley, Gunstone, and Harwood (1994), the greatest vitamin E effect is exhibited by α-tocopherol (relative activity 100), followed by β-tocopherol (50), α-tocotrienol (30), and γ-tocopherol (10). Conversely, the best in vitro antioxidant activity ranks in the order γ-tocopherol (relative antioxidant activity 100), δ-tocopherol (68), β-tocopherol (64), and α-tocopherol (35) (Pongracz, Weiser, and Matzinger, 1995). The in vitro antioxidant activities of tocotrienols are unknown, whereas it has recently been shown that plastochromanol-8 is a more powerful antioxidant than α-tocopherol (Olejnik, Gogolewski, and Nogala-Kalucka, 1997)

The tocopherol derivatives are the predominant chromanol form in vegetable oils. The most relevant exceptions are palm oil, which in addition to tocopherols contains large amounts of tocotrienols (Padley, Gunstone, and Harwood, 1994), and linseed oil, which contains both tocopherol derivatives and plastochromanol-8 (Velasco and Goffman, 2000). The average chromanol composition of vegetable oils is given in Table 12.5.

Because of their beneficial in vivo and in vitro action, the increase of total chromanol content of vegetable oils is an important objective in oilseed breeding. Also, since the chromanol derivatives differ in their relative in vivo and in vitro activity, the modification of the chromanol profile for specific end uses of the modified oil or fat is an important breeding goal. Thus, Shintani and Dellapenna (1988) focused on increasing the proportion of α-tocopherol in vegetable oils as a way of increasing their vitamin E activity. They followed a biotechnological approach, based on overexpression of γ-tocopherol methyltransferase (TMT), to convert naturally occurring γ-tocopherol of *Arabidopsis* seed oil into α-tocopherol. The resulting oil from *Arabidopsis* lines overexpressing γ-TMT was ninefold that of the oil from the wild-type line. Conversely, Demurin, Skoric, and Karlovic (1996) attempted to increase the concentration of γ-tocopherol content in sunflower oil with the aim of improving oil oxidative stability. By evaluating the existing variability in sunflower germplasm, they identified and selected lines with seed oils containing 50 percent α-tocopherol and 50 percent β-tocopherol, and lines producing 95 percent γ-tocopherol, in comparison with standard lines characterized by 95 percent α-tocopherol.

TABLE 12.5. Chromanol content (mg/kg) of oils and fats

Fats and oils	Tocopherols				Tocotrienols			P-8[a]
	α	β	γ	δ	α	γ	δ	
Canola	202	65	490	9	_[b]	–	–	–
Castor	28	29	111	310	–	–	–	–
Coconut	–	–	–	4	20	–	–	–
Cottonseed	338	17	429	3	–	–	–	–
Linseed	4	–	407	–	–	–	–	142
Maize	282	54	1034	54	–	–	–	–
Olive	93	–	7	–	–	–	–	–
Palm	89	–	18	–	128	323	72	–
Palm kernel	62	–	–	–	–	–	–	–
Peanut	178	9	213	8	–	–	–	–
Safflower	477	–	44	10	–	–	–	–
Soybean	100	8	1021	421	–	–	–	–
Sunflower	670	27	11	–	–	–	–	–

Source: Data from Padley, Gunstone, and Harwood, 1994, except for linseed from Velasco and Goffman, 2000.
[a]Plastochromanol-8
[b]Indicates absence of compound in oil

Carotenoids

Carotenoids are 40-carbon polyunsaturated hydrocarbons (carotenes) and their oxygenated derivatives (xanthophylls). Carotenes are either linear or cyclized at one or both ends of the molecule (Goodwin, 1980). Similar to chromanols, they play an important role in vivo as source of provitamin A, as well as in vitro, protecting oils from oxidation (Henry, Catignani, and Schwartz, 1998; Hirschberg, 1999). β-carotene is the most nutritionally active carotene as provitamin A (Ong and Choo, 1997). Their main in vivo protective effect has traditionally been attributed to antioxidant action (Palozza and Krinsky, 1992; Miller et al., 1996). Other benefits of carotenoids such as conversion to retinoids or effects on cell communication have also been described (Halliwell, 1996). The in vitro activity of carotenoids seems to be related to the inhibition of photooxidation, acting as a filter for light of short wavelengths (Warner and Fränkel, 1987).

Palm oil is the richest oil source of carotenoids, with a concentration of 500 to 700 ppm (Ong and Choo, 1997). The concentration of carotenoids in seed oils is considerably lower (Uppström, 1995). Some experiments directed to the manipulation of carotenoid production in plants through metabolic engineering have already been conducted (Hirschberg, 1999), though so far no attempt has been made to alter carotenoid contents of vegetable oils.

Phenolic Compounds

Polar phenolic compounds are important natural antioxidants present in olive oil, especially in extra virgin oil, which is obtained from the fruit mesocarp by mechanical pressing (Tsimidou, Papadoupoulos, and Boskpu, 1992). They include a wide variety of simple phenols (e.g., hydroxytyrosol, tyrosol), aldehydic secoiridoids (e.g., oleuropein and derivatives), flavonoids, and lignans (Owen et al., 2000). According to Aparicio and colleagues (1999), phenolic compounds are the main contributors to olive oil stability. Among them, hydroxytyrosol, oleuropein, and caffeic acid seem to be the most powerful antioxidants (Saija et al., 1998; De la Puerta, Gutierrez, and Hoult, 1999). Nevertheless, many of the phenolics present in olive oil have not yet been completely identified (Shukla, Wanasundara, and Shahidi, 1997).

In general, seed oils are devoid of phenolic compounds. One exception is sesame oil, which contains a powerful antioxidant, sesamol, as well as several bisfuranyl lignans (Potterat, 1997).

Phytosterols

Phytosterols or plant sterols are essential components of the membranes, playing an important role in the control of membrane fluidity and permeability as well as in signal transduction. Their role in plant cells is similar to that of cholesterol in mammalian cells (Piironen et al., 2000). Chemically, they are steroid alcohols (triterpenes) synthesized from squalene in the isoprenoid pathway (Benveniste, 1986). Vegetable oils are the richest natural sources of plant sterols. Among them, the highest contents are found in maize and rapeseed (Rossell and Pritchard, 1991). The predominant sterols in vegetable oils belong to the 4-desmethyl sterol class, which contributes more than 85 percent of total sterols. Within this group, sitosterol (usually above 50 percent of 4-desmethyl sterols), campesterol, sigmasterol, Δ^5-avenasterol, Δ^7-avenasterol, and Δ^7-stigmastenol are the most significant (Piironen et al., 2000). Brassicasterol is a sterol characteristic of the Brassicaceae family and therefore is present in rapeseed/canola oils (Uppström, 1995). Phytosterols of vegetable oils occur mainly as free sterols and esters of fatty acids, especially as esters of oleic and linoleic acids (Piironen et al.,

2000). Saturated plant sterols (stanols) occur in low amounts in vegetable oils (Dutta and Åppelqvist, 1996).

Plant sterols and stanols contained in vegetable oils lower total and LDL serum cholesterol in humans by inhibiting cholesterol absorption from the intestine. Stanols have greater potential to lower cholesterol than sterols because they are virtually unabsorbable (Nguyen, 1999). Plant sterol and stanol esters are currently being incorporated into food products such as margarines as a dietary ingredient for lowering serum cholesterol (Miettinen et al., 1995).

Oil Components with Anticarcinogenic Activity

As a general rule, all oil components with antioxidant action are of great value in cancer prevention, as free radicals produced through oxidative processes are directly implicated in carcinogenesis (Borek, 1993). In addition, some oil components without a marked antioxidant role have an anticancer protective effect through other mechanisms of action. One of the most relevant examples is squalene, a triterpene of the isoprenoid pathway (Benveniste, 1986). In most vegetable oils, squalene is an intermediary in the synthesis of phytosterols and its final concentration in the oil is low. In olive oil, however, there is an important accumulation of squalene (Kiritsakis, 1987), which has been found to have very weak antioxidant activity (Psomiadou and Tsimidou, 1999). Its chemopreventive efficacy seems to be through a strong inhibitory activity of certain enzymes implicated in oncogene activation (Newmark, 1999).

MEAL QUALITY

Oilseed meals are extensively used as protein supplements for use in animal feeds, as approximately between 20 and 50 percent of weight of meals is protein. Some oilmeal is further processed to produce concentrates (with 50 to 60 percent of crude protein) or isolates (nearly pure protein) for use in human food (Bell, 1989). Oilmeals are mainly valued for low fiber content, high protein content of good quality, and absence or low presence of toxic and antinutritional compounds.

Fiber Content

Fiber is not a homogenous chemical entity. It refers to the carbohydrates that are not truly digested by the animal and therefore do not contribute energy when consumed. They include cellulose, hemicellulose, pectins, gums, mucilages, and lignin-hemicellulose complexes (Vohra, 1989).

Fiber is predominantly associated with the seed hull. Some oilseeds, particularly sunflower and safflower, are characterized by a high hull proportion. In the case of sunflower, hull percentage among genotypes may vary from 10 to 60 percent of the total achene weight (Miller and Fick, 1997). In safflower, hull content ranges from 20 percent in reduced-hull genotypes to about 45 percent in white-hull types (Fernández-Martínez, 1997). In these oilseeds, dehulling is necessary to render meals useful as protein supplements (Bell, 1989). In sunflower, it has been shown that genetic variability for hullability or facility for dehulling exists (Denis, Domínguez, and Vear, 1994). Other oilseeds contain much lower hull contents, for example, between 15 and 20 percent of seed weight in rapeseed (Niewiadomski, 1990) and 7 percent to 8 percent of seed weight in soybean (Mounts, Wolf, and Martinez, 1987). In these cases the oilmeals are commercially available either with or without hulls or may be partially dehulled (Bell, 1989).

In some cases it has been possible to reduce fiber content by selecting for lower hull content. In rapeseed/canola, yellow seeds are characterized by thinner seed coats than dark seeds, which is associated with about 4 percent lower fiber content (Stringam, McGregor, and Pawlowski, 1974). This fact has encouraged breeding for yellow-seeded types as a means of improving the nutritional value of the meal (Baetzel, Friedt, and Lühs, 2000). In safflower, the identification of a thin-hulled mutant (Ebert and Knowles, 1966) allowed a reduction of crude fiber content from about 30 to 11 percent of the total seed (achene) weight (Weiss, 1983).

Protein Content and Quality

The oilmeals obtained after oil extraction from oilseeds contain high levels of protein, from about 20 percent in palm kernel meal to about 45 to 50 percent in soybean meal. Most oilseeds yield a defatted meal with about 35 to 50 percent crude protein (Table 12.2).

Proteins are polymers of amino acids. The amino acids are a group of primary amines that contain a central carbon atom to which are attached a hydrogen atom, an amino group (NH_2), and a carboxyl group (COOH). Proteins of oilseeds can be divided into three functional groups: (1) storage proteins with no enzymatic activity, which are the most abundant; (2) proteins having a structural function; and (3) enzymes (Niewiadomski, 1990). Another generalized classification of proteins is based on their solubility in various solvents. In this system, the four classical types of proteins are albumins, globulins, prolamins, and glutenins (Osborne, 1924). The globulins represent the major storage proteins in all oilseeds (Bell, 1989), accounting for about 90 percent of the seed protein in soybean (Wilson, 1987), 70 percent in rapeseed/canola (Uppström, 1995), and 60 percent in sunflower (Dorrell and Vick, 1997).

As long as the proportion of hull remains constant, oil and protein contents are negatively correlated (Röbbelen, 1981). In consequence, selection for high protein usually results in lower oil content. Because of the negative correlation between both traits, some authors have suggested separate development of high-oil and high-protein cultivars (Röbbelen, 1981). Others, however, recommended conducting selection for the sum of oil and protein while maintaining acceptable oil content, since the latter has a higher price in the market (Jímenez et al., 1985; Miller and Fick, 1997).

The quality of the protein has to be evaluated in terms of the nutritional balance of the absorbed amino acids. The amino acids of major interest are arginine, histidine, isoleucine, leucine, lysine, methionine, phenylalanine, threonine, tryptophan, and valine, which are essential amino acids; i.e., they cannot be synthesized by the human body and have to be incorporated through the diet. The amino acids cysteine, tyrosine, and glutamic acid are not essential but can partially satisfy the need for essential amino acids (Bell, 1989).

Nutritionally, most oilseed meals are deficient in some amino acids (Table 12.2). Soybean meal has a limited content of sulfur amino acids methionine and cysteine, although it has an adequate content of the other essential amino acids. It is worth noting the high lysine content of soybean meal, which complements the low lysine content of cereals (Bell, 1989). In addition to a limited content in sulfur amino acids, the other important legume oilseed, peanut, is also characterized by a low lysine content. The proteins of the nonlegume oilseeds are nutritionally adequate in sulfur amino acids and, with the exception of rapeseed/canola, nutritionally inadequate in lysine (Norton, 1989). The seed protein of rapeseed/canola has the best balanced amino acid composition of oilseeds, also comparing favorably with cereals (Rosa, 1999).

Toxic and Antinutritional Compounds in Oilmeals

Plants produce and accumulate potentially toxic compounds as a chemical protection against herbivores (Ågren and Schemske, 1993). In the case of oilseeds, most of the compounds with toxic or antinutritional properties remain in the meal after oil extraction, considerably reducing its value for human food and animal feed. Toxic compounds may have serious deleterious effects on both livestock and humans. Antinutritional compounds reduce the nutritive value of the feed, mainly by negatively affecting palatability, digestibility, or both. Most of the compounds or classes of compounds with detrimental properties present in oilseeds are toxic if ingested at high concentrations, but their effects at low concentrations are more likely to be antinutritional (Griffiths, Birch, and Hillman, 1998).

Some classes of antinutritional compounds are widely distributed in oilseeds (e.g., phytates, phenolics) while others are specific to certain plant families (e.g., glucosinolates in the Brassicaceae). Some of the most relevant are briefly described as follows.

Phytates

Phytic acid (myoinositol 1, 2, 3, 4, 5, 6-hexakis-dihydrogen phosphate) is a major component of cereals and oilseeds, constituting between 1 and 3 percent of the total seed weight. In oilseeds, phytic acid usually occurs as a mixture of calcium, magnesium, and potassium salts (phytates), in cristaloid-type globoids in the cells of the radicle and the cotyledon (Yiu, Altosaar, and Fulcher, 1983). Physiologically, phytic acid plays an important function as a primary reserve of energy, phosphorus, and myoinositol in the seed (Graf, 1983). Between 50 and 80 percent of the phosphorus of oilseeds is stored in the form of phytic acid (Lolas, Palamidis, and Markakis, 1976). This phosphorus is nutritionally unavailable to nonruminant livestock (Erdman, 1979).

Phytic acid is a strong chelating agent that can bind metal ions, reducing the availability of calcium, iron, magnesium, zinc, and other trace elements (Oberleas, Muhrer, and O'Dell, 1966). In addition, phytates form complexes with amino acids, reducing digestibility and amino acid availability (Thompson, 1990). These antinutritional properties limit the use of the oilseed protein for animal feed. In human nutrition, however, several beneficial effects have been reported. Phytic acid is believed to have a marked in vivo antioxidant effect, to decrease the risk of iron-mediated colon cancer, and to lower serum cholesterol and triglycerides (Martínez et al., 1995; Greiner and Jany, 1996).

Among oilseeds, rapeseed/canola contains the highest concentration of phytic acid. Matthäus, Lösing, and Fiebig (1995) reported a range from 2.0 to 4.0 g/100 g seed in comparison with 2.5 to 2.6 g/100 g in linseed, 1.9 g/100 g in peanut, 1.2 to 1.7 g/100 g in soybean, and 1.9 g/100 g in sunflower. Because of the chelating action of phytic acid, oilmeals for animal feed are often supplemented with the enzymes phytase and acid phosphatase, which improves their digestibility and increases the bioavailability of phosphorus and metal ions (Zyla and Korelski, 1993; Aldeola, 1995).

Another strategy is to reduce the ratio of phytate phosphorus to inorganic phosphorus in the seed by plant breeding. This has recently been achieved in soybean by Wilcox and colleagues (2000), who developed two mutants with considerably increased proportion of inorganic phosphorus in relation to phytate phosphorus.

Phenolics

Phenolic compounds are common in most oilseeds. They exert a detrimental effect on meal quality by interacting with amino acids, denaturing proteins, and inhibiting enzymes, thus lowering the nutritional value of the meal for animal feed (Sozulski, 1979). Phenolic compounds also limit the utilization of the meal as a source of human food-grade protein because they confer undesired color, bitter taste, and/or astringency to oilseed protein products (Naczk et al., 1998).

The content of phenolics in rapeseed/canola is much higher than in other oilseeds, about ten times that in peanut and cottonseed and about 30 times that in soybean (Shahidi and Naczk, 1992). The most important phenolic compound in rapeseed/canola seeds is sinapine, the choline ester of sinapic acid, which represents from 5.0 to 17.7 g•kg^{-1} total seed weight (Velasco and Möllers, 1998). Another important phenolic compound is chlorogenic acid, which is responsible for the production of yellow-green coloration following oxidation in sunflower meal. The presence of chlorogenic acid limits the broad use of sunflower meal for human consumption (Dorrell and Vick, 1997). Both sinapine and chlorogenic acid are predominantly present in the seed kernels; therefore, the dehulling of the seeds scarcely reduces the presence of these phenolics in the meal (Uppström, 1995; Pedrosa et al., 2000).

Tannins are complex phenolic compounds (polyphenolic compounds) that form complexes with proteins and reduce their availability to animals. They are present in variable proportions in most oilseeds. Matthäus (1997) reported a variation from 0.04 mg/g in soybean to 3.8 mg/g in rapeseed. Peanuts, not included in the mentioned evaluation, also have a significant amount of tannins (Bell, 1989). In peanuts, tannins are mainly concentrated in the testa, which is usually removed to improve energy digestibility of the meal (Weiss, 1983).

Glucosinolates

Glucosinolates (GSLs) are a family of secondary plant metabolites particularly abundant in seeds and green tissues of the family Brassicaceae

(Kjaer, 1976). They consist of a thioglucoside linked to a variety of side chains which are usually amino acid derivatives (Höglund et al., 1991). More than 100 different GSLs showing different side chain structure have been identified in the plant kingdom, although only around 15 or 16 occur in significant amounts in the genus *Brassica,* to which the oilseeds rapeseed and canola belong (Rosa, 1999). Both the GSLs and their degradation products are associated with antinutritive and toxic effects, limiting the usefulness of seeds and seed meals for human and animal feed (Sørensen, 1990).

Traditional rapeseed cultivars contained high levels of glucosinolates, about 110 to 150 μmol/g seed. The discovery in the middle 1960s that the Polish variety 'Bronowski' contained much lower glucosinolate content, about 10 to 12 μmoles/g seed (Josefsson and Åppelqvist, 1968; Downey, Craig, and Youngs, 1969), opened up the development of rapeseed cultivars which combined the previously developed zero erucic acid trait (Stefansson, Hougen, and Downey, 1961) with very low glucosinolate levels. Canola was the name adopted by the rapeseed industry in Canada in 1978 to designate rapeseed (*Brassica napus* and *B. rapa*) cultivars with less than 1 percent erucic acid in the seed oil and a glucosinolate content below 30 μmol/g oil-extracted, air-dried meal (Vaisey-Genser and Eskin, 1987). The drastic reduction of glucosinolates took place in the seeds but not vegetative tissues, which maintained similar high glucosinolate content as the traditional cultivars. This fact was of great value for the productive potential of the new low-glucosinolate cultivars, as glucosinolates have an important protective effect against pests and diseases (Mithen, 1992).

Other Antinutritional Factors

Among the most important antinutritional compounds present in soybean seeds are trypsin inhibitors. Trypsin inhibitors are types of protease inhibitors, which are proteins found in virtually all legume species (Liener and Kakade, 1980). Soybean trypsin inhibitors cause pancreatic lesions, particularly hypertrophy and hyperplasia, in animals fed with raw soybean meal (Liener et al., 1985). The trypsin inhibitors are heat labile, being inactivated during the toasting phase of meal production. Toasted meals do not, therefore, cause problems to animals fed with soybean meal (Rackis, Wolf, and Baker, 1985).

Cottonseed meal contains gossypol, a polyphenolic compound present in the pigment glands of both the vegetative tissues and the seeds of cotton (Kohel, 1989). Gossypol is toxic to monogastric animals and also causes discoloration in foods (Vroh-Bi et al., 1999). The identification of glandless types of cotton resulted in a great improvement of meal quality (McMichael,

1960) but also in a greater susceptibility to insect attack (Calhoun, 1997). In consequence, breeding efforts to develop glanded-plant, glandless-seed cottonseed cultivars are under way (Vroh-Bi et al., 1999).

Castor seeds contain two highly toxic endosperm proteins, ricin and *Ricinus communis* agglutin, which limit the utilization of castor meal for animal nutrition (Pinkerton et al., 1999). Seeds of linseed contain cyanogenic glycosides which, upon hydrolysis by enzymatic action, release hydrogen cyanide (HCN), a powerful inhibitor of the respiratory enzyme cytochrome oxidase. Nevertheless, the enzyme responsible for the hydrolysis of the cyanogenic glycosides is inactivated by heat during hot-pressing oil extraction (Oomah, Mazza, and Kenaschuk, 1992; Shahidi and Wanasundara, 1997).

BREEDING AND PRODUCTION STRATEGIES

Ultimately, production of oil crops aims at maximizing the profit from the harvest. Most oil crops produce two main products: oil and meal. In most cases the biggest profit is obtained from the oil, while oilmeal makes a secondary contribution to the overall economic value of the harvest. There are exceptions to this rule, the most representative being soybean products. Soybean seeds contain a relatively low oil content as compared with the other oil crops. One ton of soybean seed yields approximately 180 kg oil and 800 kg meal. The ratio of the sale value of meal to that of the oil changes frequently depending on the situation in particular markets (Hatje, 1989).

Increasing oil yield of oilseeds is achieved by increasing seed yield and/or increasing the oil content of the seed. Oil content depends on both the percentage of hull and the oil concentration in the kernel. In most cases, significant increases of seed oil content have been achieved by a reduction in hull percentage. In sunflower, it has been estimated that about two-thirds of the increase in achene oil content occurred during selection for this trait resulted from reduction in hull percentage, and about one-third from an increase in kernel oil content (Fick and Miller, 1997). Oil content in the seed kernel is considered to be a quantitative trait strongly influenced by the environment, although it shows a relatively high heritability in comparison with other quantitative traits such as yield (Grami, Baker, and Stefansson, 1977; Röbbelen, 1990; Miller and Fick, 1997). Consequently, plant breeders have been able to increase seed kernel oil content by changing the oil:protein:carbohydrate ratios in favor of oil by using conventional breeding methods (Murphy, 1995). Further increases using classical breeding approaches are becoming progressively more difficult and efforts to improve oil content by biotechnological means are under way (Töpfer, Martini, and Schell, 1995; Zou et al., 1997; Martini and Schulte, 1998).

Traditionally, the fatty acid composition of the oil has been considered to be the main factor defining oil quality. Therefore, great breeding efforts have been devoted to its modification for special purposes, including both food and nonfood uses of the oils. The concentration of a particular fatty acid in the seed oil is a qualitative trait governed by a reduced number of major genes. With few exceptions in which significant maternal effects have been reported, fatty acid concentration is determined by the genotype of the developing embryo (Velasco, Pérez-Vich, and Fernández-Martínez, 1999). This fact has considerably facilitated breeding for modified fatty acid composition, as early selection on single seeds is as effective as selection on single plants (Röbbelen, 1990). Based on this advantage, the half-seed technique for nondestructive selection for seed oil fatty acid composition was developed by Downey and Harvey (1963) and extensively used for oilseed breeding over the past forty years.

Environmental factors, especially temperature during seed development, influence the fatty acid composition of the oil. Canvin (1965) demonstrated that, in general, the level of unsaturation of the oil decreases as temperature increases. In sunflower, this effect has been extensively studied for the desaturation step from oleic acid (18:1) to linoleic acid (18:2). It has been concluded that temperature exerts a multiple effect on oleic acid desaturation, affecting the availability of substrate (oleate), the activity of the microsomal oleate desaturase (FAD2) enzyme (Garcés, Sarmiento, and Mancha, 1992), and the availability of oxygen, which in turn is also involved in the regulation of the enzyme (Martínez-Rivas, García-Díaz, and Mancha, 2000). As a result, higher temperatures promote higher levels of oleic acid content in the oil. This fact has traditionally been used as an important criterion for commercial production of sunflower in the United States, where oil obtained from warm regions has been used for specific markets requiring higher levels of oleic acid (Robertson, Morrison, and Wilson, 1978). The influence of temperature on the fatty acid profile of the oil is genotype dependent. A remarkable example is the high oleic acid sunflower mutant developed by Soldatov (1976), which is much less sensitive to temperature fluctuations than standard sunflower. In an experiment conducted in a growth chamber, Fernández-Martínez and colleagues (1986) found a range of variation for oleic acid between 38 and 62 percent in standard sunflower compared with 91 to 95 percent in high-oleic acid sunflower. The commercial production of an oil crop having a specific fatty acid profile requires its stable expression over environments, which represents one of the major goals in oilseed breeding.

Major achievements in the modification of the fatty acid composition of seed oil have been accomplished in all important oilseeds. Particularly successful has been the induction of mutations with physical or chemical mutagenizing agents, which has produced a wide variation in fatty acid profile in rapeseed/canola, sunflower, soybean, and linseed (see review in Velasco, Pérez-Vich, and Fernández-Martínez, 1999). In recent years, however, major attention has been paid to the utilization of genetic engineering for the modification of the fatty acid biosynthetic pathway (Kinney, 1999).

The elimination of potentially toxic and/or antinutritional factors has been the main objective in breeding for meal quality in most oilseed crops. One of the most spectacular achievements has been the improvement in meal quality of old rapeseed cultivars by drastically reducing the glucosinolate content. Such a modification resulted in a change in status of the crop, from low-quality to high-quality meal for animal feed (Buzza, 1995). As described for oil quality, breeding efforts on meal quality are shifting toward a greater weight of biotechnological approaches. Thus, a number of reports have described the successful application of genetic transformation for the modification of the total seed protein content (Kohno-Murase et al., 1994), amino acid composition (Clercq et al., 1990; Altenbach et al., 1992; Kohno-Murase et al., 1995), tocopherol composition (Shintani and Dellapenna, 1998), and antinutritional compounds (Chavedej et al., 1994).

A major problem that oilseed breeders have had to confront is that each change in the oil or meal quality has initially been associated with reductions in seed and oil yields, in comparison with earlier quality forms. The reason for this cannot be attributed to physiological constraints but to a transfer of selection intensity from yield to quality traits (Becker, Löptien, and Röbbelen, 1999). However, cultivars with improved oil or meal quality and similar or even better agronomic performance than earlier quality cultivars can be developed by maintaining an adequate selection intensity on yield. Current sunflower cultivars with high oleic acid content (Fernández-Martínez, Muñoz, and Gómez-Arnau, 1993), canola cultivars exhibiting a simultaneous reduction of erucic acid and glucosinolates (Röbbelen and Thies, 1980), and linseed cultivars with low linolenic acid content (Scarth, Mcvetty, and Rimmer, 1997) are illustrative examples.

Oil crops, particularly oilseeds, have traditionally been at the forefront of plant breeding. This leading position has been augmented with the advent of the biotechnological revolution. In the next several years novel breeding tools derived from biotechnology and molecular genetics will be called on to play an increasingly important role. Nevertheless, effective improvement of grain quality in oil crops will necessarily require the adequate integration of such novel tools with traditional breeding approaches.

REFERENCES

Ågren, J. and Schemske, D.W. (1993). The cost of defense against herbivores: An experimental study of trichome production in *Brassica rapa*. *American Naturalist* 141: 338-350.

Aldeola, A. (1995). Digestive utilization of minerals by weanling pigs fed copper- and phytase-supplemented diets. *Canadian Journal of Animal Science* 75: 603-610.

Altenbach, S.B., Kuo, C.C., Staraci, L.C., Pearson, K.W., Wainwright, C., Georgescu, A., and Townsend, J. (1992). Accumulation of a Brazil nut albumin in seeds of transgenic canola results in enhanced levels of seed protein methionine. *Plant Molecular Biology* 18: 235-245.

Álvarez-Ortega R., Cantisán, S., Martínez-Force, E., and Garcés, R. (1997). Characterization of polar and nonpolar seed lipid classes from highly saturated fatty acid sunflower mutants. *Lipids* 32: 833-837.

Aparicio, R., Roda, L., Albi, M.A., and Gutierrez, F. (1999). Effect of various compounds on virgin olive oil stability measured by Rancimat. *Journal of Agricultural and Food Chemistry* 7: 4150-4155.

Åppelqvist, L.-A. (1989). The chemical nature of vegetable oils. In Downey, R.K., Röbbelen, G., and Ashri, A. (Eds.), *Oil Crops of the World* (pp. 22-37). New York: McGraw-Hill.

Auld, D.L., Heikkinen, M.K., Erickson, D.A., Sernyk, J.L., and Romero, J.E. (1992). Rapeseed mutants with reduced levels of polyunsaturated fatty acids and increased levels of oleic acid. *Crop Science* 32: 657-662.

Baetzel, R., Friedt, W., and Lühs, W. (2000). Genetic modification of seed colour as means of rapeseed (*Brassica napus* L.) improvement. Proceedings of the Third International Crop Science Congress (p. 33), Hamburg, Germany, August 17-22.

Becker, H., Löptien, H., and Röbbelen, G. (1999). Breeding: An overview. In Gómez-Campo, C. (Ed.), *Biology of Brassica Coenospecies* (pp. 413-499). Amsterdam: Elsevier Science B.V.

Bell, J.M. (1989). Nutritional characteristics and protein uses of oilseed meals. In Downey, R.K., Röbbelen, G., and Ashri, A. (Eds.), *Oil Crops of the World* (pp. 192-207). New York: McGraw-Hill.

Bell, J.M. and Shires, A. (1982). Composition and digestibility by pigs of hull fractions from rapeseed cultivars with yellow or brown seed coats. *Canadian Journal of Animal Science* 62: 557-565.

Benveniste, P. (1986). Sterol biosynthesis. *Annual Review of Plant Physiology and Plant Molecular Biology* 37: 275-308.

Bewley, J.D. and Black, M. (1978). *Physiology and Biochemistry of Seeds in Relation to Germination*. Berlin: Springer-Verlag.

Bickert, C., Lühs, W., and Friedt, W. (1994). Variation for fatty acid content and triacylglycerol composition in different *Linum* species. *Industrial Crops and Products* 2: 229-237.

Borek, C. (1993). Molecular mechanisms in cancer induction and prevention. *Environmental Health Perspectives* 101(Suppl. 3): 237-245.

Bouchereau, A., Clossais-Besnard, N., Bensaoud, A., Leport, L., and Renard, M. (1996). Water stress effects on rapeseed quality. *European Journal of Agronomy* 5: 19-30.

Bracco, U. (1994). Effect of triglyceride structure on fat absorption. *American Journal of Clinical Nutrition* 60(Suppl.): 1002s-1009s.

Bubeck, D.M., Fehr, W.R., and Hammond, E.G., (1989). Inheritance of palmitic and stearic acid mutants of soybean. *Crop Science* 29: 652-656.

Buzza, G.C. (1995). Plant breeding. In Kimber, D.S. and McGregor, D.I. (Eds.), *Brassica Oilseeds: Production and Utilization* (pp. 153-175). Wallingford, UK: CAB International.

Calhoun, D.S. (1997). Inheritance of high glanding, and insect resistance trait in cotton. *Crop Science* 37: 1181-1186.

Canvin, D.T. (1963) The formation of oil in the seed of *Ricinus communis* L. *Canadian Journal of Biochemistry* 41: 1879-1885.

Canvin, D.T. (1965). The effect of temperature on the oil content and fatty acid composition of the oils from several oil seed crops. *Canadian Journal of Botany* 43: 63-69.

Chan, J., Bruce, V., and McDonald, B. (1991). Dietary alpha-linolenic acid is as effective as oleic acid and linoleic acid in lowering blood cholesterol in normolipidemic men. *American Journal of Clinical Nutrition* 53: 1230-1234.

Chavedej, S., Brisson, N., McNeil, J., and De Luca, V. (1994). Redirection of tryptophan leads to production of low indole glucosinolate canola. *Proceedings of the National Academy of Sciences, USA* 89: 184-188.

Chen, J.C.F., Tsai, C.C.Y., and Tzen, J.T.C. (1999). Cloning and secondary structure-analysis of caleosin, a unique calcium-binding protein in oil bodies of plant seeds. *Plant and Cell Physiology* 40: 1079-1086.

Clercq, A.D., Vanderwiele, M., Damme, J.V., Guerche, P., Montagu, M.V., Venderkerckhove, J., and Krebbers, E. (1990). Stable accumulation of modified 2s albumin seed storage protein with a higher methionine content in transgenic plants. *Plant Physiology* 94: 970-979.

Crapiste, G.H., Brevedan, M.I.V., and Carelli, A.A. (1999). Oxidation of sunflower oil during storage. *Journal of the American Oil Chemists' Society* 76: 1437-1443.

De la Puerta, R., Gutierrez, V.R., and Hoult, J.R.S. (1999). Inhibition of leukocyte 5-lipoxygenase by phenolics from virgin olive oil. *Biochemical Pharmacology* 57: 445-449.

Dehesh, K., Edwards, P., Hayes, T., Cranmer, A.M., and Fillatti, J. (1996). Two novel thioesterases are key determinants of the bimodal distribution of acyl-chain length of cuphea-palustris seed oil. *Plant Physiology* 110: 203-210.

Demurin, Y., Skoric, D., and Karlovic, D. (1996). Genetic variability of tocopherol composition in sunflower seeds as a basis for breeding for improved oil quality. *Plant Breeding* 115: 33-36.

Denis, L., Domínguez, J., and Vear, F. (1994). Inheritance of "hullability" in sunflower (*Helianthus annuus* L.). *Plant Breeding* 113: 27-35.

Dorrell, D.G. and Vick, B.A. (1997). Properties and processing of oilseed sunflower. In Schneiter, A.A. (Ed.), *Sunflower Technology and Production* (pp. 709-

745). Madison, WI: American Society of Agronomy, Crop Science Society of America, and Soil Science Society of America.

Downey, R.K., Craig, B.M., and Youngs, C.G. (1969). Breeding rapeseed for oil and meal quality. *Journal of the American Oil Chemists' Society* 46: 121-123.

Downey, R.K. and Harvey, B.L. (1963). Methods of breeding for oil quality in rape. *Canadian Journal of Plant Science* 43: 271-275.

Dutta, P.C. and Åppelqvist, L.-A. (1996). Saturated sterols (stanols) in unhydrogenated and hydrogenated edible vegetable oils and in cereal lipids. *Journal of the Science of Food and Agriculture* 71: 383-391.

Dybing, C.D. and Zimmerman, D.C. (1966). Fatty acid composition in maturing flax as influenced by the environment. *Plant Physiology* 41: 1465-1470.

Ebert, W.W. and Knowles, P.F. (1966). Inheritance of pericarp types, sterility, and dwarfness in several safflower crosses. *Crop Science* 6: 579-582.

Erdman, J.W., Jr. (1979). Oilseed phytates: Nutritional implications. *Journal of the American Oil Chemists' Society* 56: 736-741.

Erickson, E.A., Wilcox, J.R., and Cavins, J.F. (1988). Inheritance of altered palmitic acid percentage in two soybean mutants. *Journal of Heredity* 79: 465-468.

Fan, Y.Y. and Chapkin, R.S. (1998). Importance of dietary gamma-linolenic acid in human health and nutrition. *Journal of Nutrition* 128: 1411-1414.

Fehr, W.R., Welke, G.A., Hammond, E.G., Duvick, D.N., and Cianzio, S.R. (1991). Inheritance of reduced palmitic acid in seed oil of soybean. *Crop Science* 31: 88-89.

Fernández-Martínez, J. (1997). Update on safflower genetic improvement and germplasm resources. In *Proceedings of the Fourth International Safflower Conference* (pp. 187-195). Bari, Italy: Arti Grafiche Savaresse.

Fernández-Martínez, J., del Río, M., and de Haro, A. (1993). Survey of safflower (*Carthamus tinctorius* L.) germplasm for variants in fatty acid composition and other seed characters. *Euphytica* 69: 115-122.

Fernández-Martínez, J., Jiménez-Ramírez, A., Domínguez-Giménez, J., and Alcántara, M. (1986). Influencia de la temperatura en el contenido de ácido oleico y linoleico del aceite de tres genotipos de girasol. *Grasas y Aceites* 37: 236-331.

Fernández-Martínez, J., Muñoz, J., and Gómez-Arnau, J. (1993). Performance of near-isogenic high and low oleic acid hybrids of sunflower. *Crop Science* 33: 1158-1163.

Fick, G.N. and Miller, J.F. (1997). Sunflower breeding. In Schneiter, A.A. (Ed.), *Sunflower Technology and Production* (pp. 395-439). Madison, WI: American Society of Agronomy, Crop Science Society of America, and Soil Science Society of America.

Food and Agricultural Organization of the United Nations (FAO). (2001). *Production Yearbook 1999,* Volume 53. Rome: FAO Statistical Series.

Frandsen, G.I., Mundy, J., and Tzen, J.T.C. (2001). Oil bodies and their associated proteins, oleosin and caleosin. *Physiologia Plantarum* 112: 301-307.

Frankel, E.N. (1991). Recent advances in lipid oxidation. *Journal of the Science of Food and Agriculture* 54: 495-511.

Friedt, W. (1988). Biotechnology in breeding of industrial oil crops: The present status and future prospects. *Fat Science Technology* 90: 51-55.

Garcés, R., Sarmiento, C., and Mancha, M. (1992). Temperature regulation of oleate desaturase in sunflower (*Helianthus annuus* L.) seeds. *Planta* 186: 461-465.

Goffman, F.D. and Galletti, S. (2001). Gamma-linolenic acid and tocopherol contents in the seed oil of 47 accessions from several *Ribes* species. *Journal of Agricultural and Food Chemistry* 49: 349-354.

Goodwin, T.W. (1980). *The Biochemistry of the Carotenoids*, Volume 1. London, New York: Chapman and Hall.

Graef, G.L., Fehr, W.R., and Hammond, E.G. (1985). Inheritance of three stearic acid mutants of soybean. *Crop Science* 25: 1076-1079.

Graf, E. (1983). Applications of phytic acid. *Journal of the American Oil Chemists' Society* 60: 1861-1867.

Grami, B., Baker, R.J., and Stefansson, B.R. (1977). Genetics of protein and oil content in summer rape: Heritability, number of effective factors and correlations. *Canadian Journal of Plant Science* 57: 937-943.

Green, A.G. (1986). A mutant genotype of flax (*Linum usitatissimum* L.) containing very low levels of linolenic acid in its seed oil. *Canadian Journal of Plant Science* 66: 499-503.

Greiner, R. and Jany, K.-D. (1996). Ist Phytat ein unerwünchster Inhaltsstoff in Getreideprodkten? *Getreide Mehl Brot* 50: 368-372.

Griffiths, D.W., Birch, A.N.E., and Hillman, J.R. (1998). Antinutritional compounds in the Brassicaceae: Analysis, biosynthesis, chemistry and dietary effects. *Journal of Horticultural Science and Biotechnology* 73: 1-18.

Halliwell, B. (1996). Antioxidants in human health and disease. *Annual Review of Nutrition* 16: 33-50.

Hammond, E.G. and Fehr, W.R. (1983). Registration of A6 germplasm line of soybean. *Crop Science* 23: 192-193.

Harwood, J.L. (1996). Recent advances in the biosynthesis of plant fatty acids. *Biochimica et Biophysica Acta* 1301: 7-56.

Hatje, G. (1989). World importance of oil crops and their products. In Downey, R.K., Röbbelen, G., and Ashri, A. (Eds.), *Oil Crops of the World* (pp. 1-21). New York: McGraw-Hill.

Henry, L.K., Catignani, G.L., and Schwartz, S.J. (1998). The influence of carotenoids and tocopherols on the stability of safflower seed oil during heat-catalyzed oxidation. *Journal of the American Oil Chemists' Society* 75: 1399-1402.

Hirschberg, J. (1999). Production of high-value compounds: Carotenoids and vitamin E. *Current Opinion in Biotechnology* 10: 186-191.

Hitz, W.D., Yadav, N.S., Reiter, R.S., Mauvais, C.J., and Kinney, A.J. (1995). Reducing polyunsaturation in oils of transgenic canola and soybean. In Kader, J.C. and Mazliak, P. (Eds.), *Plant Lipids Metabolism* (pp. 506-508). Dordrecht, the Netherlands: Kluwer Academic Publishers.

Höglund, A.S., Lenman, M., Falk, A., and Rask, L. (1991). Distribution of myrosinase in rapeseed tissues. *Plant Physiology* 95: 213-221.

Horrobin, D.F. (1992). Nutritional and medical importance of gamma-linolenic acid. *Progress in Lipid Research* 31: 163-194.

Huang, A.H.C. (1996). Oleosins and oil bodies in seeds and other organs. *Plant Physiology* 110: 1055-1061.

Ichihara, K. and Noda, M. (1982). Some properties of diacylglycerol acyltransferase in a particulate fraction from maturing safflower seeds. *Phytochemistry* 21: 1895-1901.

Jaworski, J.G., Clough, R.C., and Barnum, S.R. (1989). A cerulenin insensitive short chain 3-ketoacyl-acyl carrier protein synthetase in *Spinacia oleracea* leaves. *Plant Physiology* 90: 41-44.

Jensen, S.K., Liu, Y.G., and Eggum, B.O. (1995). The influence of seed size and hull content on the composition and digestibility in rats. *Animal Feed Science and Technology* 54: 9-19.

Jiménez, A., Fernández, J., Domínguez, J., Gimeno, V., and Alcántara, M. (1985). Considerations in breeding for protein yield in sunflower. In Fernández-Martinez, J. (Ed.), *Proceedings of the Sixth Meeting of The EUCARPIA Section Oil and Protein Crops* (pp. 45-60). Córdoba, Spain, June 10-13. Cordoba: CIDA.

Jones, A., Davies, H.M., and Voelker, T.A. (1995). Palmitoyl-acyl carrier protein (ACP) thioesterase and the evolutionary origin of plant acyl-ACP thioesterases. *Plant Cell* 7: 359-371.

Josefsson, E. and Åppelqvist, L.-A. (1968). Glucosinolates in seed of rape and turnip rape as affected by variety and environment. *Journal of the Science of Food and Agriculture* 19: 564-570.

Kamal-Eldin, A. and Åppelqvist, L.A. (1996). The chemistry and antioxidant properties of tocopherols and tocotrienols. *Lipids* 31: 671-701.

Kinney. A.J. (1997). Development of genetically engineered oilseeds. In Williams, J.P., Khan, M.U., and Lem, N.W. (Eds.), *Physiology, Biochemistry and Molecular Biology of Plants Lipids* (pp. 193-200). Dordrecht, the Netherlands: Kluwer Academic Publishers.

Kinney, A.J. (1999). New and improved oils from genetically-modified oilseed plants. *Lipid Technology* 11: 36-39.

Kiritsakis, A. (1987). Olive oil: A review. *Advances in Food Research* 31: 453-482.

Kjaer, A. (1976). Glucosinolates in the Cruciferae. In Vaughan, J.G., MacLeod, A.J., and Jones, B.M.G. (Eds.), *The Biology and Chemistry of the Cruciferae* (pp. 207-219). London: Academic Press.

Knapp, S.J. (1990). New temperate oilseed crops. In Janick, J. and Simon, J.E. (Eds.), *Advances in New Crops* (pp. 203-210). Portland, OR: Timber Press.

Knowles, P.F. and Mutwakil, A. (1963). Inheritance of low iodine value of safflower selections from India. *Economic Botany* 17: 139-145.

Kohel, R.J. (1989). Cotton. In Downey, R.K., Röbbelen, G., and Ashri, A. (Eds.), *Oil Crops of the World* (pp. 404-415). New York: McGraw-Hill.

Kohno-Murase, J., Murase, M., Ichikawa, H., and Imamura, J. (1994). Effects of an antisense napin gene on storage compounds in transgenic *Brassica napus* seeds. *Plant Molecular Biology* 26: 1115-1124.

Kohno-Murase, J., Murase, M., Ichikawa, H., and Imamura, J. (1995). Improvement in the quality of seed storage protein by transformation of *Brassica napus* with an antisense gene for cruciferin. *Theoretical and Applied Genetics* 91: 627-631.

Kumar, P.R. and Tsunoda, S. (1980). Variation in oil content and fatty acid composition among seeds from the Cruciferae. In Tsunoda, S., Hinata, K., and Gómez-Campo, C. (Eds.), *Brassica Crops and Wild Allies* (pp. 235-252). Tokyo: Scientific Societies Press.

Liener, I.E., and Kakade, M.L. (1980). Protease inhibitors. In Liener, I.E. (Ed.), *Toxic Constituents of Plant Foodstuffs* (pp. 7-57). New York: Academic Press.

Liener, I.E., Nitsan, Z., Srisangnam, C., Rackis, J.J., and Gumbmann, M.R. (1985). The USDA trypsin inhibitor study: II. Time related biochemical changes in the pancreas of rats. *Plant Foods for Human Nutrition* 35: 243-257.

List, G.R., Mounts, T.L., Orthoefer, F., and Neff, W.E. (1996). Potential margarine oils from genetically modified soybeans. *Journal of the American Oil Chemists' Society* 73: 729-732.

Lobb, K. (1992). Fatty acid classification and nomenclature. In Chow, C.K. (Ed.), *Fatty Acids in Foods and Their Health Implications* (pp. 1-16). New York: Marcel Dekker.

Lolas, G.M., Palamidis, N., and Markakis, P. (1976). The phytic acid-total phosphorus relationship in barley, oats, soybeans, and wheat. *Cereal Chemistry* 53: 867-871.

Martínez, C., Ros, G., Periago, M.J., López, G., Ortuño, J., and Rincón, F. (1995). El ácido fítico en la alimentación humana. *Food Science and Technology International* 2: 201-209.

Martínez-Rivas, J.M., García-Díaz, M.T., and Mancha, M. (2000). Temperature and oxygen regulation of microsomal oleate desaturase (FAD2) from sunflower. *Biochemical Society Transactions* 28: 892-894.

Martini, N. and Schulte, W. (1998). Ansätze zur quantitativen Beeinflussung der Speicherlipide in Samen. *Vorträge für Pflanzenzüchtung* 41: 69-81.

Matthäus, B. (1997). Antinutritive compounds in different oilseeds. *Fett/Lipid* 99: 170-174.

Matthäus, B., Lösing, R., and Fiebig, H.-J. (1995). Determination of inositol phosphates IP3-IP6 in rapeseed and rapeseed meal by an HPLC method: Part 2. Investigations of rapeseed and rapeseed meal and comparison with other methods. *Fat Science and Technology* 97: 372-374.

McHughen, A. (1992). Revitalisation of an ancient crop: Exciting new developments in flax breeding. *Plant Breeding Abstracts* 62: 1031-1035.

McMichael, S.C. (1960). Combined effects of the glandless genes *gl2* and *gl3* on pigmant glands in the cotton plant. *Agronomy Journal* 46: 385-386.

Mensink, R.P. and Katan, M.B. (1989). Effect of a diet enriched with monounsaturated or polyunsaturated fatty acids on levels of low-density and high-density lipoprotein cholesterol in healthy women and men. *New England Journal of Medicine* 321: 436-441.

Mensink R.P., Temme, E.H.M., and Hornstra, G. (1994). Dietary saturated and trans fatty acids and lipoprotein metabolism. *Annals of Medicine* 56: 461-464.

Miettinen, T.A., Puska, P., Gylling, H., Vanhanen, H., and Vartianen, E. (1995). Reduction of serum cholesterol with sitostanol-ester margarine in a mildly hypercholesterolemic population. *New England Journal of Medicine*. 333: 1308-1312.

Miller, J.F. and Fick, G.N. (1997). The genetics of sunflower. In Schneiter, A.A. (Ed.), *Sunflower Technology and Production* (pp. 441-495). Madison, WI: American Society of Agronomy, Crop Science Society of America, and Soil Science Society of America.

Miller, J.F. and Vick, B.A. (1999). Inheritance of reduced stearic and palmitic acid content in sunflower seed oil. *Crop Science* 39: 364-367.

Miller, N.J., Sampson, J., Candeias, L.P., Bramley, P.M., and Rice-Evans, C.A. (1996). Antioxidant activities of carotenes and xantophylls. *FEBS Letters* 384: 240-242.

Mithen, R. (1992). Leaf glucosinolate profiles and their relationship to pest and disease resistance in oilseed rape. In Johnson, R. and. Gellis, G.J. (Eds.), *Breeding for Disease Resistance* (pp. 71-83). Dordrecht, the Netherlands: Kluwer Academic Publishers.

Mounts, T.L., Wolf, W.J., and Martinez, W.H. (1987). Processing and utilization. In Wilcox, J.R. (Ed.), *Soybeans: Improvement, Production, and Uses*, Second Edition (pp. 819-866). Madison, WI: American Society of Agronomy, Crop Science Society of America, and Soil Science Society of America.

Muggli, R. (1994). Vitamin E-Bedarf bei Zufuhr von Polyenfettsäuren. *Fat Science and Technology* 96: 17-19.

Murphy, D.J. (1995). The use of conventional and molecular genetics to produce new diversity in seed oil composition for the use of plant breeders: Progress, problems and future prospects. *Euphytica* 85: 433-440.

Naczk, M., Amarowicz, R., Sullivan, A., and Shahidi, F. (1998). Current research developments on polyphenolics of rapeseed/canola: A review. *Food Chemistry* 62: 489-502.

Naested, H., Frandsen, G.I., Jauh, G.Y., Hernandezpinzon, I., Nielsen, H.B., Murphy, D.J., Rogers, J.C., and Mundy, J. (2000). Caleosins—Ca2+-binding proteins associated with lipid bodies. *Plant Molecular Biology* 44: 463-476.

Newmark, H.L. (1999). Squalene, olive oil, and cancer risk: Review and hypothesis. *Annals of the New York Academy of Science* 889: 193-203.

Nguyen, T.T. (1999). The cholesterol-lowering action of plant sterols. *Journal of Nutrition* 129: 2109-2112.

Niewiadomski, H. (1990). *Rapeseed: Chemistry and Technology*. Amsterdam: Elsevier.

Norton, G. (1989). Nature and biosynthesis of storage proteins. In Downey, R.K., Röbbelen, G., and Ashri, A. (Eds.), *Oil Crops of the World* (pp. 165-191). New York: McGraw-Hill.

Oberleas, D., Muhrer, M.E., and O'Dell, B. (1966). Dietary metal complexing agents and zinc availability in the rat. *Journal of Nutrition* 90: 56-62.

Ohlrogge, J.B., Browse, J., and Somerville, C.R. (1991). The genetics of plant lipids. *Biochimica et Biophysica Acta* 1082: 1-26.

Olejnik, D., Gogolewski, M., and Nogala-Kalucka, M. (1997). Isolation and some properties of plastochromanol-8. *Nahrung* 41:101-104.

Ong, A.S.H. and Choo, Y.M. (1997). Carotenoids and tocols from palm oil. In Shahidi, F. (Ed.), *Natural Antioxidants. Chemistry, Health Effects, and Applica-*

tions (pp. 133-149). Champaign, IL: American Oil Chemist's Society (AOCS) Press.

Oomah, B.D., Mazza, G., and Kenaschuk, E.O. (1992). Cyanogenic compounds in flaxseed. Journal of Agricultural and Food Chemistry 40: 1346-1348.

Osborne, T.B. (1924). The Vegetable Proteins, Second Edition. London: Longman, Green and Co.

Osorio, J., Fernández-Martínez, J., Mancha, M., and Garcés, R. (1995). Mutant sunflowers with high concentration of saturated fatty acids in the oil. Crop Science 35: 739-742.

Owen, R.W., Mier, W., Giacosa, A., Hull, W.E., Spiegelhalder, B., and Bartsch, H. (2000). Phenolic compounds and squalene in olive oils: The concentration and antioxidant potential of total phenols, simple phenols, secoiridoids, lignans and squalene. Food Chemistry and Toxicology 38: 647-659.

Padley, F.B., Gunstone, F.D., and Harwood, J.L. (1994). Occurrence and characteristics of oils and fats. In Gunstone, F.D., Harwood, J.L., and Padley, F.B. (Eds.), The Lipid Handbook (pp. 47-223). London: Chapman and Hall.

Palozza, P. and Krinsky, N.I. (1992). β-carotene and α-tocopherol are synergistic antioxidants. Archives of Biochemistry and Biophysics 297: 184-187.

Pascual-Villalobos, M.J., Röbbelen, G., Correal, E., and Ehbrecht-von-Witzke, S.E. (1992). Performance test of Euphorbia lagascae Spreng., and oilseed species rich in vernolic acid, in southeast Spain. Industrial Crops and Products 1: 185-190.

Pedrosa, M.M., Muzquiz, M., García-Vallejo, C., Burbano, C., Cuadrado, C., Ayet, G., and Robredo, L.M. (2000). Determination of caffeic and chlorogenic acids and their derivatives in different sunflower seeds. Journal of the Science of Food and Agriculture 80: 459-464.

Piironen, V., Lindsay, D.G., Miettinen, T.A., Toivo, J., and Lampi, A.M. (2000). Plant sterols: Biosynthesis, biological function and their importance to human nutrition. Journal of the Science of Food and Agriculture 80: 939-966.

Pinkerton, S.D., Rolfe, R., Auld, D.L., Ghetie, V., and Lauterbach, B.F. (1999). Selection of castor for divergent concentrations of ricin and Ricinus communis agglutin. Crop Science 39: 353-357.

Pongracz, G., Weiser, H., and Matzinger, D. (1995). Tocopherole: Antioxidanten der Natur. Fat Science and Technology 97: 90-104.

Potterat, O. (1997). Antioxidants and free-radical scavengers of natural origin. Current Organic Chemistry 1: 415-440.

Psomiadou, E. and Tsimidou, M. (1999). On the role of squalene in olive oil stability. Journal of Agricultural and Food Chemistry 47: 4025-4032.

Putt, E.D. (1997). Sunflower early history. Properties and processing of oilseed sunflower. In Schneiter, A.A. (Ed.), Sunflower Technology and Production (pp. 1-19). Madison, WI: American Society of Agronomy, Crop Science Society of America, and Soil Science Society of America.

Rackis, J.J., Wolf, W.J., and Baker, E.C. (1985). Protease inhibitors in plant foods: Content and inactivation. Advances in Experimental Medicine and Biology 199: 299-347.

Rahman, S.M., Takagi, Y., Miyamoto, K., and Kawakita, T. (1995). High stearic acid soybean mutant induced by X-ray irradiation. *Bioscience Biotechnology and Biochemistry* 59: 922-923.

Rakow, G. (1973). Selektion auf Linol- und Linolensäuregehalt in Rapssamen nach mutagener Behandlung. *Zeitschrift für Pflanzenzüchtung* 69: 62-82.

Renaud, S.C., Ruf, J.C., and Petithory, D. (1995). The positional distribution of fatty acids in palm oil and lard influences their biological effect in rats. *Journal of Nutrition* 125: 229-237.

Reske, J., Siebrecht, J., and Hazebroek, J. (1997). Triacylglycerol composition and structure in genetically modified sunflower and soybean oils. *Journal of the American Oil Chemists' Society* 74: 989-998.

Röbbelen, G. (1981). Potentials and restrictions of breeding for protein improvement in rapeseed. In Bunting, E.S. (Ed.), *Production and Utilization of Protein in Oilseed Crops* (pp. 3-11). The Hague: Martinus Nijhoff Publishers.

Röbbelen, G. (1990). Mutation breeding for quality improvement. A case study for oilseed crops. *FAO/IAEA Mutation Breeding Review* 6: 1-44.

Röbbelen, G. and Nitsch, A. (1975). Genetical and physiological investigations on mutants for polyenoic fatty acids in rapeseed, *Brassica napus* L.: I. Selection and description of new mutants. *Zeitschrift für Pflanzenzüchtung* 75: 93-105.

Röbbelen, G. and Thies, W. (1980). Biosynthesis of seed oil and breeding for improved oil quality of rapeseed. In Tsunoda, S., Hinata, K., and Gómez-Campo, C. (Eds.), *Brassica Crops and Wild Allies* (pp. 254-283). Tokyo: Japan Scientific Societies Press.

Robertson, J.A., Morrison, W.H., III, and Wilson, R.L. (1978). Effect of sunflower hybrid or variety and planting location on oil content and fatty acid composition. In *Proceedings of the Eighth International Sunflower Conference* (pp. 524-532). Minneapolis, MN, July 23-27. Minneapolis, MN: International Sunflower Association.

Rosa, E.A.S. (1999). Chemical composition. In Gómez-Campo, C. (Ed.), *Biology of Brassica Coenospecies* (pp. 315-357). Amsterdam: Elsevier Science B.V.

Rossell, J.B. and Pritchard, J.L.R. (1991). *Analysis of Oilseeds, Fats, and Fatty Foods*. New York: Elsevier Applied Science.

Rowland, G.G. and Bhatty, R.S. (1991). Ethyl methanesulphonate induced fatty acid mutations in flax. *Journal of the American Oil Chemists' Society* 67: 213-214.

Rücker B. and Röbbelen, G. (1997). Mutants of *Brassica napus* with altered seed lipid fatty acid composition. In *Proceedings of the Twelfth International Symposium on Plant Lipids* (pp. 316-318). Dordrecht, the Netherlands: Kluwer Academic Publishers.

Saija, A., Trombetta, D., Tomaino, A., Locascio, R., Princi, P., Uccella, N., Bonina, F., and Castelli, F. (1998). In-vitro evaluation of the antioxidant activity and biomembrane interaction of the plant phenols oleuropein and hydroxytyrosol. *International Journal of Pharmaceutics* 166: 123-133.

Scarth, R., Mcvetty, P.B.E., and Rimmer, S.R. (1997). Allons low linolenic acid summer rape. *Canadian Journal of Plant Science* 77: 125-126.

Shahidi, F. (1997). Natural antioxidants: An overview. In Shahidi, F. (Ed.), *Natural Antioxidants: Chemistry, Health Effects, and Applications* (pp. 1-11). Champaign, IL: AOCS Press.

Shahidi, F. and Naczk, M. (1992). An overview of the phenolics of canola and rapeseed: Chemical, sensory and nutritional significance. *Journal of the American Oil Chemists' Society* 69: 917-924.

Shahidi, F. and Wanasundara, P.K.J.P.D. (1997). Cyanogenic glycosides of flaxseeds. *ACS Symposium Series* 662: 171-185.

Shintani, D. and Dellapenna, D. (1998). Elevating the vitamin-E content of plants through metabolic engineering. *Science* 282: 2098-2100.

Shukla, V.K.S., Wanasundara, P.K.J.P.D., and Shahidi, F. (1997). Natural antioxidants from oilseeds. In Shahidi, F. (Ed.), *Natural Antioxidants: Chemistry, Health Effects, and Applications* (pp. 97-132). Champaign, IL: AOCS Press.

Small, D.M. (1991). The effects of glyceride structure on absorption and metabolism. *Annual Review of Nutrition* 11: 413-434.

Smith, K.J. and Huyser, W. (1987). World distribution and significance of soybean. In Wilcox, J.R. (Ed.), *Soybeans: Improvement, Production, and Uses*, Second Edition (pp. 1-22). Madison, WI: American Society of Agronomy, Crop Science Society of America, and Soil Science Society of America.

Soldatov, K.I. (1976). Chemical mutagenesis in sunflower breeding. In *Proceedings of the Seventh International Sunflower Conference* (pp. 352-357). Krasnodar, USSR: International Sunflower Association.

Somerville, C. (1991). Prospects for genetic modification of the composition of edible oils from higher plants. *Food Biotechnology* 5: 217-228.

Somerville, C. and Browse, J. (1991). Plant lipids: Metabolism, mutants, and membranes. *Science* 252: 80-87.

Sørensen, H. (1990). Glucosinolates: Structure, properties, function. In Shahidi, F. (Ed.), *Canola and Rapeseed: Production, Chemistry, Nutrition and Processing Technology* (pp. 149-172). New York: Van Nostrand Reinhold.

Sozulski, F. (1979). Organoleptic and nutritional effects of phenolic compounds on oilseed protein products: A review. *Journal of the American Oil Chemists' Society* 56: 711-715.

St. Angelo, A.J. (1996). Lipid oxidation in foods. *Critical Reviews in Food Science and Nutrition* 36: 175-224.

Stefansson, B.R., Hougen, F.W., and Downey, R.K. (1961). Note on the isolation of rape plants with seed oil free from erucic acid. *Canadian Journal of Plant Science* 41: 218-219.

Stojšin, D., Ablett, G.R., Luzzi, B.M., and Tanner, J.W. (1998). Use of gene substitution values to quantify partial dominance in low palmitic acid soybean. *Crop Science* 38: 1437-1441.

Stojšin, D., Luzzi, B.M., Ablett, G.R., and Tanner, J.W. (1998). Inheritance of low linolenic acid level in the soybean line RG10. *Crop Science* 38: 1441-1444.

Stringam, G.R., McGregor, D.I., and Pawlowski, S.H. (1974). Chemical and morphological characteristics associated with seedcoat color in rapeseed. In *Proceedings of the Fourth International Rapeseed Conference* (pp. 99-108). Giessen, Germany.

Takagi, Y., Hossain, A.B.M.M., Yanagita, T., Matsueda, T., and Murayama, A. (1990). Linolenic acid content in soybean improved by X-ray irradiation. *Agricultural and Biological Chemistry* 54: 1735-1738.

Takagi, Y., Rahman, S.M., Joo, H., and Kawakita, T. (1995). Reduced and elevated palmitic acid mutants in soybean developed by X-ray irradiation. *Bioscience Biotechnology and Biochemistry* 59: 1778-1779.

Tatum, V. and Chow, C.K. (1992). Effects of processing and storage on fatty acids in edible oils. In Chow, C.K. (Ed.), *Fatty Acids in Foods and Their Health Implications* (pp. 337-351). New York: Marcel Dekker.

Thompson, A.E., Dierig, D.A., Johnson, E.R., Dahlquist, G.H., and Kleiman, R. (1994). Germplasm development of *Vernonia galamensis* as a new industrial oilseed crop. *Industrial Crops and Products* 3: 185-200.

Thompson, L.U. (1990). Phytates in canola/rapeseed. In Shahidi, F. (Ed.), *Canola and Rapeseed: Production, Chemistry, Nutrition and Processing Technology* (pp. 173-192). New York: Van Nostrand Reinhold.

Töpfer, R., Martini, N., and Schell, J. (1995). Modification of plant lipid synthesis. *Science* 268: 681-686.

Traber, M.G. and Sies, H. (1996). Vitamin E in humans: Demand and delivery. *Annual Review of Nutrition* 16: 321-347.

Tsimidou, M., Papadoupoulos, G., and Boskpu, D. (1992). Phenolic compounds and stability of virgin olive oil. Part I. *Food Chemistry* 45: 141-144.

Tzen, J.T.C., Cao, Y.Z., Laurent, P., Ratnayake, C., and Huang, A.H.C. (1993). Lipids, proteins, and structure of seed oil bodies from diverse species. *Plant Physiology* 101: 267-276.

Uppström, B. (1995). Seed chemistry. In Kimber, D.S. and McGregor, D.I. (Eds.), *Brassica Oilseeds: Production and Utilization* (pp. 217-242). Wallingford, UK: CAB International.

U.S. Department of Agriculture (USDA) (1992). The food guide pyramid. Home and Garden Bulletin no. 252, USDA Human Nutrition Information Service.

Vaisey-Genser, M. and Eskin, N.A.M. (1987). *Canola Oil: Properties and Performance.* Winnipeg: Canola Council of Canada.

van der Wal, R.J. (1960). Calculation of the distribution of the saturated and unsaturated acyl groups in fats from pancreatic lipase hydrolisis data. *Journal of the American Oil Chemists' Society* 37: 18-20.

Velasco, L. and Fernández-Martínez, J.M. (2000). Isolation of lines with contrasting seed oil fatty acid profiles from safflower germplasm. *Sesame and Safflower Newsletter* 15: 104-108.

Velasco, L., Fernández-Martínez, J.M., and De Haro, A. (2001). Relationship of test weight and seed quality traits in Ethiopian mustard. *Journal of Genetics and Breeding* 55: 91-94.

Velasco, L. and Goffman, F.D. (2000). Tocopherol, plastochromanol and fatty acid patterns in the genus *Linum*. *Plant Systematics and Evolution* 221: 77-88.

Velasco, L. and Möllers, C. (1998). Nondestructive assessment of sinapic acid esters in *Brassica* species: II. Evaluation of germplasm and identification of phenotypes with reduced levels. *Crop Science* 38: 1650-1654.

Velasco, L., Pérez-Vich, B., and Fernández-Martínez, J.M. (1999). The role of mutagenesis in the modification of the fatty acid profile of oilseed crops. *Journal of Applied Genetics* 40: 185-209.

Vles, R.O. and Gottenbos, J.J. (1989). Nutritional characteristics and food uses of vegetable oils. In Downey, R.K., Röbbelen, G., and Ashri, A. (Eds.), *Oil Crops of the World* (pp. 63-86). New York: McGraw-Hill.

Voelker, T.A., Jones, A., Cranmer, A.M., Davies, H.M., and Knutzon, D.S (1997). Broad-range and binary-range acyl-acyl-carrier-protein thioesterases suggest an alternative mechanism for medium-chain production in seeds. *Plant Physiology* 114: 669-677.

Vohra, P. (1989). Carbohydrate and fiber content of oilseeds and their nutritional importance. In Downey, R.K., Röbbelen, G., and Ashri, A. (Eds.), *Oil Crops of the World* (pp. 208-225). New York: McGraw-Hill.

Vroh-Bi, I., Baudoin, J.P., Hau, B., and Mergeai, G. (1999). Development of high-gossypol cotton plants with low-gossypol seeds using trispecies bridge crosses and in vitro culture of seed embryos. *Euphytica* 106: 243-251.

Warner, K. and Frankel, E.N. (1987). Effects of β-carotene on light stability of soybean oil. *Journal of the American Oil Chemists' Society* 64:213-218.

Weiss, E.A. (1983). *Oilseed Crops.* New York: Longman.

White, P.J. (1992). Fatty acids in oilseeds (vegetable oils). In Chow, C.K. (Ed.), *Fatty Acids in Foods and Their Health Implications* (pp. 237-262). New York: Marcel Dekker.

Wilcox, J.R., Cavins, J.F., and Nielsen, N.C. (1984). Genetic alteration of soybean oil composition by a chemical mutagen. *Journal of the American Oil Chemists' Society* 61: 97-100.

Wilcox, J.R., Premachandra, G.S., Young, K.A., and Raboy, V. (2000). Isolation of high seed inorganic P, low-phytate soybean mutants. *Crop Science* 40: 1601-1605.

Willett, W.C. and Ascherio, A. (1994). Trans fatty acids: Are the effects only marginal? *American Journal of Public Health* 84: 722-724.

Wilson, R.F. (1987). Seed metabolism. In Wilcox, J.R. (Ed.), *Soybeans: Improvement, Production, and Uses,* Second Edition (pp. 643-686). Madison, WI: American Society of Agronomy, Crop Science Society of America and Soil Science Society of America.

Wong, R.S.C. and Swanson, E. (1991). Genetic modification of canola oil: High oleic acid canola. In Haberstrohn, C. and Morris, C.F. (Eds.), *Fat and Cholesterol Reduced Foods: Technologies and Strategies* (pp. 153-164). The Woodlands, TX: Portfolio Publ. Co.

Yiu, S.H., Altosaar, I., and Fulcher, R.G. (1983). The effects of commercial processing on the structure and microchemical organization of rapeseed. *Food Microstructure* 2: 165-173.

Yodice, R. (1990). Nutritional and stability characteristics of high oleic sunflower seed oil. *Fat Science and Technology* 92: 121-126.

Zou, J.T., Katavic, V., Giblin, E.M., Barton, D.L., MacKenzie, S.L., Keller, W.A., Hu, X., and Taylor, D.C. (1997). Modification of seed oil content and acyl composition in the Brassicaceae by expression of a yeast *sn*-2 acyltransferase gene. *Plant Journal* 9: 909-923.

Zyla, K. and Korelski, J. (1993). In-vitro and in-vivo dephosphorylation of rapeseed meal by means of phytate-degrading enzymes derived from *Aspergillus niger*. *Journal of the Science of Food and Agriculture* 61: 1-6.

Chapter 13

The Malting Quality of Barley

Roxana Savin
Valeria S. Passarella
José Luis Molina-Cano

INTRODUCTION

Importance of Grain Quality

Grain quality comprises a group of characteristics that collectively determine the usefulness of the harvested grains for a particular end use. Frequently it is regarded by both breeders and producers to be as important as yield. Not only are these characteristics the reason why only a few plant species are used to satisfy most human requirements for food and fiber (Slafer and Satorre, 1999), but also their importance for grain trading has been increasing in recent decades (Wrigley, 1994). Therefore, to breed and manage grain crops to achieve a certain quality standard and to be able to predict the quality of a particular crop is rather important. Achievement of this objective is dependent upon our knowledge of the major factors that could modify grain composition and, consequently, its quality.

. As grain markets have become more specialized, there is growing pressure on farmers to produce grains with greater uniformity and with certain characteristics (Wrigley, 1994). Appropriate husbandry to obtain grains with high and stable quality will likely be of increasing importance in achieving economic benefits. It is well known that grain quality may be modified by environment or crop management techniques. However, the strategies and tools required to produce grains with certain quality characteristics are not as well established as those for achieving high yields. Within this context, it has become increasingly important to improve our

We wish to thank Stuart Swanston (Scottish Crop Research Institute) and Gustavo Slafer (University of Buenos Aires) for critical reading and suggestions to this chapter. We are deeply indebted to Sandy MacGregor (Canadian Grain Commission) who generously allowed us to use his original micrographs reproduced in Figure 13.2.

understanding of the factors that determine grain quality (Gooding and Davies, 1997).

Malting and Brewing Processes

The essential aim of brewing is the conversion of cereal starch into alcohol to make a palatable beverage. Two processes are involved: the starch is first converted to soluble sugars by amylolytic enzymes, and second, the sugars are fermented to alcohol by enzymes present in yeast (Kent and Evers, 1994). Only the key aspects of the malting and brewing processes are summarized here, to provide the reader with a general framework to understand the subsequent sections of this chapter. Further and more comprehensive treatment of the topic can be found in Cook (1962), Briggs (1978), Pollock (1979), Briggs and colleagues (1986), Hough and colleagues (1987), Palmer (1989), Moll (1991), MacGregor and Bhatty (1993), Hardwick (1995), Lewis and Young (1995), and Kunze (1996).

Malting can be defined as the commercial exploitation of those processes that lead to germination (Sparrow, 1970, cited by Swanston and Ellis, 2002). The malting process commences with the steeping of barley in water to achieve a moisture level sufficient to activate metabolism in the embryonic and aleurone tissues, leading in turn to the development of hydrolytic enzymes (Figure 13.1) (Bamforth and Barclay, 1993). Moisture uptake into the starchy endosperm is also critical before the food reserves of that tissue can be mobilized through the action of enzymes during the germination process. In this mobilization phase, referred to as *modification,* the cell walls and protein matrix of the starchy endosperm are degraded, exposing the starch granules and rendering the grain friable and readily milled. After a period of germination sufficient to achieve an even modification, the green malt is dried and kilned to arrest germination and stabilize the malt (Figure 13.1). The kilning process also introduces flavor and color properties, which are very important in the subsequent beer production.

Malt is similar in appearance to barley grain, but closer physical examination reveals that the embryo has developed considerably during malting and that the endosperm has become friable (Figure 13.2). Substantial changes have taken place in many of the constituents of the grain, specifically in its cell walls, starch, and protein. During the malting process a considerable synthesis of enzymes takes place, and these enzymes are in turn responsible for the major physical changes observed (Pollock, 1962).

The key process in beer production is the fermentation of the sugars contained in the wort to form alcohol and carbon dioxide (Figure 13.1) (Hough et al., 1987; Moll, 1991; Lewis and Young, 1995; Kunze, 1996). To provide

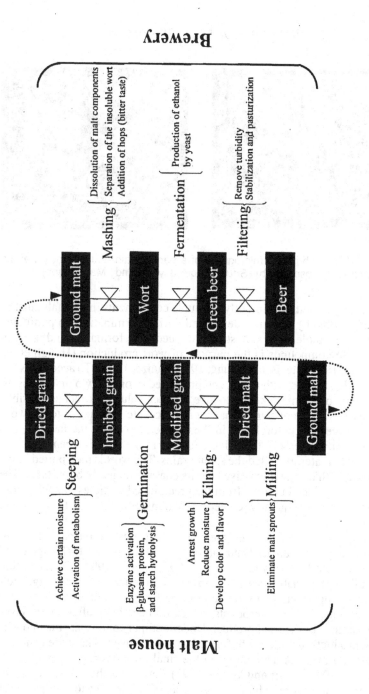

FIGURE 13.1. Diagrammatic summary of malting and brewing processes, with the key events during these processes shown.

431

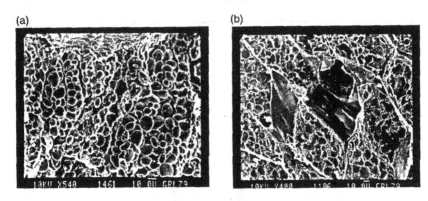

FIGURE 13.2. High-power micrograph of a barley grain endosperm (a) and a modified malt endosperm (b) (*Source:* Courtesy of Sandy MacGregor.)

the necessary conditions, initially insoluble components in the malt must be rendered soluble by enzymes developed during germination; in particular, soluble fermentable sugars must be produced. The formation and dissolution of these compounds is the purpose of wort production (Hough et al., 1987; Moll, 1991; Lewis and Young, 1995; Kunze, 1996). To accomplish it, malt is milled, mixed with water, and processed in one of two mash vessels, the mash tun or the mash kettle, to give as much soluble extract as possible. Where extra sources of starch (maize grits, rice, etc.) are added to the wort, they are processed, beforehand, in the adjunct cooker. In the lauter vessel (mash separation vessel), the soluble extract in the wort is separated from the insoluble material, called the spent grains. The wort is then boiled in the wort kettle with hops, which give beer its characteristically bitter taste. The hot wort so produced is freed from the precipitated particles in a whirlpool or centrifuge, subsequently cooled, and transferred to the fermentation tanks.

To transform wort into beer, the sugars in it must be fermented by enzymes of the yeast to ethanol and carbon dioxide (Figure 13.1) (Hough et al., 1987; Moll, 1991; Lewis and Young, 1995; Kunze, 1996). Fermentation and conditioning (maturation) are carried out in fermentation and lager cellars at low temperature, to optimize yeast performance and to avoid the formation of undesirable by-products that could negatively affect both flavor and appearance of the final product. Under optimal conditions, the undesirable by-products are either not formed or are removed. There are different fermentation systems depending on the final beer desired (Hough et al., 1987; Moll, 1991; Lewis and Young, 1995); here yeast characteristics are of paramount importance. The filtration process is intended to remove yeast and other turbidity-causing materials from beer (Moll, 1991; Kunze, 1996), and different filtration techniques are available (Kunze, 1996). The stabilization of beer, both microbiologically and colloidally, aims at keeping its quality intact for as long as possible, thus preserving its shelf life under commercial conditions. The microbiological stabilization uses different systems, e.g., pasteurization, flash pasteurization, cold sterile filling, and so on, whereas the colloidal stabilization makes use of different agents to prevent the formation of colloidal hazes (Hough et al., 1987; Hardwick, 1995; Kunze, 1996). Finally, beer is carbonated and filled in bottles or cansor kegs or is conditioned with additional yeast and filled in casks to be distributed for sale.

Objectives

All the malting and brewing processes described act on the raw material, i.e., barley grains, the quality of which is strongly dependent on the composition of these grains. This, in turn, is dependent on genotypic characteristics (conferred to them by breeding) and environmental effects (some of which may be altered by crop management). In this chapter we review the major grain structural components that affect malting quality in barley and discuss the genotypic and environmental factors that may modify it.

GRAIN STRUCTURAL COMPONENTS
THAT AFFECT MALTING QUALITY

Starch and Nonstarch Polysaccharides

Starch, protein, and β-glucan comprise around 80 percent of barley grain weight (MacGregor and Fincher, 1993). It is the products of starch (by far the main polysaccharide in grain cereals) hydrolysis that are fermented to alcohol in brewing. Therefore, genetic factors affecting the content of starch or the facility to hydrolyze it into simple sugars would be expected to have consequences for malting and brewing quality.

Starch gives the main substrates for yeast to produce alcohol during brewing fermentation. In the mature barley grain it exists in granular form, with two distinct populations of large (A-type) and small (B-type) granules, the latter accounting for 90 percent of the total number but only 10 percent of the total volume (Bathgate and Palmer, 1972). Starch is a combination of the two polysaccharides amylose (AM) and amylopectin (AP), with AP constituting approximately 75 percent of the starch in normal, cultivated barley. In spite of the straight chain of AM and its molecular simplicity, AP is more readily degraded by amylases (Ellis, 1976). The high amylose character is also associated with a reduction in the size of the A-type granules (Ellis, 1976), and the smaller the granules the more heavily they are imbedded into the surrounding protein matrix (Swanston, 1994). This gives rise to a very compact endosperm structure which is difficult to disrupt either enzymatically or mechanically (Swanston, Ellis, and Stark, 1995). Thus, B-type granules are more difficult to degrade than the larger A-type granules, and stresses reducing the proportion of A-type granules tend to reduce malt extract, beyond any additional effect these stresses may have on the amount of starch. Allan and colleagues (1995) suggested that barley samples with a greater proportion of larger diameter starch granules could potentially produce more fermentable sugars because they would have a greater percentage of starch by volume. In addition, a portion of the AM forms a complex with lipids which remains stable at temperatures above 90°C (Tester and Morrison, 1993) and is not degraded during malting, this being exacerbated in high AM genotypes (Swanston, Ellis, and Stark, 1995).

Nonstarch polysaccharides and lignin (~10 to 20 percent of grain) are structural components of cell walls and negatively affect the water uptake and germination of the barley grain (Molina-Cano, Swanston, and Ullrich, 2000). Other fiber components that are partially soluble are mixed-linked (1→3, 1→4) β-glucans (1.5 to 11.5 percent) and arabinoxylans (~4 to 8 percent), which have a tendency to increase the viscosity of solutions and colloids, thus hampering the diffusion of water and enzymes throughout the endosperm. β-glucans are the primary components in the starchy endosperm cell walls (Fincher, 1975). Arabinoxylans, also known as pentosans, are found primarily in the cell walls of the hull and aleurone (Han and Froseth, 1992).

High levels of β-glucan have long been regarded as deleterious to malting quality, as they may reduce the rate of endosperm modification (Bamforth, 1982). Residual cell walls are barriers to amylolytic and proteolytic enzymes acting during malting and mashing. In addition, high levels of β-glucan in the wort cause increases in viscosity, which may lead to filtration problems (Bamforth and Barclay, 1993). Barley β-glucans are classed as water soluble or insoluble and can be assessed separately by an enzymatic method (Aman and Graham, 1987).

Storage Proteins

The negative correlation between malt extract yield (i.e., the total extractable material that is likely to be obtained from a given malt) and barley protein content, first reported by Bishop in the 1930s (Molina-Cano, Swanston, and Ullrich, 2000), is now known to be mainly due to the hordeins, the major fraction of the endosperm storage proteins in barley grains. There are two main reasons for this negative correlation: (1) the relative increase in other grain components naturally implies a likely reduction in starch content, the main source of extract (Smith, 1990), and (2) hordein acts as a physical barrier to starch degradation due to its role as the main component of the endosperm protein matrix into which the starch granules are embedded. Increases in hordein content thus imply a restricted access to the amylolytic enzymes during malting (Palmer, 1989).

When examining cultivar and environmental effects on malting quality of Australian barleys grown under a Mediterranean-type climate (Eagles et al., 1995), it was found that seasonal differences in malt extract levels could not be explained completely by differences in protein concentration. This indicates that other factors, such as the type of protein present or the starch characteristics, influence malt extract independently of grain protein concentration.

Hordeins are soluble in alcohol/water mixtures and, on average, account for up to 60 percent of the total grain nitrogen (Brennan et al., 1998). They are classified into four groups, or families of polypeptides, called B-, C-, D-, and γ-hordeins, and are encoded by the genes *Hor2* (B-fraction), *Hor1* (C-fraction), *Hor3* (D-fraction), and *Hor5* (γ-fraction), located on barley chromosome 5(1H) (Kreis and Shewry, 1992). The B- and C-fractions account for 70 to 80 percent and 10 to 20 percent, respectively, of the total hordein, while the D and γ groups are quantitatively minor components.

There is still uncertainty on the specific effect of the different hordein subunits on malting performance (Swanston and Ellis, 2002). Furthermore, the results may differ among the various methodologies used to measure the hordein fractions (Janes and Skerritt, 1993) and/or a clear definition of the different groups that conformed a certain fraction (reviewed by Smith, 1990). For example, some researchers found a negative relationship between B-hordein (Baxter and Wainwright, 1979), D-hordein (Howard et al., 1996) or a colloidal aggregate of D- and B-hordeins linked by thiol groups, named gel protein (van den Berg et al., 1981; Smith and Lister, 1983), and malting quality. Recently, analyses of isogenic barley lines with and without D-hordeins clearly demonstrated that D-hordein was a major component of gel protein but failed to observe differences in malting performance, raising uncertainties about the deleterious influence of this hordein fraction on malting quality. As a continuation of this work, the same six pairs of near-isogenic lines, differing in the presence or absence of D-hordein, were included in genetic backgrounds with different B- and C-hordein alleles (Brennan et al., 1998). The conclusion reached by the authors was that differences in malting quality were not related to either the presence or absence of D-hordein or to gel protein levels.

Additional effects of hordein subunits on other malting attributes were found. Peltonen and colleagues (1994) evaluated the effect of D-, C-, and B-hordeins on the malting quality of northern European barleys. They found that the B-fraction had some effect on malting quality by regulating diastatic power (a complex of enzymes that hydrolyze starch in barley endosperms). In the absence of a reducing agent in the extraction solvent, a high concentration of B-hordein was correlated with an increase in extract yield. When environmental conditions favored high nitrogen uptake efficiency, a larger proportion of D-hordein disulphide bonds was synthesized, thus decreasing malting quality. It was speculated that with lower grain nitrogen content and lower D:B hordein ratio, malting quality increased (Peltonen et al., 1994). Contradictory results were obtained, however, by Molina-Cano, Swanston, and Ullrich (2000), who found that B-hordeins were associated with a decrease of malt extract, which was significantly greater in Nordic compared to Mediterranean barleys.

Study of the malting behavior of barleys grown in Spain and Scotland (Molina-Cano et al., 1995) permitted the formulation of a model stating that a lower content of insoluble β-glucans and a higher B/C-hordein ratio was linked to superior malting quality. The use of total grain protein content as a predictor of malting quality was considered to be inadequate to fully account for the different malting behavior of northern and southern European barleys (as also discussed in Savin and Molina-Cano, 2002).

Further research into genetic (G) and environmental (E) influences on hordein content was carried out by Molina-Cano and colleagues (Molina-Cano, Polo, Romera, et al., 2001; Molina-Cano, Polo, Sopena, et al., 2001), using induced mutants in the malting barley cultivar Triumph that were grown in both Spain and Scotland. The mutant TL 43 showed higher B-hordein content than 'Triumph' in Scotland but lower content in Spain; it displayed consistently higher C- and D-hordein content than 'Triumph' in both environments, i.e., there was GxE interaction for B-hordein and none for C- and D-hordein content. There were also differences in grain ultra-structure between the two lines, as TL43 showed a more dense protein matrix than 'Triumph,' together with a thinner pericarp, testa, and aleurone layers (Molina-Cano, Polo, Sopena, et al., 2001). When studying water uptake, a general conclusion was that although influenced both by genotype and environment, water uptake was more dependent on the latter. B-hordein quantity and distribution were important factors hampering water uptake, but soluble β-glucan content, previously implicated in determining differences in water uptake between grains produced in Spain and Scotland, appeared to have little effect in this study (Molina-Cano, Polo, Sopena, et al., 2001).

Endosperm modification during malting in the same genotypes as, i.e., TL43/'Triumph,' was also studied in Spain and Scotland during different seasons (Swanston and Molina-Cano, submitted). Over seasons, the mutant generally gave slightly lower extracts than 'Triumph,' but this was associated with higher protein levels in the malted grain. At given nitrogen levels, TL43 had a similar extract to 'Triumph,' but a higher Kolbach index, suggesting that the extract contained higher proportions of protein-derived material. Patterns of development of both extract and fermentability, between two and five days after the initiation of germination, were determined mainly by the genotype, but the environment influenced the levels obtained for both traits. Two days after the initiation of germination, TL43 grown at Lleida, Spain, gave higher soluble nitrogen than 'Triumph' but, thereafter, the relative rates of nitrogen solubilization were similar for both cultivars. At both experimental sites (Scotland and Spain), TL43 gave extracts with a higher proportion of nitrogenous material than 'Triumph' and, consequently, fermentability was always lower in the mutant.

GENOTYPIC AND ENVIRONMENTAL FACTORS AFFECTING MALTING QUALITY

The environment during grain filling may produce important changes in grain composition and quality in cereals (see Chapter 11; Gooding and Davies, 1997; Savin and Molina-Cano, 2002). However, genotypic differences in the response to the environment can also be important (Stone and Nicolas, 1994; Savin and Nicolas, 1996; Wallwork et al., 1998b). Moreover, genotype × environment interaction is one of the causes of unpredictable variation found in quantitative traits such as malting quality (Molina-Cano, Polo, Romera, et al., 2001). In this section, the major environmental factors involved in the determination of malting quality are discussed, considering also genetic variability in the response to those factors.

High Temperatures

It is well known that temperature can have a profound impact on crop development and grain yield (e.g., Slafer and Rawson, 1994). For example, it is well established that optimum temperature for maximum grain weight in temperate cereals is around 15 to 18°C (e.g., Chowdhury and Wardlaw, 1978). However, in most barley-growing areas, mean temperature during grain filling is higher than this optimum.

High temperatures can also have an important effect on grain composition and quality. The responses to high temperatures have been divided into two categories (Stone and Nicolas, 1994, Wardlaw and Wrigley, 1994): (1) those resulting from sustained periods of moderately high temperature (25 to 30 to 32°C) and (2) those originated from brief periods (three to five days) of very high temperature (ca. >35°C). This division presumes that the type of responses and the mechanisms involved differ between these two thermal ranges (Wardlaw and Wrigley, 1994). Plant responses to moderately high temperature result largely from changes in the rate and duration of existing processes. In contrast, under very high temperatures some processes are severely retarded while other physiological processes may be induced or intensified.

Responses to Moderately High Temperature

The temperature range from 15 to 32°C involves a progressive decrease in grain size with temperature increase (Chowdhury and Wardlaw, 1978; Wardlaw and Wrigley, 1994). Although grain growth rate increases with temperature, the duration of grain growth is reduced to a greater extent (e.g., Chowdhury and Wardlaw, 1978). In this range, wheat yield has been shown to decline approximately 3 to 4 percent for each 1°C rise in average temperature above 15°C (Wardlaw and Wrigley, 1994) under both controlled and field conditions. Barley shows the same trend but with an apparently reduced sensitivity. For instance, Savin and colleagues (1997a) showed a reduction in grain weight with increases in temperature of only 1 percent/°C (Figure 13.3a).

It is commonly accepted that accumulation of starch is more sensitive to high temperature than protein accumulation (Jenner, Ugalde, and Aspinall, 1991), resulting in a reduction in relative starch content (Figure 13.3a). In some experiments, there were reductions in the number of both A- and B-type starch granules due to heat stress (MacLeod and Duffus, 1988). However, Savin and colleagues (1997a) showed that moderately high temperatures (27 or 30°C) commencing 20 days after anthesis and lasting until maturity reduced the final size of the grains and their starch content with no clear trends in starch granule number.

In general, grain nitrogen content per grain was unchanged by moderately high temperatures (Figure 13.3a), but grain nitrogen percentage increased (Glennie-Holmes and Jacobsen, 1994; Savin et al., 1997a). In addition, protein composition may change under elevated temperature. Swanston and colleagues (1997) found changes in the B:C-hordein ratio when comparing results obtained in Scotland and Spain with the same cultivar. B:C-hordein was lower in the hotter environment of Spain.

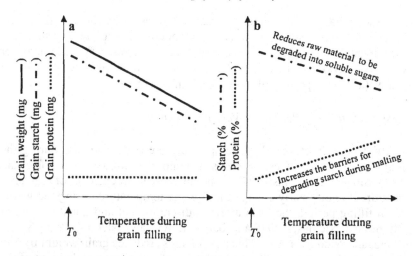

FIGURE 13.3. Theoretical reduction of grain weight, starch, and protein absolute content (a) and starch and protein relative content (b) as temperature increases from an initial temperature (T_0) during grain filling. (*Source:* References for these data can be found in Jenner, Ugalde, and Aspinall, 1991; Savin et al., 1997a; Savin and Molina-Cano, 2002 [a]; and Glennie-Holmes and Jacobsen, 1994; Savin and Molina-Cano, 2002; Passarella, Savin, and Slafer, 2002 [b].)

Differences between sites in temperature during grain filling may also change β-glucan deposition. Swanston and colleagues (1997) found differences in the pattern of β-glucan deposition during grain filling between barley crops grown in Spain and Scotland, with Spanish-grown samples having higher levels of total β-glucan but lower levels of the insoluble fraction. In a study aimed to characterize the differences between the malting behavior of Scandinavian and Iberian barleys, Molina-Cano and colleagues (2002) pointed out that total and insoluble β-glucans are of paramount importance. They were an effective barrier to extract development in the North, but were associated with an increase in the South. A conclusion was that β-glucans in the Iberian barleys contributed to extract as a consequence of the capacity of these barleys to synthesize and release β-glucan hydrolazes during germination.

A general consequence of exposure to moderately high temperatures during grain filling under controlled conditions was a reduction in malt extract (Figure 13.3b) (Glennie-Holmes and Jacobsen, 1994). The effect is evident, even after exposure for only a few days to moderately high temperatures. For instance, when the spikes (but not the whole canopy) were exposed to brief periods (five days) of high temperatures (Passarella, Savin, and Slafer, 2002), malt extract was reduced by ca. 2 percent. This figure, although apparently small, is rather important considering the very slight environmental change (the spike temperature was about 30°C at midday for only five days) and there was also a reduction in the amount of maltable grains (as the proportion of grains <2.5 mm increased).

Responses to Brief Periods of Very High Temperature

Brief periods of very high temperature are quite common during the grain-filling phase of cereal crops in temperate areas (MacNicol et al., 1993; Stone and Nicolas 1994). Although these short episodes of high temperature do not greatly change the average temperature during the whole grain-filling period, they can affect both grain yield and quality in wheat (Stone and Nicolas, 1994) and barley (Savin and Nicolas, 1996; Savin, Stone, and Nicolas, 1996). This type of stress reduces grain weight by 5 to 30 percent depending on the cultivar, time of exposure, and duration of the stress (Savin and Molina-Cano, 2002).

Reductions in grain weight are, in general, closely matched with starch content per grain (Savin and Molina-Cano, 2002) and are more related to the reduction in the number of B-type than A-type starch granules (Savin and Molina-Cano, 2002). This reduction may be caused by an irreversible effect of heat stress on the activity of soluble starch synthase, a key enzyme for the synthesis of starch (Wallwork et al., 1998a).

Similar to the effects of continuous moderately high temperatures, grain protein percentage is frequently increased when grains are exposed to brief periods of very high temperature. The increase in protein percentage is primarily due to a reduction in grain starch content, as nitrogen accumulation is comparatively unresponsive to brief episodes of very high temperatures (Savin et al., 1997b; Savin and Nicolas, 1996, 1999; Wallwork et al., 1998a).

β-glucan accumulation in the grains begins approximately 15 days after anthesis (Aman, Graham, and Tilly, 1989). Heat stress early in the accumulation phase could reduce β-glucan content more severely than a later stress (Savin et al., 1997b). For example, when heat stress commenced 15 days after anthesis, β-glucan content was significantly reduced in heat-stressed plants (Savin et al., 1997b; Wallwork et al., 1998b), whereas when the stress commenced later it was not significantly affected (Savin and Nicolas, 1996; Savin, Stone, and Nicolas, 1996).

Exposure to brief periods of very high temperature reduced hot water extract in some experiments (Wallwork et al., 1998b; Savin, Stone, and Nicolas, 1996; Savin et al., 1997b) but not in others (MacNicol et al., 1993; Savin et al., 1997b). This could be an indication that individual components such as starch, nitrogen, or β-glucan content cannot individually explain malt extract values when barley plants are exposed to short periods of high temperature. It is likely that interactions among these quality attributes may be responsible for the limited effect of those stresses on malt extract in most experiments (Savin and Molina-Cano, 2002).

Drought

In most of the rainfall regions of the world, small cereal grains are subjected to water stress which may occur at different stages during the life cycle. Traditionally, barley is grown in places with less availability of resources than wheat. However, barley is usually the highest-yielding temperate cereal in low rainfall areas where there is a Mediterranean-type climate (López-Castañeda and Richards, 1995). The yield advantage of barley is particularly evident under dry conditions, but it may disappear or be reversed when water is not a limiting factor (Araus, Slafer, and Romagosa, 1999). Little is known about the effects of postanthesis drought on grain quality in cereals, as only a few experiments have been performed to assess these effects. In addition, a problem with many drought experiments is that the intensity and timing of the stress, relative to grain growth, are often poorly defined, so that results are difficult to interpret or compare (Savin and Molina-Cano, 2002).

Grain filling depends partly on current photosynthesis and partly on the transfer of assimilate accumulated before flowering (Bonnett and Incoll, 1992). The amount of assimilate coming from photosynthesis after flowering depends on the efficiency with which the plants can use the limited water available during grain filling (Passioura, 1994). On the other hand, water-limited plants may translocate considerable amounts of preanthesis assimilates to the grain. The proportion of grain weight that comes from this source varies widely with species and environments and depends strongly on the pattern of drought (Passioura, 1994).

Only five reported studies appear to have examined the effects of drought on malting quality (reviewed in Savin and Molina-Cano, 2002). Experiments have been conducted in greenhouses (Savin and Nicolas, 1996, 1999), growth chambers (Morgan and Riggs, 1981; MacNicol et al., 1993), and under field conditions (Coles, Jamieson, and Haslemore, 1991). Grain weight reduction varied between 3 and 30 percent compared to the well-watered control, depending on the intensity and timing of exposure as well as the genotype (Savin and Molina-Cano, 2002). This grain weight reduction seems to have occurred primarily because starch accumulation had been reduced by drought. Apparently, the final number of the endosperm cells was not altered by water stress (Brooks, Jenner, and Aspinall, 1982), but the size or number of the A-type or B-type starch granules in the endosperm decreased, depending on the timing of the water stress (Brooks, Jenner, and Aspinall, 1982; Savin and Nicolas, 1999).

Morgan and Riggs (1981) found that barley extract viscosity (an indicator of β-glucan content) increased with drought applied from anthesis onward. However, when severe (Savin and Nicolas, 1996) or mild (Savin and Nicolas, 1999) droughts were applied, there was a tendency for β-glucan content in the grains to be reduced. This is in agreement with results reported under field (Stuart, Loi, and Fincher, 1988; Coles, Jamieson, and Haslemore, 1991) and growth chamber conditions (MacNicol et al., 1993). Savin and Molina-Cano (2002), reviewing results from several drought experiments, showed a linear relationship between malt extract and β-glucan degradation (i.e., difference between the percentage of β-glucans in grain and in malt, when the latter is given as a percentage, Stuart, Loi, and Fincher, 1998).

Nutrient Availability

Nitrogen is the nutrient that most often limits crop yield. Nitrogen plays numerous key roles in plant biochemistry and is an essential constituent of enzymes, chlorophyll, nucleic acids and storage proteins. Consequently, a deficiency in the supply or availability of nitrogen may have a significant influence on crop yield and grain quality. In malting barley, nutrient management is essential because high nitrogen availability may increase yield but could also be detrimental to quality, in contrast to the situation in breadmaking wheat (see Chapter 11). As discussed earlier, a high grain nitrogen content is inversely related to malting quality (Smith, 1990; Howard et al., 1996). Therefore, the amount of soil nitrogen required to maximize yield and quality would differ for each particular combination of genotype and environment.

However, grain yield and grain protein concentration are generally negatively correlated in most of the cropping systems (Smith et al., 1999; Stone and Savin, 1999). The amount of carbohydrates accumulated in a small grain cereal is most often sink limited during grain filling, whereas the amount of nitrogen is usually source limited under normal field conditions (Dreccer, Grashoff, and Rabbinge, 1997). The final protein concentration will thus depend on the nitrogen availability during the crop cycle (Stone and Savin, 1999). For instance, several authors reported an increase of C-hordeins under conditions of high fertility (Shewry et al., 1983).

ACHIEVING BARLEY-GRAIN QUALITY TARGETS

Breeding

*The Analytical Assessment of Malting Quality
in Breeding Programs*

The analytical methods used to evaluate malting quality in samples from breeding programs must be tailored for such purposes. The high number of samples to be analyzed, their small size, and the short time available to produce the results (Molina-Cano, 1991) greatly condition the methodology, which should also provide the lowest possible operating costs. As yet, sufficiently rapid methods have not been designed, since malting is essential, nor are the available ones cheap enough (the average cost of a fully analyzed malt sample is between 50 to 150 U.S. dollars, depending on the number of analytical parameters measured). Furthermore, the greater variability of nitrogen content between grains derived from ear-to-row progenies, than in grains harvested from dense plots, makes the quality data obtained during the visual selection phase of the breeding program (F_2 to F_5) rather unreliable. Other complicating factors during this phase of the program are GxE interaction and heterozygosity.

The predictive methods developed in the first half of the twentieth century, analyzing unmalted grain, by the Bendelow and Meredith team (reviewed in Bendelow, 1981) determined the potential extract and the amylolytic activity in previously digested barley flour, in the first case with a solution of malt enzymes and in the second with papain. Finally, it was determined that malting is essential if one wants an accurate quality assessment (Bendelow, 1981, personal communication).

Other nonmalting methods include milling energy (Allison et al., 1979), the measure of the viscosity of an acid extract of barley flour, as a predictor of β-glucan content (Aastrup, 1979), sedimentation (Reeves et al., 1979), falling time (Morgan, 1977), and barley hordein content (Baxter and Wainwright, 1979). Though useful under certain conditions, none of these methods is widely applicable (Bendelow, 1981).

The development of enzymatic (McCleary and Glennie-Holmes, 1985), high performance liquid chromatography (HPLC) (Pérez-Vendrell et al., 1995), and fluorimetric (Jørgensen and Aastrup, 1988) methods has enabled breeders to readily select for low levels of β-glucan in barley grain.

The first micromalting devices were developed in the 1950s, such as those of the Plant Breeding Institute in Cambridge, United Kingdom (Whitmore and Sparrow, 1957), or the Canadian Grain Research Laboratory (Atkinson and Bendelow, 1976). They had capacity for a reasonable number of very small samples, but they were manually operated and thus expensive and cumbersome, besides having a limited level of precision. Completely automated and computer-controlled micromalting models have now been developed, with an enlarged capacity of 100 to 300 samples, depending on sample size (Gothard, Morgan, and Smith, 1980; Takeda et al., 1981; Glennie-Holmes, Moon, and Cornish, 1990). Other devices include those of Carlsberg and Abed of Denmark and Phoenix and Joe White of Australia. An interesting development was the micromalting equipment of the Scottish Crop Research Institute, which allows assessment of experimental data laid out with complex statistical designs (Swanston, 1997).

Although evaluation of a commercial malt can be based on a large number of parameters (European Brewery Convention [EBC], 1987), the assessment of breeding samples relies on a smaller number of quality parameters. They can be summarized in a single figure or index, such as the EBC quality index Q (Molina-Cano, 1987) or the Carlsberg index (Ingversen, Englyst, and Jorgensen, 1989), thus supplying the breeder with a single score with likely predictive value for quality selection.

Breeding Barley for Malting in the Molecular Era

The classical barley breeding programs were conceived and carried out with the aid of the methods described previously which have been reliable and widely accepted over the years. New tools offered by DNA methodologies will have an increasing role in present and future malting barley breeding efforts, mainly because with these methods it is possible to directly observe the genotype, rather than the phenotype, in contrast to the classical methods used. As Swanston and Ellis (2002) pointed out, barley breeders are now able to exploit the knowledge that has been gained on factors affecting quality and to use DNA markers, such as restriction fragment length polymorphisms (RFLPs), random amplified polymorphic DNA (RAPDs), and amplified fragment length polymorphisms (AFLPs), to identify the genetic basis of desired traits to be selected in breeding.

The development of different genetic maps locating quantitative trait loci (QTLs) affecting quality traits has brought about a wealth of knowledge to aid malting barley breeders. Updated reviews on this topic include Kleinhofs (2000), Hayes and Jones (2000), Meyer and colleagues (2000), Molina-Cano, Swanston, and Ullrich (2000), Barr and colleagues (2000), Swanston and Ellis (2002), and Ullrich (2002). The rapid development experienced in this field should oblige both breeders and geneticists to update their knowledge through the World Wide Web. The main sites that we are aware of at the time of writing this chapter include the following:

- Washington State University Barley Genomics
  http://barleygenomics.wsu.edu
- Institute of Plant Genetics and Crop Plant Research—Gatersleben
  http://www.ipk-gatersleben.de/en/
- *Barley Genetics Newsletter*
  http://wheat.pw.usda.gov/ggpages/bgn
- U.S. Department of Agriculture GrainGenes Database
  http://wheat.pw.usda.gov/ggpages/outline.html
- *Molecular Breeding* (journal)
  http://www.kluweronline.com/issn/1380-3743
- North American Barley Genome Project
  http://barleyworld.org
- Scottish Crop Research Institute
  http://www.scri.sari.ac.uk/

Barley breeders have made significant changes and improvements in the crop, and today the malting industry has access to high quality raw material. This has been achieved largely as a result of phenotypic selection (Swanston and Ellis, 2002), making use of increasingly sophisticated testing procedures. Malting, however, comprises a highly complex series of interrelated biochemical pathways occurring simultaneously, so the underlying genetic control is equally complex. In recent years, major steps have been taken toward locating some of the critical genes and assessing their role. This is likely to continue, making use of techniques such as expressed sequence tags (ESTs), where the messenger RNA associated with activated genes is extracted to clone sequences of cDNA. One possible application of ESTs to malting barley has already been outlined briefly (Swanston et al., 1999). The vast improvements in information technology of recent years also make it possible to compare cDNA sequences with those already known and held in databases; around 70 percent of the ESTs from malted barley showed homology with known sequences (Swanston et al., 1999). Not all sequences that have been determined have a known function, but it has been possible to detect sequences associated with both carbohydrate and amino acid metabolism. Future developments are likely to include mapping of ESTs and comparison of map locations with those of known genes or QTLs.

The future, beyond any doubt, will arise from the interaction of classical breeding, genomics, and proteomics, although genetic engineering will also have an important role. Updated accounts on these topics were recently offered by Kleinhofs (2000), Waugh (2000), Lörz and colleagues (2000) and Fincher (2000).

Crop Management

Although both grain yield and quality are determined throughout the growing season, important decisions that will strongly affect them should be taken before sowing (Calderini and Dreccer, 2002). Among others, the choice of the genotype and the amount of nitrogen available are central for successfully combining the genotype potential for yield and quality with the environmental availability of resources. Final grain quality is the result of the interaction between the genotype, the natural environment, and crop management practices (Gooding and Davies, 1997). In extensive production systems it is not possible to provide each stage of the crop cycle with the optimal combination of environmental factors to reach the highest possible yield and quality; therefore, a trade-off is to make presowing decisions to ensure that critical crop stages for the definition of yield and quality are given a preferential environment (Calderini and Dreccer, 2002). Nevertheless, knowledge of the effects of environment and GxE interaction is still rather imprecise, so management strategies with the objective of increasing yields, while obtaining high malting quality barley, are difficult to design. In this section only a brief discussion on how crop management may modify barley malting quality is presented.

Grain yield is more strongly related to grain number than to grain size, with individual grain weight being the most stable component of yield (Smith et al., 1999). Thus, in general, grain yield is more responsive to the preanthesis period conditions than to those occurring after anthesis. However, grain quality may be determined by responses during both the pre- and postanthesis periods. A strong relationship can occur between plant protein content at anthesis (which is the result of crop nitrogen absorption before anthesis) and final grain protein content (Molina-Cano, Gracia, and Ciudad, 2001). The synthesis of the different grain components and the final grain composition can be altered by high temperatures, drought, and nitrogen availability during grain filling as discussed previously (MacNicol et al., 1993; Savin and Nicolas, 1996, 1999; Wallwork et al., 1998a,b) in relation to the length of the grain growth period as an environmental factor (Savin and Molina-Cano, 2002). Therefore, choosing both the appropriate cultivar for the environment and the amount of nitrogen fertilizer to apply are key aspects of crop management.

Choosing Cultivars

Malting barley is a crop sold on the basis of cultivar; therefore, cultivar choice strongly determines whether the grains will be accepted by the industry. Moreover, it is a common situation that contracts with the malting companies include the obligation of sowing certain cultivars, although acceptance of the grain is subjected to several additional requirements set by the company.

Genotypes are generally grouped into different classes according to (1) time of sowing: winter, Mediterranean, and spring cultivars and/or (2) spike type: two- and six-row cultivars. The use of a particular group will be associated with climatic considerations, mainly the length of the growing season and the likelihood of freezing temperatures during winter (Calderini and Dreccer, 2002). The first step toward choosing the best genotype from the broad range of possibilities defined by the combination of these two groupings (winter-spring barley or two-six rowed barley) is to clearly and precisely define the agroclimatic characteristics of the crop production system (Calderini and Dreccer, 2002).

Nitrogen Fertilizer

Addition of nitrogen fertilizer is one of the most frequently used methods for altering grain yield and quality (Stone and Savin, 1999). Starting from a low level of nitrogen availability, the first increment of nitrogen fertilizer increases the amounts of both starch and protein in the grain, but the response of starch is usually the greater (Figure 13.4).

Therefore, it tends to increase yield but decrease protein percentage, resulting in the frequently reported negative relationship between grain yield and protein percentage (see Stone and Savin, 1999). Before the critical level of nitrogen is attained, the response of starch and protein accumulation enters a second region of response (Figure 13.4), in which additional nitrogen fertilizer will often have a reduced (but still positive) effect on starch accumulation and a proportionally greater impact on protein accumulation. The net effect of nitrogen in the second region of response is therefore a small increase in yield and a comparatively large increase in protein percentage (Figure 13.4). As greater amounts of nitrogen are added, the crop may reach the third region of response (Figure 13.4), at which maximum yield has been attained. In this region, additional fertilizer does not affect the amount of starch in the grain, but it does increase the amount of grain protein. As a result, protein percentage is highly responsive to nitrogen in this "luxury consumption" region of nitrogen addition. Thus, addition of nitrogen to a

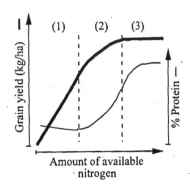

FIGURE 13.4. Diagrammatic representation of the response of yield (━━) and protein percentage (──) to nitrogen fertilizer (*Source:* From P. J. Stone and R. Savin, 1999, Grain quality and its physiological determinants. In E. H. Satorre and G. A. Slafer (Eds.), *Wheat: Ecology and Physiology of Yield Determination,* pp. 85-120. Reproduced with permission of The Haworth Press, Inc.)

soil that is rather poor in terms of nitrogen availability increases yield, whereas fertilizing a crop when the soil has a relatively high nitrogen level increases protein percentage (Stone and Savin, 1999). In barley crops for malting purposes the decision to add nitrogen fertilizer is more critical than in wheat crops for bread making, since the objective is to have high grain yields, so sufficient nitrogen must be present to achieve this; on the other hand, nitrogen should not lead to grain protein levels high enough to cause a negative relationship between malt extract and protein content (Molina-Cano, Polo, Romera, et al., 2001). Therefore, the final decision on the amount of nitrogen fertilizer to add should come from the expected yield responses at each site and also depend on the temperature, water availability, and the type of malt required by the local industry.

CONCLUSIONS

Malting quality is more than the sum of the contributions of starch, protein, and β-glucans to malt extract or any other measure of quality. It is not just the presence of a given constituent that determines quality, but rather the interaction between different constituents. In addition, the final composition of the barley grains may be modified by environment and crop management. Understanding the influence of the major environmental factors on grain composition and how these changes affect malting performance is a major factor in obtaining the quality required.

In recent years important improvements have been achieved in techniques for measuring the different components of grain and malt and also in the selection of high quality raw material through different breeding techniques. This is likely to continue in the future. However, a better understanding of how particular environmental factors may modify the composition of the barley grain and its subsequent transformation into malt and beer is still lacking. A better understanding of how these factors interact is also required to predict the final quality of the barley crop under different crop management regimes and environmental conditions.

REFERENCES

Aastrup, S. (1979). The relationship between the viscosity of an acid flour extract of barley and its beta-glucan content. *Carlsberg Research Communications* 44: 289-304.

Allan, G.R., Chrevatidis, A., Sherkat, F., and Stuart, I.M. (1995). The relationship between barley starch and malt extract for Australian barley varieties. *Proceedings of the Fifth Scientific and Technical Convention of the Institute of Brewing, Victoria Falls, Zimbabwe:* 70-79.

Allison, M.J., Cowe, I.A., Borzucki, R., Bruce, F.M., and MacHale, R. (1979). Milling energy of barley. *Journal of the Institute of Brewing* 85: 262-264.

Aman, P. and Graham, H. (1987). Analysis of total and insoluble mixed-linked (1-3), (1-4) β-D-glucan in barley and oats. *Journal of Agricultural and Food Chemistry* 35: 704-709.

Aman, P., Graham, H., and Tilly, A.C. (1989). Content and solubility of mixed linked (1-3, 1-4)-β-D-glucan in barley and oats during kernel development and storage. *Journal of Cereal Science* 10: 45-50.

Araus, J.L., Slafer, G.A., and Romagosa, I. (1999). Durum wheat and barley yields in antiquity estimated from ^{13}C discrimination of archaeological grains: A case study from the Western Mediterranean Basin. *Australian Journal of Plant Physiology* 26: 345-352.

Atkinson, J.M. and Bendelow, V.M. (1976). Automatic malting equipment for quality selection in barley breeding programs. *Canadian Journal of Plant Science* 56: 1007-1010.

Bamforth, C.W. (1982) Barley β-glucans: Their role in malting and brewing. *Brewers Digest* 57: 22-27.

Bamforth, C.W. and Barclay A.H.P. (1993). Malting technology and the uses of malt. In MacGregor, A.W. and Bhatty, R.S. (Eds.), *Barley: Chemistry and Technology* (pp. 297-354). St. Paul, MN: American Association of Cereal Chemists.

Barr, A.R., Jefferies, S.P., Warner, P., Moody, D.B., Chalmers, K.J., and Langridge, P. (2000). Marker assisted selection in theory and practice. In Logue, S. (Ed.), *Barley Genetics VIII: Proceedings of the Eighth International Barley Genetics*

Symposium, Volume I, *Invited Papers* (pp. 167-178). Glen Osmond: Adelaide University.

Bathgate, G.N. and Palmer, G.H. (1972). A reassessment of the chemical structure of barley and wheat starch granules. *Die Starke* 24: 336-341.

Baxter, E.D. and Wainwright, T. (1979). Hordein and malting quality. *Journal of the American Society of Brewing Chemists* 37: 8-12.

Bendelow, V.M. (1981). Selection for quality in malting barley breeding. In Asher, M.J.C., Ellis, R.P., Hayter, A.M., and Whitehouse, R.N.H. (Eds.), *Barley Genetics IV: Proceedings of the Fourth International Barley Genetics Symposium* (pp. 181-185). Edinburgh: Edinburgh University Press.

Bonnett, G.D. and Incoll, L.D. (1992). The potential preanthesis and postanthesis contributions of stem internodes to grain yield in crops of winter barley. *Annals of Botany* 69: 219-225.

Brennan, C.S., Smith, D.B., Harris, N., and Shewry, P.R. (1998). The production and characterisation of Hor 3 null lines of barley provides new information on the relationship of D hordein to malting performance. *Journal of Cereal Science* 28: 291-299.

Briggs, D.E. (1978). *Barley.* London: Chapman and Hall.

Briggs, D.E., Hough, J.S., Stevens, R., and Young, T.W. (1986). *Malting and Brewing Science,* Volume I: *Malt and Sweet Wort.* London: Chapman and Hall.

Brooks, A., Jenner, C.F., and Aspinall, D. (1982). Effects of water deficit on endosperm starch granules and on grain physiology of wheat and barley. *Australian Journal of Plant Physiology* 9: 423-436.

Calderini, D.F. and Dreccer, M.F. (2002). Choosing genotype, sowing date, and plant density. In Slafer, G.A., Molina-Cano, J.L., Savin, R., Araus, J.L., and Romagosa, I. (Eds.), *Barley Science: Recent Advances from Molecular Biology to Agronomy of Yield and Quality* (pp. 413-444). Binghamton, NY: Food Products Press.

Chowdhury, S.I. and Wardlaw, I.F. (1978). The effect of temperature on kernel development in cereals. *Australian Journal of Agricultural Research* 29: 205-223.

Coles, G.D., Jamieson, P.D., and Haslemore, R.M. (1991). Effect of moisture stress on malting quality in Triumph barley. *Journal of Cereal Science* 14: 161-177.

Cook, A.H. (1962). *Barley and Malt.* New York: Academic Press.

Dreccer, M.F., Grashoff, C., and Rabbinge, R. (1997). Source-sink ratio in barley (*Hordeum vulgare* L.) during grain filling: Effects on senescence on grain nitrogen concentration. *Field Crops Research* 49: 269-277.

Eagles, H.A., Bedggod, A.G., Panozzo, J.F., and Martin, P.J. (1995). Cultivar and environmental effects on malting quality in barley. *Australian Journal of Agricultural Research* 46: 831-844.

Ellis, R.P. (1976). The use of high amylose barley for the production of whisky worts. *Journal of the Institute of Brewing* 82: 280-281.

European Brewery Convention (1987). *Analytica EBC,* Fourth Edition. Brauerei- und Getranke- Rundschau. Zurich, Switzerland: European Brewery Convention.

Fincher, G.B. (1975). Morphology and chemical composition of barley endosperm cell walls. *Journal of the Institute of Brewing* 81: 116-122.

Fincher, G.B. (2000). Potential applications for protein engineering in barley improvement. In Logue, S. (Ed.), *Barley Genetics VIII: Proceedings of the Eighth International Barley Genetics Symposium*. Volume I, *Invited Papers* (pp. 194-195). Glen Osmond: Adelaide University.

Glennie-Holmes, M. and Jacobsen, J. (1994). Effect of temperature during grain filling on malting quality of barley: II. Different diurnal temperature ranges in the phytotron. *Proceedings of the Workshop on Heat Tolerance in Temperate Cereals*. Hawaii.

Glennie-Holmes, M., Moon, R.L., and Cornish, P.B. (1990). A computer controlled micro-malter for assessment of the malting potential of barley cross-breds. *Journal of the Institute of Brewing* 96: 11-16.

Gooding, M.J. and Davies, W.P. (1997). *Wheat Production and Utilization: Systems, Quality and the Environment*. Gloucestershire, UK: CAB International, University Press.

Gothard, P.G., Morgan, A.G., and Smith, D.B. (1980). Evaluation of a micro-malting procedure used to aid a plant breeding programme. *Journal of the Institute of Brewing* 86: 69-73.

Han, M.S. and Froseth, J.A. (1992). Composition of pearling fractions of barleys with normal and waxy starch. *Proceedings Western Section American Society Animal Science* 43: 155-158.

Hardwick, W.A. (1995). *Handbook of Brewing*. New York: M. Dekker.

Hayes, P.M. and Jones, B.L. (2000). Malting quality from a QTL perspective. In Logue, S. (Ed.), *Barley Genetics VIII: Proceedings of the Eighth International Barley Genetics Symposium*, Volume I, *Invited Papers* (pp. 107-114). Glen Osmond: Adelaide University.

Hough, J.S., Briggs, D.E., Stevens, R., and Young, T.W. (1987). *Malting and Brewing Science*, Volume II, *Hopped Wort and Beer*. London: Chapman and Hall.

Howard, K.A., Gayler, K.R., Eagles, H.A., and Halloran, G.M. (1996). The relationship between D hordein and malting quality in barley. *Journal of Cereal Science* 24: 47-53.

Ingversen, J., Englyst, A., and Jorgensen, K.G. (1989). Evaluation of malting quality in a barley breeding programme: Use of alpha-amylase and beta-glucan levels in malt as preselection tools. *Journal of the Institute of Brewing* 95: 99-103.

Janes, P.W. and Skerritt, J.H. (1993). High performance liquid chromatography of barley proteins: Relative quantities of hordein fractions correlate with malt extract. *Journal of the Institute of Brewing* 99: 77-84.

Jenner, C.F., Ugalde, D.T., and Aspinall, D. (1991). The physiology of starch and protein deposition in the endosperm of wheat. *Australian Journal of Plant Physiology* 18: 211-226.

Jørgensen, K.G. and Aastrup, S. (1988). Determination of β-glucan in barley, malt, wort and beer. In Linskens, H.F. and Jackson, J.F. (Eds.), *Modern Methods of Plant Analysis, New Series: Beer Analysis*, Volume 12 (pp. 88-108). Berlin, Heidelberg, New York: Springer.

Kent, N.L. and Evers, A.D. (1994). *Technology of Cereals*. Oxford, UK: Pergamon.

Kleinhofs, A. (2000). The future of barley genetics. In Logue, S. (Ed.), *Barley Genetics VIII: Proceedings of the Eighth International Barley Genetics Symposium*, Volume I, *Invited Papers* (pp. 6-12). Glen Osmond: Adelaide University.

Kreis, M. and Shewry, P.R. (1992). The control of protein synthesis in developing barley seeds. In Shewry, P.R. (Ed.), *Barley: Genetics, Biochemistry, Molecular Biology and Biotechnology* (pp. 319-333). Wallingford, UK: CAB International.

Kunze, W. (1996). *Technology of Brewing and Malting.* Berlin: VLB.

Lewis, M.J. and Young, T.W. (1995). *Brewing.* London: Chapman and Hall.

López-Castañeda, C. and Richards, R.A. (1995). Variation in temperate cereals in rainfed environments: I. Grain yield, biomass and agronomic characteristics. *Field Crops Research* 37: 51-62.

Lörz, H., Serazetdinova, L., Leckband, G., and Lütticke, S. (2000). Transgenic barley: A journey with obstacles and milestones. In Logue, S. (Ed.), *Barley Genetics VIII: Proceedings of the Eighth International Barley Genetics Symposium*, Volume I, *Invited Papers* (pp. 189-193). Glen Osmond: Adelaide University.

MacGregor, A.W. and Bhatty, R.S. (1993). *Barley: Chemistry and Technology.* St. Paul, MN: American Association of Cereal Chemists, Monograph Series.

MacGregor, A.W. and Fincher, G.B. (1993). Carbohydrates of the barley grain. In MacGregor, A.W. and Bhatty, R.S. (Eds.), *Barley: Chemistry and Technology* (pp. 73-130). St. Paul, MN: American Association of Cereal Chemists.

MacLeod, L.C. and Duffus, C.M. (1988). Temperature effects on starch granules in developing barley grains. *Journal of Cereal Science* 8: 29-37.

MacNicol, P.K., Jacobsen, J.V., Keys, M.M., and Stuart, I.M. (1993). Effects of heat and water stress on malt quality and grain parameters of Schooner barley grown in cabinets. *Journal of Cereal Science* 18:61-68.

McCleary, B.V. and Glennie-Holmes, M. (1985). Enzymic quantification of (1-3), (1-4)-β-D-glucan in barley and malt. *Journal of the Institute of Brewing* 91: 285-295.

Meyer, R.C., Swanston, J.S., Young, G.R., Lawrence, P.E., Bertie, A., Richtie, J., Wilson, A., Brosnan, J., Pearson, S., Bringhurst, T., et al. (2000). Genetic approaches to improving distilling quality in barley. In Logue, S. (Ed.), *Barley Genetics VIII: Proceedings of the Eighth International Barley Genetics Symposium*, Volume I, *Invited Papers* (pp. 115-117). Glen Osmond: Adelaide University.

Molina-Cano, J.L. (1987). The EBC barley and malt committee index for the evaluation of malting quality in barley and its use in breeding. *Plant Breeding* 98: 249-256.

Molina-Cano, J.L. (1991). Breeding barley for malting and feeding quality. In Molina-Cano, J.L. and Brufau, J. (Eds.), *New Trends in Barley Quality for Malting and Feeding, Options Méditerranéenes, Serie Seminaires, No. 20* (pp. 35-44). Zaragoza, Spain: IAMZ-CIIHEAM.

Molina-Cano, J.L., Gracia, P., and Ciudad, F. (2001). Prediction of final grain protein content with plant protein content of malting barley under Mediterranean conditions. *Journal of the Institute of Brewing* 107(6): 359-366.

Molina-Cano, J.L., Polo, J.P., Romera, E., Araus, J.L., Zarco, J., and Swanston, J.S. (2001). Relationships between barley hordeins and malting quality in a mutant of cv. Triumph I. Genotype by environment interaction of hordein content. *Journal of Cereal Science* 34: 285-294.

Molina-Cano, J.L., Polo, J.P., Sopena, A., Voltas, J., Pérez-Vendrell, A.M., and Romagosa, I. (2001). Mechanisms of malt extract development in barleys from different european regions: II. Effect of barley hordein fractions on malt extract yield. *Journal of the Institute of Brewing* 106(2): 117-123.

Molina-Cano, J.L., Ramo, T., Ellis, R.P., Swanston, J.S., Bain, H., Uribe-Echeverría, T., and Pérez-Vendrell, A.M. (1995). Effect of grain composition on water uptake by malting barley: A genetic and environmental study. *Journal of the Institute of Brewing* 101: 79-83.

Molina-Cano, J.L., Romera, E., Aikasalo, R., Pérez-Vendrell, A.M., Larsen, J., and Rubió, A. (2002). A reappraisal of the differences between Scandinavian and Spanish barleys: Effect of β-glucan content and degradation to malt extract yield in the cv. Scarlett. *Journal of the Institute of Brewing* 108(2): 221-226.

Molina-Cano, J.L., Swanston, J.S., and Ullrich, S.E. (2000). Genetic and environmental effects on malting and feed quality: A dynamic field of study. In Logue, S. (Ed.), *Barley Genetics VIII: Proceedings of the Eighth International Barley Genetics Symposium*, Volume I, *Invited Papers* (pp. 127-134). Glen Osmond: Adelaide University.

Moll, M. (1991). *Beers and Coolers*. Andover, UK: Intercept Ltd.

Morgan, A.G. (1977). The relationship between barley extract curves and malting ability. *Journal of the Institute of Brewing* 83: 231-234.

Morgan, A.G. and Riggs, T.J. (1981). Effects of drought on yield and on grain and malt characters in spring barley. *Journal of Science Food Agriculture* 32: 339-346.

Palmer, G.H. (1989). *Cereal Science and Technology*. Aberdeen: Aberdeen University Press.

Passarella, V.S., Savin, R., and Slafer, G.A. (2002). Grain weight and malting quality in barley as affected by brief periods of increased spike temperature under field conditions. *Australian Journal of Agricultural Research* 53: 1219-1227.

Passioura, J.B. (1994). The yield of crops in relation to drought. In Boote, K.J., Bennett, J.M., Sinclair, T.R., and Paulsen, G.M. (Eds.), *Physiology and Determination of Crop Yield* (pp. 343-364). Madison, WI: American Society of Agronomy.

Peltonen, J., Rita, H., Aikasalo, R., and Home, S. (1994). Hordein and malting quality in northern barleys. *Hereditas* 120: 231-239.

Pérez-Vendrell, A.M., Guasch, J., Francesch, M., Molina-Cano, J.L., and Brufau, J. (1995). Determination of β-(1-3),(1-4)-D-glucans in barley by reversed-phase high performance liquid chromatography. *Journal of Chromatography A* 718: 291-297.

Pollock, J.R.A. (1962). The nature of the malting process. In Cook, A.H. (Ed.), *Barley and Malt* (pp. 303-399). New York and London: Academic Press.

Pollock, J.R.A. (1979). *Brewing Science*, Volume I. London: Academic Press.

Reeves, S.G., Baxter, E.D., Martin, H.L., and Wainwright, T. (1979). Prediction of the malting quality of barley by a modified Zeleny sedimentation test. *Journal of the Institute of Brewing* 85: 141-143.

Savin, R. and Molina-Cano, J.L. (2002). Changes in malting quality and its determinants in response to abiotic stresses. In Slafer, G.A., Molina-Cano, J.L., Savin, R., Araus, J.L., and Romagosa, I. (Eds.), *Barley Science: Recent Advances from Molecular Biology to Agronomy of Yield and Quality* (pp. 523-550). Binghamton, NY: Food Products Press.

Savin, R. and Nicolas, M.E. (1996). Effects of short periods of drought and high temperature on grain growth and starch accumulation of two malting barley cultivars. *Australian Journal of Plant Physiology* 23: 201-210.

Savin, R. and Nicolas, M.E. (1999). Effects of timing of heat stress and drought on grain growth and malting quality of barley. *Australian Journal of Agricultural Research* 50: 357-364.

Savin, R., Stone, P.J., and Nicolas, M.E. (1996). Responses of grain growth and malting quality of barley to short periods of high temperature in field studies using portable chambers. *Australian Journal of Agricultural Research* 47: 465-477.

Savin, R., Stone, P.J., Nicolas, M.E., and Wardlaw, I.F. (1997a). Effects of heat stress and moderately high temperature on grain growth and malting quality of barley. *Australian Journal of Agricultural Research* 48: 615-624.

Savin, R., Stone, P.J., Nicolas, M.E., and Wardlaw, I.F. (1997b). Effects of the temperature regime before heat stress on grain growth and malting quality of barley. *Australian Journal of Agricultural Research* 48: 625-634.

Shewry, P.R., Franklin, J., Parmar, S., Smith, S.J., and Miflin, B.J. (1983). The effects of sulphur starvation on the amino acid and protein composition of barley grain. *Journal of Cereal Science* 1: 21-31.

Slafer, G.A. and Rawson, H.M. (1994). Sensitivity of wheat phasic development to major environmental factors: A re-examination of some assumptions made by physiologists and modelers. *Australian Journal of Plant Physiology* 21: 393-426.

Slafer, G.A. and Satorre, E.H. (1999). An introduction to the physiological-ecological analysis of wheat yield. In Satorre, E.H. and Slafer, G.A. (Eds.), *Wheat: Ecology and Physiology of Yield Determination* (pp. 3-12). Binghamton, NY: Food Products Press.

Smith, D.B. (1990). Barley seed protein and its effects on malting and brewing quality. *Plant Varieties and Seeds* 3: 63-80.

Smith, D.B. and Lister, P.R. (1983). Gel forming proteins in barley grain and their relationship with malting quality. *Journal of Cereal Science* 1: 229-239.

Smith, D.L., Dijark, M., Bulman, P., Ma, B.L., and Hamel, C. (1999). Barley: Physiology of yield. In Smith, D.L. and Hamel, C. (Eds.), *Crop Yield: Physiology and Processes* (pp. 67-107). Berlin, Heidelberg: Springer-Verlag.

Stone, P.J. and Nicolas, M.E. (1994). The effects of short periods of high temperature during grain filling on grain yield and quality vary widely between wheat cultivars. *Australian Journal of Plant Physiology* 21: 887-900.

Stone, P.J. and Savin, R. (1999). Grain quality and its physiological determinants. In Satorre, E.H. and Slafer, G.A. (Eds.), *Wheat: Ecology and Physiology of Yield Determination* (pp. 85-120). Binghamton, NY: Food Products Press.

Stuart, I.M., Loi, L., and Fincher, G.B. (1988). Varietal and environmental variations in (1-3, 1-4)-β-glucan levels and (1-3, 1-4)-β-glucanase potential in barley: Relationships to malting quality. *Journal of Cereal Science* 7: 61-71.

Swanston, J.S. (1994). Malting performance of barleys with altered starch composition. PhD thesis. Edinburgh, Heriot-Watt University.

Swanston, J.S. (1997). Assessment of malt. *Ferment* 10: 29-34.

Swanston, J.S. and Ellis, R.P. (2002). Genetics and breeding of malt quality attributes. In Slafer, G.A., Molina-Cano, J.L., Savin, R., Araus, J.L., and Romagosa, I. (Eds.), *Barley Science: Recent Advances from Molecular Biology to Agronomy of Yield and Quality* (pp. 85-114). Binghamton, NY: Food Products Press.

Swanston, J.S., Ellis, R.P., Pérez-Vendrell, A., Voltas, J., and Molina-Cano, J.L. (1997). Patterns of barley grain development in Spain and Scotland and their implication for malting quality. *Cereal Chemistry* 74: 456-461.

Swanston, J.S., Ellis, R.P., and Stark, J.R. (1995). Effects on grain and malting quality of genes altering barley starch composition. *Journal of Cereal Science* 22: 265-273.

Swanston, J.S. and Molina-Cano, J.L. The relationship between protein solubilisation and malting quality in a mutant of cv. Triumph. *Journal of the Institute of Brewing* (submitted).

Swanston, J.S., Thomas, W.T.D., Powell, W., Meyer, R., Machray, G.C., and Hedley, P.E. (1999). Breeding barley for malt whisky distilling—A genome based approach. In Campbell, I. (Ed.), *Proceedings of the Fifth Aviemore Conference on Malting, Brewing and Distilling*. London: Institute of Brewing.

Takeda, G., Sekiguchi, T., Kurai, K., and Seko, H. (1981). New procedure for micro-malting and its application for quality selection in barley breeding. *Japanese Journal of Breeding* 31: 414-422.

Tester, R.F. and Morrison, W.R. (1993). Swelling and gelatinization of cereal starches: VI. Starches from waxy Hector and Hector barleys at four stages of grain development. *Journal of Cereal Science* 17: 11-18.

Ullrich, S.E. (2002). Genetics and breeding of barley feed quality attributes. In Slafer, G.A., Molina-Cano, J.L., Savin, R., Araus, J.L., and Romagosa, I. (Eds.), *Barley Science: Recent Advances from Molecular Biology to Agronomy of Yield and Quality* (pp. 115-142). Binghamton, NY: Food Products Press.

van den Berg, R., Muts, G.C.J., Drost, B.W., and Graveland, A. (1981). Proteins from barley to wort. In *Proceedings European Brewery Convention Congress* (pp. 461-469). Copenhagen.

Wallwork, M.A.B., Logue, S.J., MacLeod, L.C., and Jenner, C.F. (1998a). Effect of high temperature during grain-filling on starch synthesis in developing barley grain. *Australian Journal of Plant Physiology* 25: 173-181.

Wallwork, M.A.B., Logue, S.J., MacLeod, L.C., and Jenner, C.F. (1998b). Effects of a period of high temperature during grain filling on the grain growth charac-

teristics and malting quality of three australian malting barleys. *Australian Journal of Agricultural Research* 49: 1287-1296.
Wardlaw, I.F. and Wrigley, C.W. (1994). Heat tolerance in temperate cereals: An overview. *Australian Journal of Plant Physiology* 21: 695-703.
Waugh, R. (2000). Current perspectives in barley genomics. In Logue, S. (Ed.), *Barley Genetics VIII: Proceedings of the Eighth International Barley Genetics Symposium,* Volume I, *Invited Papers* (pp. 205-211). Glen Ormond: Adelaide University.
Whitmore, E.T. and Sparrow, D.H.B. (1957). Laboratory micro-malting technique. *Journal of the Institute of Brewing* 63: 397-398.
Wrigley, C.W. (1994). Developing better strategies to improve grain quality for wheat. *Australian Journal of Agricultural Research* 45: 1-17.

Index

Page numbers followed by the letter "f" indicate figures; those followed by the letter "t" indicate tables.

Printed in the United States
by Baker & Taylor Publisher Services